Studies in Classification, Data Analysis, and Knowledge Organization

AF815165

For further volumes:
http://www.springer.com/series/1564

Francesco Palumbo · Carlo Natale Lauro
Michael J. Greenacre
Editors

Data Analysis and Classification

Proceedings of the 6th Conference of the Classification and Data Analysis Group of the Società Italiana di Statistica

Springer

Editors

Professor Francesco Palumbo
Department of Institution
in Economics and Finance
Università di Macerata
Via Crescimbeni, 20
62100 Macerata
Italy
francesco.palumbo@unimc.it

Professor Carlo Natale Lauro
Department of Mathematics and Statistics
Università Federico II di Napoli
Via Cinthia - Complesso Universitario
di Monte Sant'Angelo
80126 Napoli
Italy
clauro@unina.it

Professor Michael J. Greenacre
Department of Economics and Business
Universitat Pompeu Fabra
Ramon Trias Fargas, 25–27
08005 Barcelona
Spain
michael.greenacre@gmail.com

ISSN 1431-8814
ISBN 978-3-642-03738-2 e-ISBN 978-3-642-03739-9
DOI: 10.1007/978-3-642-03739-9
Springer Heidelberg Dordrecht London New York

Library of Congress Control Number: 2009936001

Cover design: SPi Publisher Services

Printed on acid-free paper

Springer is part of Springer Science+Business Media (www.springer.com)

Preface

This volume contains revised versions of selected papers presented at the biennial meeting of the Classification and Data Analysis Group (CLADAG) of the Italian Statistical Society, which was held in Macerata, September 12–14, 2007. Carlo Lauro chaired the Scientific Programme Committee and Francesco Palumbo chaired the Local Organizing Committee.

The scientific programme scheduled 150 oral presentations and one poster session. Sessions were organised in five plenary sessions, 10 invited paper specialised sessions and 24 solicited paper sessions. Contributed papers and posters were 54 and 12, respectively.

Five eminent scholars, who have given important impact in the Classification and Data Analysis fields, were invited as keynote speakers, they are *H. Bozdogan, S.R. Masera, G. McLachlan, A. Montanari, A. Rizzi*.

Invited Paper Specialised Sessions focused on the following topics:

- Knowledge extraction from temporal data models
- Statistical models with errors-in-covariates
- Multivariate analysis for microarray data
- Cluster analysis of complex data
- Educational processes assessment by means of latent variables models
- Classification of complex data
- Multidimensional scaling
- Statistical models for public policies
- Classification models for enterprise risk management
- Model-based clustering

It is worth noting that two of the ten specialised sessions were organised by the French (*Classification of complex data*) and Japanese (*Multidimensional scaling*) classification societies. The SPC is grateful to professors Okada (Japan) and Zighed (France), who took charge of the Japanese and French specialised session organisation, respectively. The SPC is grateful to the Italian statisticians who actively cooperated in the organisation of the specialised and solicited sessions: they were mainly responsible for the success of the conference.

On the occasion of the ClaDAG conference in Macerata, the SPC decided to have two sessions dedicated to young researchers who had finished their PhD programme during the year before the conference.

Thus, the conference provided a large number of scientists and experts from home and abroad with an attractive forum for discussion and mutual exchange of knowledge. Plenary and specialised sessions topics were agreed, aiming at fitting the mission of ClaDAG within the fields of Classification, Data Analysis and Multivariate Statistics.

All papers published in the present volume have been reviewed by the most qualified scholars from many countries, for each specific topic. The review process was quite long but very accurate to meet the publisher's standard of quality and the prestige of the series.

The more methodologically oriented papers focus on developments in clustering and discrimination, multidimensional data analysis, data mining. Many papers also provide significant contributions in a wide range of fields of application. This suggested the presentation of the 51 selected papers in nine parts, one more section consists of the keynote lectures. Section names are listed below:

1. Keynote lectures
2. Cluster analysis
3. Multidimensional scaling
4. Multivariate analysis and applications
5. Classification and classification trees
6. Statistical models
7. Latent variables
8. Knowledge extraction from temporal data
9. Statistical methods for financial and economics data
10. Missing values

We wish to express our gratitude to the other members of the Scientific Programme Committee Andrea Cerioli (Università degli Studi di Parma) Paolo Giudici (Università degli Studi di Pavia) Antonio Giusti (Università degli Studi di Firenze) Pietro Mantovan (Università degli Studi "Cà Foscari" di Venezia) Angelo Marcello Mineo (Università degli Studi di Palermo) Domenico Piccolo (Università degli Studi di Napoli Federico II) Marilena Pillati (Università degli Studi di Bologna) Roberto Rocci (Università degli Studi di Roma "Tor Vergata") Sergio Zani (Università degli Studi di Parma).

We gratefully acknowledge the University of Macerata and its Departments of *Istituzioni Economiche e Finanziarie* and *Studi sullo Sviluppo Economico* for financial support. We are also indebted to SISTAR Marche who has partially supported the publishing of the present volume. We thank all the members of the Local Organizing Committee: D. Bruzzese, C. Davino M. Gherghi, G. Giordano L. Scaccia, G. Scepi, for their excellent work in managing the organisation of the sixth ClaDAG conference. We desire to express our special thanks to Cristina Davino, for her skilful accomplishment of the duties of Scientific Secretary of ClaDAG 2007, and to Dr. Rosaria Romano for her assistance in producing this volume.

Finally, we would like to thank Dr. Martina Bihn of Springer-Verlag, Heidelberg, for her support and dedication to the production of this volume.

Macerata Francesco Palumbo
Naples Carlo N. Lauro
Barcelona Michael J. Greenacre
June 2009

List of Referees

We are indebted with our colleagues who kindly accepted to revise one or more papers. Their work has been essential to the quality of the present volume.

T. Aluja Banet, J. Antoch, E. Beccalli, D. Blei, S.A. Blozis, D. Bruzzese, M. Chavent, D. Dorn, G. Elliott, V. Esposito-Vinzi, A. Flores-Lagunes, L.C. Freeman, G. Giampaglia, Z. Huang, F. Husson, S. Ingrassia, C. Kascha, H.A.L. Kiers, S. Klink, I. Lerman, P.G. Lovaglio, A.H. Marshall, G. McLachlan, S. Mignani, M. Misuraca, A. Morineau, I. Moustaki, F. Murtagh, A. Nasraoui, L. Lebart, L. Norden, T. Poibeau, M. Riani, F. Rijmen, J. Sander, G. Saporta, Y. Sheng, F.P. Schoenberg, T.A.B. Snijders, R. Turner, L. Trinchera, A. Uhlendorff, J.K. Vermunt, B.Y.Yeap T.P. York, N.L. Zhang, J. Zhuang, D. Zighed

Contents

Part VI Statistical Models

Part VII Latent Variables

Part VIII Knowledge Extraction from Temporal Data

Part XI Missing Values

Contributors

Rafik Abdesselam ERIC EA 3038, University of Lyon 2, 69676, Bron, France rafik.abdesselam@univ-lyon2.fr

Giada Adelfio Department of Statistical and Mathematical Sciences, University of Palermo, viale delle Scienze, ed 13, 90128, Palermo, Italy, adelfio@unipa.it

Valentina Adorno Department of Economics, University of Bologna, Piazza Scaravilli, 2 Bologna, valentina.adorno@unibo.it

Marco Alfò Dipartimento di Statistica, Probabilità e Statistiche Applicate, Piazzale Aldo Moro, 5 - 00185 Roma, marco.alfo@uniroma1.it

Alessandra Amendola Di.S.E.S. Università degli Studi di Salerno, Via Ponte Don Melillo 84084 Fisciano (SA), Italy, alamendola@unisa.it

Balzanella Antonio Università degli Studi di Napoli Federico II, Via Cinthia I-80126 Napoli, Italy, balzanella2@alice.it

Anthony C. Atkinson London School of Economics, London WC2A 2AE, UK, a.c.atkinson@lse.ac.uk

Laura Attardi Dip.to di Progettazione Aeronautica, Università di Napoli, Italy, attardi@unina.it

Luigi Augugliaro Dipartimento di Scienze Statistiche e Matematiche, Università di Palermo, Viale delle Scienze, Edificio 13, 90128, Palermo, Italy, augugliaro@dssm.unipa.it

Silvia Bacci Department of Statistics "G. Parent", Viale Morgagni 59, 50134 Firenze, Italy, s.bacci@ds.unifi.it

Roberto Baragona Department of Sociology and Communication, Sapienza University of Rome, Via Salaria 113, 00198 Rome, Italy, roberto.baragona@uniroma1.it

Cristina Bernini Department of Statistics, University of Bologna, Via Belle Arti 41, Bologna, Italy, cristina.bernini@unibo.it

Silvia Bianconcini Department of Statistics, University of Bologna, Via Belle Arti, 41 - 40126 Bologna, Italy, silvia.bianconcini@unibo.it

Matilde Bini Department of Statistics "G. Parenti", Viale Morgagni, 59, 50134 Firenze, Italy, bini@ds.unifi.it

Sergio Bolasco Dipartimento di Studi Geoeconomici, Linguistici, Statistici, Storici per l'Analisi Regionale, Sapienza, University of Rome, Via del Castro Laurenziano 9, Roma, sergio.bolasco@uniroma1.it

Elvio Bonafede University of Pavia, Corso Strada Nuova 65, Italy, concetto.bonafede@unipv.it

Anna Lisa Bondi Department of Statistical and Mathematical Sciences "S. Vianelli" University of Palermo,viale delle Scienze - ed. 13, 90128 Palermo, Italy, bondi@dssm.unipa.it

Giuseppe Bove Dipartimento di Scienze dell'Educazione, Università degli Studi Roma Tre, Italy, bove@uniroma3.it

Adriana Brogini Department of Statistics, University of Padova, via Cesare Battisti 241, 35121, Padova, Italy, brogini@stat.unipd.it

Bruno Bertaccini Department of Statistics, Università degli Studi di Firenze "G. Parenti", Viale Morgagni, 59, 50134 Firenze, Italy, bertaccini@ds.unifi.it

Silvia Cagnone Department of Statistics, University of Bologna, Via Belle Arti, 41 - 40126 Bologna, Italy, silvia.cagnone@unibo.it

Alessio Canzonetti Dipartimento Studi Geoeconomici, Linguistici, Statistici, Storici per l'Analisi regionale - Facolta' di Economia - Sapienza Universita' di Roma, Via del Castro Laurenziano 9, Roma, alessio.canzonetti@uniroma1.it

Rosella Castellano DIEF, Università di Macerata, Via Crescimbeni, 20, 62100 Macerata, Italy, castellano@unimc.it

Paola Cerchiello University of Pavia, Corso Strada Nuova 65, Italy, paola.cerchiello@unipv.it

Andrea Cerioli Dipartimento di Economia, University of Parma, Via Kennedy 6, Italy, andrea.cerioli@unipr.it

Bruno Chiandotto Università degli Studi di Firenze, Dip.to di Statistica 'G. Parenti', Italy, chiandotto@ds.unifi.it

Marcello Chiodi Department of Statistical and Mathematical Sciences, University of Palermo, viale delle Scienze, ed 13, 90128, Palermo, Italy, chiodi@unipa.it

Paolo Chirico Dipartimento di Statistica e Matematica applicata, Via Maria Vittoria 38, 10100, Torino, Italy, paolo.chirico@unito.it

Claudio Conversano Department of Economics, University of Cagliari, Viale Fra Ignazio 17, I-09123, Cagliari, Italy, conversa@unica.it

Marcella Corduas Dipartimento di Scienze Statistiche, Università di Napoli Federico II, Via L.Rodino, 80138, Napoli(I), Italy, corduas@unina.it

Cristina Davino University of Macerata, Dipartimento di Studi sullo sviluppo economico, Italy, cdavino@unimc.it

Alfonso Iodice D'Enza Dipartimento di Scienze Economiche e Finanziarie Università di Cassino, Rome, iodicede@unina.it

Matteo Dimai Department of Economics and Statistics, University of Trieste, P.le Europa 1, 34127 Trieste, Italy, matteo.dimai@gmail.com

Elise Dusseldorp TNO Quality of Life, Department of Statistics, Leiden, the Netherlands, elise.dusseldorp@tno.nl

Paolo Eusebi Dipartimento di Economia, Finanza e Statistica; Università degli Studi di Perugia, Italy, peusebi@stat.unipg.it

Luigi Fabbris Statistics Department, University of Padua, Via C. Battisti 241, 35121 Padova, Italy, luigi.fabbris@unipd.it

Silvia Figini Department of Statistics and Applied Economics L. Lenti, University of Pavia, Italy, silvia.figini@unipv.it

Carlo Gaetan Department of Statistics, University Ca' Foscari, Venice, Italy, gaetan@unive.it

Marta Giorgino EURES, Via Col di Nava, 3 - 00141 Roma, Italy, m.giorgino@eures.it

Luca Greco Department PE.ME.IS - Section of Statistics, University of Sannio, Benevento, Italy, luca.greco@unisannio.it

Marco Grisoli Project Manager - Area Excelencia y Marketing Estratégico France Telecom España marco.grisoli@orange-ftgroup.com

Thorsten Hermes Universitat Bremen, Am Fallturm 1, D-28359 Bremen, Germany, hermes@tzi.de

Domenica Fioredistella Iezzi Università degli Studi di Roma "Tor Vergata", Italy, stella.iezzi@uniroma2.it

Antonio Irpino Dipartimento di Studi Europei e Mediterranei, Second University of Naples, Via del Setificio, 15, Belvedere di San Leucio, 81100 Caserta, Italy, antonio.irpino@unina2.it

Arne Jacobs Universitat Bremen, Am Fallturm 1, D-28359 Bremen, Germany, jarne@tzi.de

Dario Luzio Dipartimento di Chimica e Fisica della Terra, University of Palermo, via Archirafi, 26, 90123, Palermo, Italy, luzio@unipa.it

Pietro Mantovan Department of Statistics, University Ca Foscari, S Giobbe, Cannaregio, 873 -I-30121 Venezia, Italy, pietro.mantovan@unive.it

Antonello Maruotti Dipartimento di Statistica, Probabilità e Statistiche Applicate, Piazzale Aldo Moro, 5 - 00185 Roma, antonello.maruotti@uniroma1.it

Geoffrey J. McLachlan Department of Mathematics and Institute for Molecular Bioscience, University of Queensland, Brisbane, QLD 4072, Australia, gjm@maths.uq.edu.au

Giovanna Menardi Department of Economics and Statistics, P.le Europa, 1 Trieste, Italy, Giovanna.Menardi@econ.units.it

Angelo M. Mineo Dipartimento di Scienze Statistiche e Matematiche, Università di Palermo, Viale delle Scienze, Edificio 13, 90128, Palermo, Italy, elio.mineog@dssm.unipa.it

Simona Caterina Minotti Dipartimento di Statistica, Università degli Studi di Milano-Bicocca, Via Bicocca degli Arcimboldi 8, 20126 Milano, Italy, simona.minotti@unimib.it

Francesco Mola Department of Economics, University of Cagliari, Viale Fra Ignazio 17, I-09123, Cagliari, Italy, mola@unica.it

Giorgio E. Montanari Dipartimento di Economia, Finanza e Statistica; Università degli Studi di Perugia, Italy, giorgio@stat.unipg.it

Isabella Morlini DSSCQ, Università di Modena e Reggio Emilia, Modena, Italy, isabella.morlini@unimore.it

Marcella Niglio Di.S.E.S. Università degli Studi di Salerno, Via Ponte Don Melillo 84084 Fisciano (SA), Italy, mniglio@unisa.it

S.K. Ng Department of Mathematics, University of Queensland Brisbane, QLD 4072, Australia, skn@maths.uq.edu.au

Akinori Okada Graduate School of Management and Information Sciences, Tama University, 4-1-1 Hijirigaoka, Tama-shi, Tokyo 206-0022, Japan, okada@tama.ac.jp

Edoardo Otranto Dipartimento di Economia, Impresa e Regolamentazione, Via Torre Tonda 34, 07100 Sassari, Italy, eotranto@uniss.it

Massimo Pagani "Luigi Sacco" Hospital, University of Milan, Via G.B. Grassi 74, 20157 Milan, Italy, mp@ctnv.unimi.it

Francesco Palumbo Dipartimento di Istituzioni Economiche e Finanziarie Università di Macerata, Faculty of Economics, Macerata, Italy, palumbo@unimc.it

Andrea Pastore Department of Statistics, University Ca Foscari, S Giobbe, Cannaregio, 873 -I-30121 Venezia, Italy, andrea.pastore@unive.it

Pasquale Pavone Dipartimento di Studi Geoeconomici, Linguistici, Statistici, Storici per l'Analisi Regionale, Sapienza, University of Rome, Via del Castro Laurenziano 9, Roma, pavone03@libero.it

Guido Pellegrini Department of Economic Theory and Quantitative Methods for Political Choices, Sapienza University of Rome, Piazzale Aldo Moro 5, Roma, Italy, guido.pellegrini@uniroma1.it

Domenico Perrotta European Commission (EC), Joint Research Centre (JRC), Institute for the Protection and Security of the Citizens (IPSC), Global Security and Crisis Management (GSCM), Via Enrico Fermi 2749, Ispra, Italy, domenico.perrotta@ec.europa.eu

Antonella Plaia Department of Statistical and Mathematical Sciences "S. Vianelli", University of Palermo, viale delle Scienze - ed. 13, 90128 Palermo, Italy, plaia@unipa.it

Mariano Porcu Dip. Ric. Economiche e Sociali - Univ. di Cagliari, Viale S. Ignazio 78, Italy, mrporcu@unica.it

M. Giovanna Ranalli Dipartimento di Economia, Finanza e Statistica; Università degli Studi di Perugia, Italy, giovanna@stat.unipg.it

M. Riani Dipartimento di Economia, University of Parma, Via Kennedy 6, Italy, mriani@unipr.it

Giulia Rivellini Università Cattolica del Sacro Cuore, Largo Gemelli 1, 20123 Milano, Italy, giulia.rivellini@unicatt.it

Alfredo Rizzi Dipartimento di Statistica, Probabilità e Statistiche Applicate, Università di Roma "La Sapienza" P.le A.Moro, 5 - 00185 Roma, alfredo.rizzi@uniroma1.it

Roberta Varriale Università degli Studi di Firenze, Dip.to di Statistica 'G. Parenti', Italy, roberta.varriale@ds.unifi.it

Elvira Romano Seconda Università degli Studi di Napoli, via Del Setificio 81100 Caserta, Italy, elvroman@unina.it

Yasmin H. Said Isaac Newton Institute for Mathematical Sciences, Cambridge University, Cambridge, CB3 0EH UK, ysaid99@hotmail.com
and
Department of Computational and Data Sciences, George Mason University MS 6A2, Fairfax, VA 22030, USA

Luisa Scaccia DIEF, Università di Macerata, Via Crescimbeni, 20, 62100 Macerata, Italy, scaccia@unimc.it

Germana Scepi University of Naples, Via Cinthia, Monte Sant'Angelo (NA), Italy, scepi@unina.it

Sergio Scippacercola Dipartimento di Matematica e Statistica - Università degli studi di Napoli Federico II - Via Cinthia, 80126 – Napoli, Italy, sergio.scippacercola@unina.it

Luca Scrucca Dipartimento di Economia, Finanza e Statistica, Università degli Studi di Perugia, Perugia, Italy, luca@stat.unipg.it

Walid K. Sharabati Department of Statistics, Purdue University, West Lafayette, IN 47907, USA, sharabati@gmail.com

Debora Slanzi Department of Statistics, University Ca' Foscari, San Giobbe Canareggio 873, 30121, Venezia, Italy, debora.slanzi@unive.it

Nadia Solaro Department of Statistics, University of Milan-Bicocca, Via Bicocca degli Arcimboldi 8, 20126 Milan, Italy, nadia.solaro@unimib.it

Isabella Sulis Dip. Ric. Economiche e Sociali - Univ. di Cagliari, Viale S. Ignazio 78, Italy, isulis@unica.it

Nicola Torelli Department of Economics and Statistics, University of Trieste, P.le Europa 1, 34127 Trieste, Italy, nicola.torelli@econ.units.it

Francesca Torti Università Milano Bicocca, Facoltà di Statistica, Milano, Italy, f.torti1@campus.unimib.it, francesca.torti@jrc.it

Rosanna Verde Dipartimento di Studi Europei e Mediterranei, Second University of Naples, Via del Setificio, 15, Belvedere di San Leucio, 81100 Caserta, Italy, rosanna.verde@unina2.it

Domenico Vistocco Dip.to di Scienze Economiche, Università di Cassino, Italy, vistocco@unicas.it

Cosimo Vitale Di.S.E.S. Università degli Studi di Salerno, Via Ponte Don Melillo 84084 Fisciano (SA) Italy, vitale@unina.it

Salvatore Vitrano Statistical Office, Ministry for Cultural Heritage and Activities, Collegio Romano 27, 00186 Rome, Italy, svitrano@beniculturali.it

Giorgio Vittadini Dipartimento di Metodi Quantitativi per le Scienze Economiche ed Aziendali, Università degli Studi di Milano-Bicocca, Via Bicocca degli Arcimboldi 8, 20126 Milano, Italy, giorgio.vittadini@unimib.it

K. Wang Department of Mathematics, University of Queensland Brisbane, QLD 4072, Australia, kwang@maths.uq.edu.au

Adalbert F.X. Wilhelm Jacobs University Bremen, P.O. Box 75 05 61, D-28725 Bremen, Germany, a.wilhelm@jacobs-university.de

Edward J. Wegman Department of Computational and Data Sciences, George Mason University, Fairfax, VA, USA, ewegman@gmail.com

Susanna Zaccarin Università di Trieste, Piazzale Europa 1, 34127 Trieste, Italy, susanna.zaccarin@econ.units.it

Sergio Zani Dipartimento di Economia, Università di Parma, Italy, sergio.zani@unipr.it

Part I
Key-note

Clustering of High-Dimensional and Correlated Data

Geoffrey J. McLachlan, Shu-Kay Ng, and K. Wang

Abstract Finite mixture models are being commonly used in a wide range of applications in practice concerning density estimation and clustering. An attractive feature of this approach to clustering is that it provides a sound statistical framework in which to assess the important question of how many clusters there are in the data and their validity. We consider the applications of normal mixture models to high-dimensional data of a continuous nature. One way to handle the fitting of normal mixture models is to adopt mixtures of factor analyzers. However, for extremely high-dimensional data, some variable-reduction method needs to be used in conjunction with the latter model such as with the procedure called EMMIX-GENE. It was developed for the clustering of microarray data in bioinformatics, but is applicable to other types of data. We shall also consider the mixture procedure EMMIX-WIRE (based on mixtures of normal components with random effects), which is suitable for clustering high-dimensional data that may be structured (correlated and replicated) as in longitudinal studies.

1 Introduction

Clustering procedures based on finite mixture models are being increasingly preferred over heuristic methods due to their sound mathematical basis and to the interpretability of their results. Mixture model-based procedures provide a probabilistic clustering that allows for overlapping clusters corresponding to the components of the mixture model. The uncertainties that the observations belong to the clusters are provided in terms of the fitted values for their posterior probabilities of component membership of the mixture. As each component in a finite mixture model corresponds to a cluster, it allows the important question of how many clusters there are in the data to be approached through an assessment of how many components are

G.J. McLachlan (✉)
Department of Mathematics and Institute for Molecular Bioscience, University of Queensland, Brisbane, QLD 4072, Australia
e-mail: gjm@maths.uq.edu.au

F. Palumbo et al. (eds.), *Data Analysis and Classification*,
Studies in Classification, Data Analysis, and Knowledge Organization,
DOI 10.1007/978-3-642-03739-9_1,

needed in the mixture model. These questions of model choice can be considered in terms of the likelihood function.

Scott and Symons (1971) were one of the first to adopt a model-based approach to clustering. Assuming that the data were normally distributed within a cluster, they showed that their approach is equivalent to some commonly used clustering criteria with various constraints on the cluster covariance matrices. However, from an estimation point of view, this approach yields inconsistent estimators of the parameters. This inconsistency can be avoided by working with the mixture likelihood formed under the assumption that the observed data are from a mixture of classes corresponding to the clusters to be imposed on the data, as proposed by Wolfe (1965) and Day (1969). Finite mixture models have since been increasingly used to model the distributions of a wide variety of random phenomena and to cluster data sets; see, for example, McLachlan and Peel (2000).

2 Definition of Mixture Models

We let $\boldsymbol{Y}$ denote a random vector consisting of p feature variables associated with the random phenomenon of interest. We let $\boldsymbol{y}_1, \ldots, \boldsymbol{y}_n$ denote an observed random sample of size n on $\boldsymbol{Y}$. With the finite mixture model-based approach to density estimation and clustering, the density of $\boldsymbol{Y}$ is modelled as a mixture of a number (g) of component densities $f_i(\boldsymbol{y})$ in some unknown proportions $\pi_1, \ldots, \pi_g$. That is, each data point is taken to be a realization of the mixture probability density function (p.d.f.),

$$f(\boldsymbol{y}; \Psi) = \sum_{i=1}^{g} \pi_i f_i(\boldsymbol{y}), \tag{1}$$

where the mixing proportions π_i are nonnegative and sum to one. In density estimation, the number of components g can be taken sufficiently large for (1) to provide an arbitrarily accurate estimate of the underlying density function. For clustering purposes, each component in the mixture model (1) corresponds to a cluster. The posterior probability that an observation with feature vector y_j belongs to the ith component of the mixture is given by

$$\tau_i(y_j) = \pi_i f_i(y_j)/f(y_j) \tag{2}$$

for $i = 1, \ldots, g$. A probabilistic clustering of the data into g clusters can be obtained in terms of the fitted posterior probabilities of component membership for the data.

An outright partitioning of the observations into g (nonoverlapping) clusters $C_1, \ldots, C_g$ is effected by assigning each observation to the component to which it has the highest estimated posterior probability of belonging. Thus the ith cluster C_i contains those observations $\boldsymbol{y}_j$ with $\hat{z}_{ij} = 1$, where $\hat{z}_{ij} = 1$ if $i = h^*$, and zero otherwise, and

$$h^* = \arg\max_h \hat{\tau}_h(y_j); \tag{3}$$

$\hat{\tau}_i(y_j)$ is an estimate of $\tau_i(y_j)$. As the notation implies, $\hat{z}_{ij}$ can be viewed as an estimate of z_{ij} which, under the assumption that the observations come from a mixture of g groups $G_1, \ldots, G_g$, is defined to be one or zero according as the jth observation $\boldsymbol{y}_j$ does or does not come from G_i ($i = 1, \ldots, g$; $j = 1, \ldots, n$).

3 Maximum Likelihood Estimation

On specifying a parametric form $f_i(y_j; \theta_i)$ for each component density, we can fit this parametric mixture model

$$f(y_j; \Psi) = \sum_{i=1}^{g} \pi_i f_i(y_j; \theta_i) \tag{4}$$

by maximum likelihood (ML). Here $\Psi = (\omega^T, \pi_1, \ldots, \pi_{g-1})^T$ is the vector of unknown parameters, where ω consists of the elements of the θ_i known a priori to be distinct. In order to estimate Ψ from the observed data, it must be identifiable. This will be so if the representation (4) is unique up to a permutation of the component labels. The maximum likelihood estimate (MLE) of Ψ, $\hat{\Psi}$, is given by an appropriate root of the likelihood equation,

$$\partial \log L(\Psi)/\partial\Psi = \mathbf{0}, \tag{5}$$

where $L(\Psi)$ denotes the likelihood function for Ψ,

$$L(\Psi) = \prod_{j=1}^{n} f(y_j; \Psi).$$

Solutions of (5) corresponding to local maximizers of $\log L(\Psi)$ can be obtained via the expectation-maximization (EM) algorithm of Dempster et al. (1977); see also McLachlan and Krishnan (1997). Let $\hat{\Psi}$ denote the estimate of Ψ so obtained.

4 Choice of Starting Values for the EM Algorithm

McLachlan and Peel (2000) provide an in-depth account of the fitting of finite mixture models. Briefly, with mixture models the likelihood typically will have multiple maxima; that is, the likelihood equation will have multiple roots. Thus the EM algorithm needs to be started from a variety of initial values for the parameter vector Ψ or for a variety of initial partitions of the data into g groups. The latter can be obtained by randomly dividing the data into g groups corresponding to the g components of the mixture model. With random starts, the effect of the central limit theorem tends to have the component parameters initially being similar at least in large samples. Nonrandom partitions of the data can be obtained via some clustering procedure such as k-means.

The choice of root of the likelihood equation in the case of homoscedastic normal components is straightforward in the sense that the ML estimate exists as the global maximizer of the likelihood function. The situation is less straightforward in the case of heteroscedastic normal components as the likelihood function is unbounded. Usually, the intent is to choose as the ML estimate of the parameter vector Ψ the local maximizer corresponding to the largest of the local maxima located. But in practice, consideration has to be given to the problem of relatively large local maxima that occur as a consequence of a fitted component having a very small (but nonzero) variance for univariate data or generalized variance (the determinant of the covariance matrix) for multivariate data. Such a component corresponds to a cluster containing a few data points either relatively close together or almost lying in a lower-dimensional subspace in the case of multivariate data. There is thus a need to monitor the relative size of the fitted mixing proportions and of the component variances for univariate observations, or of the generalized component variances for multivariate data, in an attempt to identify these spurious local maximizers.

5 Clustering via Normal Mixtures

Frequently, in practice, the clusters in the data are essentially elliptical, so that it is reasonable to consider fitting mixtures of elliptically symmetric component densities. Within this class of component densities, the multivariate normal density is a convenient choice given its computational tractability.

Under the assumption of multivariate normal components, the ith component-conditional density $f_i(y;\, \theta_i)$ is given by

$$f_i(y;\, \theta_i) = \phi(y;\, \mu_i, \Sigma_i), \tag{6}$$

where θ_i consists of the elements of μ_i and the $\frac{1}{2}p(p+1)$ distinct elements of Σ_i $(i = 1, \ldots, g)$. Here

$$\phi(y;\, \mu_i, \Sigma_i) = (2\pi)^{-\frac{p}{2}} |\Sigma_i|^{-1/2} \exp\{-\tfrac{1}{2}(y - \mu_i)^T \Sigma_i^{-1} (y - \mu_i)\}. \tag{7}$$

One attractive feature of adopting mixture models with elliptically symmetric components such as the normal or t-densities, is that the implied clustering is invariant under affine transformations of the data; that is, invariant under transformations of the feature vector y of the form,

$$y \to C y + a, \tag{8}$$

where C is a nonsingular matrix. If the clustering of a procedure is invariant under (8) for only diagonal C, then it is invariant under change of measuring units but not rotations.

It can be seen from (7) that the mixture model with unrestricted component-covariance matrices in its normal component distributions is a highly parameterized one with $\frac{1}{2}p(p+1)$ parameters for each component-covariance matrix Σ_i $(i = 1, \ldots, g)$. As an alternative to taking the component-covariance matrices to be the same or diagonal, we can adopt some model for the component-covariance matrices that is intermediate between homoscedasticity and the unrestricted model, as in the approach of Banfield and Raftery (1993). They introduced a parameterization of the component-covariance matrix Σ_i based on a variant of the standard spectral decomposition of Σ_i.

The mixture model with normal components (7) is sensitive to outliers since it adopts the multivariate normal family for the distributions of the errors. An obvious way to improve the robustness of this model for data which have longer tails than the normal or atypical observations is to consider using the multivariate t-family of elliptically symmetric distributions; see McLachlan and Peel (1998, 2000). It has an additional parameter called the degrees of freedom that controls the length of the tails of the distribution. Although the number of outliers needed for breakdown is almost the same as with the normal distribution, the outliers have to be much larger.

6 Factor Analysis Model for Dimension Reduction

As remarked earlier, the g-component normal mixture model with unrestricted component-covariance matrices is a highly parameterized model with $\frac{1}{2}p(p+1)$ parameters for each component-covariance matrix Σ_i $(i = 1, \ldots, g)$. As discussed above, Banfield and Raftery (1993) introduced a parameterization of the component-covariance matrix Σ_i based on a variant of the standard spectral decomposition of $\Sigma_i (i = 1, \ldots, g)$. However, if p is large relative to the sample size n, it may not be possible to use this decomposition to infer an appropriate model for the component-covariance matrices. Even if it is possible, the results may not be reliable due to potential problems with near-singular estimates of the component-covariance matrices when p is large relative to n.

A common approach to reducing the number of dimensions is to perform a principal component analysis (PCA). But as is well known, projections of the feature data y_j onto the first few principal axes are not always useful in portraying the group structure. A global nonlinear approach to dimension reduction can be obtained by postulating a finite mixture of linear submodels for the distribution of the full observation vector Y_j given the (unobservable) factors. see Hinton et al. (1997), McLachlan and Peel (2000), and McLachlan et al. (2003). The mixture of factor analyzers model is given by

$$f(\boldsymbol{y}_j;\,\Psi) = \sum_{i=1}^{g} \pi_i \phi(\boldsymbol{y}_j;\,\boldsymbol{\mu}_i,\,\Sigma_i), \tag{9}$$

where the ith component-covariance matrix $\boldsymbol{\Sigma}_i$ has the form

$$\Sigma_i = B_i B_i^T + D_i \quad (i = 1, \ldots, g) \tag{10}$$

and where B_i is a $p \times q$ matrix of factor loadings and D_i is a diagonal matrix ($i = 1, \ldots, g$). The parameter vector Ψ now consists of the mixing proportions π_i and the elements of the μ_i, the B_i, and the D_i. With this approach, the number of free parameters is controlled through the dimension of the latent factor space. By working in this reduced space, it allows a model for each component-covariance matrix with complexity lying between that of the isotropic and full covariance structure models without any restrictions on the covariance matrices. The mixture of factor analyzers model can be fitted by using the alternating expectation–conditional maximization (AECM) algorithm of Meng and van Dyk (1997).

A formal test for the number of factors can be undertaken using the likelihood ratio λ, as regularity conditions hold for this test conducted at a given value for the number of components g. For the null hypothesis that $H_0 : q = q_0$ vs. the alternative $H_1 : q = q_0 + 1$, the statistic $-2\log\lambda$ is asymptotically chi-squared with $d = g(p - q_0)$ degrees of freedom. However, in situations where n is not large relative to the number of unknown parameters, we prefer the use of the BIC criterion. Applied in this context, it means that twice the increase in the log likelihood ($-2\log\lambda$) has to be greater than $d \log n$ for the null hypothesis to be rejected.

The mixture of factor analyzers model is sensitive to outliers since it uses normal errors and factors. Recently, McLachlan et al. (2007) have considered the use of mixtures of t analyzers in an attempt to make the model less sensitive to outliers.

7 Some Recent Extensions for High-Dimensional Data

The EMMIX-GENE program of McLachlan et al. (2002) has been designed for the normal mixture model-based clustering of a limited number of observations that may be of extremely high-dimensions. It was called EMIX-GENE as it was designed specifically for problems in bioinformatics that require the clustering of a relatively small number of tissue samples containing the expression levels of possibly thousands of genes. But it is applicable to clustering problems outside the field of bioinformatics involving high-dimensional data. In situations where the sample size n is very large relative to the dimension p, it might not be practical to fit mixtures of factor analyzers to data on all the variables, as it would involve a considerable amount of computation time. Thus initially some of the variables may have to be removed. Indeed, the simultaneous use of too many variables in the cluster analysis may serve only to create noise that masks the effect of a smaller number of variables. Also, the intent of the cluster analysis may not be to produce a clustering of the observations on the basis of all the available variables, but rather to discover and study different clusterings of the observations corresponding to different subsets of the variables; see, for example, Soffritti (2003) and Galimberti and Soffritti (2007).

Therefore, the EMMIX-GENE procedure has two optional steps before the final step of clustering the observations. The first step considers the selection of a subset of relevant variables from the available set of variables by screening the variables on an individual basis to eliminate those which are of little use in clustering the observations. The usefulness of a given variable to the clustering process can be assessed formally by a test of the null hypothesis that it has a single component normal distribution over the observations. A faster but ad hoc way is to make this decision on the basis of the interquartile range. Even after this step has been completed, there may still remain too many variables. Thus there is a second step in EMMIX-GENE in which the retained variables are clustered (after standardization) into a number of groups on the basis of Euclidean distance so that variables with similar profiles are put into the same group. In general, care has to be taken with the scaling of variables before clustering of the observations, as the nature of the variables can be intrinsically different. Also, as noted above, the clustering of the observations via normal mixture models is invariant under changes in scale and location. The clustering of the observations can be carried out on the basis of the groups considered individually using some or all of the variables within a group or collectively. For the latter, we can replace each group by a representative (a metavariable) such as the sample mean as in the EMMIX-GENE procedure.

8 Mixtures of Normal Components with Random Effects

Up to now, we have considered the clustering of data on entities under two assumptions that are commonly adopted in practice; namely:

(a) There are no replications on any particular entity specifically identified as such.
(b) All the observations on the entities are independent of one another.

These assumptions should hold for the clustering of, say, tissue samples consisting of the expression levels of many (possibly thousands) of genes, although the tissue samples have been known to be correlated for different tissues due to flawed experimental conditions. However, condition (b) will not hold for the clustering of gene profiles, since not all the genes are independently distributed, and condition (a) will generally not hold either as the gene profiles may be measured over time or on technical replicates. While this correlated structure can be incorporated into the normal mixture model (9) by appropriate specification of the component-covariance matrices Σ_i, it is difficult to fit the model under such specifications. For example, the M-step may not exist in closed form.

Accordingly, Ng et al. (2006) have developed the procedure called EMMIX-WIRE (**EM**-based **MIX**ture analysis **W**ith **R**andom **E**ffects) to handle the clustering of correlated data that may be replicated. They adopted conditionally a mixture of linear mixed models to specify the correlation structure between the variables and to allow for correlations among the observations. It also enables covariate information to be incorporated into the clustering process.

To formulate this procedure, we consider the clustering of n gene profiles $\boldsymbol{y}_j$ ($j = 1, \ldots, n$), where we let $\boldsymbol{y}_j = (\boldsymbol{y}_{1j}^T, \ldots, \boldsymbol{y}_{mj}^T)^T$ contain the expression values for the jth gene profile and

$$\boldsymbol{y}_{tj} = (y_{1tj}, \ldots, y_{r_t tj})^T \qquad (t = 1, \ldots, m)$$

contains the r_t replicated values in the tth biological sample ($t = 1, \ldots, m$) on the jth gene. The dimension p of $\boldsymbol{y}_j$ is given by $p = \sum_{t=1}^{m} r_t$.

With the EMMIX-WIRE procedure, the observed p-dimensional vectors $\boldsymbol{y}_1, \ldots, \boldsymbol{y}_n$ are assumed to have come from a mixture of a finite number, say g, of components in some unknown proportions $\pi_1, \ldots, \pi_g$, which sum to one. Conditional on its membership of the ith component of the mixture, the profile vector $\boldsymbol{y}_j$ for the jth gene ($j = 1, \ldots, n$), follows the model

$$\boldsymbol{y}_j = X\beta_i + Ub_{ij} + Vc_i + \varepsilon_{ij}, \tag{11}$$

where the elements of β_i are fixed effects (unknown constants) modelling the conditional mean of $\boldsymbol{y}_j$ in the ith component ($i = 1, \ldots, g$). In (11), b_{ij} (a q_b-dimensional vector) and c_i (a q_c-dimensional vector) represent the unobservable gene- and tissue-specific random effects, respectively. These random effects represent the variation due to the heterogeneity of genes and samples (corresponding to $b_i = (b_{i1}^T, \ldots, b_{in}^T)^T$ and c_i, respectively). The random effects b_i and c_i, and the measurement error vector $(\varepsilon_{i1}^T, \ldots, \varepsilon_{in}^T)^T$ are assumed to be mutually independent, where X, U, and V are known design matrices of the corresponding fixed or random effects, respectively. The presence of the random effect c_i for the expression levels of genes in the ith component induces a correlation between the profiles of genes within the same cluster.

With the LMM, the distributions of $\boldsymbol{b}_{ij}$ and $\boldsymbol{c}_i$ are taken, respectively, to be multivariate normal $N_{q_b}(0, H_i)$ and $N_{q_c}(0, \theta_{ci} I_{q_c})$, where H_i is a $q_b \times q_b$ covariance matrix and I_{q_c} is the $q_c \times q_c$ identity matrix. The measurement error vector ε_{ij} is also taken to be multivariate normal $N_p(0, A_i)$, where $A_i = \text{diag}(W\xi_i)$ is a diagonal matrix constructed from the vector $(W\xi_i)$ with $\xi_i = (\sigma_{i1}^2, \ldots, \sigma_{iq_e}^2)^T$ and W a known $p \times q_e$ zero-one design matrix.

We let $\Psi = (\psi_1^T, \ldots, \psi_g^T, \pi_1, \ldots, \pi_{g-1})^T$ be the vector of all the unknown parameters, where ψ_i is the vector containing the unknown parameters β_i, the distinct elements of H_i, θ_{ci}, and ξ_i of the ith component density ($i = 1, \ldots, g$). The estimation of Ψ can be obtained by the ML approach via the EM algorithm, proceeding conditionally on the tissue-specific random effects c_i as formulated in Ng et al. (2006). The E- and M-steps can be implemented in closed form. In particular, an approximation to the E-step by carrying out time-consuming Monte Carlo methods is not required. A probabilistic or an outright clustering of the genes into g components can be obtained, based on the estimated posterior probabilities of component membership given the profile vectors and the estimated tissue-specific random effects $\hat{c}_i$ for $i = 1, \ldots, g$; see Ng et al. (2006).

References

Banfield, J., & Raftery, A. (1993). Model-based gaussian and non-gaussian clustering. *Biometrics, 49*, 803–821.

Day, N. (1969). Estimating the components of a mixture of two normal distributions. *Biometrika, 56*, 463–474.

Dempster, A., Laird, N., & Rubin, D. (1977). Maximum likelihood from incomplete data via the EM algorithm (with discussion). *Journal of the Royal Statistical Society B, 39*, 1–38.

Galimberti, G., & Soffritti, G. (2007). Model-based methods for identifying multiple cluster structures in a data set. *Computational Statistics and Data Analysis, 52*, 520–536.

Hinton, G., Dayan, P., & Revow, M. (1997). Modeling the manifolds of images of handwritten digits. *IEEE Transactions on Neural Networks, 8*, 65–73.

McLachlan, G., Bean, R., & Ben-Tovim Jones, L. (2007). Extension of the mixture of factor analyzers model to incorporate the multivariate t distribution. *Computational Statistics and Data Analysis, 51*, 5327–5338.

McLachlan, G., Bean, R., & Peel, D. (2002). A mixture model-based approach to the clustering of microarray expression data. *Bioinformatics, 18*, 413–422.

McLachlan, G., & Krishnan, T. (1997). *The EM algorithm and extensions*. New York: Wiley.

McLachlan, G., & Peel, D. (1998). Robust cluster analysis via mixtures of multivariate t-distributions. In: A. Amin, D. Dori, P. Pudil, & H. Freeman (Eds.), *Lecture notes in computer science* (Vol. 1451, pp. 658–666). Berlin: Springer.

McLachlan, G., & Peel, D. (2000). *Finite mixture models*. New York: Wiley.

McLachlan, G., Peel, D., & Bean, R. (2003). Modelling high-dimensional data by mixtures of factor analyzers. *Computational Statistics and Data Analysis, 41*, 379–388.

Meng, X., & van Dyk, D. (1997). The EM algorithm – an old folk song sung to a fast new tune (with discussion). *Journal of the Royal Statistical Society B, 59*, 511–567.

Ng, S., McLachlan, G., Wang, K., Ben-Tovim Jones, L., & Ng, S. (2006). A mixture model with random-effects components for clustering correlated gene-expression profiles. *Bioinformatics, 22*, 1745–1752.

Scott, A., & Symons, M. (1971). Clustering methods based on likelihood ratio criteria. *Biometrics, 27*, 387–397.

Soffritti, G. (2003). Identifying multiple cluster structures in a data matrix. *Communications in Statistics – Simulation and Computation, 32*, 1151–1177.

Wolfe, J. (1965). *A computer program for the computation of maximum likelihood analysis of types* (Technical Report SRM 65-112). US Naval Personnel Research Activity, San Diego.

Statistical Methods for Cryptography

Alfredo Rizzi

Abstract In this note, after recalling certain results regarding prime numbers, we will present the following theorem of interest to cryptography: Let two discrete s.v.'s (statistical variable) X, Y assume the value: $0, 1, 2, \ldots, m-1$. Let X be uniformly distributed, that is, it assumes the value $i(i = 0, 1, \ldots, m-1)$ with probability $1/m$ and let the second s.v. Y assume the value i with probability $(p_i : \sum_{i=1}^{m-1} p_i = 1, p_i \geq 0)$. If the s.v. $Z = X + Y$ (mod m) is uniformly distributed and m is a prime number, at least one of the two s. v. X and Y is uniformly distributed.

1 Introduction

In today's world the need to protect vocal and written communication between individuals, institutions, entities and commercial agencies is ever present and growing. Digital communication has, in part, been integrated into our social life. For many, the day begins with the perusal of e-mail and the tedious task of eliminating spam and other messages we do not consider worthy of our attention. We turn to the internet to read newspaper articles, to see what's on at the cinema, to check flight arrivals, the telephone book, the state of our checking account and stock holdings, to send and receive money transfers, to shop on line, for students' research and for many other reasons. But the digital society must adequately protect communication from intruders, whether persons or institutions which attack our privacy. Cryptography (from $\kappa\rho\upsilon\pi\tau o\varsigma$, hidden), the study and creation of secret writing systems in numbers or codes, is essential to the development of digital communication which is absolutely private insofar as being impossible to be read by anyone to whom it is not addressed. Cryptography seeks to study and create systems for ciphering and to verify and authenticate the integrity of data. One must make the distinction between

A. Rizzi
Dipartimento di Statistica, Probabilità e Statistiche Applicate,
Università di Roma "La Sapienza" P.le A.Moro, 5 - 00185 Roma
e-mail: alfredo.rizzi@uniroma1.it

F. Palumbo et al. (eds.), *Data Analysis and Classification*,
Studies in Classification, Data Analysis, and Knowledge Organization,
DOI 10.1007/978-3-642-03739-9_2,

cryptoanalysis, the research of methods an "enemy" might use to read the messages of others and cryptography. Cryptography and cryptoanalysis are what make up cryptology.

Until the 1950s cryptography was essentially used only for military and diplomatic communication. The decryption of German messages by the English and of Japanese messages by the Americans played a very important role in the outcome of the Second World War. The great mathematician Alan Touring made an essential contribution to the war effort with his decryption of the famous Enigma machine which was considered absolutely secure by the Germans. It was the Poles, however, who had laid the basis for finding its weak link. Cryptography also played a vital role in the Pacific at the battle of Midway Regarding Italy, the naval battles of Punta Stilo and of Capo Matapan were strongly influenced by the interception and decryption of messages.

1.1 Different disciplines in cryptography

There are four disciplines which have important roles in cryptography:

1. Linguistics, in particular Statistical Linguistics
2. Statistics, in particular the Theory of the Tests for the Analysis of Randomness and of Primality and Data Mining
3. Mathematics, in particular Discrete Mathematics
4. The Theory of Information

The technique of Data Mining seems to be of more use in the analysis of a great number of data which are exchanged on a daily basis such as satellite data. Technical developments are largely inter-disciplinary. This suggests that new applications will be found which will, in turn, lead to new queries and problems for the scholars of Number Theory, Modular Arithmetic, Polynomial Algebra, Information Theory and Statistics to apply to cryptography.

Until the 1950s the decryption of messages was based exclusively on statistical methods and specific techniques of cryptography. In substance, the working instruments of cryptography, both for the planning of coding systems and for reading messages which the sender intended remain secret, were statistical methods applied to linguistics. The key to decoding systems using poly-alphabetic substitution and simple and double transposition has always been the analysis of the statistical distribution of graphemes (letters, figures, punctuation marks, etc.). Mathematics was not fundamental to the work of the cryptoanalyst.

Today, with the advent of data processing technology, coding of messages is done by coding machines. The structure of reference is the algebra of Galois (GF(q)). The search for prime numbers, in particular tests of primality, are of notable interest to modern cryptology.

In this note, after recalling certain results regarding prime numbers, we will present a theorem of interest to cryptography.

2 Prime Numbers

The questions regarding prime numbers have interested many scholars since the dawn of mathematics. We need only recall Euclid in ancient times and Fermat, Eulero, Legendre, Gauss and Hilbert in the last four hundred years. Gauss, in 1801, in *Disquisitiones Arithmeticae*, stated that the problem of distinguishing prime numbers from composite numbers and that of the factorization of these composite numbers were among the most important and useful in arithmetics. Moreover, he added, the very dignity of science itself seemed to require that such an elegant problem be explored from every angle which might help clarify it.

The great calculation resources which are today available to scholars all over the world have led many to deal with questions relative to primes and some to try and *falsify* certain conjectures. Numerous are the web sites devoted to these numbers. The most noteworthy fact of this situation is that information arrives on the web in real time, not only in print and these are among the most frequented sites. This leads many to confront questions regarding primes which are of limited importance. A form of emulation is stimulated in which we see many universities in the entire world, but particularly the United States, make great efforts to find a new prime and so become the "leader of the pack", if only for a short while as with setting a record in a sport. This happened, and is happening in the efforts to find the largest known prime to which some universities devote massive calculation resources for many years as occurred with the confirmation of the famous theorem of four colors in postal zones at the University of Illinois and the proof that the 23rd Mersenne number is prime.

When speaking of research in prime numbers reference is often made to possible applications in cryptography and in particular cryptographic systems with an RSA public key. The RSA system is based on the choice of two primes of sufficient size and on the relations introduced by Eulero in 1700. This is the source of interest in basic research in prime numbers which could, in some way, have operative results in various coding systems.

2.1 *Tests of primality*

The theoretical basis for the tests of primality, whether deterministic or probabilistic, has its origin in the research of the Swiss mathematician Leonardo Eulero (1707–1783) and the Frenchman Pierre de Fermat (1601–1665). Let $\mathbf{Z_n}$, be the set $[1, 2, \ldots, n]$. Let $\mathbf{Z}_n^*$ be the set of the integers prime with n. The cardinality of $\mathbf{Z}_n$ is indicated by $\phi(n)$. This is known as Eulero's function.

Theorem 1. *The number of primes with n is equal to:*

$$\phi(n) = n \prod \left(1 - \frac{1}{p_j}\right),$$

where p_j varies in all the primes which are divisors of n (including n if it is prime).

This demonstration can be seen in texts of the Theories of Numbers. If n is a prime number Eulero's function $\phi(n)$ is reduced to:

$$\phi(n) = n(1 - \frac{1}{n}) = n - 1.$$

If n is a composite number it is reduced to: $\phi(n) < n - 1$.

Theorem 2 (Eulero's). *For any* $n > 2$ *and* $a : (a, n) = 1$

$$a^{\phi(n)} \equiv 1(\text{mod} n) \; \forall a \in \mathbf{Z_n}.$$

With Fermat's so-called Little Theorem one is able to consider a particular case as Eulero did whenever n is prime. In essence Fermat, had formulated a concept which was completely demonstrated to be a particular case of the preceding theorem. This was also demonstrated in various ways during the eighteenth century.

Theorem 3 (Fermat's). *If n is prime then:*

$$a^{n-1} \equiv 1(\text{mod} n) \; \forall a \in \mathbf{Z_n^*}.$$

2.2 Deterministic tests

Those procedures which allow the determination of prime numbers through the application of a certain algorithm are called *deterministic tests*. The theory of complexity, an important branch of Computer Science, allows one to quantify the computational difficulty of a specific procedure. In general, complexity is measured by the processing resources necessary for the implementation of the algorithms in terms of memory capacity used, time taken for their execution, etc. For the problem of determining the primality of an integer it is enough to refer to the time taken for the execution of the algorithm. The simplest deterministic test of primality for a number n is based on the successive division of n by all primes inferior to the square root of n. Naturally this test is not applicable to very large integers. There are many valid deterministic tests of primality in numbers smaller than a particular n. For example (Pomerance et al. 1980):

1. If $n < 1.373.653$ and satisfies Fermat's relation (par. 2.1) for base 2 and 3, then n is prime.
2. If $n < 25.326.001$ and satisfies Fermat's relation (par. 2.1) for base 2, 3 and 5 then n is prime.
3. If $n < 2.152.3002.898.747$ and satisfies Fermat's relation (par. 2.1) for base 2, 3, 5, 7 and 11 then n is prime.
4. If $n < 341.550.071.728.321$ and satisfies Fermat's relation (par. 2.1) for base 2, 3, 5, 7, 11, and 13 then n is prime.

2.3 Some deterministic tests

The important results of M. Agrawal, N. Kayal and N. Saxena appear in "Annals of Mathematics", where they have proposed a deterministic test based on the following:

Theorem 4. *p is prime if and only if*

$$(x-a)^p \equiv (x^p - a)(\text{mod}\, p),$$

where a is a number prime with p.

The simple demonstration is based on the fact that, if i comes between 0 and p the coefficients $\binom{n}{p}$ calculated on *modulo p*, in the binomial development of the first member of the preceding relation are null and, furthermore, $a^p \equiv a(\text{mod}\, p)$.

Vice-versa, if p is not prime one of its factors does not divide $\binom{n}{p}$ mod *(p)* and therefore, the indicated relation is not valid. The algorithm supplied by the authors, carried out in only 13 lines, allows one to discover whether a number is prime or composite.

The result of greatest theoretic interest, demonstrated by the authors in the work cited, is the following:

Theorem 5. *The asymptotic complexity of the algorithm is* $O^{\sim}(\log^{21/2} n)$ *where the symbol* $O^{\sim}(f(n))$ *is for* $O^{\sim}(f(n)\ poly(\log f(n))$, *where f(n) is any function of n.*

In practice, however, the authors recall that in many cases this is faster than indicated. Therefore one deals with an algorithm P, or actually an algorithm in which the time of execution is the function of a polynomial which depends on n. The other algorithms for the analysis of primality noted in literature are *NP*, or rather, their execution in a *polynomial* time depends on non-deterministic procedures. In 1986 Goldwasser and Kilian proposed a randomized algorithm, based on an elliptical curve which works in very wide hypotheses, in polynomial time for *quasi* all inputs. The algorithm certifies primality. The probabilistic tests of primality verify the null hypothesis $\mathbf{H}_0 : n$ is a prime number. If the hypothesis is not verified the number is surely composite. This is a statistical test in which the probability of errors of the second type, or rather of accepting a false hypothesis, is a number other than zero. Very little attention has been paid by scientific literature to these, very particular statistical tests.

The most noted statistical test of primality is that of Miller and Rabin, proposed in 1976. We define as *witness* a number which meets the requirements of Fermat's so-called Little Theorem to be a composite number. The test in question is based on the following:

Theorem 6. *If n is an odd composite number then the number of witnesses of which it is composed will be at least:* $(n-1)/2$.

Theorem 7. *Considering an odd integer and an integer s, the probability that a composite number is considered to be prime is less than* 2^s.

Empirical evidence shows that the probability that a composite number is found to be prime is actually, in the long term, less than that indicated. There have been shown to exist only 21,253 composite numbers in base 2 which are inferior to 25 billion and which satisfy Fermat's Little Theorem. These are called pseudo-primes. There is, therefore a probability of about 8×10^{-6} that a composite number n will satisfy the relation $2^{n-1} \equiv 1(\text{mod} n)$.

The problems of factorizing a number and of determining if a number is prime are by their nature diverse. In many processing procedures, however, these are treated together. In every case it is easier to determine whether a number is prime than to find all of its factors. Today, even with the super computers available and the improved algorithms which are known, it is not possible to factorize numbers having more than a few hundred digits.

3 The Sum Modulo *m* of Statistical Variables

The deterministic and non-deterministic methods co-exist, at times in the same procedure. Algorithms are being found which are always more efficient and easier to use. But there is no doubt that probabilistic tests of primality are the only ones applicable when n is *particularly* high and one hasn't the fortune to find oneself in a very particular situation. For instance, if the number is composite and divisible by one of the initial primes. Deterministic tests of primality can be applied, relatively quickly, to numbers having a very large amount of digits. There is, however, a limit on the number of digits as we learn from the *theory of complexity*. Probabilistic tests of primality furnish results which are completely acceptable in situations which are very general. They require negligible time to process and are those applicable in research situations Rizzi (1990), Scozzafava (1991).

Theorem 8. *Let two discrete s.v. (statistical variable) X, Y assume the value: $0, 1, 2, \ldots, m-1$. Let X be uniformly distributed, that is, it assumes the value $i\,(i = 0, 1, \ldots, m-1)$ with probability $1/m$, and let the second s.v. Y assume the value i with probability $(p_i, \sum_{i=1}^{m-1} p_i = 1, p_i \geq 0)$. Then, if the two s.v. are independent, it follows that the s.v. Z obtained as a sum modulo m: Z = X + Y(mod m) is uniformly distributed.*

Proof. If the s.v. X assumes the value i, then the s.v. Z can assume the values:

$$i, i+1, i+2, \ldots, m-1, 0, 1, 2, \ldots, i-1$$

respectively with probabilities:

$$p_0, p_1, p_2, \ldots, p_{m-1} - i, \ldots, p_{m-1}$$

assumed by the Y. If we let i assume the values $0, 1, 2, \ldots, m-1$, it follows that the s.v. Z assumes the general value $h(h = 0, 1, \ldots, m-1)$ with probability

$$\begin{cases} \frac{1}{m} p_h & \text{if } X = 0 \quad Y = h \\ \frac{1}{m} p_{h-1} & \text{if } X = 1 \quad Y = h-1 \\ \vdots & \vdots \qquad \vdots \\ \frac{1}{m} p_0 & \text{if } X = h \quad Y = 0 \\ \vdots & \vdots \qquad \vdots \\ \frac{1}{m} p_{h-1} & \text{if } X = m-1 \quad Y = h+1 \end{cases}$$

It follow immediately by summation:

$$P(Z = h) = \frac{1}{m} \sum_{i=0}^{h} p_i + \frac{1}{m} \sum_{i=h+1}^{m-1} p_i = \frac{i}{m}.$$

The above theorem can be easily generalized to the sum (mod m) of n s.v., one of which will be uniformly distributed.

Theorem 9. *Let two independent s.v. X and Y assume the values:* $0, 1, 2, \check{E}m - 1$ *respectively with probabilities*

$$p_0, p_1, \ldots, p_{m-1} \left(p_i \geq 0, \sum_{i=1}^{m-1} p_i = 1 \right)$$

$$q_0, q_1, \ldots, q_{m-1} \left(q_i \geq 0, \sum_{i=0}^{m-1} q_i = 1 \right)$$

Then, if the s.v. $Z = X + Y (\text{mod } m)$ is uniformly distributed and m is a prime number, at least one of the two s. v. X and Y is uniformly distributed.

Proof. The table of the sum *(mod m)* of the s.v. X and Y is as follows: As the X and Y are independent, the s. v. Z assumes the value 0 with probability

$$p_0 q_0 + \cdots + p_2 q_{m-2} + p_1 q_{m-1}$$

	0	1	2	$\ldots$	i	$\ldots$	$m-2$	$m-1$	
0	0	1	2		i		$m-1$	$m-1$	p_0
1	1	2	3		$i+1$		$m-1$	0	p_1
2	2	3	4		$i+2$		0	1	p_2
$\vdots$									
j	j	$j+1$			$i+j$		$j+m-2 \pmod m$	$j+m-1 \pmod m$	p_j
$\vdots$									
$m-1$	$m-1$	0			$i-1$		$m-3$	$m-2$	p_{m-1}
	q_0	q_1			q_i		q_{m-2}	q_{m-1}	1

Such probability, according to the hypothesis of uniform distribution of Z, must be $1/m$. By the same token, in order to compute the probabilities that the s.v. Z assumes the values $1, 2, \ldots, m-1$, the following system can be written:

$$\begin{cases} p_0q_0 & +p_{m-1}q_1 + \cdots + p_2q_{m-2} + p_1q_{m-1} = 1/m \\ p_1q_0 & +p_0q_1 \quad + \cdots + p_3q_{m-2} + p_2q_{m-1} = 1/m \\ p_2q_0 & +p_1q_1 \quad + \cdots + p_4q_{m-2} + p_3q_{m-1} = 1/m \\ \vdots & \\ p_{m-1}q_0 & +p_{m-2}q_1 + \cdots + p_1q_{m-2} + p_0q_{m-1} = 1/m \end{cases}.$$

If the p_i are known (the reasoning is the same if the q_i are known) the above system is a system of m equations in the m unknowns q_i. It follows:

$$q_i = \frac{1}{\Delta} \begin{vmatrix} p_0\, p_{m-1} & \ldots\, 1/m \,\ldots\, p_1 \\ p_1\, p_0 & \ldots\, 1/m \,\ldots\, p_2 \\ \vdots & \\ p_{m-1}\, p_{m-2} & \ldots\, 1/m \,\ldots\, p_0 \end{vmatrix},$$

where Δ is the determinant of the p_i coefficients, and it can be easily seen to be $\neq 0$ if at least one $p_i \neq 1/m (i = 0, 1, \ldots, m-1)$. (In the opposite case the s.v. X is uniformly distributed and the theorem is proved). In fact it is a "circulating" determinant. In order to show the theorem in general, it is sufficient to show that

$$q_i = q_0 \quad \forall i = 1, 2, \ldots, m-1.$$

In this case, as $\sum_{i=o}^{m-1} q_i \geq 0$ we have $q_i = 1/m$. Then, it is enough to show that

$$\begin{vmatrix} \frac{1}{m} & p_{m-1} & \cdots & p_1 \\ \frac{1}{m} & p_0 & \cdots & p_2 \\ \vdots & & & \\ \frac{1}{m} & p_{m-2} & \cdots & p_0 \end{vmatrix} = \begin{vmatrix} p_0 & p_{m-1} & \cdots & \frac{1}{m} & \cdots & p_1 \\ p_1 & p_0 & \cdots & \frac{1}{m} & \cdots & p_2 \\ \vdots & & & & & \\ p_{m-1} & p_{m-2} & \cdots & \frac{1}{m} & \cdots & p_0 \end{vmatrix}.$$

The two determinants are equal because, in order to transform the second into the first one, it is necessary to perform, due to the circulating nature of the permutation of the p_i, $m-2$ rows' inversions and $m-2$ inversions over the columns, that is $2(m-2)$ inversions over rows and columns in all. If an even number of inversions is performed, the sign of the determinant is unchanged. In this way, for instance, if $m-1=3$ we have that the number of q_0 is

$$\begin{vmatrix} \frac{1}{3} & p_1 & p_2 \\ \frac{1}{3} & p_2 & p_0 \\ \frac{1}{3} & p_0 & p_1 \end{vmatrix} = \begin{vmatrix} p_0 & \frac{1}{3} & p_2 \\ p_1 & \frac{1}{3} & p_0 \\ p_2 & \frac{1}{3} & p_1 \end{vmatrix}.$$

References

Agrawal, M., Kayal, N., & Saxena, N. (2004). Primes in p. *Annals of Mathematics, 160*, 781–793.

Goldwasser, S., & Kilian, J. (1986) Almost all primes can be quickly certified. In *Proceedings of the eighteenth annual ACM symposium on Theory of Computing* (pp. 316–329). New York: ACM Press.

Pomerance, C., Selfridge, J. L., & Wagstaff, Jr., S. S. (1980). The pseudoprimes to 25×10^9. *Mathematics of Computation, 35*, 1003–1026.

Rizzi, A. (1990). Some theorem on the sum modulo m of two independent random variables. *Metron, 48*, 149–160.

Scozzafava, P. (1991). Sum and difference modulo m between two independent random variables. *Metron, 49*, 495–511.

Part II
Cluster Analysis

An Algorithm for Earthquakes Clustering Based on Maximum Likelihood

Giada Adelfio, Marcello Chiodi, and Dario Luzio

Abstract In this paper we propose a clustering technique set up to separate and find out the two main components of seismicity: the background seismicity and the triggered one. We suppose that a seismic catalogue is the realization of a non homogeneous space–time Poisson clustered process, with a different parametrization for the intensity function of the Poisson-type component and of the clustered (triggered) component. The method here proposed assigns each earthquake to the cluster of earthquakes, or to the set of independent events, according to the increment to the likelihood function, computed using the conditional intensity function estimated by maximum likelihood methods and iteratively changing the assignment of the events; after a change of partition, MLE of parameters are estimated again and the process is iterated until there is no more improvement in the likelihood.

1 Introduction

A basic description of seismic events provides the distinction of earthquakes in foreshocks, aftershocks, mainshocks and isolated events. A cluster of earthquakes is formed by the main event of each sequence, its foreshocks and its aftershocks, that could occur before and after the mainshock, respectively. Isolated events are spontaneous earthquakes that do not trigger a sequence of aftershocks and because of this characteristic, space–time features of principal earthquakes (main and isolated events) are close to those of a Poisson process that is stationary in time, since the probability of occurrence of future events is constant in time irrespectively of the past activity, even if nonhomogeneous in space. Therefore, the seismogenic features controlling the kind of seismic release of background and clustered seismicity are not similar (Adelfio et al. 2006b), and to describe the seismicity of an area in space, time and magnitude domains, sometimes it is useful to study separately the features

G. Adelfio (✉)
Department of Statistical and Mathematical Sciences, University of Palermo,
viale delle Scienze, ed 13, 90128, Palermo, Italy
e-mail: adelfio@unipa.it

F. Palumbo et al. (eds.), *Data Analysis and Classification*,
Studies in Classification, Data Analysis, and Knowledge Organization,
DOI 10.1007/978-3-642-03739-9_3,

of *independent* events and *triggered* ones. Indeed, to estimate parameters of phenomenological laws useful for the description of seismicity, a reasonable definition of "earthquake cluster" is required; furthermore the prediction of the occurrence of large earthquakes (related to the assessment of seismic risk in space and time) is complicated by the presence of clusters of aftershocks, that are superimposed to the background seismicity, according to some (unknown) mixing parameter, and shade its principal characteristics. For these purposes the preliminary subdivision of a seismic catalog in background seismicity (represented by isolated events, that do not trigger any further event, and the mainshock of each seismic sequence) and clustered events, is sometimes required. At this regard, a seismic sequences detection technique is presented; it is based on MLE of parameters that identify the conditional intensity function of a model describing seismic activity as a clustering-process, which represents a slight modification of the ETAS model (Epidemic Type Aftershocks-Sequences model; Ogata, 1988, Ogata et al., 2004). Diagnostics for these models is discussed in Adelfio and Chiodi (2008).

In Sect. 2 conditional intensity function of point processes is introduced, focusing on the description of ETAS model and related models. In Sect. 3 the features of the proposed method are defined. Finally in Sect. 4 an example of application is proposed and some conclusive remarks for future works are reported.

2 Conditional Intensity Function of the Clustering Procedure

A seismic catalogue, assumed as realization of a space–time point process, contains information about seismic events occurred in a region, in a given time interval. In particular, given a seismic catalogue of n events, the i-th row of the catalogue gives quantitative information about the estimated latitude, longitude and depth (x_i, y_i, z_i), the time of occurrence (t_i) and the magnitude (m_i) of the seismic event U_i, $(i = 1, \ldots, n)$. In this paper the depth z will not be considered, since the high level of its measurement error.

To describe the features of the seismic activity of a space–time area the definition of a conditional intensity function is required. The conditional intensity function of a space–time point process can be defined as

$$\lambda(t, \mathbf{x}|\mathcal{H}_t) = \lim_{dt, d\mathbf{x} \to 0} \frac{E[N([t, t + dt) \times [\mathbf{x}, \mathbf{x} + d\mathbf{x})|\mathcal{H}_t)]}{\ell(dt)\ell(d\mathbf{x})}, \tag{1}$$

where $\ell(x)$ is the Lebesgue measure of x; $\mathcal{H}_t$ is the space–time occurrence history of the process up to time t, i.e. the σ-algebra of events occurring at times up to but not including t; $dt, d\mathbf{x}$ are time and space increments respectively, and $E[N([t, t + t) \times [\mathbf{x}, \mathbf{x} + d\mathbf{x})|\mathcal{H}_t)]$ is the history-dependent expected value of occurrence in the volume $\{[t, t + dt) \times [\mathbf{x}, \mathbf{x} + d\mathbf{x})\}$. The conditional intensity function completely identifies the features of the associated point process (Daley and Vere-Jones 2003) (i.e. if it is independent of the history but dependent only on the current time and the spatial locations (1) supplies a nonhomogeneous Poisson process; a constant conditional intensity provides a homogeneous Poisson process).

For a space–time point process, the log likelihood function is defined by

$$\log L = \sum_{i=1}^{n} \log \lambda(x_i, y_i, t_i) - \int_{T_0}^{T_{max}} \int_{\Omega_{xy}} \lambda(x, y, t) dx\, dy\, dt, \tag{2}$$

where (x_i, y_i, t_i) are the space–time coordinates of the $i-th$ event $(i = 1, 2, \ldots, n)$, $(T_0 - T_{max})$ is the observed period of time and Ω_{xy} is the space region.

2.1 The ETAS Model

The conditional intensity function used in our procedure is a variation of ETAS model, a self-exciting point process describing earthquakes catalogs as a realization of a branching or epidemic-type point process. The conditional intensity function of the ETAS model in a point x, y, t, m of the space–time-magnitude domain, conditioned to the space–time occurrence history of the process up to time t, denoted by $\mathcal{H}_t$, is defined by

$$\lambda(x, y, t, m|\mathcal{H}_t) = J(m) \left[\mu(x, y) + \sum_{t_j < t} g(t - t_j) f(x - x_j, y - y_j|m_j) \right], \tag{3}$$

where x_j, y_j, t_j, m_j are the space–time-magnitude coordinates of the observed events up to time t, $J(m)$ is the density of magnitude (Gutenberg and Richter 1944); inside the squared brackets there is the sum of the spontaneous activity $\mu(x, y)$ and the triggered one, given by the product of the time and space (conditioned to magnitude) probability distributions. The main hypothesis of the model states that all events, both a mainshock or an aftershock, have the possibility of generating offsprings.

In ETAS model, background seismicity $\mu(x, y)$ is assumed stationary in time, while time triggering activity is represented by a non stationary Poisson process according to the modified Omori's formula (Utsu, 1961). In this model, the occurrence rate of aftershocks at time t following the earthquake of time τ, is described by

$$g(t - \tau) = \frac{K}{(t - \tau + c)^p}, \quad \text{with } t > \tau \tag{4}$$

with K a normalizing constant, c and p characteristic parameters of the seismic activity of the given region; p is useful for characterizing the pattern of seismicity, indicating the decay rate of aftershocks in time.

In (3), $f(\cdot,\cdot)$ is the spatial distribution, conditioned to magnitude of the generating event; a number of its formulations are proposed in Ogata (1998), where the occurrence rate of aftershocks is related to the mainshock magnitude.

2.2 Intensity Function for a Particular Clustered Inhomogeneous Poisson Process

In our procedure, we assume that the seismic catalog is the realization of a clustered inhomogeneous Poisson process, assuming that the events of the background seismicity come from a space–time Poisson process (spatially inhomogeneous) and that among these there is a number k of mainshocks that can generate aftershocks sequences, inhomogeneous both in space and times, as a function of the magnitude of the main event. Differently from ETAS model we do not assume that each event can generate an offspring. Therefore, in our procedure we consider the following intensity function:

$$\lambda(x, y, t; \boldsymbol{\theta}) = \lambda_t \mu(x, y) + K_0 \sum_{\substack{j=1 \\ (t_j < t)}}^{k} g_j(x, y) \frac{\exp[\alpha(m_j - m_0)]}{(t - t_j + c_j)^{p_j}}, \tag{5}$$

where $\boldsymbol{\theta} = (\lambda_t, K_0, c_j, p_j, \alpha)$. In (5) t_j and m_j are time of the first event and magnitude of the mainshock of the cluster j, $g_j(x, y)$ is the space intensity of the cluster j and $\mu(x, y)$ is the background one; K_0 and λ_t are the weights of the clustered seismicity and of the background one, respectively. Background seismicity is assumed stationary in time, while time aftershock activity is represented by the modified Omori formula (Utsu 1961), with parameters c_j and p_j, relating the occurrence rate of aftershocks to the mainshock magnitude m_j, with α measuring the influence on the relative weight of each sequence, m_0 the completeness threshold of magnitude, i.e. the lower bound for which earthquakes with higher values of magnitude are surely recorded in the catalog.

In our approach space intensity, both of background seismicity $\mu(x, y)$ and of each cluster $g_j(x, y)$, $j = 1, \ldots, k$, is estimated by a bivariate kernel estimator: it is computed either using only the independent events (isolated and mainshocks) or points belonging to the cluster, including the mainshock, respectively.

The used bivariate kernel estimator is

$$\hat{f}(x, y) = \frac{1}{n h_x h_y} \sum_{i=1}^{n} K\left(\frac{x - X_i}{h_x}, \frac{y - Y_i}{h_y}\right), \tag{6}$$

where $K(\cdot,\cdot)$ is a generic bivariate kernel function and $\mathbf{h} = (h_x, h_y)$ is the vector of the smoothing constants. In both cases the smoothing constant is evaluated with Silverman's rule, that for the one-dimensional case (Silverman 1986) is

$h_{opt} = 1.06An^{-1/5}$, with A = min{standard deviation, range-interquartile/1.34}, which optimizes the estimator asymptotic behavior in terms of mean integrated square error and provides valid results on a wide range of distributions. In the evaluation of (5), different kinds of parametrization are considered to take into account for different assumptions on the seismicity of an area, (e.g. Omori's law parameters of the k clusters can be assumed equal or distinct in each cluster). The choices can be compared at the end of the procedure on the basis of the final likelihood values.

3 The Proposed Clustering Method

In our clustering procedure we assume that a catalog of n events may be partitioned in $k + 1$ sets, one relative to the background seismicity and k relative to clusters, according to partition $\mathcal{P}_{k+1}$. Then, on the basis of this partition, three types of events are identified: n_0 isolated points, k mainshocks and n_j points belonging to the j-th cluster ($j = 1, 2, \ldots, k$), where $\sum_{j=0}^{k} n_j = n$. The goal of this method is to find a good partition of events, according to likelihood function maximization, with respect the vector of parameters θ of the intensity function and the partition $\mathcal{P}_{k+1}$, by an iterative procedure. Briefly, given partition $\mathcal{P}_{k+1}$, we compute the estimate $\hat{\theta}$ which maximizes the likelihood function (2) and, on the basis of the estimated value $\hat{\theta}$, we look for a better partition moving single points from their current position to a new subset (a new cluster or the set of main events) such that the likelihood function increases, until a convergence criterion is achieved.

In our approach the likelihood is not complete, but it is conditioned on the assumption that a subset of primary events is known following a version of the original trigger model (see Ogata, 2001). The complete likelihood is simulated alternating the cluster identification and the maximization steps, as in E-M algorithm structure.

3.1 Finding a Candidate Cluster and Likelihood Changes

At each iteration s, given $\hat{\theta}^{(s)}$, the cluster r_h which maximizes the conditional intensity function is found for each unit U_h, ($h = 1, \ldots, n$) either an isolated or a clustered point; approximately this is obtained comparing the k contributions to the sum:

$$\sum_{\substack{j=1 \\ (t_{0_j} < t_h)}}^{k} g_j(x_h, y_h) \frac{\exp\left[\alpha(m_j - m_0)\right]}{(t_h - t_{0_j} + c)^p}$$

and assigning temporally each unit U_h to the cluster r that maximizes

$$I(t_h > t_{0_r}) g_r(x_h, y_h) \frac{\exp\left[\alpha(m_r - m_0)\right]}{(t_h - t_{0_r} + c)^p} \quad r = 1, 2, \ldots, k. \tag{7}$$

If the partition changes (from $\mathcal{P}_{k+1}^{(s)}$ to $\mathcal{P}_{k+1}^{*(s)}$) because of a movement of a single unit, we examine the change in the log-likelihood function $\log L(\boldsymbol{\theta}; \mathbf{x}, \mathbf{y}, \mathbf{t}, \mathcal{P}_{k+1}^{(s)})$. Schematically, kinds of change of partition are due to three different types of movement of units: unit U_h moves from background seismicity to cluster r (refereed as type A), unit U_h moves from cluster r to the set of background seismicity (type B) and unit U_h moves from cluster r to cluster q (type C).

We compute the variation in the log-likelihood function for each kind of movement (A, B and C) and for each possible change on the current partition induced by the movement of a unit U_h, $h = 1, 2, \ldots, n$, assuming that $\hat{\boldsymbol{\theta}}^{(s)}$ does not change in each iteration.

3.2 The Algorithm of Clustering

The technique of clustering that we propose leads to an intensive computational procedure, implemented by software R (R Development Core Team 2007). The main steps can be summarized as follows:

1. Iteration $s = 1$. The algorithm starts from a partition $\mathcal{P}_{k+1}^{(s)}$ found by a window-based method (similar to a single-linkage procedure) or other clustering hierarchical methods. Briefly, the starting method puts events in a cluster if in it there is at least an event inside a window of δ_s units of space and δ_t units of time. δ_s, δ_t are given as input.
2. Clusters with a minimum fixed number of elements are found out: the number k of clusters is determined.
3. Partition $\mathcal{P}_{k+1}^{(s)}$ is then completed with the set of isolated points, constituted by the n_0 points not belonging to clusters.
4. Estimation of the space seismicity (6) both for background and k clusters.
5. Maximum Likelihood Estimation of parameters: in (5) it is possible to assume either common Omori law parameters c and p over all cluster or varying c_j and p_j in each cluster (this could depend on the available catalog): as default, we consider the second type parametrization. An iterative simplex method is used for the maximization of the likelihood (2). $\hat{\boldsymbol{\theta}}^{(s)}$ is the value of the MLE.
6. Finding a better partition $\mathcal{P}_{k+1}^{(s)}$: for each unit U_h, either an isolated or a clustered point, the *best candidate cluster* r_h is found, according to the rule in (7).
7. Different kinds of movements are tried (type A, B or C, as in Sect. 3.1).
8. Points are assigned to the *best set* of events (best in the sense of likelihood).
9. Points are moved from clusters to background (and viceversa) if their movement increases the current value of the likelihood (2).
10. If no point is moved the algorithm stops, otherwise $\mathcal{P}_{k+1}^{(s)}$ is updated, $s = s + 1$ and the algorithm come back to step 2.

In the last steps (6–9), the likelihood (2), is computed using the current value of $\hat{\boldsymbol{\theta}}^{(s)}$. On the basis of the final partition and the final values of the estimates, the vector of estimated intensities for each point is computed.

4 Application to a Real Catalog and Final Remarks

The proposed method could be the basis to carry out an analysis of the complexity of the seismogenic processes relative to each sequence and to the background seismicity, separately. It has been applied to a catalog of 4,295 seismic events occurred in the Southern Tyrrhenian Sea from February 2000 to December 2005, providing a plausible separation of the different components of seismicity and clusters that have a good interpretability. In Figs. 1 and 2 the found clusters and some of their features are shown. The algorithm identified eight clusters, with a total of 964 events; the remaining 3,331 events were assigned to the background seismicity and the estimated parameters are $\hat{\alpha} = 0.2061$, $\hat{\lambda}_t = 0.000016$ and $\hat{K}_0 = 0.000389$. No relevant dependence of estimated parameters on the magnitude values has been observed.

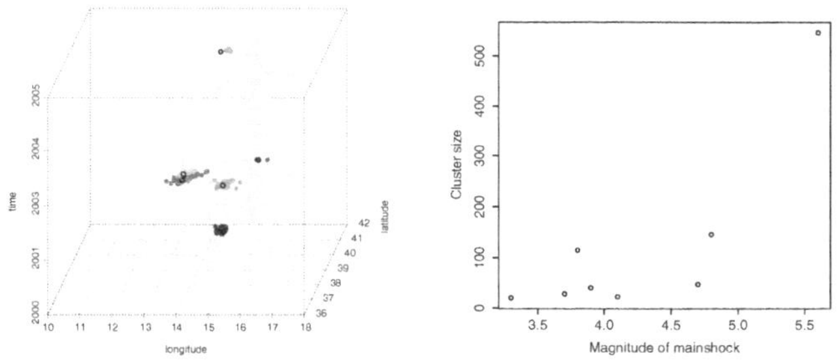

Fig. 1 *On the left:* space–time plot of clusters (*filled circles*) and isolated events (*asterisks*) of the Southern Tyrrhenian catalog from 2000 to 2005. (*Open circle* is used for main event of each cluster.) *On the right:* plot of clusters size vs. mainshocks magnitude

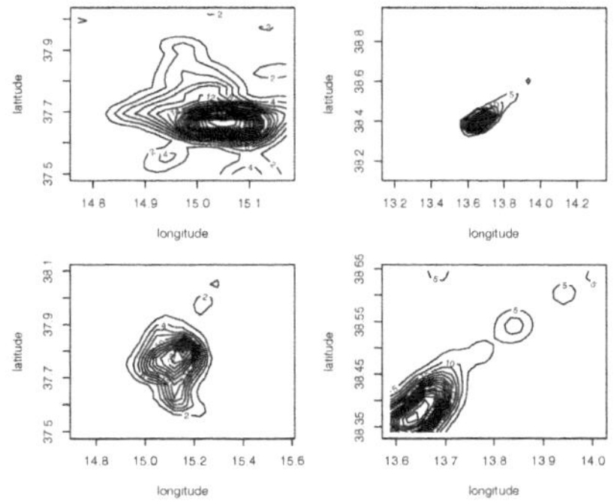

Fig. 2 Contour plot of main clusters spatial distribution

Comparing the current version of the clustering proposed method to its first version (Adelfio et al. 2006a) some extensions have been introduced. In this improved version the moving of points from their current position to a better set (in sense of likelihood) does not require the definition of fixed thresholds and, as described in Sect. 2.2, different kinds of parametrization are introduced allowing to take into account for different assumptions about the seismicity of an area (e.g. Omori law parameters). On the other hand, the optimization steps can be improved in the future, for instance, minimizing the computational burden of the algorithm, reducing the dependence of the convergence of the iterative algorithm on some initial choices.

References

Adelfio, G., Chiodi, M., De Luca, L, & Luzio, D. (2006a). Nonparametric clustering of seismic events. In S. Zani, A. Cerioli, M. Riani, & M. Vichi (Eds.), *Data analysis, classification and the forward search* (pp. 397–404). Berlin: Springer.

Adelfio, G., Chiodi, M., De Luca, L., Luzio, D., & Vitale, M. (2006b). Southern-tyrrhenian seismicity in space–time-magnitude domain. *Annals of Geophysics, 49*(6), 1139–1151.

Adelfio, G., & Chiodi, M. (2008). Second-order diagnostics for space–time point processes with application to seismic events. *Environmetrics*, doi:10.1002/env.961.

Daley, D. J., & Vere-Jones, D. (2003). *An introduction to the theory of point processes* (2nd edition). New York: Springer.

Gutenberg, B., & Richter, C. F. (1944). Frequency of earthquakes in California. *Bulletin of the Seismological Society of America, 34*, 185–188.

Ogata, Y. (1988). Statistical models for earthquake occurrences and residual analysis for point processes. *Journal of the American Statistical Association, 83*(401), 9–27.

Ogata, Y. (1998). Space–time point-process models for earthquake occurrences. *Annals of the Institute of Statistical Mathematics, 50*(2), 379–402.

Ogata, Y. (2001). Exploratory analysis of earthquake clusters by likelihood-based trigger models. *Journal of Applied Probability, 38*(A), 202–212.

Ogata, Y., Zhuang, J., & Vere-Jones, D. (2004). Analyzing earthquake clustering features by using stochastic reconstruction. *Journal of Geophysical Research, 109*(B05301), 1–17.

R Development Core Team. (2005). *R: A language and environment for statistical computing*. Vienna, Austria: R Foundation for Statistical Computing.

Silverman, B. W. (1986). *Density estimation for statistics and data analysis*. London: Chapman and Hall.

Utsu, T. (1961). A statistical study on the occurrence of aftershocks. *Geophysical Magazine, 30*, 521–605.

A Two-Step Iterative Procedure for Clustering of Binary Sequences

Francesco Palumbo and A. Iodice D'Enza

Abstract Association Rules (AR) are a well known data mining tool aiming to detect patterns of association in data bases. The major drawback to knowledge extraction through AR mining is the huge number of rules produced when dealing with large amounts of data. Several proposals in the literature tackle this problem with different approaches. In this framework, the general aim of the present proposal is to identify patterns of association in large binary data. We propose an iterative procedure combining clustering and dimensionality reduction techniques: each iteration involves a quantification of the starting binary attributes *and* an agglomerative algorithm on the obtained quantitative variables. The objective is to find a quantification that emphasizes the presence of groups of co-occurring attributes in data.

1 Introduction

Association rules (AR) mining aims to detect patterns of association in large transaction data bases. Transactions are binary sequences recording the presence/absence of a finite set of attributes or items. Let $\mathcal{A} \subseteq \mathcal{I}$ and $\mathcal{B} \subseteq \mathcal{I}$ be two disjoint subsets of the set $\mathcal{I}$ of binary attributes, the expression $(A \Longrightarrow B)$ (to be read *if* A *then* B) represents a general association rule, where $A \in \mathcal{A}$ and $B \in \mathcal{B}$. In the simplest case, both A and B refer to the presence of a single attribute (whereas $\bar{A}$ and $\bar{B}$ refer to the absence).

In other words, an AR is a logical relation: A refers to the antecedent part or *body* and B is termed consequent part or *head*. The association strength of a rule is often measured by the indexes

A. Iodice D'Enza (✉)
Dipartimento di Scienze Economiche e Finanziarie Università di Cassino, Rome
e-mail: iodicede@unina.it

F. Palumbo et al. (eds.), *Data Analysis and Classification*,
Studies in Classification, Data Analysis, and Knowledge Organization,
DOI 10.1007/978-3-642-03739-9_4,

$$\begin{aligned} support &= P(A \cap B), \\ confidence &= P(B|A) \equiv \frac{P(A \cap B)}{P(A)}. \end{aligned} \tag{1}$$

The *support* represents the empirical probability that A and B occur together: it expresses the association strength. The confidence represents the logical dependence of B from A: in other words it is the conditional probability of B given A.

Although the support/confidence framework is the most commonly used in AR mining, Lenca et al. (2007) show how most association indexes can be suitably adopted in this context.

A general formalization for a simple rule is then:

$$A \Longrightarrow B : \{support = P(A \cap B), confidence = P(B \mid A)\}.$$

The aim of association rule mining is to identify interesting rules revealing the presence of items that systematically occur together. Since huge amounts of rules may prevent the detection of truly interesting patterns of association, most of the AR mining procedures deal with support and confidence thresholds to define the rules to extract or discard. It is hard to identify associations characterizing low-occurring attributes: To study low co-occurrences low supports thresholds have to be set; however, the lower the support threshold, the larger the number of generated AR will be.

In the literature, different proposals tackle the problem of detecting patterns of association among low occurring attributes through the combined use of clustering and AR mining procedures. Plasse et al. (2007) propose an approach characterized by a former clustering of the starting attributes (item) and a latter AR mining within the obtained groups. Iodice D'Enza et al. (2007) propose another approach involving clustering techniques: in this case, clustering is on binary records (transactions, individuals) characterizing data; thus AR are mined separately in the groups and compared to global association patterns.

The aim of the present proposal is to identify patterns of association in large binary data. In particular, we propose an iterative procedure combining clustering and factorial (dimensionality reduction) techniques: each iteration consists of (1) a quantification of the starting binary attributes and (2) an agglomerative algorithm on the obtained quantitative variables. The objective is to find a quantification that better emphasizes the presence of groups of co-occurring attributes in data.

The next section is dedicated to the definition of data structure and to the description of the proposed procedure; in the last section we present an example of application of the procedure to a famous benchmark data set (BMS-Webview).

2 Clustering and Dimensionality Reduction

The starting data structure $\mathbf{S}$ is a $(n \times p)$ Boolean matrix characterized by n binary vectors (objects) considered with respect to p Boolean variables (e.g. *presence* or *absence* of an attribute). The disjunctive coded version of $\mathbf{S}$ is indicated by the $(n \times 2p)$ matrix $\mathbf{Z}$. Let us indicate with K the number of clusters and let $\underline{\pi} = \pi_1, \pi_2, \ldots, \pi_K$ be a random vector containing the probabilities for a sequence to belong to the kth group. We aim at identifying groups of sequences such that $\sum_j \left(E\left[P(G_k \mid A_j)\right] - E\left[P(G_k)\right]\right)$ is maximized, it corresponds to estimate the vector $\underline{\pi}$ given $\mathbf{Z}$.

The aim is to partition the sequences in homogeneous groups, each characterized by a set of highly co-occurring attributes.

Let us consider the matrix $F_{K \times P}$ with general element f_{kj} being the frequency of the jth attribute in the kth group. It is easy to verify the following relation holds:

$$\sum_{j=1}^{p} \left(E\left[P(G_k \mid A_j)\right] - E\left[P(G_k)\right]\right) = \sum_{k=1}^{K} \left(\sum_{j=1}^{p} f_{kj} P(G_k \mid A_j) - f_{k.} P(G_k) \right)$$
$$= \sum_{k=1}^{K} \left(\sum_{j=1}^{p} f_{kj} \frac{P(G_k \cap A_j)}{P(A_j)} - f_{k.} P(G_k) \right). \tag{2}$$

The quantities $\sum_{j=1}^{p} f_{kj} P(G_k|A_j) = \sum_{j=1}^{p} \frac{f_{kj}^2}{f_{.j}}$ and $f_{k.} P(G_k) = \sum_{j=1}^{p} \frac{f_{kj}^2}{n}$, where $P(G_k \mid A_j)$ and $P(G_k)$ indicate the empirical probabilities. According to Lauro and Balbi (1999) the following relation results:

$$\frac{1}{n} \sum_{k=1}^{K} \left(\sum_{j=1}^{p} \frac{f_{kj}^2}{f_{.j}} - \frac{f_{k.}^2}{n} \right) = \frac{1}{n} \sum_{k=1}^{K} \sum_{j=1}^{p} f_{.j} \left(\frac{f_{kj}}{f_{.j}} - f_{k.} \right)^2. \tag{3}$$

Considering the identity in (3), it is worth to notice that the problem in (2) can be rewritten as a special case of Non-Symmetrical Correspondence Analysis (NSCA) (Palumbo and Verde 1996).

Clustering huge and sparse datasets is a challenging task. Non-hierarchical methods, like k-means algorithm (MacQueen 1967), or one of its versions, seem to be preferable when dealing with large amounts data. In the framework of transaction datasets, several algorithms ensure good results, fast computation and scalability. However, the a priori choice of the number of clusters affects the solution considerably. Hierarchical methods do not require as input the number of clusters and they provide a dendrogram representation of the clustering structure. The computational effort is in higher hierarchical than in non-hierarchical procedures. Recent advances in numerical methods and the more computational capability of modern computers suggests that we reconsider hierarchical algorithms. We propose the application of a non-hierarchical agglomerative procedure to linear combinations (factors) of the starting binary variables, since this ensures a two-fold advantage: the number of variable (factors) to consider is critically reduced; furthermore factors are orthogonal and quantitative. Thus we can use the agglomerative algorithm on Euclidean distances and the Ward linkage method.

An algebraic formalization of the problem leads to the following expression

$$\frac{1}{n}\left[\mathbf{X}^{\mathsf{T}}\mathbf{Z}(\Delta)^{-1}\mathbf{Z}^{\mathsf{T}}\mathbf{X} - \frac{p}{n}\left(\mathbf{X}^{\mathsf{T}}\mathbf{1}\mathbf{1}^{\mathsf{T}}\mathbf{X}\right)\right]\mathbf{U} = \Lambda\mathbf{U}, \tag{4}$$

where $\Delta = diag(\mathbf{Z}^{\mathsf{T}}\mathbf{Z})$ and $\mathbf{1}$ is a n-dimensional vector of ones. The trace of the target matrix

$$\left[\mathbf{X}^{\mathsf{T}}\mathbf{Z}(\Delta)^{-1}\mathbf{Z}^{\mathsf{T}}\mathbf{X} - \frac{p}{n}\left(\mathbf{X}^{\mathsf{T}}\mathbf{1}\mathbf{1}^{\mathsf{T}}\mathbf{X}\right)\right] \tag{5}$$

corresponds to quantity in (3). Furthermore, we have denoted $\mathbf{X}$ the $n \times K$ disjunctive coded matrix that assigns each sequence to one of the K groups. Λ and $\mathbf{U}$ are the eigenvalues diagonal matrix and the eigenvector matrix, respectively. Note that the expression (5) is that same quantity minimized in the NSCA (Palumbo and Verde 1996) and that it corresponds to maximize the sum of squared Euclidean distances between group profiles and the marginal profile.

In expression (5), since matrices $\mathbf{X}$, Λ and $\mathbf{U}$ are unknown, a direct solution is not possible. Then, according to Vichi and Kiers (2001), we adopt an iterative two-step procedure involving factorial analysis and clustering. Thus, two steps are required to alternately maximize the relation

$$\sum_j \left(E\left[P\left(G_k|A_j\right)\right] - E\left[P(G_k)\right]\right).$$

More specifically, the first step aims to find the optimal quantification of items, given the partition of sequences defined by matrix $\mathbf{X}$: the quantity to maximize is

$$\sum_j \left(E\left[P(G_k|A_j^*)\right] - E\left[P(G_k)\right]\right)$$

with $A_j^* = A_j u_j$ being the quantified version of the starting binary variable A_j. Note that in this step $\sum_j (E\left[P(G_k)\right]$ is fixed.

The second step maximizes the quantity $\sum_j (E\left[P(G_k)\right]$, since the n binary sequences are re-assigned to the K groups, according to the items quantification obtained in step 1: the changing matrix in this step is $\mathbf{X}$, while the quantification of items A_j, $(j = 1, \ldots, p)$, is fixed.

The algorithm proceeds as follows:

- *Step 0*: pseudo-random generation of matrix $\mathbf{X}$.
- *Step 1*: a singular value decomposition is performed on the matrix resulting from (5), obtaining the matrix Ψ, such that

$$\Psi = \mathbf{Z}(\Delta)^{-1}\mathbf{Z}^{\mathsf{T}}\mathbf{X}\mathbf{U}\Lambda^{\frac{1}{2}}; \tag{6}$$

Rows of Ψ contain the elements of the linear combination of the starting binary sequences with weights u_j resulting from the quantification of items A_j.

- *Step 2*: a Euclidean squared distance based agglomerative algorithm on the projected sequences (Ψ matrix) obtaining an update of the matrix **X**.

Steps 1 and 2 are repeated until the clusters become stable. Notice that both factorial analysis and the clustering are based on the squared Euclidean distance, which ensures the consistency of the whole procedure: in fact clustering of sequences is not performed on original sequences but on their factorial coordinates. In addition, the two alternate steps lead to satisfy the same criterion.

3 Example

In order to illustrate an application of the proposed procedure we refer to the data set $BMS - Webview1$, already used in several proposals in the literature. In particular, after a pre-processing phase the dimensions of the **S** matrix are $n = 3,289$ and $p = 160$.

The number of clusters is set to $K = 5$, Fig. 1 shows the dendrogram representation of the hierarchy resulting from the agglomerative clustering of the starting binary sequences. All the resulting clusters but the central one have a similar size: the larger cluster represented in the center of Fig. 1 represents the null centroid cluster and it contains all the sequences with almost all null values. Furthermore, the figure shows that a little higher cutoff point determines four groups; a little lower cutoff level determines six groups. The choice is on $K = 5$ since the solution for $K = 6$ just splits the null centroid cluster; in the $K = 4$ solution the first two clusters starting from the left of the figure are merged together and it is not a desirable effect. The random initialization of **X** matrix is performed.

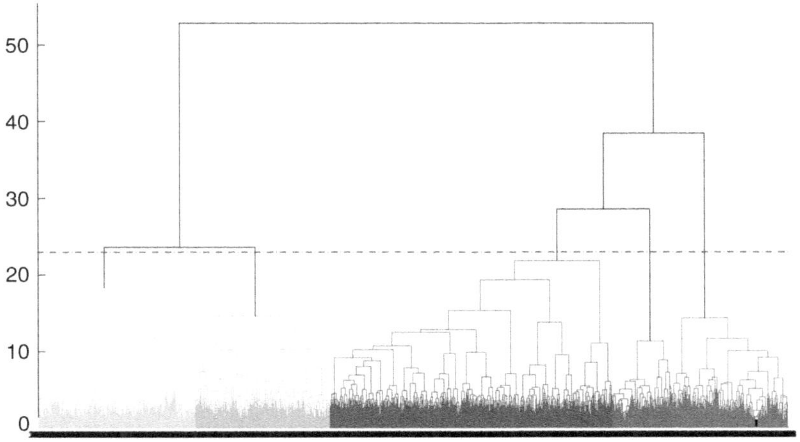

Fig. 1 Dendrogram representation resulting from the agglomerative clustering of starting sequences

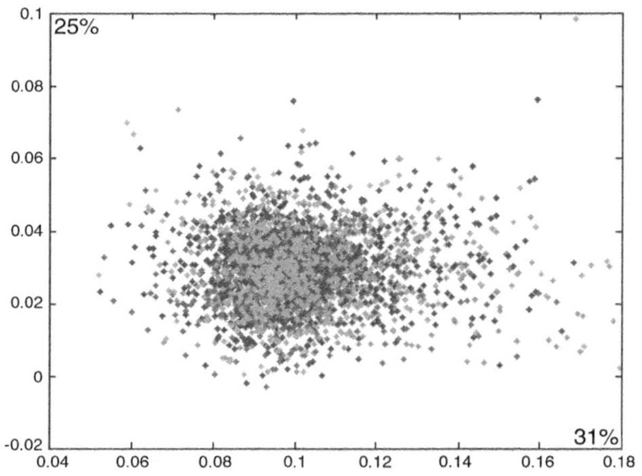

Fig. 2 Factorial representation of the sequences: randomized assignment to clusters

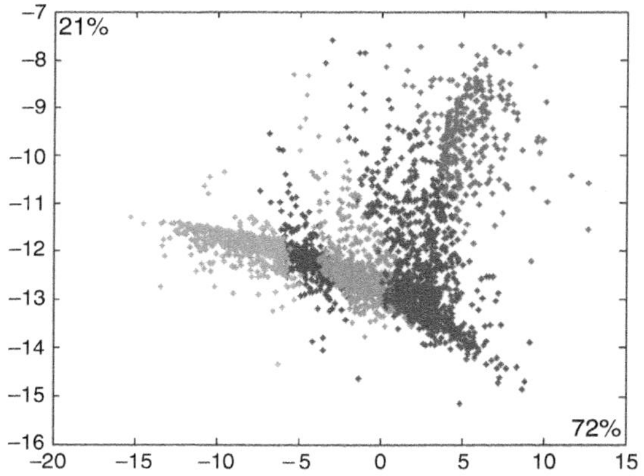

Fig. 3 Factorial representation of the sequences: procedure-based assignment to clusters

A further input parameter of the procedure is the number q, indicating the dimensionality of the solution. Obtained dimensions are a combination of the original p dimensions ($q \ll p$). Although choosing the optimal number of dimensions is an open issue, it is possible to choose the first q dimensions that preserve a satisfying percentage of the original variability characterizing data. In this application, choice is to consider the first $q = 4$ dimensions, enough to provide a suitable approximation of starting data structure.

Figure 2 shows the starting configuration of points, corresponding to the random matrix $\mathbf{X}$. The two steps are repeated five times before convergence. The solution

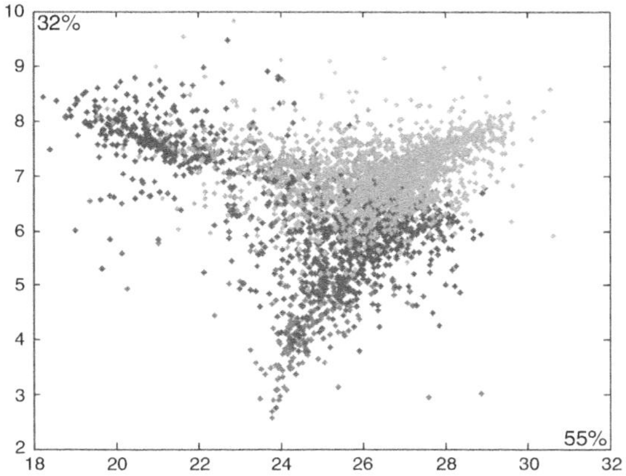

Fig. 4 Factorial representation of the sequences: procedure-based assignment to clusters

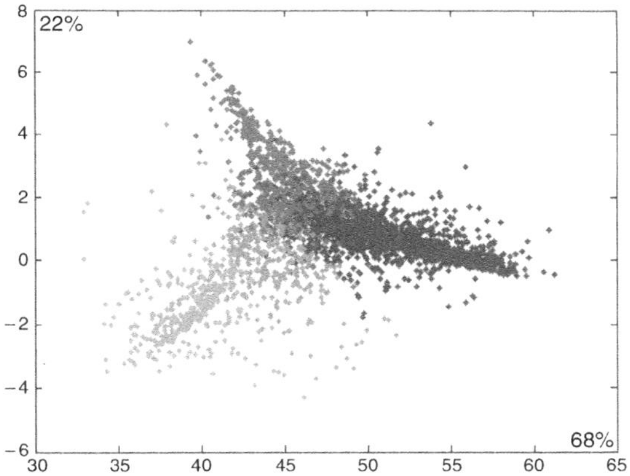

Fig. 5 Factorial representation of the sequences: procedure-based assignment to clusters

is represented in Fig. 3. Figures 4 and 5 show the evolving configuration of points: it clearly emerges that the points belonging to different clusters are step-by-step better separated. In the last configuration of points it is evident that they characterize they characterize different areas of the map. The different areas of the factorial map are characterized by different attributes: this means that it is possible to describe the obtained clusters of sequences in terms of sets of relevant attributes. A final consideration is on the number of clusters: we assumed the number of clusters to be fixed in the different iterations; use of an agglomerative algorithm determines a hierarchy at each iteration. Then the user may choose a different number of clusters modifying the cutoff point of the corresponding dendrogram.

References

Iodice D'Enza, A., Palumbo, F., & Greenacre, M. (2007). Exploratory data analysis leading towards the most interesting simple association rules. *Computational Statistics and Data Analysis*, doi:10.1016/j.csda.2007.10.006.

Lauro, C. N., & Balbi, S. (1999). The analysis of structured qualitative data. *Applied Stochastic Models and Data Analysis, 15*(1), 1–27.

Lenca, P., Vaillant, B., Meyer, P., & Lallich, S. (2007). Association rule interestingness measures: Experimental and theoretical studies. In G. Guillet & H. J. Hamilton (Eds.), *Quality measures in data mining*. Berlin: Springer.

MacQueen, J. (1967). Some methods for classification and analysis of multivariate observations. In *Proceedings of the Fifth Berkeley Symposium on Mathematical Statistics and Probability*, University of California Press.

Palumbo, F., & Verde, R. (1996). Analisi Fattoriale Discriminante Non-Simmetrica su Predittori Qualitativi (in italian). In *Atti della XXXVIII Riunione scientifica della Societ Italiana di Statistica*, Rimini, Italy.

Plasse, M., Niang, N., Saporta, G., Villeminot, A., & Leblond, L. (2007). Combined use of association rules mining and clustering methods to find relevant links between binary rare attributes in a large data set. *Computational Statististics Data Analysis*, doi:10.1016/j.csda.2007.02.020.

Vichi, M., & Kiers, H. A. L. (2001). Factorial k-means analysis for two way data. *Computational Statistics and Data Analysis, 37*, 49–64.

Clustering Linear Models Using Wasserstein Distance

Antonio Irpino and Rosanna Verde

Abstract This paper deals with the clustering of complex data. The input elements to be clustered are linear models estimated on samples arising from several sub-populations (typologies of individuals). We review the main approaches to the computation of metrics between linear models. We propose to use a Wasserstein based metric for the first time in this field. We show the properties of the proposed metric and an application to real data using a dynamic clustering algorithm.

1 Introduction

Complex data arise ever more frequently. Indeed, the overwhelming growth of data is pushing the development of new data mining techniques. The output of these techniques is in general the description of patterns of data expressed as descriptions of clusters of individuals (as in market segmentation techniques), and of models describing sets of individuals (time series, linear regressions, etc.). More broadly, the output may be viewed as statistical models describing the synthesis of sets of individuals or the causal relationships among the descriptors of such sets of individuals. The amount of acquired knowledge is growing rapidly. It is therefore urgent to develop techniques allowing the analysis of this kind of information. Much work has already done using several approaches: symbolic, functional, fuzzy and compositional data analysis can be used concurrently to analyze data not presented in a standard way. This paper aims to contribute to the field of model classification by proposing a new metric for comparing linear models.

According to McCullagh (2007), a statistical model is a set of probability distributions associated with the sample space S, while a parameterized statistical model is a parameter set Θ together with a function $P : \Theta \rightarrow \mathcal{P}(S)$, which assigns to each

A. Irpino (✉)
Dipartimento di Studi Europei e Mediterranei, Second University of Naples, Via del Setificio, 15, Belvedere di San Leucio, 81100 Caserta, Italy
e-mail: antonio.irpino@unina2.it

F. Palumbo et al. (eds.), *Data Analysis and Classification*,
Studies in Classification, Data Analysis, and Knowledge Organization,
DOI 10.1007/978-3-642-03739-9_5,

parameter point $\theta \in \Theta$ a probability distribution P_θ on S. Here $\mathcal{P}(S)$ is the set of all probability distributions on S.

In previous contributions (Irpino et al. 2006; Irpino and Verde 2006; Irpino and Romano 2007) we showed how Wasserstein distance is particularly efficient for clustering data described by multi-valued variables or distributions, as for the statistical models considered by McCullagh (2007). In this paper, we introduce this distance in order to compare linear models by considering their parameters as estimates obtained from samples of sub-populations (typologies) of individuals. In particular, we refer to the simplest linear model, the classical linear regression model, as a particular case of the parameterized statistical model.

According to McCullagh (2007), in the standard linear regression model $Y \sim N(X\beta, \sigma^2 I_n)$ on $\mathcal{R}^n$, each parameter β is defined in $\mathcal{R}^p$ and σ^2 in $[0, \infty)$.

In some proposals, models are compared with respect to the point estimates of the parameters. The clustering of modeled data is infrequently mentioned in the literature. Some methods have been developed for clustering time series (Piccolo 1990). In the context of consumer preferences analysis, a new approach has recently been presented by Romano et al. (2006) where a distance is built as the convex linear combination of two Euclidean distances embedding information both of the estimated parameters and of model fitting. Finally, in the framework of functional data analysis, another contribute was proposed by Ingrassia et al. (2003) to identify homogenous typologies of curves.

The new distance we introduce permits comparison with the interval estimates of parameters, instead of point estimates. It also takes into consideration the point estimates, and the size and shape of the variability of the distributions of estimators. It is worth observing that Wasserstein distance is commonly related to the errors of the models and to the sizes of samples used for estimating. In that sense, the Wasserstein distance appears to be consistent with the definition of the parameterized statistical model, taking into consideration the distribution of the estimators.

This paper is organized in the following way. Section 2 introduces the notation of the input data for the proposed clustering algorithm. Section 3 shows the general schema of the dynamic clustering algorithm, the Wasserstein distance used as allocation function (Sect. 3.1) and for the construction of a representation function that is consistent with the criterion optimized in the algorithm (Sect. 3.2). In Sect. 4, we show an application using Bank of Italy dataset from the 2004 Households Survey. Section 5 gives some perspectives and concluding remarks.

2 Input Data and the Clustering Algorithm

Given a set of typologies of individuals identifying sub-populations (market segments, strata of a population, grouped individuals, etc.) on which a regression analysis has been performed using the same set of variables related by the same causal relationship. We want to study the linear dependency structure of the response variable y_i from the set of p predictors $\{X_{1i}, \ldots, X_{pi}\}$. From each typology a linear

Table 1 Input data table

Typology	$\widehat{\beta}_0$	$\widehat{\beta}_1$	...	$\widehat{\beta}_p$	s_{β_0}	s_{β_1}	...	s_{β_p}	n	s	R^2
...	...	...	...	...	...	...	...	...	...	...	...
i	$\widehat{\beta}_{0i}$	$\widehat{\beta}_{1i}$	...	$\widehat{\beta}_{pi}$	$s_{\beta_{0i}}$	$s_{\beta_{1i}}$	...	$s_{\beta_{pi}}$	n_i	s_i	R_i^2
...	...	...	...	...	...	...	...	...	...	...	...

model is estimated. A typology i is usually described by the structured information containing at least: the point estimates of the $p+1$ parameters $\widehat{\beta}_{ji}$, the standard error of the parameters $s_{\beta_{ji}}$, the sample size n_i, the standard error of the model s_i and the goodness of fit index R^2 (Table 1).

In this paper, for the models we use the OLS (Ordinary Least Squares) method. Under the classical hypothesis it is known that the statistics:

$$T = \frac{\frac{B_{ji}-\beta_{ji}}{\sqrt{Var(B_{ji})}}}{\sqrt{\frac{s_i^2}{\sigma_i^2}}}$$

is a Student's T pdf with $(n_i - p - 1)$ degrees of freedom. It is usually used as pivotal quantity for the interval estimates of parameters. Thus, we assume this structured information in order to cluster typologies similar association structure. In order to compare two linear models, we assume they are similar if they generate the same interval estimates for the parameters.

3 Dynamic Clustering of Linear Models

The dynamic clustering algorithm (DCA) (Diday 1971) represents a general reference for unsupervised non hierarchical iterative clustering algorithms. Let E be a set of elements. The general DCA looks for the partition $P \in P_k$ of E in k classes C_h (for $h = 1, \ldots k$), among all the possible partitions P_k, and the vector $L \in L_k$ of k prototypes G_h (for $h = 1, \ldots k$) representing the classes in P, such that, the following Δ fitting criterion between L and P is minimized

$$\Delta(P^*, L^*) = Min\{\Delta(P, L) \mid P \in P_k, L \in L_k\}. \tag{1}$$

Such a criterion is defined as the sum of dissimilarity or distance measures $\delta(\cdot)$ of fitting between each object belonging to a class $C_h \in P$ and the class representation (prototype) $G_h \in L$. In our case, we propose using an L_2 Wasserstein based distance as a criterion function.

3.1 Wasserstein Distance for Distributions

If F and G are the distribution functions of two random variables f and g respectively, the Wasserstein L2 metric is defined as (Gibbs and Su 2002)

$$d_W(F,G) := \left(\int_0^1 \left(F^{-1}(t) - G^{-1}(t)\right)^2 dt\right)^{1/2}, \tag{2}$$

where F^{-1} and G^{-1} are the quantile functions of the two distributions. Irpino and Romano (2007) proved that the proposed distance can be decomposed as

$$d_W^2 = \underbrace{\left(\mu_f - \mu_g\right)^2}_{Location} + \underbrace{\left(\sigma_f - \sigma_g\right)^2}_{Size} + \underbrace{2\sigma_f\sigma_g\left(1 - \rho_{QQ}(F,G)\right)}_{Shape}, \tag{3}$$

where

$$\rho_{QQ}(F,G) = \frac{\int_0^1 \left(F^{-1}(t) - \mu_f\right)\left(G^{-1}(t) - \mu_g\right) dt}{\sigma_i \sigma_j} \tag{4}$$

is the correlation of the quantiles of the two distributions as represented in a classical QQ plot. It is worth noting that $0 < \rho_{QQ} \leq 1$ differs from the classical range of variation of the Bravais–Pearson's ρ.

Let us consider the estimates of two linear models from two samples (h and k) on the same set of p variables. Considering the interval estimation of model parameters, the squared Wasserstein distance between two estimated models is

$$\begin{aligned} d_W^2(y_h, y_k) = & \sum_{j=0}^{p} \left(\hat{\beta}_{jh} - \hat{\beta}_{jk}\right)^2 \\ & + \sum_{j=0}^{p} \left(s_{\beta_{jh}} \sqrt{\frac{n_h - p - 1}{n_h - p - 3}} - s_{\beta_{jk}} \sqrt{\frac{n_k - p - 1}{n_k - p - 3}}\right)^2 \\ & + 2\sum_{j=0}^{p} s_{\beta_{jh}} s_{\beta_{jk}} \sqrt{\frac{n_h - p - 1}{n_h - p - 3}\frac{n_k - p - 1}{n_k - p - 3}} \\ & \left[1 - \rho_{QQ}(T_{n_h-p-1}, T_{n_k-p-1})\right]^2. \end{aligned} \tag{5}$$

If the two samples have the same size $n_h = n_k$, the squared distance can be simplified as

$$d_W^2(y_h, y_k) = \sum_{j=0}^{p} \left(\hat{\beta}_{jh} - \hat{\beta}_{jk}\right)^2 + \frac{n_h - p - 1}{n_h - p - 3} \sum_{j=0}^{p} \left(s_{\beta_{jh}} - s_{\beta_{jk}}\right)^2. \tag{6}$$

For large samples, T_{n_h-p-1} and T_{n_k-p-1} can be approximated by the normal standard distribution, then (6) is equal to zero and the distance can be simplified as

$$d_W(y_h, y_k) = \sqrt{\sum_{j=0}^{p} \left(\hat{\beta}_{jh} - \hat{\beta}_{jk}\right)^2 + \sum_{j=0}^{p} \left(s_{\beta_{jh}} - s_{\beta_{jk}}\right)^2}. \tag{7}$$

This is also verified in the case of normal assumption of the errors, where the β's are estimated using the maximum likelihood method.

3.2 Representation and Allocation Functions

Considering a dynamic clustering of I linear models associated with I typologies into k classes, we use the Wasserstein distance as a criterion function. A prototype G_h associated to a class C_h is an element of the space of description of E, and it should be represented as a linear model. The algorithm is initialized by generating k random clusters. Generally, the criterion $\Delta(P, L)$ is based on an additive distance on the p descriptors. The criterion function can be written as

$$\Delta(P, L) = \sum_{h=1}^{k} \sum_{i \in C_h} d_W^2(y_i, G_h) = \sum_{h=1}^{k} \sum_{i \in C_h} \sum_{j=0}^{p} \left[\int_0^1 \left(F_{ji}^{-1}(t) - F_{jh}^{-1*}(t)\right)^2 dt\right]. \tag{8}$$

The k prototypes of the k clusters are obtained by minimizing the criterion in (8). The prototype of the generic cluster C_h, for the j-th parameter, is a distribution where the t-th quantile is the mean of the t-th quantiles of the distribution for the j-th parameter computed for the typologies belonging to the cluster. Under the condition of applicability of (7), the criterion function can be written as

$$\Delta(P, L) = \sum_{h=1}^{k} \sum_{i \in C_h} d_W^2(y_i, G_h) = \sum_{h=1}^{k} \sum_{i \in C_h} \sum_{j=0}^{p} \left[\left(\hat{\beta}_{ji} - \hat{\beta}_{jh}^*\right)^2 + \left(s_{\beta_{ji}} - s_{\beta_{jh}}^*\right)^2\right]. \tag{9}$$

We obtain a prototypal linear model where, for the generic cluster h, the prototypal parameters $\hat{\beta}_{jh}^*$ are the means of the point estimates of the parameters $\hat{\beta}_{ji}$ (of $i \in C_h$) and the prototypal standard deviations $s_{\beta_{jh}}^*$ are the means of the standard deviations $s_{\beta_{ji}}$ of the models belonging to the cluster C_h.

It is interesting to note that the Wasserstein distance can be considered for the definition of an inertia measure that satisfies the Huygens theorem of decomposition of inertia. Indeed, we showed (Irpino and Verde 2006; Irpino et al. 2006) that it can be considered as an extension of the Euclidean distance between quantile functions and it is consistent with the within criterion minimized in the dynamical clustering algorithm. Once the prototypes are determined, the algorithm runs until convergence upon a stationary point. Each linear model of E is allocated to the cluster according to the minimal Wasserstein distance to the prototype. The set of prototypes that minimize the within criterion is recomputed.

4 An Application Using Bank of Italy Household Survey Data

The Survey on Household Income and Wealth (SHIW) began in the 1960s with the aim of gathering data on the incomes and savings of Italian households. The sample used in the most recent surveys (2004) comprises about 8,000 households (24,000 individuals), distributed over about 300 Italian municipalities. We have extracted 80 typologies of households broken down by region and size (single person households, couple households, medium sized household with three to four members, large sized households with more than four members). For each sub-sample we have estimated the classic consumption–income relation, or as J. M. Keynes called it, the "consumption function":

$$\text{NDCONSUMP} = \text{fi}_0 + \text{fi}_1 \times \text{INCOME} + \text{error}, \quad (10)$$

where NDCONS is the consumption of non durable goods and services, β_0 is the "autonomous (non durable) consumption" of a household, β_1 is the "marginal propensity to consume" non durable goods and services and INCOME is the disposable income (after taxes and transfer payments). The analysis using 7,851 households (257 were dropped because they were considered anomalous) gives the following model:

$$\begin{array}{lllll} \text{NDCONSUMP} = & 6,879.17 & + 0.473 \times \text{INCOME} + & \text{ERROR} \\ & (145.166) & (0.0043) & (6,678.0) \end{array},$$

where the standard errors are reported in the parentheses. The R^2 statistics is equal to 0.6024. In order to give the same role to the parameters for the computation of the distance, we standardize the original data using the global mean and the global standard deviation of "INCOME" and "CONSUMPTION", resulting in the following model

$$\begin{array}{lll} \text{NDCONSUMPst} = 0.776162 \times \text{INCOMEst} + & \text{ERROR} \\ (0.007117) & (0.63061) \end{array}.$$

Then we estimated the same model for 73 typologies of households (seven of which have been added in the most similar typology for region and size because the sample size was less than ten households). We then performed a DCA choosing eight clusters, described in Table 2. Figure 1 is the graph of the lines related to the eight prototypes and the line computed using all households. Avoiding comments on the economic aspects of the consumption function, we may note that the consumption behaviors of the eight clusters are not in general related to the geographic location of households. Rather, it seems more related to household size. Analysing only the

Table 2 Clusters and prototypes obtained from the Dynamic Clustering ($k = 8$)

Cluster (#)	Members Prototypal model parameters CONSUMPTIONst$_c$ = fi$_{0c}$ + fi$_{1c}$ × INCOMEst$_c$ (in brackets the prototypal standard errors)
1	Marche_Couple; Marche_Medium; Sicily_Large; Trentino AA_Couple; Umbria_Couple; Umbria_Large
(6)	$\beta_0 = 0.21935(0.00852)$ $\beta_1 = 0.92368(0.00761)$
2	Abruzzo_Large; Molise_Large; Abruzzo_Medium; Calabria_Medium; Campania_Large; Campania_Medium; Emilia R_Couple; Emilia R_Large; Emilia R_Medium; Emilia R_Single; Friuli VG_Single; Lazio_Couple; Lazio_Medium; Liguria_Couple; Liguria_Medium; Lombardy_Medium; Molise_Medium; Piedmont_Medium; Val d'Aosta_Medium; Sardinia_Medium; Trentino AA_Medium; Tuscany_Couple; Tuscany_Large; Tuscany_Medium; Veneto_Couple; Veneto_Medium
(24)	$\beta_0 = 0.04801(0.01240)$ $\beta_1 = 0.66531(0.00723)$
3	Basilicata_Large; Calabria_Large; Basilicata_Medium; Calabria_Single; Marche_Single; Piedmont_Couple; Val d'Aosta_Couple; Puglia_Large; Puglia_Medium
(7)	$\beta_0 = -0.20084(0.00591)$ $\beta_1 = 0.55390(0.00663)$
4	Basilicata_Couple; Basilicata_Single; Calabria_Couple; Campania_Couple; Lazio_Single; Molise_Single; Puglia_Single; Sardinia_Couple; Sardinia_Single; Veneto_Single
(10)	$\beta_0 = 0.03414(0.02003)$ $\beta_1 = 1.03627(0.02035)$
5	Friuli VG_Large; Trentino AA_Large; Friuli VG_Medium; Lazio_Large; Sardinia_Large; Umbria_Medium
(5)	$\beta_0 = 0.28947(0.03252)$ $\beta_1 = 0.44637(0.01944)$
6	Liguria_Large; Piedmont_Large; Val d'Aosta_Large; Lombardy_Large; Marche_Large; Veneto_Large
(4)	$\beta_0 = 0.49887(0.03842)$ $\beta_1 = 0.70009(0.02037)$
7	Abruzzo_Couple; Abruzzo_Single; Campania_Single; Friuli VG_Couple; Liguria_Single; Lombardy_Couple; Lombardy_Single; Piedmont_Single; Val d'Aosta_Single; Puglia_Couple; Sicily_Couple; Sicily_Medium; Tuscany_Single
(12)	$\beta_0 = -0.07468(0.00272)$ $\beta_1 = 0.83643(0.00345)$
8	Molise_Couple; Sicily_Single; Trentino AA_Single; Umbria_Single
(4)	$\beta_0 = 0.29868(0.01198)$ $\beta_1 = 1.21674(0.01929)$

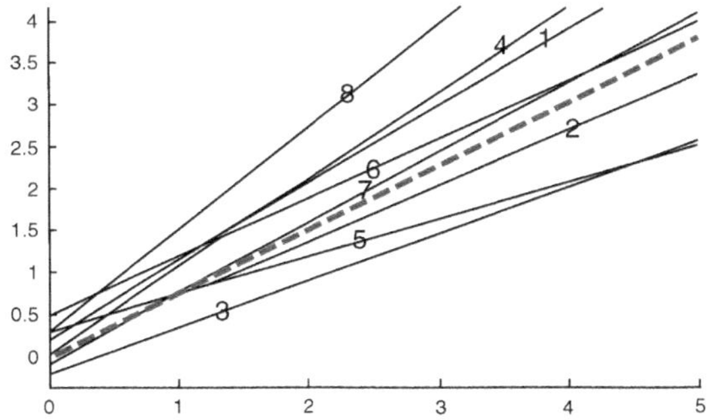

Fig. 1 The eight prototype lines and the line estimated from all the households (*dashed and bold line*)

distribution of income and consumption in general, the clusters are generated on the basis of the geographic location of households, displaying in general lower income (and consumption) in the south of Italy compared to the north.

5 Conclusions and Perspectives

This paper introduces a new metric for the comparison of linear models based on the comparison of their parameters. The satisfaction of the Huygens theorem allows its use with several techniques such as metric multidimensional scaling, hierarchical clustering based on the Ward criterion, and it permits extending the classical measure for the evaluation of the clustering output based on the comparison of the within and the between inertia. On the other hand, it is important to consider a correct data preprocessing in order to give the same role to the parameters to be compared. A great effort is needed to manage the correlation structure of the estimators of the parameters within and between the models to be compared. In this case, the Minkowski extension of the distance represents an approximation of excess for the distance between two models described by p parameters. The question of a multivariate version of Wasserstein distance is still open for discussion and a general and analytic solution has not been offered until now (Cuesta-Albertos et al. 1997).

References

Cuesta-Albertos, J. A., Matrán, C., & Tuero-Diaz, A. (1997). Optimal transportation plans and convergence in distribution. *Journal of Multivariate Analysis, 60*, 72–83.

Diday, E. (1971). La méthode des Nueées dynamiques. *Revue de statistique appliquée, 19*(2), 19–34.

Gibbs, A. L., & Su, F. E. (2002). On choosing and bounding probability metrics. *International Statistical Review, 7*(3), 419–435.

Ingrassia, S., Cerioli, A., & Corbellini, A. (2003). Some issues on clustering of functional data. In: M. Shader, W. Gaul, & M. Vichi (Eds.), *Between data science and applied data analysis* (pp. 49–56). Berlin: Springer.

Irpino, A., & Romano, E. (2007). Optimal histogram representation of large data sets: Fisher vs piecewise linear approximations. *RNTI, E-9*, 99–110.

Irpino, A., & Verde, R. (2006). A new Wasserstein based distance for the hierarchical clustering of histogram symbolic data. In: V. Batanjeli, H. H. Bock, A. Ferligoj, & A. Ziberna, (Eds.), *Data science and classification, IFCS 2006* (pp. 185–192). Berlin: Springer.

Irpino, A., Verde, R., & Lechevallier, Y. (2006). Dynamic clustering of histograms using Wasserstein metric. In: A. Rizzi, & M. Vichi (Eds.), *COMPSTAT 2006 – Advances in computational statistics* (pp. 869–876). Berlin: Physica.

McCullagh, P. (2007). What is a statistical model? *The Annals of Statistics, 30*(5), 1225–1310.

Piccolo, D. (1990). A distance measure for classifying ARIMA models. *Journal of Time Series Analysis, 11*, 153–164.

Romano, E., Giordano, G., & Lauro, C. N. (2006). An inter model distance for clustering utility function. *Statistica Applicata, 18*(3), 521–533.

Comparing Approaches for Clustering Mixed Mode Data: An Application in Marketing Research

Isabella Morlini and Sergio Zani

Abstract Practical applications in marketing research often involve mixtures of categorical and continuous variables. For the purpose of clustering, a variety of algorithms has been proposed to deal with mixed mode data. In this paper we apply some of these techniques on two data sets regarding marketing problems. We also propose an approach based on the consensus between partitions obtained by considering separately each variable or subsets of variables having the same scale. This approach may be applied to data with many categorical variables and does not impose restrictive assumptions on the variable distribution. We finally suggest a summarizing fuzzy partition with membership degrees obtained as a function of the classes determined by the different methods.

1 Introduction

Clustering mixed feature-type data is a task frequently encountered in marketing research. It may occur, for instance, in the field of segmentation, when mining descriptive competitor's products is aimed at grouping competitors according to the characteristics of their product. The purpose for clustering is to allow the marketing and the sales program of a company to focus on contrasting the subset of products that are most likely to compete its offering. In addition to continuous variables such as price and technical characteristics, products may be described by presence or absence of various optional accessories and by other categorical or nominal variables. Such mixed mode data are often encountered in other disciplines like, for example, psychiatry and medicine. Although a number of studies have provided guidelines for other clustering problems, there appear to be few recommendations about the best strategy to use with this type of data (Everitt and Merette 1990). Some recent studies have extended the k-means algorithm to cluster mixed numerical and categorical variables (see, e.g., Ahhmad and Dey 2007). The k-means algorithm designed to analyze data with categorical variables proposed by Zhang et al. (2006)

I. Morlini (✉)
DSSCQ, Università di Modena e Reggio Emilia, Modena, Italy
e-mail: isabella.morlini@unimore.it

F. Palumbo et al. (eds.), *Data Analysis and Classification*,
Studies in Classification, Data Analysis, and Knowledge Organization,
DOI 10.1007/978-3-642-03739-9_6,

is also applicable to cluster mixed mode data. In this paper other approaches are considered and applied to a real problem. The first one involves the application of classical standard techniques like the k-means or the self organizing map (SOM) algorithm (Kohonen 1984) to the data standardized in some ways. The second is the application of the mixture model for large data sets implemented in the package SPSS and described in Chiu et al. (2001) and Zhang et al. (1996), based on the EM algorithm. We then propose a further approach based on the consensus between partitions obtained considering separately each variable or considering subsets of continuous and categorical variables. Since there is no evidence in the literature of the superiority of one method over the others, we then suggest to find a final fuzzy partition whose membership function is obtained by considering the classes reached with the different techniques.

2 Obtaining Partitions with Mixed Mode Data

There are a variety of methods which might be used for clustering mixed mode data. Some of these are based on the application of standard hierarchical or non hierarchical techniques to the Gower's dissimilarity index or to the data standardized in some ways (see, e.g., Milligan and Cooper 1988). Other methods are based on mixture models assuming suitable distributions for the quantitative and the categorical variables. Some of these models, for example, assume that categorical variables arise from underlying unobservable, continuous variables (see, e.g., Everitt 1988; Everitt and Merette 1990; Coleman and Woodruff 2000), some others impose a multinomial distribution for the qualitative variables. The main drawbacks of this second class of models is that they impose very restrictive assumptions on the distributions (such as, for example, the independency of variables) and, in general, they can be applied to data set with few categorical variables.

The third way to proceed which we suggest in this work consists in finding a consensus partition which summarizes partitions separately obtained for subsets of variables. Each subset may consist in a single variable or in variables having the same scale.

If we denote by P_k a partition of a set on n objects and by $\mathbf{P} \equiv \{P_1, P_2, \ldots, P_m\}$ the set of m different partitions of the same objects (which may be obtained by changing the clustering algorithm or by choosing different variables), the consensus partition C is that minimizing a loss function between the set $\mathbf{P}$ and C, subject to some constrains. Constrains may regard the number of groups in C or the state of being a member of $\mathbf{P}$ or belonging to the set $\mathbf{A}$ of all possible partitions in g groups of n objects. Let $\Delta(P_k, P_l)$ be a measure of the difference between P_k and P_l. A common definition for $\Delta(P_k, P_l)$ is the number of pairs of objects which belong to different classes in P_k and P_l:

$$\Delta(P_k, P_l) = \sum_{1 \leq i \leq j \leq n} (c_{ijk} - c_{ijl})^2, \tag{1}$$

where $c_{ijk} = 1$ if object i and object j belong to the same class in P_k and $c_{ijk} = 0$ otherwise, $(i, j = 1, \ldots, n; k = 1, \ldots, m)$. The consensus partition is found by solving:

$$\min \sum_{k=1,\ldots,m} \Delta(P_k, C). \tag{2}$$

If (2) is subject to $C \in \mathbf{P}$ then the consensus partition is called *medoid* (see Gordon and Vichi 1998). If (2) is subject to $C \in \mathbf{A}$, then C is called *median* (Kaufman and Rousseeuw 1990). The medoid partition is the one maximizing the sum of the Rand indexes between all the other partitions. The algorithm for finding the median partition is described in Gordon and Vichi (1998). For clustering n objects with q quantitative variables and c categorical variables, we find a primary partition P_q using the q quantitative variables and a primary partition P_c using the c categorical variables. We then find the median partition of P_q and P_c. In order to reduce as much as possible the influence of other factors except for the variable choice, in both primary partitions we use variables standardized in the interval [0, 1] and we choose the same clustering method and the same number of groups.

An alternative consensus partition is found for the set of $q + c$ primary partitions obtained by considering each single variable. Here we find the medoid partition, which is computationally less demanding and may be regarded as providing an approximate solution to the problem of obtaining the median partition.

Even though simulation studies aimed at comparing clustering techniques are quite common in the literature, examining differences in algorithms and assessing their performance is nontrivial and conclusions depend on the data structure and on the simulation study itself. For these reasons, in this paper, we only apply our consensus method and different techniques to the same real data sets and we try to reach some insights about the characteristics of the different methods by looking at the Rand index computed for each couple of partitions. We then suggest not to choose one technique over the others, but to apply several algorithms to the same data and then find a final fuzzy partition whose membership function depends on all clusters obtained. This way to proceed is particularly convenient in marketing research, since it allows to discover secondary segments of clients or competitors, besides the primary segments.

3 Some Illustrative Applications

The methodology described earlier is first illustrated by application to the analysis of partitions of 25 home theater models of different brands. Seven variables are considered to identify clusters of models which might be the closest competitors. Three are quantitative features: price (in euros), power (in kilowatts) and number of speakers. The other four are dichotomic variables: the presence of DVD recorder, of wireless technology, of DVX player, of a radio data system. We standardize quantitative variables to the range [0,1]. In Table 1, within each column,

Table 1 Clusters obtained with different techniques and final fuzzy partition

Model	k-Means (Gower)	SOM	Mixt. model	Consensus		Membership degrees		
				Median	Medoid	1	2	3
Akai 4200	1	1	1	1	1	1	0	0
Hitachi K180	1	1	1	1	1	1	0	0
Lg DAT200	1	1	1	1	1	1	0	0
Waitec HTXE	1	1	1	1	1	1	0	0
Genesis AV3	1	1	1	1	2	0.8	0.2	0
Kenwood SLIM1	1	1	1	1	1	1	0	0
Orion HTS2965	1	1	1	1	1	1	0	0
Pioneer DCS323	1	1	3	2	1	0.6	0.2	0.2
Samsung UP30	1	1	1	1	1	1	0	0
Samsung TWP32	1	1	1	1	1	1	0	0
Sharp AT1000	1	1	1	1	1	1	0	0
Teac PLD2100	1	1	1	1	1	1	0	0
Jbl DSC1000	3	3	3	3	2	0	0.2	0.8
Kenwood 5.1D	2	3	2	3	1	0.2	0.4	0.4
Panasonic HT88	3	3	3	3	2	0	0.2	0.8
Philips RGB500	1	1	2	2	1	0.6	0.4	0
Pioneer HTW	2	3	2	3	1	0.2	0.4	0.4
Sony PALPRO	1	1	1	1	1	1	0	0
Technics DV290	3	3	3	2	2	0	0.4	0.6
Thomson DPL943	3	3	3	3	1	0.2	0	0.8
Jvc THR1	2	2	2	2	2	0	1	0
Kenwood CIN5.4	2	3	2	3	2	0	0.6	0.4
Pioneer RCS9H	2	3	2	3	3	0	0.4	0.6
Sony RH7000	2	3	2	3	3	0	0.4	0.6
Yamaha YHT941	2	3	2	3	2	0	0.6	0.4

models assigned to the same label belong to the same cluster. We set the number of clusters equal to 3 since an explorative analysis with hierarchical clustering techniques show that the optimal "cut" of all dendrograms obtained with different linkages is in three groups. The first column of Table 1 reports results obtained with the k-means cluster analysis using the Gower's coefficient of similarity. The second column reports results reached with SOM. The third column reports clusters obtained with the mixture model implemented in SPSS. The fourth column reports the median consensus between partition P_q, obtained with the k-means algorithm applied to the three quantitative variables, and P_c, obtained with the k-means algorithm applied to the four dichotomic variables (considering as numerical the values 0 and 1). The fifth column reports the medoid partition among the sevens obtained with the k-means algorithm applied to each single variable. Of course, for each binary variable, groups reached are simply the ones with models having the optional accessories and the ones with models without them. Labels reported for each partition are not the original ones. Groups have been re-labelled in order to indicate with the same label the closest group in the other partitions. This has been done by analyzing the contingency tables for each couple of partitions and the centroids. Once the groups have been re-labelled, the final fuzzy partition which we propose in this paper is easily obtained by computing, for each object, the proportion of labels in each row. The fuzzy partition is reported in the last three columns of Table 1: in each column there is the membership degree for each group. The membership degrees of an object sum up to one and thus each object has the same total influence. Furthermore, degrees formally resemble the probabilities of the object of being a member of the corresponding cluster. It is clear that some models have very similar characteristics since they are grouped together by all methods. These models belong to a single - homogeneous - cluster also in the fuzzy partition. Some other models have different behaviors with the different clustering techniques. For these objects it is coherent to assign a membership degree to different clusters, which shows the relationships with more than one group and detects market sub-segments of competitors, besides primary segments. A second observation regards the structure of segments shown by the fuzzy partition. In cluster 3 there are only models with a membership degree less than one (no one model belongs exclusively to this segment). In cluster 1 there are many objects with a membership function equal to 1, while in cluster 2 there is only one model belonging exclusively to this group. Due to this behavior, this last model seems to be an outlier and the assumption is supported by the analysis of the row profile later shown in Table 3.

In order to characterize the three groups in the fuzzy partition, we may compute the cluster weighted means. Means reported in Table 2 show that the three clusters highlight three specific segments of home theater models. Cluster 1 groups the cheapest models with the smallest power, the fewest number of speakers and, in general, no optional accessories at all. Cluster 3, on the contrary, groups the most expensive models with high power, a great number of speakers and quite all accessories. Cluster 2 groups models in an intermediate situation: they have high power and a great number of speakers, but they are not so expensive and may not have all optional accessories considered. Some models may have primary competitors in one group and secondary competitors in other groups. Let us consider, for example, models reported in Table 3. The first one (Genesis AV3) has the main competitors in group 1 because it does not have optional accessories. In a less degree, however, it has secondary competitors in group 2 because of the price, the power and the number of speakers. On the contrary, the second model (Philips RGB500) primary belongs to group 1 for the cheap price and the small number of speakers, but it may

Table 2 Clusters weighted means in the fuzzy partition

Cluster	1	2	3
Price	280	483	479
Power	434	895	849
No. of speakers	4.9	6.5	6.5
DVD recorder	0.1	0.8	0.5
Wireless technology	0.0	0.6	0.8
DVX	0.0	0.6	1.0
Radio data system	0.7	0.8	1.0

Table 3 Row profile of some models

Model	Price	Power	Speakers	DVD rec.	Wireless	DVX	Radio d.s.
Genesis AV3	520	600	6	0	0	0	0
Philips RGB500	330	600	5	1	0	0	1
Panasonic HT88	550	900	6	0	1	1	1
Jvc THR1	494	810	7	1	1	0	0

Table 4 Rand index between couples of partitions

	P_c	P_p	k-Means	SOM	Mixture	Median cons	Medoid cons
P_c	1.00	0.59	0.93	0.83	0.95	0.83	0.66
P_p	0.59	1.00	0.62	0.60	0.61	0.60	0.61
k-Means	0.93	0.62	1.00	0.90	0.88	0.81	0.71
SOM	0.83	0.60	0.90	1.00	0.78	0.87	0.69
Mixture	0.95	0.61	0.88	0.78	1.00	0.87	0.62
Median cons	0.83	0.60	0.81	0.87	0.87	1.00	0.61
Medoid cons	0.66	0.61	0.71	0.69	0.62	0.61	1.00

have competitors in group 2 due to the presence of some optional accessories. The third model (Panasonic HT88) strongly belongs to group 3 for the presence of many optional accessories but it has a secondary class of competitors in group 2 for the price not so expensive. As mentioned before, the fourth model (Jvc THR1) behaves like an outlier, being the only one exclusively belonging to group 2. This model, indeed, despite the high desirable technical characteristics (power and speakers), does not have fundamental accessories like the DVX and the radio data system.

Table 4 reports the Rand index computed for each couple of partitions. Results may give an insight about the characteristics of the different methods. The median and the medoid consensus lead to partitions quite similar to those obtained with the other techniques. Using Gower's coefficient and the mixture model, the obtained partitions have the greatest similarity index with the partition reached with only categorical variables and the smallest index with the one reached with only quantitative variable. SOM shows a similar behavior. The result that using Gower's index may be reduced the importance of the quantitative variables in the classification is not new in literature. The partition which seems less depending on categorical variables is the medoid consensus between the seven partitions reached with each single variable.

The second data set (http://www.pubblicitaitalia.it/audiweb.asp) regards 70 internet domains and contains four continuous variables describing the visits in the month of April 2006 and three dichotomic variables regarding the presence of some features. Variables are: reach, in % (Reach), pages per person (Person), time per person (Time), number of visits per person (Visit), the option of registration (Registration), the presence of newsletters (News) and the presence of RSS (Rss). Even if the optimal cuts in dendrograms obtained with hierarchical techniques seem to be in more than four groups, we set a number of clusters equal to 4, in order to hold down the number of segments and better characterize the competitors of each

Table 5 Clusters weighted means in the fuzzy partition

cluster	1	2	3	4
Reach	6.19	5.13	7.73	5.03
Person	25.17	17.55	30.26	11.88
Time	648	499	822	334
Visit	2.99	2.26	3.74	1.93
Registration	0.77	0.83	0.80	0.27
News	0.51	0.07	0.46	0.20
Rss	0.29	0.10	0.49	0.06

Table 6 Rand index between couples of partitions

	P_c	P_p	k-Means	SOM	Mixture	Median cons	Medoid cons
P_c	1.00	0.51	0.55	0.56	0.58	0.79	0.67
P_p	0.51	1.00	0.55	0.47	0.51	0.64	0.47
k-Means	0.55	0.55	1.00	0.82	0.86	0.48	0.58
SOM	0.56	0.82	0.82	1.00	0.77	0.48	0.57
Mixture	0.58	0.86	0.86	0.77	1.00	0.47	0.61
Median cons	0.79	0.48	0.48	0.48	0.47	1.00	0.55
Medoid cons	0.67	0.58	0.58	0.57	0.61	0.55	1.00

domain. Table 5 reports the weighted means of each variable in the final fuzzy partition. Here again there are two clusters grouping objects in two opposite situations: domains highly visited and, in general, with the presence of optional features (group 3) and domains less visited with no optional features (group 4). Clusters 2 and 3 contain domains in intermediate situations. Here again there are no objects exclusively belonging to the top segment (group 3) but there are some domains exclusively belonging to the worst segment (group 4). Then the market of internet domains, as well as the market of home theaters, seems to be fuzzier for objects belonging to the top segment of competitors and less indistinct for objects belonging to the worst segment. In group 2 and 3 all membership degrees are less than one. This means that there are no typical row profiles for explaining intermediate situations and domains in these groups have different characteristics.

Table 6 reports the Rand index computed for each couple of partitions. In this example the consensus partitions show less degrees of similarity with the other techniques and present the highest similarity indexes with the primary partitions of continuous and categorical variables. All methods do not seem to reduce the importance of the continuous variables in the classification. Indeed, it must be considered that in this example continuous variables are more numerous than dichotomic ones. It must be also considered that the number of groups has been kept small to be suitable for marketing purposes and is not the optimal one, as shown by previous explorative analysis.

4 Discussion

Although it is clearly impossible to generalize from the results presented, it does appear that the method proposed is likely to give results comparable to those obtained with other well-established techniques. The major practical advantages of the method formulated here is that it does not impose restrictive assumptions on the distributions, it may be applied to data sets involving many categorical variables and it does not seem to reduce the importance of the continuous variables in the classification. In the examples we only have quantitative and binary variables, but the procedure may be applied also to categorical and ordinal variables. Indeed, for obtaining the median consensus, we may find more than two primary partitions of variables having the same scale.

In this work we only suggest a possible, simple and non parametric way to proceed for clustering mixed mode data in marketing research. Our aim is not to show the superiority of one method over the others but to enlarge the set of techniques to be applied in explorative analysis. Indeed, rather than choosing one method over the others, we have shown that the final fuzzy partitions obtained by considering all methods applied are able to determine sub-segments of competitors besides primary segments and so to be useful, in practice, in marketing research.

Future research involves the application of different data recording. Rather than standardized in the interval [0 1] quantitative variables can be transformed to a continuous pair of doubled variables, using their z-scores (see Greenacre 2007, p. 184). For clustering data with a different scale, a transformation that may work well may be obtained by computing for each quantitative variables their respective z scores (by subtracting the mean and dividing by the standard deviation) and then creating the following doubled versions: positive value $= \frac{1+z}{2}$ and negative value $= \frac{1-z}{2}$. Even though it has some negative values, the range is limited.

References

Ahhmad, A., & Dey, L. (2007). A k-means clustering algorithm for mixed numeric and categorical data. *Data and Knowledge Engineering, 63*(2), 503–527.

Chiu, T., Fang, D., Chen, J., Wang, Y., & Jeris, C. (2001). A robust and scalable clustering algorithm for mixed type attributes in large database environment. In *Proceedings of the seventh ACM SIGKDD international conference on knowledge discovery and data mining* 263–268. San Francisco, CA.

Coleman, D. A., & Woodruff, D. L. (2000). Cluster analysis for large datasets: An effective algorithm for maximizing the mixture likelihood. *Journal of Computational and Graphical Statistics, 9*(4), 672–688.

Everitt, B. S. (1988). A finite mixture model for the clustering of mixed mode data. *Statistics and Probability Letters, 6*, 305–309.

Everitt, B. S., & Merette, C. (1990). The clustering of mixed-mode data: A comparison of possible approaches. *Journal of Applied Statistics, 17*(3), 284–297.

Gordon, A. D., & Vichi, M. (1998). Partitions of partitions. *Journal of Classification, 15*, 265–285.

Greenacre, M. (2007). *Correspondence analysis in practice*. New York: Chapman and Hall.

Kaufman, L., & Rousseeuw, P. J. (1990). *Finding groups in data: An introduction to cluster analysis*. New York: Wiley.
Kohonen, T. (1984). *Self-organization and associative memory*. London: Springer.
Milligan, G. W., & Cooper, M. C. (1988). A study of standardization of variables in cluster analysis. *Journal of Classification, 5*, 181–204.
Zhang, T., Ramakrishnan, R., & Livny, M. (1996). BIRCH: An efficient data clustering method for very large databases. In: *Proceedings of the ACM SIGMOD Conference on Management of Data* 103–114. Montreal, Canada.
Zhang, P., Wang, X., & Song, P. X. (2006). Clustering categorical data based on distance vectors. *JASA, 101*(473), 355–367.

The Progressive Single Linkage Algorithm Based on Minkowski Ultrametrics

Sergio Scippacercola

Abstract This paper focuses on the problem to find an ultrametric whose distortion is close to optimal. We introduce the Minkowski ultrametric distances of the n statistical units obtained by a hierarchical Cluster method (single linkage). We consider the distortion matrix which measures the difference between the initial dissimilarity and the ultrametric approximation. We propose an algorithm which by the application of the Minkowski ultrametrics reaches a minimum approximation. The convergence of the algorithm allows us to identify when the ultrametric approximation is at the local minimum. The validity of the algorithm is confirmed by its application to sets of real data.

1 Introduction

Cluster analysis is designed to detect hidden *groups* or *clusters* in a set of objects which are described by data such that the members of each cluster are similar to each other while groups are hopefully well separated (Bock 1996; Gordon 1996). We define a partition P_s of a population P of n statistical units consisting in s, non-empty, classes $C_1, C_2, \ldots, C_s$(clusters), such that:

$$C_i \neq \emptyset \quad \forall i; \quad C_i \bigcap C_j = \emptyset \quad \forall\, i, j = 1, 2, \ldots, s; \quad P = \bigcup C_i .$$

The statistical units in a cluster are highly similar to each other. Each cluster must also be sharply dissimilar from the other clusters. Each statistical unit must belong to a single cluster. The clustering methods are distinguished in *non-hierarchical* and *hierarchical* procedures.

The non-hierarchical clustering methods lead to a partition of the n statistical units into k classes defined a priori. Hierarchical methods produce a sequence of partitions (from *1* to n clusters) that can be ordered by nested increasing levels to

S. Scippacercola
Dipartimento di Matematica e Statistica - Università degli studi di Napoli Federico II - Via Cinthia, 80126 – Napoli, Italy
e-mail: sergio.scippacercola@unina.it

F. Palumbo et al. (eds.), *Data Analysis and Classification*,
Studies in Classification, Data Analysis, and Knowledge Organization,
DOI 10.1007/978-3-642-03739-9_7,

become a single cluster (Rizzi 1985). The single linkage clustering algorithm is one of several hierarchical clustering algorithms. This algorithm merges clusters based on the distance between the two closest observations in each cluster (Sebert 1998). Ultrametrics are a class of metrics that occur in applications involving hierarchical clustering. The ultrametrics produce a distortion of the initial distances. Low-distortion embeddings of ultrametrics have been a subject of mathematical studies (Bădoiu et al. 2006).

Our paper focuses on the problem of finding an ultrametric whose distortion is close to optimal. More precisely, our main purpose is to introduce a succession of distance matrices to obtain the matrix of minimum distortion from the initial data. Section 2 is dedicated to hierarchical methods which affect the initial dissimilarity by ultrametric approximations. In Sect. 3, we propose an algorithm that by means of a succession of Minkowski ultrametrics reaches a minimum approximation. Finally (Sect. 4), we highlight the validity of the algorithm by its application to sets of multidimensional data.

2 Ultrametric Approximations

In the present paper, we consider a matrix ***X***, of p quantitative variables observed on n statistical units. Let $D^{\lambda} = (d_{ij}^{\lambda})$ be the λ order Minkowski distance matrix defined by the real-valued function (Borg and Lingoes 1987):

$$d_{ij}^{\lambda} = \left[\sum_{h=1,p} \left| x_{ih} - x_{jh} \right|^{\lambda} \right]^{1/\lambda} \quad (i, j = 1, 2, \ldots, n; \ \lambda \text{ integer} \geq 1). \qquad (1)$$

Let $U^{\lambda} = (u_{ij}^{\lambda})$ be the Minkowski ultrametric distances (Scippacercola 2003) obtained by a hierarchical Cluster method (single linkage) (Mardia et al. 1989). Let

$$\Delta^{\lambda} = \left\{ \delta_{ij}^{\lambda} \right\} = D^{\lambda} - U^{\lambda} \qquad (\forall \text{ i, j}; \lambda \text{ integer} \geq 1) \qquad (2)$$

be the distortion matrix which measures the difference between the initial dissimilarity d_{ij}^{λ} and the ultrametric approximation u_{ij}^{λ}.

In the following, we focus our attention onto the ultrametric approximation u_{ij}^{λ} obtained by the single-linkage algorithm thanks to its various analytical properties (Jain et al. 1999; Mardia et al. 1989). Many authors (Chandon et al. 1980; De Soëte 1988) have solved the minimisation of (2) when $\lambda = 2$, by means of a global minimum obtained by generalising group-average clustering (Scozzafava 1995). As an alternative to this approach, we hereby suggest this problem should be solved by an algorithm (*Progressive Single Linkage Algorithm*) which obtains a local minimum of (2).

3 The Progressive Single Linkage Algorithm

By the Jensen inequality (Hardy et al. 1964), the following inequalities are valid for the Minkowski metrics (Rizzi 1985):

$$d_{ij}^{1} \geq d_{ij}^{2} \geq \cdots \geq d_{ij}^{\lambda} \qquad (\mathrm{i,j} = 1,2,\ldots,\mathrm{n})\ (\lambda \text{ integer } \geq 1)\,. \tag{3}$$

If λ increases, the distance between i and j does not increase. By the single linkage the u_{ij}^{λ} meets the ultrametric inequality:

$$u_{ij}^{\lambda} \leq \max\,(u_{ik}^{\lambda}, u_{jk}^{\lambda}) \qquad (\forall\ \mathrm{i,j,k})\ (\lambda \text{ integer } \geq 1)\,. \tag{4}$$

Theorem 1. *The sequence of ultrametric approximation matrices* $\Delta^{\lambda} = (\delta_{ij}^{\lambda})$ *for* $\lambda = 1, 2, \ldots$ *converges to the matrix* Δ^{∞}.

Proof. Indeed, by (3) and (4) if we assume

$$u_{ij}^{\lambda} \leq \mathrm{d}_{ij}^{\lambda} \qquad \lambda \text{ integer } \geq 1\,, \tag{5}$$

and

$$u_{ij}^{1} \geq \mathrm{u}_{ij}^{2} \geq \cdots \geq \mathrm{u}_{ij}^{\lambda} \quad \forall\, i,j; \lambda\text{integer } \geq 1, \tag{6}$$

it follows that the sequence of scalars

$$(d_{ij}^{1} - \mathrm{u}_{ij}^{1}), (d_{ij}^{2} - \mathrm{u}_{ij}^{2}), \ldots, (d_{ij}^{\lambda} - \mathrm{u}_{ij}^{\lambda}) \qquad \forall\, i,j; \lambda \text{ integer } \geq 1, \tag{7}$$

i.e.,

$$\delta_{ij}^{1}, \delta_{ij}^{2}, \ldots, \delta_{ij}^{\lambda} \qquad (\lambda \text{ integer } \geq 1). \tag{8}$$

converges to $\left(\delta_{ij}^{\infty}\right)$ $\forall i,j = 1,2,\ldots,n$. Therefore, the sequence of the Δ^{λ} converges to the Δ^{∞} matrix. □

According to this theorem, the $\left(\delta_{ij}^{\lambda}\right)$ convergence allows us to identify when the ultrametric approximation is at the local Minimum (the algorithm *stopping criterion*), i.e.,

$$\delta_{ij}^{\lambda-1} \approx \delta_{ij}^{\lambda} \qquad (\lambda \text{ integer } \geq 1) \Rightarrow \Delta^{\lambda-1} \approx \Delta^{\lambda}. \tag{9}$$

Therefore the λ value becomes an implicit measure of ultrametric approximation. The ultrametric approximation can be evaluated, for each λ, by means of the following indices:

1. Total ultrametric approximation: $T^{\lambda} = \sum \frac{\delta_{ij}^{\lambda}}{2} \quad \forall i,j.$
2. Absolute distortion index: $\mathrm{ABSIND}^{\lambda} = \frac{\sum \left|\delta_{ij}^{\lambda}\right|}{\frac{n(n-1)}{2}} \quad \forall i,j.$

3. Squared distortion index: $\text{SQRIND}^{\lambda} = \frac{\Sigma(\delta_{ij}^{\lambda})^2}{\frac{n(n-1)}{2}} \quad \forall i, j.$

In the Appendix, we describe the main steps of the progressive single linkage algorithm (PSLA). The computation is fast also as concerns large matrix D and requires the matrix $\boldsymbol{X}$ of the initial data as input.

4 Some Applications to Real Data

In this section, we briefly describe the results obtained by applying the progressive single linkage algorithm to real data.

The first application is referred to living conditions and welfare in ten european countries for the year 2001. We make an international comparison between ten countries with four standardized variables (rooms per person,[1] clothing and footwear,[2] employment growth,[3] meat consumption per capita[4]) (Table 1) (Eurostat n.d.).

We apply the progressive single linkage algorithm to analyse and classify the countries with regard to the living conditions and welfare. We obtain a succession of D and U matrices. The distortion indices computed are shown in Table 2. Figure 1 shows the Minkowski (d) and ultrametrics distances (u) between Greece and Netherland for some λ values referred to living conditions and welfare data. It is easy to verify (Table 2) that a local minimum is reached when $\lambda = 5$ (ABSIND = 0.60; SQRIND = 1.35). Finally, in Fig. 2 we show the dendrogram computed for $\lambda = 5$.

In the second application we consider a sub-sample (15 units) of Iris Flower Data[5] (Table 3) having four observed variables (sepal length, sepal width, petal length, petal width in centimeter). The data is standardized.

By PSLA, we obtain a succession of D and U matrices. In Table 4 we have some Minkowski inter-distances (d) and relative approximations (δ) when $\lambda = 1, 2, \ldots, 6$. It is easy to verify that the minimum is already reached when $\lambda = 6$ (ABSIND = SQRIND = 0.25). Also, we highlight (Table 4 – in bold) the inter-distances that

[1] This indicator shows the number of rooms that each person in a household has in his disposal by tenure status of the household.

[2] At current prices (% of total household consumption expenditure). Household final at current prices consumption expenditure consists of the expenditure, including imputed expenditure, incurred by resident households on individual consumption goods and services, including those sold at prices that are not economically significant.

[3] The indicator gives the change in percentage from one year to another of the total number of employed persons on the economic territory of the country or the geographical area.

[4] Apparent human consumption per capita is obtained by dividing human consumption by the number of inhabitants (resident population stated in official statistics as at 30 June).

[5] A data set with 150 random samples of flowers from the iris species setosa, versicolor, and virginica collected by Anderson (1935). From each species there are 50 observations for sepal length, sepal width, petal length, and petal width in centimeter.

Table 1 Living conditions and welfare – year 2001 (Eurostat n.d.)

Country	Room per person	Clothing and footwear	Employment growth	Meat consumption per capita
Austria	2.1	7.1	0.6	98
Denmark	2.0	5.0	0.8	114
Finland	1.6	4.6	1.5	69
France	2.0	5.1	1.8	108
Germany	1.9	6.0	0.4	88
Greece	1.4	10.6	0.3	91
Italy	1.6	8.8	2.0	91
Netherland	2.6	6.0	2.1	87
Spain	1.9	6.1	3.2	130
Sweden	2.0	5.4	1.9	73

Table 2 Distortion indices computed by the progressive single linkage algorithm relative to the living conditions and welfare data

$\lambda =$	1	2	3	4	5	6	7	8	9	10	20	50	100
T	60.03	30.94	28.32	27.57	27.19	27.32	27.47	27.62	27.75	27.86	28.34	28.57	28.64
ABSIND	1.33	0.68	0.63	0.61	0.60	0.61	0.61	0.61	0.62	0.62	0.63	0.63	0.63
SQRIND	6.69	1.85	1.50	1.40	1.35	1.35	1.35	1.36	1.36	1.37	1.40	1.41	1.41

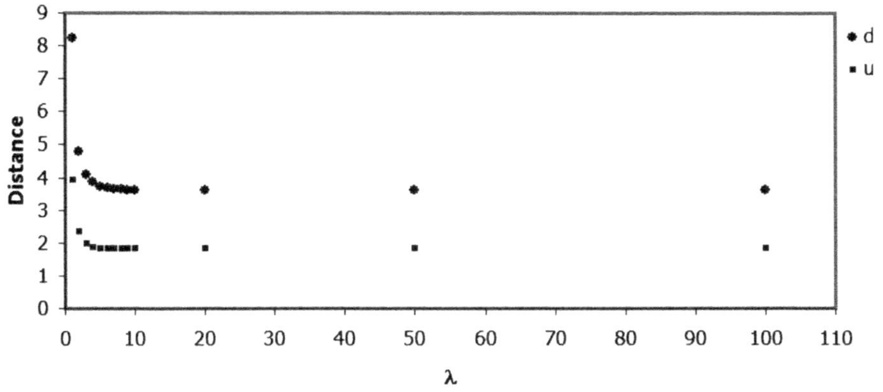

Fig. 1 Minkowski distances (d) and ultrametrics (u) from Greece to Netherland for some λ values referred to living conditions and welfare data

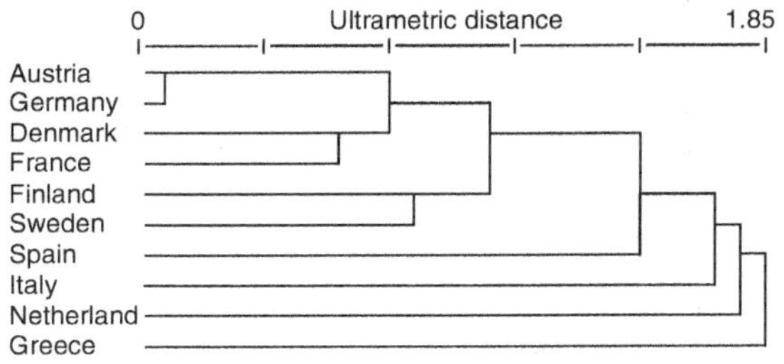

Fig. 2 Living and welfare data dendrogram by progressive single linkage algorithm (for $\lambda = 5$)

Table 3 A sub-sample of iris data

Unit	Sepal length	Sepal width	Petal length	Petal width	Type
1	5.10	3.50	1.40	0.20	Iris-setosa1
2	4.90	3.00	1.40	0.20	Iris-setosa2
3	4.70	3.20	1.30	0.20	Iris-setosa3
4	4.60	3.10	1.50	0.20	Iris-setosa4
5	5.00	3.60	1.40	0.20	Iris-setosa5
6	5.70	3.00	4.20	1.20	Iris-versic1
7	5.70	2.90	4.20	1.30	Iris-versic2
8	6.20	2.90	4.30	1.30	Iris-versic3
9	5.10	2.50	3.00	1.10	Iris-versic4
10	5.70	2.80	4.10	1.30	Iris-versic5
11	6.30	3.30	6.00	2.50	Iris-virgin1
12	5.80	2.70	5.10	1.90	Iris-virgin2
13	7.10	3.00	5.90	2.10	Iris-virgin3
14	6.30	2.90	5.60	1.80	Iris-virgin4
15	6.50	3.00	5.80	2.20	Iris-virgin5

have zero distortion (**1–3**, **1–5**, **6–7**, **8–10**, **12–14**, **12–15**). Finally, Fig. 3 shows the dendrogram computed when $\lambda = 6$.

5 Conclusions

In this paper we have introduced an extension of the single-linkage clustering algorithm. Our main purpose has been to suggest a family of Minkowski distances as a tool for measuring the distortion vis a vis the initial distance matrix of data. The present approach has used the λ variations to fastly obtain the minimum approximation. The convergence of the algorithm allows us to identify when the ultrametric approximation is at the local minimum. Then, we have obtained a dendrogram with Minkowski ultrametric distances. The suggested algorithm is computationally efficient and can be used in a lot of applications. The validity of the algorithm has been confirmed by examples of Cluster Analysis as applied to sets of real data.

Table 4 Some Minkowski inter-distances (d) and relative approximations (δ) when $\lambda = 1, 2, \ldots, 6$ by the progressive single linkage algorithm

Units	d^1	δ^1	d^2	δ^2	d^3	δ^3	d^4	δ^4	d^5	δ^5	d^6	δ^6
1–3	**0.021**	**0.000**	**0.013**	**0.000**	**0.011**	**0.000**	**0.011**	**0.000**	**0.011**	**0.000**	**0.010**	**0.000**
12–15	**0.086**	**0.000**	**0.044**	**0.000**	**0.036**	**0.000**	**0.033**	**0.000**	**0.031**	**0.000**	**0.030**	**0.000**
8–10	**0.091**	**0.000**	**0.048**	**0.000**	**0.040**	**0.000**	**0.037**	**0.000**	**0.035**	**0.000**	**0.035**	**0.000**
7–8	0.197	0.090	0.100	0.047	0.08	0.038	0.072	0.034	0.068	0.033	0.066	0.031
1–5	**0.102**	**0.000**	**0.063**	**0.000**	**0.056**	**0.000**	**0.053**	**0.000**	**0.052**	**0.000**	**0.052**	**0.000**
6–7	**0.116**	**0.000**	**0.070**	**0.000**	**0.062**	**0.000**	**0.060**	**0.000**	**0.059**	**0.000**	**0.058**	**0.000**
12–14	**0.133**	**0.000**	**0.069**	**0.000**	**0.057**	**0.000**	**0.052**	**0.000**	**0.050**	**0.000**	**0.049**	**0.000**
11–12	0.235	0.021	0.155	0.027	0.138	0.026	0.107	0.001	0.126	0.024	0.124	0.024
1–2	0.258	0.021	0.155	0.030	0.138	0.035	0.132	0.037	0.13	0.039	0.130	0.041
6–9	0.619	0.109	0.373	0.073	0.331	0.063	0.318	0.059	0.313	0.057	0.310	0.055
6–11	1.157	0.391	0.616	0.203	0.514	0.166	0.476	0.153	0.457	0.146	0.446	0.142
1–6	1.765	0.573	1.143	0.370	1.017	0.331	0.964	0.316	0.936	0.31	0.919	0.307
...	...	...	...	...	...	...	...	...	...	...	...	...
T	–	**49.19**	–	**32.24**	–	**28.66**	–	**27.22**	–	**26.48**	–	**26.05**
ABSIND		**0.47**	–	**0.31**	–	**0.27**	–	**0.26**	–	**0.25**	–	**0.25**
SQRIND		**0.88**	–	**0.38**	–	**0.30**	–	**0.27**	–	**0.26**	–	**0.25**

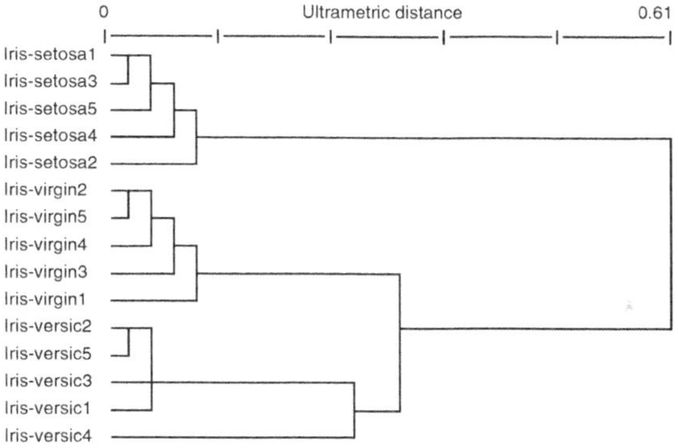

Fig. 3 Iris sub-sample dendrogram by progressive single linkage algorithm ($\lambda = 6$)

Appendix

The progressive single linkage algorithm iteratively computes the Minkowski inter-point distances by varying λ from 1 until the minimum approximation is reached. The algorithm develops in seven steps:

1. Let ϵ be the approximation.
2. Consider an initial value of λ (Minkowski parameter).
3. Repeat the step 4, 5 and 6 while the distortion indices are greater than ε.
4. Apply the single linkage algorithm (Kruskal 1956; Prim 1957; Gower and Ross 1969) to obtain the Minkowski ultrametrics distances with λ parameter.

5. Compute the distortion indices.
6. $\lambda = \lambda + 1$.
7. Build the dendrogram by the Minkowski ultrametrics distances with $\lambda - 1$ parameter.

References

Anderson, E. (1935). The irises of the Gaspé peninsula. *Bulletin of the American Iris Society, 59*, 2–5.

Bădoiu, M., Chuzhoy, J., Indyk, P., & Sidiropoulos, A. (2006). Embedding ultrametrics into low-dimensional spaces. In *Proceedings of twenty-second annual symposium on Computational Geometry SCG'06* (pp. 187–196), Sedona, AZ: ACM Press.

Bock, H. H. (1996). Probabilistic models in cluster analysis. *Computational Statistics and Data Analysis, 23*(1), 6–28.

Borg, I., & Lingoes, J. (1987). *Multidimensional similarity structure analysis*. Berlin: Springer.

Chandon, J. L., Lemaire, J., & Pouget, J. (1980). Construction de l'ultrametrique la plus proche d'une dissimilarité au sens des moindres carrés. *R.A.I.R.O. Recherche Operationelle, 14*, 157–170.

De Soëte, G. (1988). Tree representations of proximity data by least squares methods. In H. H. Bock (Ed.), *Classification and related methods of data analysis* (pp. 147–156). Amsterdam: North Holland.

Eurostat. (n.d.). *General and regional statistics*. http://epp.eurostat.ec.europa.eu

Gordon, A. D. (1996). A survey of constrained classification. *Computational Statistics and Data Analysis, 21*(1), 17–29.

Gower, J. C., & Ross, J. S. (1969). Minimum spanning trees and single linkage cluster analysis. *Applied Statistics, 18*, 54–64.

Hardy, G. H., Littlewood, J. E., & Polya, G. (1964). *Inequalities*. Cambridge: Cambridge University Press.

Jain, A. K., Murty, M. N., & Flynn, P. J. (1999). Data clustering: A Review. *ACM Computing Survey, 31*(3), 264–323.

Kruskal, J. B. (1956). On the shortest spanning subtree of a graph and the traveling salesman problem. *Proceedings of the Mathematical Society, 7*, 48–50.

Mardia, K. V., Kent, J. T., & Bibby, J. M. (1989). *Multivariate analysis*. New York: Academic.

Prim, R. C. (1957). Shortest connection network and some generalizations. *Bell System Technical Journal, 36*, 1389–1401.

Rizzi, A. (1985). *Analisi dei dati*. Rome: La Nuova Italia Scientifica.

Scippacercola, S. (2003). Evaluation of clusters stability based on minkowski ultrametrics. *Statistica Applicata – Italian Journal of Applied Statistics, 15*(4), 483–489.

Scozzafava, P. (1995). Ultrametric spaces in statistics. In A. Rizzi (Ed.), *Some relations between matrices and structures of multidimensional data analysis*. Pisa: Giardini.

Sebert, D. M., Montgomery, D. C., & Rollier, D. A. (1998). A clustering algorithm for identifying multiple outliers in linear regression. *Computational Statistics and Data Analysis, 27*(4), 461–484.

Visualization of Model-Based Clustering Structures

Luca Scrucca

Abstract Model-based clustering based on a finite mixture of Gaussian components is an effective method for looking for groups of observations in a dataset. In this paper we propose a dimension reduction method, called MCLUSTSIR, which is able to show clustering structures depending on the selected Gaussian mixture model. The method aims at finding those directions which are able to display both variation in cluster means and variations in cluster covariances. The resulting MCLUSTSIR variables are defined as a linear mapping method which projects the data onto a suitable subspace.

1 Introduction

Suppose that the observed data comes from a finite mixture with K components, each representing the probability distribution for a different group or cluster: $f(\boldsymbol{x}) = \sum_{k=1}^{K} \pi_k f_k(\boldsymbol{x}|\boldsymbol{\theta}_k)$, where π_k are the mixing probabilities ($\pi_k \geq 0$, $\sum_{k=1}^{K} \pi_k = 1$), $f_k(\cdot)$ and $\boldsymbol{\theta}_k$ are the density and the parameters of the k-th component of the mixture. With continuous data, we often take the density to be the multivariate Gaussian $\phi_k(\boldsymbol{x}|\boldsymbol{\mu}_k, \boldsymbol{\Sigma}_k)$ with parameters $\boldsymbol{\theta}_k = (\boldsymbol{\mu}_k, \boldsymbol{\Sigma}_k)$. Clusters are ellipsoidal, centered at the means $\boldsymbol{\mu}_k$, and with other geometric features, such as volume, shape and orientation, determined by $\boldsymbol{\Sigma}_k$. Parsimonious parametrization of the covariance matrix for each cluster can be adopted (Banfield and Raftery 1993; Celeux and Govaert 1995) through an eigenvalue decomposition in the form $\boldsymbol{\Sigma}_k = \lambda_k \boldsymbol{D}_k \boldsymbol{A}_k \boldsymbol{D}_k^{\top}$, where λ_k is a scalar value controlling the volume of the ellipsoid, $\boldsymbol{A}_k$ is a diagonal matrix specifying the shape of the density contours, and $\boldsymbol{D}_k$ is an orthogonal matrix which determines the orientation of the ellipsoid (see Table 1 in Fraley and Raftery, 2006). Maximum likelihood estimation for finite Gaussian mixture models is performed via the EM algorithm (Fraley and Raftery 2002; McLachlan and Peel 2000), while model selection could be based on the Bayesian Information Criterion (BIC) (Fraley and Raftery 1998).

L. Scrucca
Dipartimento di Economia, Finanza e Statistica, Università degli Studi di Perugia, Perugia, Italy
e-mail: luca@stat.unipg.it

F. Palumbo et al. (eds.), *Data Analysis and Classification*,
Studies in Classification, Data Analysis, and Knowledge Organization,
DOI 10.1007/978-3-642-03739-9_8,

In this paper we propose a dimension reduction approach which is able to show clustering structures depending on the particular Gaussian mixture model fitted. In the next section we present the methodology, then we discuss visualization issues, and we end with some applications on both simulated and real datasets.

2 Dimension Reduction for Model-Based Clustering

Suppose we describe a set of n observations on p variables through a K components Gaussian mixture model of the form $f(\boldsymbol{x}) = \sum_{k=1}^{K} \pi_k \phi_k(\boldsymbol{x}|\boldsymbol{\mu}_k, \boldsymbol{\Sigma}_k)$.

We would like to find those directions where the cluster means $\boldsymbol{\mu}_k$ vary as much as possible, provided each direction being orthogonal to the others. This amounts to solve the following optimization problem: $\text{argmax}_{\beta}\, \boldsymbol{\beta}^{\top}\boldsymbol{\Sigma}_B\boldsymbol{\beta}$ subject to $\boldsymbol{\beta}^{\top}\boldsymbol{\Sigma}\boldsymbol{\beta} = \boldsymbol{I}$, where $\boldsymbol{\Sigma}_B = \sum_{k=1}^{K} \pi_k(\boldsymbol{\mu}_k - \boldsymbol{\mu})(\boldsymbol{\mu}_k - \boldsymbol{\mu})^{\top}$ is the between-cluster covariance matrix, $\boldsymbol{\Sigma} = n^{-1}\sum_{i=1}^{n}(\boldsymbol{x}_i - \boldsymbol{\mu})(\boldsymbol{x}_i - \boldsymbol{\mu})^{\top}$ is the covariance matrix with $\boldsymbol{\mu} = \sum_{k=1}^{K} \pi_k \boldsymbol{\mu}_k$, $\boldsymbol{\beta} \in \mathbb{R}^{p \times d}$ is the spanning matrix, and $\boldsymbol{I}$ is the $(d \times d)$ identity matrix. The solution to this constrained optimization is given by the eigendecomposition of the kernel matrix $\boldsymbol{M}_I = \boldsymbol{\Sigma}_B$ with respect to $\boldsymbol{\Sigma}$. The eigenvectors, corresponding to the first d largest eigenvalues provide a basis for the subspace $\mathcal{S}(\boldsymbol{\beta})$ which shows the maximal variation among clusters means. There are at most $d = \min(p, K-1)$ directions which span this subspace.

This procedure is similar to the SIR (*Sliced Inverse Regression*) algorithm (Li 1991), but here conditioning is on the cluster memberships. It has been shown that SIR directions span at least a part of the dimension reduction subspace (Cook 1998, Chap. 6), and they may miss relevant structures in the data when within-cluster covariances are different.

The SIR_{II} method (Li 1991, 2000) exploits the information coming from the differences in the class covariance matrices. The kernel matrix is now defined as: $\boldsymbol{M}_{II} = \sum_{k=1}^{K} \pi_k(\boldsymbol{\Sigma}_k - \bar{\boldsymbol{\Sigma}})\boldsymbol{\Sigma}^{-1}(\boldsymbol{\Sigma}_k - \bar{\boldsymbol{\Sigma}})^{\top}$, where $\bar{\boldsymbol{\Sigma}} = \sum_{k=1}^{K} \pi_k \boldsymbol{\Sigma}_k$ is the pooled within-cluster covariance matrix, and directions are found through the eigendecomposition of $\boldsymbol{M}_{II}$ with respect to $\boldsymbol{\Sigma}$. Albeit the corresponding directions allow to show differences in group covariances, they are usually not able to show also location differences.

The proposed approach, called *MCLUSTSIR*, aims at finding those directions which, depending on the selected Gaussian mixture model, are able to display both variation in cluster means and variations in cluster covariances.

Definition 1. Consider the following kernel matrix

$$\boldsymbol{M} = \boldsymbol{M}_I\boldsymbol{\Sigma}^{-1}\boldsymbol{M}_I + \boldsymbol{M}_{II}. \qquad (1)$$

The basis of the dimension reduction subspace $\mathcal{S}(\boldsymbol{\beta})$ is the solution of the following constrained optimization: $\text{argmax}_{\beta}\, \boldsymbol{\beta}^{\top}\boldsymbol{M}\boldsymbol{\beta}$, subject to $\boldsymbol{\beta}^{\top}\boldsymbol{\Sigma}\boldsymbol{\beta} = \boldsymbol{I}$. This is solved through the generalized eigendecomposition

$$\boldsymbol{M}\boldsymbol{v}_i = l_i \boldsymbol{\Sigma}\boldsymbol{v}_i \qquad \boldsymbol{v}_i^\top \boldsymbol{\Sigma}\boldsymbol{v}_j = 1 \text{ if } i = j, \text{ and 0 otherwise,} \\ l_1 \geq l_2 \geq \cdots \geq l_d > 0. \tag{2}$$

The kernel matrix (1) contains information from variation on both cluster means and cluster covariances. For mixture models which assume constant within-cluster covariance matrices (i.e. E, EII, EEI, and EEE; Fraley and Raftery, 2006, Table 1), the subspace spanned by $\boldsymbol{M}$ is equivalent to that spanned by $\boldsymbol{M}_I$. In all the other cases, the second term in (1) adds further information for the identification of the dimension reduction subspace.

We give now some properties and remarks of the proposed method with proofs omitted for reasons of conciseness.

Remark 1. The eigenvectors corresponding to the first d largest eigenvalues from (2), $\boldsymbol{\beta} \equiv [\boldsymbol{v}_1, \ldots, \boldsymbol{v}_d]$, provide the *MCLUSTSIR directions*, which are the basis of the subspace $\mathcal{S}(\boldsymbol{\beta})$. However, these coefficients are uniquely determined up to a multiplication by a scalar, whereas the associated directions from the origin are unique. Hence, we can adjust their length such that they have unit length, i.e. each direction is normalized as $\boldsymbol{\beta}_j \equiv \boldsymbol{v}_j / \parallel \boldsymbol{v}_j \parallel$ for $j = 1, \ldots, d$.

Proposition 1. *MCLUSTSIR directions are invariant under affine transformation* $\boldsymbol{x} \mapsto \boldsymbol{C}\boldsymbol{x} + \boldsymbol{a}$, *for* $\boldsymbol{C}$ *a nonsingular matrix and* $\boldsymbol{a}$ *a vector of real values. Thus, MCLUSTSIR directions in the transformed scale are given by* $\boldsymbol{C}^{-1}\boldsymbol{\beta}$.

Proposition 2. *Each eigenvalue of the eigendecomposition in* (2) *can be decomposed in the sum of the contributions given by the squared variance of the between group means and the average of the squared differences among within group variances along the corresponding direction of the projection subspace, i.e.*

$$l_i = \mathbb{V}(\mathbb{E}(z_i|Y))^2 + \mathbb{E}(\mathbb{V}(z_i|Y)^2),$$

where $z_i = \beta_i^\top \boldsymbol{x}$, *for* $i = 1, \ldots, d$.

Remark 2. Let $\boldsymbol{X}$ be the $(n \times p)$ sample data matrix which we assume, with no loss of generality, having zero mean column-vectors. The sample version $\widehat{\boldsymbol{M}}$ of (1) is computed using the corresponding estimates obtained from the fit of a Gaussian mixture model, so in practice we calculate the eigendecomposition of $\widehat{\boldsymbol{M}}$ with respect to $\widehat{\boldsymbol{\Sigma}}$.

3 Visualization of Clustering Structures

MCLUSTSIR directions may help to visualize the clustering structure on a dimension reduced subspace. To this purpose we need to project the observed data and the parameters of the fitted finite mixture onto the estimated subspace.

Definition 2. Let $\boldsymbol{X}$ be the $(n \times p)$ matrix of n observations on p variables, $\widehat{\boldsymbol{\mu}}_k$ and $\widehat{\boldsymbol{\Sigma}}_k$ be the estimated mean vector and covariance matrix, respectively, for the k-th cluster, and $\mathcal{S}(\widehat{\boldsymbol{\beta}})$ be the estimated subspace spanned by the $(p \times d)$ matrix $\widehat{\boldsymbol{\beta}}$. The projection of the observed data onto the subspace $\mathcal{S}(\widehat{\boldsymbol{\beta}})$ is computed as $\widehat{\boldsymbol{Z}} = \boldsymbol{X}\widehat{\boldsymbol{\beta}}$, and we call these *MCLUSTSIR variables*. The projection of the finite mixture parameters onto the subspace $\mathcal{S}(\widehat{\boldsymbol{\beta}})$ are, respectively, $\widehat{\boldsymbol{\beta}}^{\top}\widehat{\boldsymbol{\mu}}_k$ and $\widehat{\boldsymbol{\beta}}^{\top}\widehat{\boldsymbol{\Sigma}}_k\widehat{\boldsymbol{\beta}}$.

Once observed data and parameters are expressed in the new coordinate system, we may employ several graphical tools to visualize the clustering information. Some graphics we found most useful in our experience are described in the following:

1. One-dimensional plots may be employed to display the marginal distribution for each estimated direction. For example, side-by-side box plots conditioning on cluster membership allows to see the ability of each direction to separate clusters (see diagonal panels of Fig. 2), while density estimates computed for each mixture component easily show the location and dispersion parameters along any direction (see diagonal panels of Fig. 1).
2. With two-dimensional plots we have a variety of interesting plots. Scatterplots of pairs of MCLUSTSIR variables with points marked by cluster membership are very useful to display clusters in a 2D subspace (see off-diagonal panels of Fig. 3), particularly if the directions associated with the largest eigenvalues are used. This basic plot can be enhanced by adding several other graphical tools, such as contours of the estimated densities (see off-diagonal panels of Fig. 1), classification regions according to maximum a posteriori (MAP) probability (see

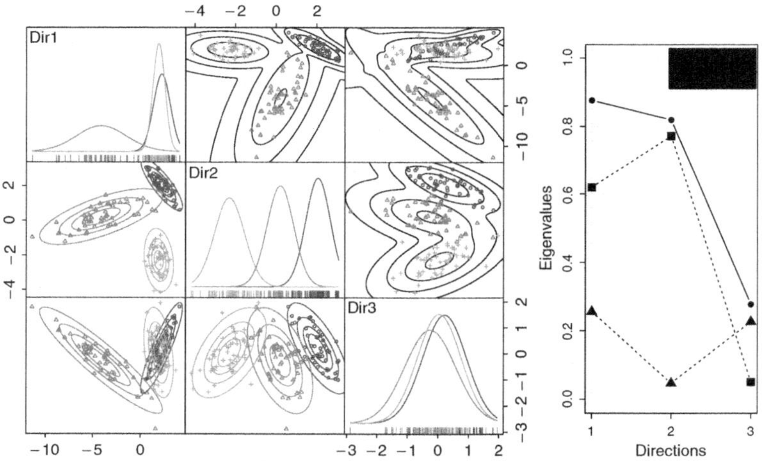

Fig. 1 Scatter plots of MCLUSTSIR variables (*left panel*): mixture (*above diagonal*) and within-cluster (*below diagonal*) bivariate density contours, with marginal univariate component densities (*diagonal*). Plot of eigenvalues (*right panel*) for each estimated direction, with contributions from means and variances differences among clusters

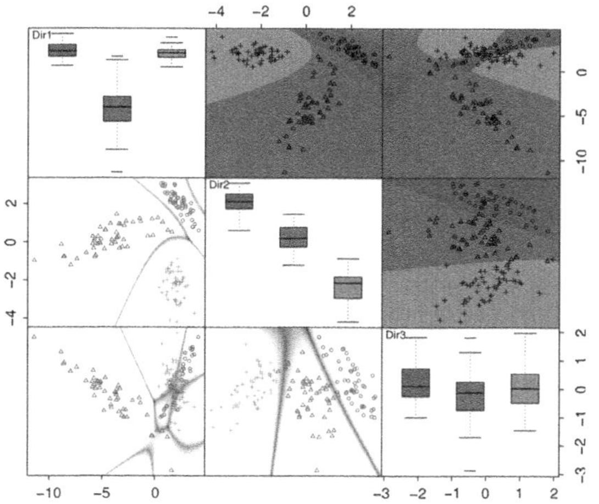

Fig. 2 Clustering plots for the estimated MCLUSTSIR variables: side-by-side box-plots along the *diagonal*, MAP regions *above the diagonal*, and uncertainty boundaries *below the diagonal*

above-diagonal panels of Fig. 2), and uncertainty boundaries (see below-diagonal panels of Fig. 2).

3. One and two-dimensional plots are perhaps the most useful and easy to interpret graph, but in principle the same ideas could be extended to *higher dimensions*. For example, 3D spinning plot and 3D contour plot showing hyperellipsoids of constant density can be visualized in a dynamic graphic device.

4 Examples

In this section we discuss the implementation of the ideas introduced in the previous sections by examples on simulated and real datasets.

4.1 Overlapping Clusters with Unconstrained Covariances

We simulated a dataset with three overlapping clusters of dimension $n_k = 50$ for $k = 1, 2, 3$. Each cluster was generated from a Gaussian distribution with means $\boldsymbol{\mu}_1 = [0, 0, 0]^\top$, $\boldsymbol{\mu}_2 = [4, -2, 6]^\top$, $\boldsymbol{\mu}_3 = [-2, -4, 2]^\top$, and covariances $\boldsymbol{\Sigma}_1 = \begin{pmatrix} 1 & 0.9 & 0.9 \\ 0.9 & 1 & 0.9 \\ 0.9 & 0.9 & 1 \end{pmatrix}$, $\boldsymbol{\Sigma}_2 = \begin{pmatrix} 2 & -1.8 & -1.8 \\ -1.8 & 2 & -1.8 \\ -1.8 & -1.8 & 2 \end{pmatrix}$, $\boldsymbol{\Sigma}_3 = \begin{pmatrix} 0.5 & 0 & 0 \\ 0 & 0.5 & 0 \\ 0 & 0 & 0.5 \end{pmatrix}$.

The Gaussian mixture model with the highest BIC is the VVV model (ellipsoidal, varying volume, shape, and orientation) with 3 components. For this model the MCLUSTSIR variables are defined as follows:

$$
\begin{aligned}
Z_1 &= +0.8443X_1 - 0.3788X_2 + 0.3791X_3 \\
Z_2 &= -0.6952X_1 - 0.3964X_2 + 0.5997X_3 \\
Z_3 &= +0.5413X_1 - 0.5523X_2 - 0.6340X_3
\end{aligned}
$$

and the corresponding eigenvalues are plotted in the right panel of Fig. 1. This graph indicates that the first direction reflects differences in means and, to a less extent, differences in variances. The following direction mainly shows differences in means, while the contribution of the last direction is negligible. These aspects are also apparent in the left panel of Fig. 1, where projections of the estimated marginal within-cluster univariate densities are reported along the diagonal. The off-diagonal panels show contours of the bivariate mixture densities (above the diagonal) and contours of the bivariate within-cluster densities (below the diagonal). Clusters appear to be clearly separated in the first two directions, with elliptical contours reflecting the geometric characteristics of the fitted model.

Finally, Fig. 2 shows some clustering plots: panels above the diagonal display MAP regions, while panels below the diagonal the corresponding uncertainty boundaries. It is confirmed that the first two MCLUSTSIR directions are able to separate the clusters with small uncertainty.

4.2 *High Dimensional Mixture of Two Normal Distributions*

Here we simulated $n = 300$ data points from a 15-dimensional mixture model. Let $\boldsymbol{X} = 0.5 \times d + d \times Y + \boldsymbol{Z}$, where $d = 0.95 - 0.05i$ $(i = 1, \ldots, 15)$, $Y \sim$ Bernoulli(0.2), $\boldsymbol{Z} \sim \mathrm{N}(\boldsymbol{\mu}, \boldsymbol{\Sigma})$ with mean $\boldsymbol{\mu}_{15\times 1} = [0, \ldots, 0]^{\top}$ and covariance matrix $\boldsymbol{\Sigma}_{15\times 15} = [\sigma_{ij}]$, $\sigma_{ii} = 1, \sigma_{ij} = -0.13 f_i f_j$, where the first eight elements of f are -0.9 and the last seven are 0.5. With this scheme the first eight variables can be considered roughly as a block of variables with the same correlations, while the rest of the variables form another block.

Chang (1983) used this setting to show the failure of principal components as a method for reducing the dimension of the data before clustering. In fact, as it can be seen from Fig. 3, the first and the last principal components are needed to separate the two clusters. On the contrary, MCLUSTSIR from a two components EEE model only requires one direction to clearly separate the clusters. Furthermore, MCLUSTSIR coefficients clearly highlight the blocking structure used to simulate the variables (see right panel of Fig. 3).

4.3 *Wisconsin Diagnostic Breast Cancer Data*

In their review on model-based clustering, Fraley and Raftery (2002) analyzed this dataset on breast tumor diagnosis. They used 3 out of 30 available features to cluster the data of 569 women diagnosed of breast cancer, among which 357 were benign and 212 malignant. The fitted model assumed unconstrained covariances (VVV)

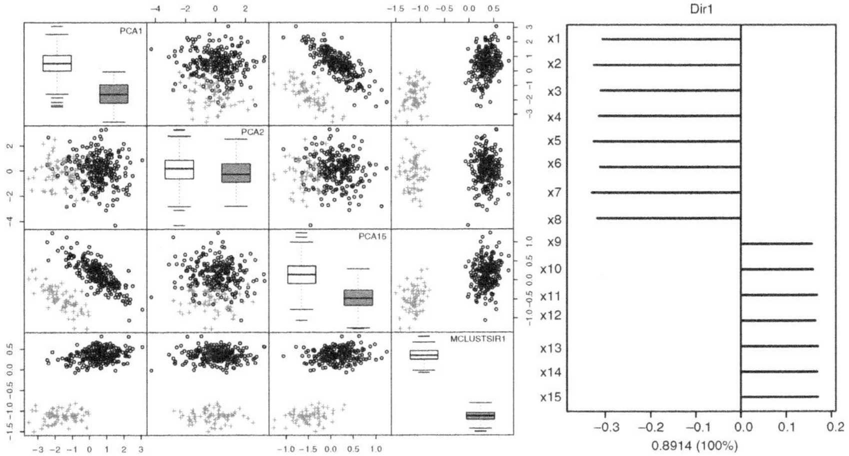

Fig. 3 Scatterplot matrix of 1st, 2nd and 15th PC, and 1st MCLUSTSIR direction for the Chang data with points marked according to cluster membership (*left panel*), and estimated coefficients defining the MCLUSTSIR variate (*right panel*)

with two components, and correctly classified 94.55% of the units. Applying the MCLUSTSIR procedure, we may plot the data onto the estimated subspace as shown in the left panel of Fig. 4. On the first direction individuals suffering from malignant cancer have a larger mean and a more dispersed distribution than the benign group. The second direction adds further information to separate the groups, both in terms of means and variances, while the last direction does not show any difference among groups, with the estimated densities which largely overlap. Thus, a plot of the data projected onto the first two MCLUSTSIR directions provide almost all the relevant clustering information. This can be contrasted with the plots of the two selected features (extreme area and mean texture) provided by Fraley and Raftery (2002, Figs. 1 and 3) which are not capable of fully representing such information.

The graph located in the first row and second column on the left panel of Fig. 4 shows the contours of the within-cluster bivariate densities projected onto the first two MCLUSTSIR directions; they appear different in orientation, shape and volume, with the group of benignant having a more compact distribution. For the same directions, the plot located in the second row and first column shows the uncertainty boundaries defining the MAP regions; the shaded areas represent regions of higher uncertainty, which are located near the overlap of the two clusters.

Finally, the plot of the corresponding eigenvalues is reported on the right panel of Fig. 4. From this graph we find confirmation that only two directions are needed, the first showing mainly differences in variances between the groups, while the second direction showing both differences in means and variances.

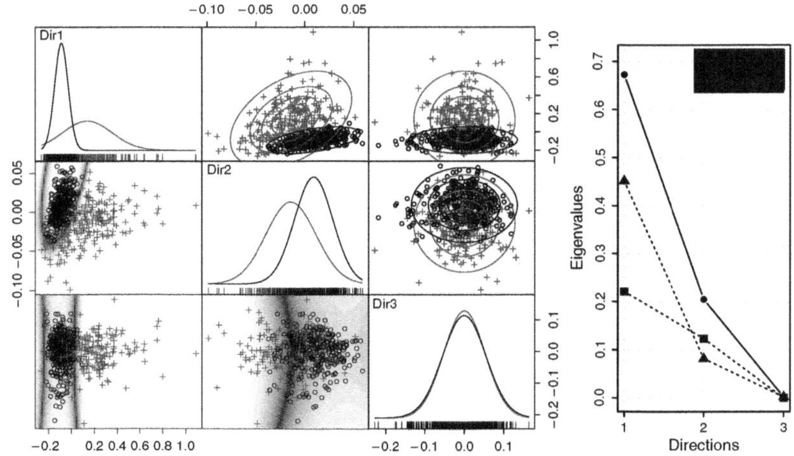

Fig. 4 Scatter plots of MCLUSTSIR variables for the WDBC data (*left panel*): bivariate within-cluster density contours (*above diagonal*), uncertainty boundaries (*below diagonal*) and marginal univariate component densities (*diagonal*). Plot of eigenvalues (*right panel*) for each estimated direction, with contributions from means and variances differences among clusters

5 Comments and Extensions

MCLUSTSIR variables are defined as a linear mapping method which projects the data onto a suitable subspace. This can be viewed as a form of (soft) feature extraction, where the components are reduced through a set of linear combinations of the original variables. The dimension of the subspace may be assessed informally through graphical exploration, or more formally by reformulating the subset selection problem as a model comparison problem (Raftery and Dean 2006).

Finally, the MCLUSTSIR approach is readily applicable to model-based discriminant analysis, both if a single Gaussian component is used for any class, or a mixture of several components are used to describe each class.

References

Banfield, J., & Raftery, A. E. (1993). Model-based Gaussian and non-Gaussian clustering. *Biometrics, 49*, 803–821.

Celeux, G., & Govaert, G. (1995). Gaussian parsimonious clustering models. *Pattern Recognition, 28*, 781–793.

Chang, W. (1983). On using principal components before separating a mixture of two multivariate normal distributions. *Applied Statistics, 32*(3), 267–275.

Cook, R. D. (1998). *Regression graphics: Ideas for studying regressions through graphics*. New York: Wiley.

Fraley, C., & Raftery, A. E. (1998). How many clusters? which clustering method? answers via model-based cluster analysis. *The Computer Journal, 41*, 578–588.

Fraley, C., & Raftery, A. E. (2002). Model-based clustering, discriminant analysis, and density estimation. *Journal of the American Statistical Association, 97*(458), 611–631.

Fraley, C., & Raftery, A. E. (2006). *MCLUST version 3 for R: Normal mixture modeling and model-based clustering* (Technical Report 504). Department of Statistics, University of Washington.

Li, K. C. (1991). Sliced inverse regression for dimension reduction (with discussion). *Journal of the American Statistical Association, 86*, 316–342.

Li, K. C. (2000). *High dimensional data analysis via the SIR/PHD approach*. Unpublished manuscript. Retrieved from http://www.stat.ucla.edu/~kcli/sir-PHD.pdf.

McLachlan, G., & Peel, D. (2000). *Finite mixture models*. New York: Wiley.

Raftery, A. E., & Dean, N. (2006). Variable selection for model-based clustering. *Journal of the American Statistical Association, 101*(473), 168–178.

Part III
Multidimensional Scaling

Models for Asymmetry in Proximity Data

Giuseppe Bove

Abstract Geometrical models to explore and represent asymmetric proximity data are usually classified in two classes: *distance models* and *scalar product models*. In this paper we focalize on scalar product models, emphasizing some relationships and showing possibilities to incorporate external information that can help the analysis of proximities between rows and columns of data matrices. In particular it is pointed out how some of these models apply to the analysis of skew-symmetry with external information.

1 Introduction

Proximities (e.g. similarity ratings), preferences (e.g. socio-matrices), flow data (e.g. import–export, brand switching) are examples of one-mode two-way data that we can represent in low-dimensional spaces by scalar product or Euclidean distance models. The difference between the two types of models from a data-analytic perspective consists in the geometrical entity (scalar product or distance) chosen to represent the entries of the data matrix. When not random asymmetry is present in the data these models have to be suitably modified by increasing the number of parameters (see, e.g. Zielman and Heiser 1996). In the next section some scalar product models for asymmetric proximities are reviewed emphasizing their relationships. Possible approaches to take into account external information in the analysis of skew-symmetry are considered in the third section.

2 A Class of Scalar Product Models

A *general* scalar product model to represent asymmetric one-mode two-way data is the *biplot model* (Gabriel 1971). Approximate *r-dimensional* biplot of a square data

G. Bove
Dipartimento di Scienze dell'Educazione, Università degli Studi Roma Tre, Italy
e-mail: bove@uniroma3.it

F. Palumbo et al. (eds.), *Data Analysis and Classification*,
Studies in Classification, Data Analysis, and Knowledge Organization,
DOI 10.1007/978-3-642-03739-9_9,

matrix $\mathbf{X} = [x_{ij}]$, whose rows and columns correspond to the same set of n objects, is based on the approximate factorization

$$\mathbf{X} = \mathbf{AB}' + \boldsymbol{\xi} \tag{1}$$

with r, the number of columns of $\mathbf{A}$ and $\mathbf{B}$, less than the rank of $\mathbf{X}$. The rows of matrices $\mathbf{A}$ and $\mathbf{B}$ provide coordinate vectors respectively for the n rows and columns of the data matrix and their scalar products approximate the entries of $\mathbf{X}$. $\boldsymbol{\xi}$ is a residual term. Non-uniqueness given to non-singular transformations of the two coordinate matrices $\mathbf{A}$ and $\mathbf{B}$ is removed assuming column-wise orthogonality and setting to 1 the norm of each column of one of the two matrices. The *r-dimensional* biplot is obtained by minimizing the sum of squared residuals $\|\boldsymbol{\xi}\|^2$, i.e., by the singular value decomposition of the data matrix $\mathbf{X}$. The method allows to represent asymmetric proximities x_{ij} by $2n$ points in a low-dimensional space (usually bidimensional). This direct representation of the data matrix can also be used to analyze symmetry $s_{ij} = \frac{1}{2}(x_{ij} + x_{ji})$ and skew-symmetry $k_{ij} = \frac{1}{2}(x_{ij} - x_{ji})$ by sum and difference of the two scalar products corresponding to the entries x_{ij} and x_{ji}, but for large n, this become quite complicated.

A particular case of the previous model is the *non-spatial model* DEDICOM (DEcomposition into DIrectional COMponents), proposed by Harshman (1978), represented by

$$\mathbf{X} = \mathbf{ARA}' + \boldsymbol{\xi}, \tag{2}$$

where $\mathbf{A}$ is a matrix of coefficients to relate the n objects to "basic concepts" underlying the objects, $\mathbf{R}$ is an $r \times r$ matrix containing relationship measures between the concepts and $\boldsymbol{\xi}$ is a residual term. DEDICOM can be obtained by the biplot model constraining $\mathbf{B}$ to have the particular form: $\mathbf{B} = \mathbf{AR}'$. DEDICOM does not provide a graphical representation of the objects, however Kiers and Takane (1994) proved that a method for a graphical representation can be obtained when the symmetric component $(\frac{1}{2}(\mathbf{R} + \mathbf{R}'))$ of matrix $\mathbf{R}$ is positive definite. In order to obtain graphical representations for the general case Kiers and Takane (1994) proposed the following constrained version of DEDICOM, named Generalized GIPSCAL (GG),

$$\mathbf{X} = \mathbf{A}(a\mathbf{I} + b\Delta)\mathbf{A}' + \boldsymbol{\xi}, \tag{3}$$

where $\mathbf{I}$ is the identity matrix, $a \geq 0$ and b are unknown constants, $\boldsymbol{\xi}$ is a residual term and Δ is a block diagonal matrix with 2×2 matrices

$$\begin{bmatrix} 0 & \delta_l \\ -\delta_l & 0 \end{bmatrix} \tag{4}$$

along the diagonal and, if n is odd, a zero element in the last diagonal position. It is easy to observe that GG is a constrained version of DEDICOM where $\mathbf{R}$ has a particular form. A graphical representation for GG is obtained reformulating the model as a constrained version of the biplot model where $\mathbf{B} = \mathbf{AT}$, with $\mathbf{T}$ a particular columnwise orthogonal matrix. This means that the graphical representation of the columns of the data matrix in each plane (dimension) is constrained to be a rotation of the row configuration. As for *general biplot* the representation in this case can be difficult to analyze for large n.

To obtain a direct representation of symmetry and skew-symmetry, others interesting scalar product models can be applied. They can be considered constrained version of DEDICOM by imposing on $\mathbf{R} = \mathbf{A}(a\mathbf{I} + b\Delta)\mathbf{A}'$ particular forms (see Rocci 2004): GIPSCAL by Chino (1978, 1990) and Generalized Escoufier and Grorud (GEG) by Rocci and Bove (2002), when $\delta_l = 1$ for each l, Escoufier and Grorud model (EG) by Escoufier and Grorud (1980), when $a = b = 1$ and $\delta_l = 1$ for each l, Singular Value Decomposition of SKew-symmetry (SVDSK) by Gower (1977), when $a = 0$, $b = 1$ and only the skew-symmetry of $\mathbf{X}$ is analysed. For these models, scalar products describe symmetry, while the area of the triangles having two object-points and the origin as vertices describe the absolute value of the skew-symmetry, whose algebraic sign is associated with the orientation of the plane. All the previous models are examples of metric multidimensional scaling (MDS), but when nonmetric data are available methods like the one proposed by Okada and Imaizumi (1987) in the context of distance models should be preferred.

When a full column rank matrix of *external variables* $\mathbf{E} = [\mathbf{e}_1, \mathbf{e}_2, \ldots, \mathbf{e}_p]$ containing additional information on the n objects is available, we can try to incorporate in the analysis the external information in order to improve data interpretation (e.g. data theory compatible MDS). This problem is considered in Bove (2006), where some proposals to incorporate external information in *general biplot*, GEG and EG models are provided along with an application to Morse code confusion data. In that paper it is pointed out that when methods for joint analysis of symmetry and skew-symmetry, like GEG and EG, fail to reveal theory-consistent explanations of the asymmetric proximity (e.g. because the symmetric component is very relevant), separate external analyzes of the two components should be preferred. In the next section proposals for the external analysis of skew-symmetry will be presented.

3 The Analysis of Skew-Symmetry with External Information

Now we show a method to incorporate external information in SVDSK (Gower 1977) following the same approach presented in Bove (2006). The model in matrix notation can be formulated as

$$\mathbf{K} = \mathbf{A}\Delta\mathbf{A}' + \boldsymbol{\xi}, \tag{5}$$

where $\mathbf{K} = [k_{ij}]$ is the skew-symmetric component of $\mathbf{X}$ and $\mathbf{A}$, Δ and $\boldsymbol{\xi}$ are defined as in the previous section. In order to incorporate external information we want the columns of $\mathbf{A}$ to be in the subspace spanned by the columns of $\mathbf{E}$. In matrix notation $\mathbf{A} = \mathbf{EC}$ where $\mathbf{C}$ is a matrix of unknown weights, so that

$$\mathbf{K} = \mathbf{A}\Delta\mathbf{A}' + \boldsymbol{\xi} = \mathbf{E}\mathbf{C}\Delta\mathbf{C}'\mathbf{E}' + \boldsymbol{\xi}. \tag{6}$$

The least squares estimate for $\mathbf{C}$ and Δ is obtained by minimizing

$$h(\mathbf{C}, \Delta) = \left\| \mathbf{K} - \mathbf{E}\mathbf{C}\Delta\mathbf{C}'\mathbf{E}' \right\|^2 \tag{7}$$

that, if we rewrite $\mathbf{E} = \mathbf{P}\mathbf{G}$, where $\mathbf{P}'\mathbf{P} = \mathbf{I}$ and $\mathbf{G}$ is a square full rank matrix, is equivalent to

$$h'(\mathbf{C}, \Delta) = \|\mathbf{P}'\mathbf{K}\mathbf{P} - \mathbf{G}\mathbf{C}\Delta\mathbf{C}'\mathbf{G}'\|^2 \tag{8}$$

as it is shown in Bove (2006, p. 70) in general for asymmetric matrices. So the minimum of $h'(\mathbf{C}, \Delta)$ (and $h(\mathbf{C}, \Delta)$) is reached when $\mathbf{C} = \mathbf{G}^{-1}\mathbf{U}$ and $\Delta = \Sigma$ where $\mathbf{U}\Sigma\mathbf{U}'$ is the *r-dimensional* singular value decomposition of the skew-symmetric matrix $\mathbf{P}'\mathbf{K}\mathbf{P}$.

An important advantage of this method of external analysis is that we need only n points in the graphical representation. On the other hand, in some applications it can happen that even the separate external analysis of skew-symmetry could cause a strong reduction of fit respect to the unconstrained approach. In these cases it is worthwhile to check if the external information is able to explain at least the size of skew-symmetry, disregarding its algebraic signs. To this aim, in Bove (2006) it was suggested to perform externally constrained symmetric MDS of the matrix $\mathbf{M}$ obtained with the absolute values of the entries of $\mathbf{K}$. Metric or ordinal scaling methods can be applied to the symmetric matrix $\mathbf{M}$, whose diagonal elements are equal to zero. These external analyses by distance models can be applied even with standard statistical software (e.g. PROXSCAL, SPSS-Categories).

The size of skew-symmetry can be also analyzed by scalar product models. In this case the diagonal entries are taken into account by assuming $\mathbf{M}$ a dissimilarity-like matrix and applying classical scaling with linear constraints discussed, for instance, in De Leeuw and Heiser (1982) and Ter Braak (1992). We remark that classical scaling perform an indirect application of a scalar product model (eigendecomposition) to the matrix $\mathbf{M}$, because a preliminary transformation of the dissimilarities into scalar products is carried out. So the pairs of objects having large skew-symmetry are represented by distant points. However, if we want to represent these pairs of points close to each other, scalar product models should be directly applied to $\mathbf{M}$.

A method to represent $\mathbf{M}$ by scalar products between only n points can be based on the approximate factorization

$$\mathbf{M} = \mathbf{A}\mathbf{A}' + \boldsymbol{\xi} \tag{9}$$

with $\mathbf{A}$ and $\boldsymbol{\xi}$ defined as previously. The external information can be incorporated in the analysis if we rewrite in the usual manner,

$$\mathbf{M} = \mathbf{A}\mathbf{A}' + \boldsymbol{\xi} = \mathbf{E}\mathbf{C}\mathbf{C}'\mathbf{E}' + \boldsymbol{\xi}. \tag{10}$$

The least squares estimation problem is based on the function

$$h(\mathbf{C}) = \left\|\mathbf{M} - \mathbf{ECC'E'}\right\|^2 \tag{11}$$

or, as shown before, equivalently on

$$h'(\mathbf{C}) = \|\mathbf{P'MP} - \mathbf{GCC'G'}\|^2. \tag{12}$$

The minimum of function $h'(\mathbf{C})$ is obtained for $\mathbf{C} = \mathbf{G}^{-1}\mathbf{U}$, where $\mathbf{UU'}$ is the best positive semi-definite (Gramian) matrix approximation of rank r of the symmetric matrix $\mathbf{P'MP}$. The computation of matrix $\mathbf{U}$ is straightforward if $\mathbf{P'MP}$ is positive semidefinite, but this usually does not hold in the applications because $\mathbf{M}$ is not positive semidefinite (diagonal entries are equal to zero). We can substitute all diagonal entries, that are not informative, with some constant positive value that makes matrix $\mathbf{M}$ positive semidefinite (e.g. the absolute value of its lowest negative eigenvalue). This transformation of $\mathbf{M}$ makes $\mathbf{P'MP}$ positive semidefinite, so $\mathbf{U}$ is easily obtained from the eigendecomposition of this last matrix.

When taking into account the diagonal entries of $\mathbf{M}$ is considered too constraining we can follow a different approach. In fact, the problem of approximating a symmetric matrix by a positive semi-definite (Gramian) matrix of lower rank is considered in Bailey and Gower (1990) and in Ten Berge and Kiers (1993). In the second paper is proposed an alternating least squares method that allows nonunit weights for the diagonal entries. This feature is particularly useful in our case because by this method we can fit only the non-diagonal entries of matrix $\mathbf{M}$ by using zero weights for the diagonal elements.

4 Conclusions

A class of scalar product models for asymmetric MDS was presented emphasizing some relationships between the models and showing possibilities to incorporate external information for the analysis of skew-symmetry. The hierarchy of the models in the class could suggest strategies in their application on real data. The choice between the models can also depends on the importance given to the symmetric and the skew-symmetric components respect to a direct analysis of the entries of the data matrix. Separate analyses of the two components seem preferable when symmetry is much more relevant in the data (see Bove 2006) or when we want to represent separately skew-symmetric residuals of statistical models (e.g. symmetry or quasi-symmetry). Future developments on this research line could consider comparative applications of the different methods proposed for the analysis of the size of skew-symmetry.

References

Bailey, R. A., & Gower J. C. (1990). Approximating a symmetric matrix. *Psychometrika, 55*, 665–675.

Bove, G. (2006). Approaches to asymmetric multidimensional scaling with external information. In S. Zani, A. Cerioli, et al. (Eds.), *Data analysis, classification and the forward search* (pp. 69–76). Berlin: Springer.

Chino, N. (1978). A graphical technique for representing asymmetric relationships between N objects. *Behaviormetrika, 5*, 23–40.

Chino, N. (1990). A generalized inner product model for the analysis of asymmetry. *Behaviormetrika, 27*, 25–46.

De Leeuw, J., & Heiser, W. J. (1982). Theory of multidimensional scaling. In P. R. Krishnaiah & L. N. Kanal (Eds.), *Handbook of statistics* (Vol. 2, pp. 285–316). Amsterdam: North Holland.

Escoufier, Y., & Grorud, A. (1980). Analyse factorielle des matrices carrées non-symétriques. In E. Diday et al. (Eds.), *Data analysis and informatics* (pp. 263–276). Amsterdam: North Holland.

Gabriel, K. R. (1971). The biplot graphic display of matrices with application to principal component analysis. *Biometrika, 58*, 453–467.

Gower, J. C. (1977). The analysis of asymmetry and orthogonality. In J. R. Barra et al. (Eds.), *Recent developments in statistics* (pp. 109–123). Amsterdam: North Holland.

Harshman, R. A. (1978). Models for analysis of asymmetrical relationships among N objects or stimuli. In *Paper presented at the first joint meeting of the Psychometric Society and the Society for Mathematical Psychology*, McMaster University, Hamilton, Ontario.

Kiers, H. A. L., & Takane, Y. (1994). A generalization of GIPSCAL for the analysis of nonsymmetric data. *Journal of Classification, 11*, 79–99.

Okada, A., & Imaizumi, T. (1987). Nonmetric multidimensional scaling of asymmetric proximities. *Behaviormetrika, 21*, 81–96.

Rocci, R. (2004). A general algorithm to fit constrained DEDICOM models. *SMA-Journal of the Italian Statistical Society, 13*, 139–150.

Rocci, R., & Bove, G. (2002). Rotation techniques in asymmetric multidimensional scaling. *Journal of Computational and Graphical Statistics, 11*, 405–419.

Ten Berge, J. M. F., & Kiers, H. A. L. (1993). An alternating least squares method for the weighted approximation of a symmetric matrix. *Psychometrika, 58*, 115–118.

Ter Braak, C. J. F. (1992). Multidimensional scaling and regression. *Statistica Applicata, 4*, 577–586.

Zielman, B., & Heiser, W. J. (1996). Models for asymmetric proximities. *British Journal of Mathematical and Statistical Psychology, 49*, 127–146.

Intimate Femicide in Italy: A Model to Classify How Killings Happened

Domenica Fioredistella Iezzi

Abstract Women's homicide (femicide) is the most serious form of violence against women. The aim of this paper is to propose a method to classify the underlying mechanism of killing. We analysed 1,125 cases of femicides in Italy from 2000 to 2005, 764 of these occurring in a domestic setting. The most important information about the mechanism of femicide is textual data, referring to newspaper articles. We propose a method to describe the killer profile on the basis of the analysis of crime scenes, possible suspect information and killer's modus operandi.

1 Introduction

Women's homicide (femicide) by intimate partners is the most serious form of violence against women. Statistics show that when a woman is killed, the perpetrator is often a man who has been intimately involved with her (EURES 2006; Steen and Hunskaar 2004). Domestic femicide is very frequent and intimate partner is the single largest category of femicide, with women most often killed by their husband, lover, ex-husband, or ex lover (Landau and Rolef 1998).

The present study analyzes the data on femicides occurred in Italy from 2000 to 2005 (EURES 2006). Since 2000 Economic and Social Research Centre (EURES) collects data on murders in Italy, and integrates this information with DEA DB (The Data Bank of the National Agency of Press - ANSA) and data from the Department of Criminal Police-Service and Analysis. These data cover all cases of intentional murders of women occurring in Italy. In Italy, the National Institute of Statistics (ISTAT) recently collected data about harassment and physical and psychological violence. It has been estimated that 6,000,743 women aged 16–70 years are victims of physical or sexual abuse during their lives. 23.7% of the female population suffers from sexual violence, 18.8% are subjected to physical violence and 4.8% are victims of rape or attempted rape. 17.3% of the violence against women is attributed to

D. F. Iezzi
Università degli Studi di Roma "Tor Vergata", Italy
e-mail: stella.iezzi@uniroma2.it

F. Palumbo et al. (eds.), *Data Analysis and Classification*,
Studies in Classification, Data Analysis, and Knowledge Organization,
DOI 10.1007/978-3-642-03739-9_10,

partners or ex-partners (Iezzi and Corradi 2007; ISTAT 2007). The aim of this paper is to propose a method to classify the mechanism of killing. We apply classification trees to explain and predict the belonging of femicides to a group (femicide within or out the family), on the basis of explanatory quantitative and qualitative variables. We define "femicide within the family" when the perpetrator is a partner, an ex-partner or a member of the family (son, daughter, cousin, uncle, etc.), while "femicide out the family" is carried out by a friend, an acquaintance or by unknown people or strangers.

2 National and International Scenario

Let us first look at the comparison of incidence of femicides with the background of homicides in Italy and in other countries. The Spanish Centre "Reina Sofia" has collected data from several countries (Fig. 1), finding that countries with the highest rates of femicides are several South American countries, in particular Guatemala with about 123 femicides per million women. Among European countries, UK, Spain, Germany, Austria, and Denmark report higher femicide rates than Italy. In fact, Italy has only 6.5 femicides per million women. The European countries with the highest rate of femicide are the former Soviet Union republics, such as Estonia, Hungary, Romania, Slovakia. The countries with the highest rate of *domestic* femicide (domestic femicide: killing of a woman by a partner or a member of her family) are Hungary, Slovenia and Finland. Many studies in United States, Canada, Australia and in other countries show that domestic homicides are the largest category of femicides.

In Italy from 1990 to 2006 the yearly number of total femicides was about 180–190 cases.[1] In 2005, the rate of femicides strongly diminished, dropping to 4.39 per million women (Table 1). During the period considered for this study (2000–2005), EURES data-base collected 1,125 femicides committed in Italy, with a mean number of 187.5 femicides per year. About 68% of femicides (764 in absolute value) are committed *within the family*. EURES, according to other relevant field studies (Mouzos and Rushforth 2003), defines Domestic Homicide when the murder happens within a familiar or intimate context (even if the relationship is finished). Therefore, we consider Domestic Homicide if the killer is:

- A relative (a member of family)
- A partner (husband, wife, fiancé, live-in partner)
- An ex partner

The majority of victims of domestic femicides (42.2) are killed by their husband or live-in partner; 15.8% by their ex husband or ex partner, 12.7% by their son or daughter. Totally Couple-femicides (or intimate femicides), that are women killed

[1] Source: homicide recorded by police.

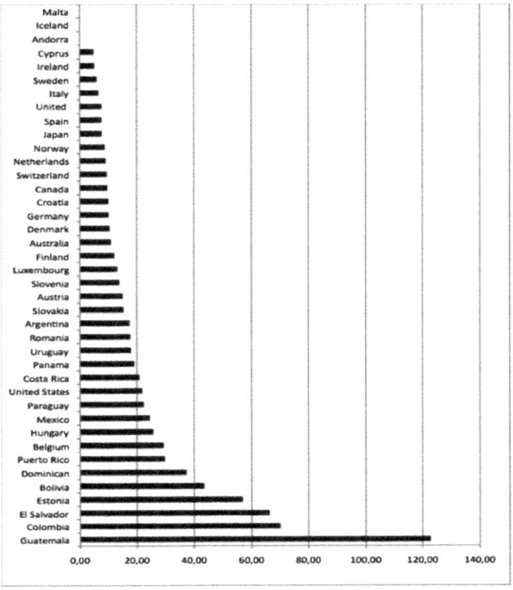

Fig. 1 Prevalence rates of femicides in 2003 (source: Second International Report "Partner Violence Against Women", Centro Reina Sofia 2003)

Table 1 Prevalence rates of femicides in Italy from 1990 to 2006 (source: EURES 2006)

	V.A.	Rate per million women
1990	184	6.31
1995	190	6.49
2000	199	6.78
2001	186	6.33
2002	187	6.36
2003	192	6.50
2004	185	6.20
2005	132	4.39
2006	181	5.99

by their partner, husband, ex partner or ex husband are 66.7% of total domestic femicides. 92.4% of victims of a domestic femicide are killed by a man and only the 7.6% by another woman.

Femicides are committed in a domestic context according to various causes: the majority of victims are killed for passion, jealousy, or as a consequence of a feeling of possession, or for the inability to accept the loss of the partner. Many are also femicides called "pietatis causa" (6.9%) that happens through the will of the author (and sometimes also of the victim) to end the sufferings of the victim.

In both family homicides and femicides occurring in Italy, the killer is commonly a male (respectively 80.5% and 92.8%).

3 Data and Method

We analysed 1,125 cases of femicides, of which 764 were domestic, occurring in Italy from 2000 to 2005 and have been collected by the EURES DB. The most important information about the mechanism of femicide is textual data, referring to newspaper articles. The victims are mainly Italian (80%). Most of these femicides were committed in a domestic setting ($p < 0.000$). Recently, the average number per year is 128 cases of femicides within the family (one every 3 days) and 60 out of the family.

There is no relationship between the size of a city and where a woman is killed ($p = 0.283$).

For each femicide, we analysed nine illustrative variables (type of femicide "in family" or "out of family", relationship between victim and murderer, profession of victim, region of murder, geographical area, size of city, citizenship of victim, age and year of murder). Moreover, we explored the textual descriptions of mechanisms of femicide (MF). The corpus of MF is composed of 18,624 words, of which 2,438 are words type. The method is composed of the following sequential steps:

(A) Pre-processing: we cleaned and normalised the *corpora* (description how killing happened).
(B) Lexical analysis: we took out the *bag of words*.
(C) Information extraction: we calculated TFIDF index (1).
(D) Building a three-way matrix: we computed the cosine distance for each key-topic.
(E) Exploration of latent dimensions: we detected the latent dimensions of femicide mechanics.
(F) Classification of cases: we wanted to assess the adequacy of classification of femicide mechanics in two classes: in and out the family.

After the pre-processing and lexical analysis (steps A and B), we build matrix X that has a dimension ($n \times p$), where n is the number of cases of femicide and p the keywords to describe the mechanics. We divided X into the following six key topics: (1) place of the crime, (2) weapon used, (3) part of the body injured, actions of killer, (4) before, (5) during and (6) after femicide. On each key-topic, we calculated term frequency inverse document frequency index (TFIDF) as the method to categorize the mechanics of killing (step C).

$$TFIDF = TF \times Ln \frac{N}{DF}, \tag{1}$$

where TF is the number of times that a keyword appears in a case of femicide, N is the number of femicide cases and DF is the document frequency.

The TFIDF measure assigns a weight to each unique word in a document representing how topic specific that word is to its document or femicide case. On each key topic of **X**, we computed the similarity between a keyword and a case of femicide by using the cosine distance (step D). The cosine of the angle formed by two document vectors (2), that describe two mechanics of killing, **A** and **B** is

$$\cos(\alpha) = \frac{< A, B >}{|A||B|}. \tag{2}$$

Cases with many common terms will have vectors closer to each other, than documents with fewer overlapping terms. In this way, we built a three-way proximity matrix $\Delta = [_{ijk}]$, where $\{ijk\}(i, j = 1, \ldots, N$ and $k = 1, \ldots, K)$ is the cosine distance between two cases of femicide in $k^{\{th\}}$ occasion (key-topic).

In step E, we applied **IN**dividual **D**ifferences **SCAL**ing (INDSCAL) algorithm on Δ to detect latent dimensions of femicide mechanics (Borg and Groenen 2005). In step F, we classified the relationship victim and offender by mechanics of killing (latent dimensions obtained in step E) and some characteristics of the victims (age, profession of victim, region of murder, geographical area, size of city, citizenship of victim and age) by Exhaustive CHAID (Chi-squared Automatic Interaction Detector, Biggs et al. 1991). At each step, Exhaustive CHAID chooses the independent (predictor) variable that has the strongest interaction with the dependent variable (femicide in the family and out of the family). Categories of each predictor are merged if they are not significantly different compared to the dependent variable. Exhaustive CHAID is a non parametrical method based on classification tree procedures are a useful tool for the analysis of large data sets characterised by high dimensionality and non standard structure, where no hypothesis can be made on the underlying distribution (Fig. 2). In the end, we used Generalization of the Logistic Regression (GLR) to assess the contribution of a risk factor to femicide and controlled for covariates (latent dimensions).

4 The Main Results

The INDSCAL on the matrix Δ detected two dimensions for location of femicide: the majority of intimate femicides happen at home and especially in the bedroom or kitchen; while other femicides occur outside home and, generally, in cars or hotels. The mechanics of these killings are very similar, e.g. the most used weapons are pistols or guns, followed by knives and the killer hits the woman repeatedly while the victim is already dead. In the family, the killer generally strikes the victim on her head or strangles her. Out of the family, he hits her on the head, chest and pubes. The latent dimensions are: location, weapon and part of the body hurt. Exhaustive CHAID trees showed that the most important variable is the profession of victims. In particular, profession of victims can classified into four clusters: (1) blue collar worker, farmer, teacher or dependent worker; (2) businesswoman, student and unemployed; (3) housewife, and nurse; (4) prostitute. The victims belong to classes no. 1, no. 2 and no. 3 were killed by relatives; while prostitutes (no. 4 class) by strangers. In the class no. 1 the killer operated in two different ways: (1) he hit

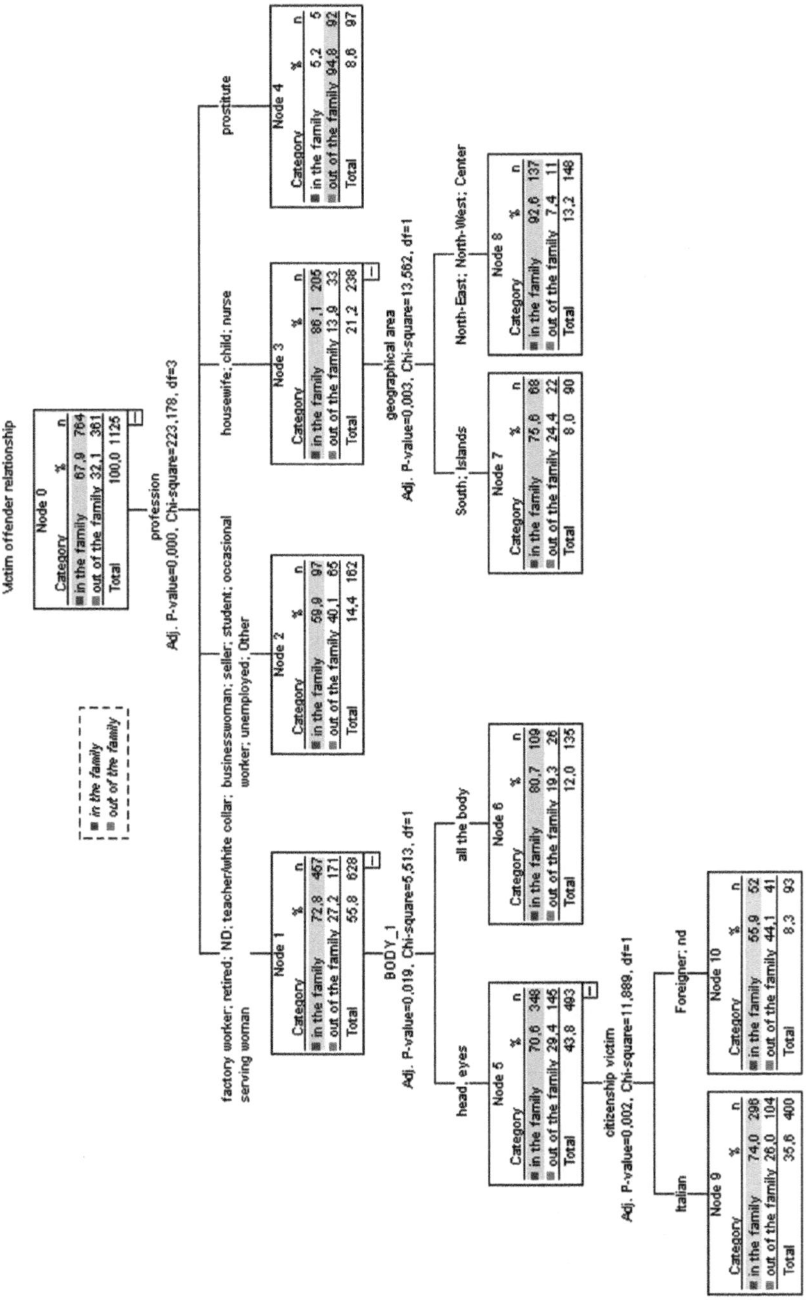

Fig. 2 Classification tree of Italian femicides

the superior part of the body, in particular head, eyes and face; (2) he beat and beat to death or struck all parts of the victim's body.

In Southern Italy, the killers use weapons that require close contact with the victims, such as knives or hands; in North-East, North-West and Centre, they prefer gun and blunt instruments. Moreover, when the victims are aged between 25 and 35, the assassin was above all an ex-partner, more than 35 was a husband. Women murder by men who are strangers are very young (from 19 to 24 years old). The risk estimate of 0.244 indicates that the category predicted by the model (femicides by relatives and by strangers) is wrong for 24.4% of the cases.

The classification showed that the model classified approximately 75.6% of the femicides correctly. The results of GLR confirmed that the dimensions latent significant are part of the body injured, actions of killer before femicide. The most important factors of risk are the citizenship of victim, the profession of murdered and area of crime (North-East, North-West, Centre, South and Islands). Exhaustive CHAID chooses the independent (predictor) variable that has the strongest interaction with the dependent variable (femicide in the family and out of the family). Categories of each predictor are merged if they are not significantly different compared to the dependent variable. Exhaustive CHAID is a non parametrical method based on classification tree procedures are a useful tool for the analysis of large data sets characterised by high dimensionality and non standard structure, where no hypothesis can be made on the underlying distribution.

According to this, we classified the femicides and predicted future cases, based on the profile of victims and scene of crime.

References

Borg, I., & Groenen, P. (2005). *Modern multidimensional scaling, theory and applications.* New York: Springer.

Biggs, D., B. de ville, & E, Suen (1991), (1991), 'A method of choosing multiway partitions for classification and decision trees,' J. of Appl. Statist., vol. 18, pp 49–62.

Centro Reina Sofia Para El Estudio De La Violencia (2007). *Violencia contra la mujer en las relaciones de pareja. Estadisticas y legislacin* (2nd International Report, 2007 Serie Documentos, Vol. 11). Valencia: Author.

EURES (2006). *L'omicidio volontario in Italia* (Rapporto Eures-Ansa 2006). Rome: Author.

Iezzi, D. F., & Corradi, C. (2007). Violence against women: How to estimate the risk? In *Risk and Prediction – Atti della Riunione Scientifica SIS* (pp. 519–520), Venice, 6–8 June 2007.

ISTAT (2007). *La violenza e i maltrattamenti contro le donne dentro e fuori la famiglia* (Rapporto). Rome: Author.

Landau, F. S., & Rolef, S. H. (1998). Intimate femicide in Israel: Temporal, social and motivational patterns. In *European Journal on Criminology Policy and Research, 6*, 75–90.

Mouzos, C., & Rushforth, C. (2003). *Family homicide in Australia* (Trends and Issues in Crime and Criminal Justice, no. 255). Australian Institute of criminology.

Steen, K., & Hunskaar, S. (2004). Gender and physical violence. *Social Science and Medicine, 59*, 567–571.

Two-Dimensional Centrality of Asymmetric Social Network

Akinori Okada

Abstract The purpose of the present study is to introduce a procedure to derive the centrality of an asymmetric social network, where the relationships among actors are asymmetric. The procedure is based on the singular value decomposition of an asymmetric matrix of friendship relationships among actors. Two kinds of the centrality are introduced; one is the centrality of extending friendship relationships from the actor to the other actors, and the other is the centrality of accepting friendship relationships from the other actors to the actor. The present procedure is based on the two largest singular values not only on the largest singular value. Each actor has two sets of the centrality; each consists of the centrality of extending and of the centrality of accepting friendship relationships. An application to help or advice relationships among managers in a company is shown.

1 Introduction

The centrality of an actor in a social network represents the importance, popularity, attractiveness, power, significance, or salience of the actor in forming friendship relationships with the other actors in a social network. By knowing the centrality of each actor, we can understand the characteristics of the social network. Several concepts of the centrality of an actor, and the corresponding procedures to derive the centrality have been introduced (Wasserman and Faust 1994). In defining and deriving the centrality, utilizing characteristic values and corresponding characteristic vectors have played the important role. Bonacich (1972) introduced a procedure to derive the centrality of an actor in a social network. The procedure assumes that the relationships among actors are symmetric; the relationship from actors j to k is equal to that from actors k to j. The procedure utilizes the characteristic vector which corresponds to the largest characteristic value of the

A. Okada
Graduate School of Management and Information Sciences, Tama University, 4-1-1 Hijirigaoka, Tama-shi, Tokyo 206-0022, Japan
e-mail: okada@tama.ac.jp

F. Palumbo et al. (eds.), *Data Analysis and Classification*,
Studies in Classification, Data Analysis, and Knowledge Organization,
DOI 10.1007/978-3-642-03739-9_11,

friendship relationship matrix among actors. Each element of the characteristic vector represents the centrality of the corresponding actor, and each actor has one centrality. The procedure (Bonacich 1972) has been extended to deal with the social network which is asymmetric where the matrix of friendship relationships is square but asymmetric (Bonacich and Lloyd 2001). While their formulation uses the characteristic vectors corresponding not only to the largest characteristic value, the centrality is derived from the characteristic vector corresponding only to the largest characteristic value. Bonacich (1972) has also been extended to deal with the social network between two different sets of actors where the matrix of friendship relationships is rectangular (Bonacich 1991) by using the singular value decomposition. The extension is based on the singular vector corresponding to the largest singular value. The purpose of the present study is to introduce a procedure to derive the centrality (a) which can represent both directions of asymmetric relationships, and (b) based on the singular vectors corresponding not only to the largest singular value but also to the second largest singular value. Each actor has two sets of the centrality; one is based on the singular vector corresponding to the largest singular value, and the other is based on the singular vector corresponding to the second largest singular value. Each set consists of two centralities, one represents the centrality of extending friendship relationships from an actor to the other actors, and the other represents the centrality of accepting friendship relationships from the other actors to the actor. The singular value represents the salience of the centrality based on the corresponding singular vector.

2 The Procedure

The present procedure is based on the singular value decomposition of a matrix of friendship relationships among actors of a social network where the relationships among actors are not necessarily symmetric. Let $\mathbf{A}$ be the matrix of friendship relationships among n actors. The (j, k) element of $\mathbf{A}$, a_{jk}, represents the friendship relationship from actors j to k. When the relationship from actors j to k is friendly, $a_{jk} = 1$, and when the relationship from actors j to k is not friendly, $a_{jk} = 0$. Two conjugate elements a_{jk} and a_{kj} are not necessarily equal, and $\mathbf{A}$ is not necessarily symmetric. The singular value decomposition of $\mathbf{A}$ is given by

$$\mathbf{A} = \mathbf{UDV}', \tag{1}$$

where $\mathbf{U}$ is the $n \times n$ matrix of left singular vectors ($\mathbf{U}'\mathbf{U} = \mathbf{I}$), $\mathbf{D}$ is the $n \times n$ diagonal matrix of singular values at its diagonal, and $\mathbf{V}$ is the $n \times n$ matrix of right singular vectors ($\mathbf{V}'\mathbf{V} = \mathbf{I}$).

The left singular vector represents the centrality corresponding to the row of $\mathbf{A}$, and the right singular vector represents the centrality corresponding to the column of $\mathbf{A}$. The jth element of the left singular vector represents the centrality of extending friendship relationship from actor j to the other actors. The kth element of the right singular vector represents the centrality of accepting friendship relationships from the other actors to actor k. While the centrality is derived by the singular or characteristic vector corresponding to the largest singular or characteristic value in the past studies (Bonacich 1972, 1991; Bonacich and Lloyd 2001), the centrality in the present study is derived by the singular vectors corresponding to the two largest singular values (cf. Okada 2008). Two left singular vectors give two centralities of extending friendship relationship of an actor, and two right singular vectors give two centralities of accepting friendship relationships of an actor. There are two sets of the centrality for an actor where each set consists of two kinds of the centrality; the centrality represented by the left singular vector, and the centrality represented by the right singular vector.

3 The Data

In the present study the matrix of friendship relationships among 21 managers at a company (Wasserman and Faust 1994, Table B.1, p. 740) was analyzed. The data were collected at an entrepreneurial small manufacturing organization on the west coast of the United States (Krackhardt 1987). The 21 managers consist of the president of the company (Manager 7), four vice-presidents (Managers 2, 14, 18, and 21) each heads up a department, and 16 supervisors from four departments. The data consist of responses to the question "Who would you go to for help or advice at work?" (Krackhardt 1987; Wasserman and Faust 1994, pp. 60–61). The friendship relationships in the present data mean help or advice relationships at work. The matrix of friendship relationships of the present study is a 21×21 asymmetric matrix. When manager j responded that she/he goes to managers k for help or advice at work, the (j, k) element of the matrix is 1. And when manager j responded that she/he does not go to manager k for help or advice at work, the (j, k) element of the matrix is 0.

4 The Analysis

The singular value decomposition of the matrix of friendship relationships among 21 managers was done. The five largest singular values were 140.8, 18.9, 9.7, 7.8, and 6.2. In the present study two left singular vectors and two right singular vectors corresponding to the largest and the second largest singular values were used to derive two sets of the centrality of 21 managers. Each manager has two sets of the centrality; the first set is given by the singular vector corresponding to the largest singular value, and the second set is given by the singular vector corresponding to the second largest singular value. Each of the two sets of the centrality for a manager consist of the two centralities; one given by the left singular vector representing the strength of going to the other manager for help or advice at work from the manager, and the other is given by the right singular vector representing the strength of being asked help or advice at work from the other managers to the manager.

5 Result

The result is shown graphically in Fig. 1 where each actor is represented as a point. The two graphical representations show two kinds of the centrality schematically and simultaneously; one is the strength of asking help or advice at work from a manager to the other managers, and the other is the strength of being asked help or advice at work from the other managers to a manager. Figure 1a consists of the first left (horizontal dimension) and right (vertical dimension) singular vectors corresponding to the largest singular value. The first left singular vector represents the strength of asking help or advice to the other managers from a manager along the aspect of the first left singular vector. The horizontal dimension is represented as Dimension 1 Out, because it represents the outward tendency from a manager. The first right singular vector represents the strength of accepting or being asked help or advice from the other managers to a manager along the aspect of the first right singular vector. The vertical dimension is represented as Dimension 1 In, because it represents the inward tendency of a manager.

The first left and right singular vectors represent the overall strength of asking or being asked help or advice among these 21 managers. All managers have the non-negative strength along the horizontal as well as the vertical dimensions. The president (Manager 7), represented as a large solid rhombus, has the larger strength along the vertical dimension than along the horizontal dimension. Of the four vice-presidents, represented as solid squares, the three (Managers 2, 14, and 21) also have the larger strength along the vertical dimension than along the horizontal dimension. This suggests that upper ranked managers (except Manager 18) have the larger strength of being asked help or advice from the other managers than of asking help or advice to the other lower ranked managers. On the contrary, nine of the 16 supervisors have the larger strength of asking help or advice than the strength of being asked help or advice.

Figure 1b consists of the second left (horizontal dimension) and the right (vertical dimension) singular vectors corresponding to the second largest singular value. The second left singular vector represents the strength of asking help or advice to the other managers from a manager along the aspect of the second left singular vector. The horizontal dimension is represented as Dimension 2 Out, because of the same reason as in the case of Fig. 1a. The second right singular vector represents the strength of accepting or being asked help or advice from the other managers to a manager along the aspect of the second right singular vector. The vertical dimension is represented as Dimension 2 In.

While in Fig. 1a the strength along Dimension 1 Out as well as along Dimension 1 In is non-negative, in Fig. 1b the strength along Dimension 2 Out as well as along Dimension 2 In is not always non-negative. Some managers have the negative strength along the horizontal as well as the vertical dimensions. The president (Manager 7), and the three vice-presidents (Managers 2, 14, and 21) have the negative strength along both the vertical and the horizontal dimensions. The meaning of the positive strength along the second left and right singular vectors in Fig. 1b is the same as that along the first left and right singular vectors in Fig. 1a. The meaning of the negative strength in Fig. 1b will be discussed in the next section, but as will be mentioned later, it can say that the larger absolute value represent the larger strength. The higher ranked managers except Manager 21 have the larger (absolute) strength along the horizontal dimension than along the vertical dimension, suggesting that these higher ranked managers have the larger strength of asking help or advice to other managers than of being asked help or advice from the other managers. Eleven supervisors of the 16 have the smaller strength of asking help or advice to the other managers than of being asked the help or advice from the other managers. This is opposite with the case of the first left and right singular vectors shown in Fig. 1a.

6 Discussion

In the present study, a procedure of giving two sets of the centrality for each actor in an asymmetric social network was introduced. Each of the two sets of the centrality has two kinds of the centrality; the strength of the outward tendency from an actor and that of the inward tendency to an actor. The present procedure was successfully applied help or advice relationships among managers at a company.

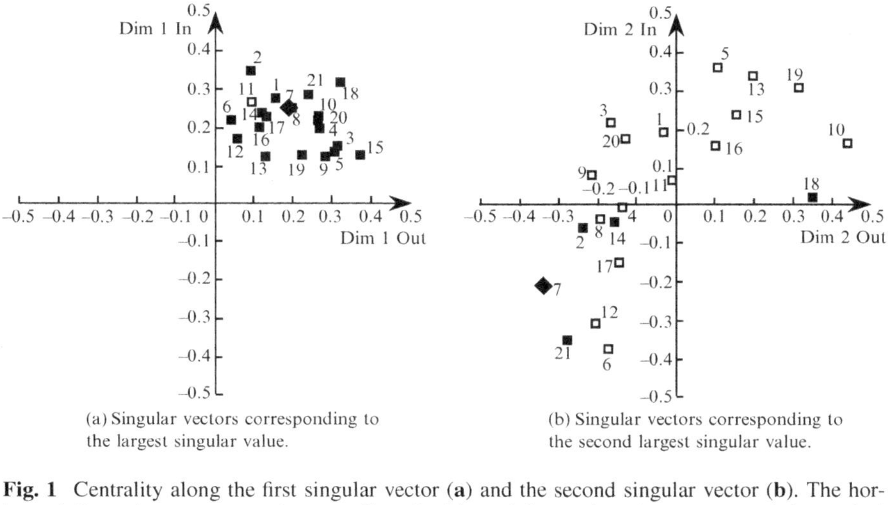

(a) Singular vectors corresponding to the largest singular value.

(b) Singular vectors corresponding to the second largest singular value.

Fig. 1 Centrality along the first singular vector (**a**) and the second singular vector (**b**). The horizontal dimension represents the centrality of asking advice to the other actors, and the vertical dimension represents the centrality of being asked advice from the other actors.

The strength of friendship relationship or the strength of the tie of asking and being asked help or advice at work from managers j to k is represented as the sum of two terms; (a) the product of the strength of asking help or advice of manager j along the first left singular vector and the strength of being asked help or advice to manager k along the first right singular vector, and (b) the product of that along the second left singular vector and that along the second right singular vector. As mentioned earlier, two singular values show the salience of the strength along the singular vectors corresponding to two largest singular values. The salience of the strength along the first singular vectors is more than seven times important than that along the second singular vectors.

In the case of Fig. 1a, the product of the elements of the left and the right singular vectors is always non-negative, because in Fig. 1a the strength along the horizontal as well as along the vertical dimensions is non-negative. The strength of friendship relationship from managers j to k along the second left and right singular vectors (Fig. 1b) is also represented as the product of two elements. But the product of the two elements can be negative in the case of the second left and right singular vectors. This should be discussed more thoroughly. There are three cases of the product of two elements along the horizontal and the vertical dimensions:

(a) The product of two positive elements is positive.
(b) The product of two negative elements is positive.
(c) The product of a positive and a negative elements is negative.

In the case of (a), the meaning of the element along the horizontal and vertical dimensions is the same as that of the first singular vectors. But in the cases of (b) and (c), it is difficult to interpret the meaning of the elements by the same manner as that for case (a).

The larger value or the positive value of the product of elements of the second left and right singular vectors shows the larger or positive friendship relationship from one manager to the other manager. The smaller value or the negative value shows the smaller or negative (friendship) relationship from one manager to the other manager. The product of two negative elements of the second left and right singular vectors is positive, showing the positive friendship relationship from one manager to the other manager. The product of the positive and the negative elements of the second left and right singular vectors is negative, showing the negative friendship relationship from one manager to the other manager. When managers j and k have the same sign along the left and right singular vectors, the product of the two corresponding elements of the second right and left singular vectors is positive, suggesting the positive friendship relationship from managers j to k. When two managers j and k have the elements of different signs for the second left and right singular vectors, the product of the two corresponding elements is negative, suggesting the negative friendship relationship from managers j to k. For example, when two managers are in the first quadrant, the relationship from a manager to the other manager is positive. This is also true when two managers are in the third quadrant. When two managers are in the second quadrant (or in the fourth quadrant), the relationship from a manager to the other manager is negative.

When two managers are in different quadrants, whether the relationship from one manager to the other is positive or negative is determined by which quadrant each of the two managers is. When a manager is in a quadrant where the element of the left singular vector is positive (the first and the fourth quadrants) and the other manager is in a quadrant where the element of the right singular vector is positive (the first and the second quadrants), the relationship from the former to the latter is positive. But the relationship from the latter in the second quadrant to the former is positive when the former is in the fourth quadrant, and is negative when the former is in the first quadrant, and the relationship from the latter in the first quadrant to the former in the fourth quadrant is negative. These tell that the second singular vectors classify managers into two groups where the relationship within the same group is positive and the relationship between two groups is negative; one consists managers in the first quadrant, and the other consists of those in the third quadrant. While the relationships from managers in the first/third quadrant to those in the second/fourth quadrant is positive, the relationship of the opposite direction is negative. While the relationship from managers in the first/third quadrant to those in the fourth/second quadrant is negative, the relationship of the opposite direction is positive. This suggests that the two groups have relationships of opposite characteristics with managers in the second and the fourth quadrants. The second singular vectors seem to represent the difference between these two groups.

The data, to which the present procedure was applied, consist of 1 and 0 (go to or do not go to another manager for help or advice), showing the relationship is binary. It seems necessary to apply the present procedure to non-binary relationship data, e.g., the frequency of communications among actors, to see the characteristics of the present procedure in analyzing these data.

The present procedure deals with an asymmetric social network or an asymmetric matrix of friendship relationships among actors. The asymmetric matrix of friendship relationships can be analyzed by asymmetric multidimensional scaling (Borg and Groenen 2005, Chap. 23) or by correspondence analysis (Greenacre 2000). While the objective of asymmetric multidimensional scaling and corresponding analysis is to summarize the asymmetric relationships among actors and visually represent the summarized relationships, the objective of the present procedure is not to summarize and visually represent the relationships but to derive the strength representing the outward and the inward tendency of each actor. In the present application, each actor has two sets of the outward and the inward tendency having two different meaning; the strength of overall tendency showing the hierarchy of the organization of the company, and the differentiation between two groups which are not friendly each other and have opposite relationships with the other actors.

References

Bonacich, P. (1972). Factoring and weighting approaches to clique identification. *Journal of Mathematical Sociology, 2*, 113–120.

Bonacich, P. (1991). Simultaneous group and individual centralities. *Social Networks, 13*, 155–168.

Bonacich, P., & Lloyd, P. (2001). Eigenvector-like measures of centrality for asymmetric relations. *Social Networks, 23*, 191–201.

Borg, I., & Groenen, P. J. K. (2005). *Modern multidimensional scaling: Theory and applications* (2nd edition). New York: Springer.

Greenacre, M. (2000). Correspondence analysis of square asymmetric matrices. *Applied Statistics, 49*, 297–310.

Krackhardt, D. (1987). Cognitive social structures. *Social Networks, 9*, 109–134.

Okada, A. (2008). Two-dimensional centrality of a social network. In C. Preisach, L. Burkhardt, & L. Schmidt-Thieme (Eds.), *Data analysis, machiine learning and applications* (pp. 381–388). Heidelberg: Springer.

Wasserman, S., & Faust, K. (1994). *Social network analysis: Methods and applications*. Cambridge, UK: Cambridge University Press.

The Forward Search for Classical Multidimensional Scaling When the Starting Data Matrix Is Known

Nadia Solaro and Massimo Pagani

Abstract This work provides an extension of the Forward Search to classical multidimensional scaling according to a double perspective: first, as a diagnostic tool in order to detect outlying units and monitor their influences on the main analysis results. Second, as a comparative tool when two or more solutions need to be compared. A case study from a clinical setting is then considered.

1 Introduction

Multidimensional Scaling (MDS) methods are a set of multivariate analysis techniques that attempt to represent proximity data into a low-dimensional Euclidean space by recovering the coordinates of points through some optimal procedure. Input data are frequently represented by dissimilarity matrices, their elements measuring to which extent points in an original, observational space are unlike. Depending on the nature of input data, MDS techniques can be divided into metric and non-metric methods. The former can be applied when input data are at least at interval-scale level, while the latter assume them at ordinal level. For a complete reference on MDS see the monographs by Cox and Cox (2001) and Borg and Groenen (2005).

As it frequently happens with most of statistical techniques, the presence of outlying units or substructures in the data, such as groups of units, might markedly influence the results of an MDS analysis. In order to prevent biased conclusions from analyses it is advisable therefore to rely on robust diagnostic tools capable to detect potential perturbations in data. On this matter, we propose to extend a specific robust methodology, that is the Forward Search (FS), to classical MDS (CMDS), which is probably the most popular MDS of metric-type. As known, the FS consists of a set of robust diagnostic tools developed in order to detect multivariate outliers especially and monitor their influence on statistics and analysis results (Atkinson et al. 2004).

N. Solaro (✉)
Department of Statistics, University of Milano-Bicocca, Via Bicocca degli Arcimboldi 8, 20126 Milan, Italy
e-mail: nadia.solaro@unimib.it

F. Palumbo et al. (eds.), *Data Analysis and Classification*, Studies in Classification, Data Analysis, and Knowledge Organization,
DOI 10.1007/978-3-642-03739-9_12,

Given that the FS is a very effective tool in exploring multidimensional data, in this work we extend the FS to CMDS by assuming that a starting data matrix of quantitative variables is known. An input dissimilarity matrix can then be set up straightforwardly. Here the purpose will be twofold:

1. To apply the FS in the context of a single MDS analysis as a pure diagnostic tool, in order to monitor the effects of perturbations in data on the principal results, such as dimension scores.
2. To apply the FS as a comparative tool, when two or more MDS solutions, deriving for instance from different input dissimilarity matrices, need to be compared.

Finally, to show the potentiality of the extension of the FS to CMDS a case study from a clinical setting concerning the Metabolic Syndrome will be considered.

2 Classical Multidimensional Scaling and the Forward Search

Given a set of n units, let: $\mathbf{\Delta} = [\delta_{ij}]_{i,j=1,\dots,n}$ be an observed dissimilarity matrix, $\mathbf{X}$ a column-centered configuration of the n points in q dimensions ($q < n$) and $\mathbf{D(X)} = [d_{ij}(\mathbf{X})]_{i,j=1,\dots,n}$ the matrix of Euclidean distances from $\mathbf{X}$. With the objective of reproducing dissimilarities δ_{ij} through distances d_{ij} in a Euclidean space, CMDS derives a solution $\mathbf{X}$ for the n points by means of the spectral decomposition theorem (SDT). Formally, if $\mathbf{\Delta}$ is Euclidean, the matrix: $\mathbf{B} = \mathbf{HAH}$ is positive semi-definite (p.s.d.) with rank $q \leq n - 1$, where $\mathbf{A} = \left[-\frac{1}{2}\delta_{ij}^2\right]$, $\mathbf{H} = \mathbf{I} - \mathbf{11}^t/n$ and $\mathbf{1}$ is a n-dimensional vector of ones. Then, the SDT can be applied to $\mathbf{B}$ to give the solution $\mathbf{X}$ in q dimensions: $\mathbf{X} = \mathbf{\Gamma\Lambda}^{1/2}$, where $\mathbf{\Lambda}$ is a diagonal matrix of eigenvalues of $\mathbf{B}$, in non-increasing order, and $\mathbf{\Gamma}$ is the corresponding matrix of normalized eigenvectors, (see, e.g. Cox and Cox, 2001). If $\mathbf{\Delta}$ is not Euclidean, $\mathbf{B}$ is not longer p.s.d., thus requiring the above procedure to be slightly modified. Two alternatives can be invoked: if the magnitude of negative eigenvalues is relatively small, then these may be simply ignored, otherwise it is advisable to add a constant to the off-diagonal elements of $\mathbf{\Delta}$ to make $\mathbf{B}$ p.s.d. In the literature this is known as "The Additive Constant Problem" (Cox and Cox 2001).

One of the crucial points is the choice of number q of dimensions, which strictly implies how well a configuration $\mathbf{X}$ fits input data. A criterion to establish how reasonably small q should be is usually based on these two goodness-of-fit indices: $\text{GOF}_1 = \sum_{i=1}^{q} \lambda_i \Big/ \sum_{i=1}^{n-1} |\lambda_i|$ and $\text{GOF}_2 = \sum_{i=1}^{q} \lambda_i \Big/ \sum \lambda_i^+$, where λ_i and λ_i^+ denotes, respectively, eigenvalues and positive eigenvalues of matrix $\mathbf{B}$. It always holds: $\text{GOF}_2 \geq \text{GOF}_1$, where the equality is attained when the λ_is are all non negative, that is matrix $\mathbf{B}$ is p.s.d. or, equivalently, $\mathbf{\Delta}$ is Euclidean. To assess the relative contribution of single dimensions to the whole fit, relative eigenvalues can be computed: $\tilde{\lambda}_s^{(1)} = \lambda_s \Big/ \sum_{i=1}^{n-1} |\lambda_i|$ or: $\tilde{\lambda}_s^{(2)} = \lambda_s \Big/ \sum \lambda_i^+$, $(s = 1, \dots, q)$.

Moreover, in order to compare two MDS solutions $\mathbf{X}$ and $\mathbf{Z}$ the so-called Procrustes statistic can be computed. It is a measure of how well a configuration $\mathbf{X}$ matches to another configuration $\mathbf{Z}$ by taking into account any possible rotation, reflection, dilation and translation (Borg and Groenen 2005; Cox and Cox 2001). In the normalized version it is given by: $R^2 = 1 - \{\mathrm{tr}[(\mathbf{X}^t\mathbf{Z}\mathbf{Z}^t\mathbf{X})^{\frac{1}{2}}]\}^2 / \{\mathrm{tr}(\mathbf{X}^t\mathbf{X})\mathrm{tr}(\mathbf{Z}^t\mathbf{Z})\}$, being $R^2 = 0$ when $\mathbf{X}$ perfectly matches with $\mathbf{Z}$, whereas $R^2 = 1$ when $\mathbf{X}$ completely mismatches with $\mathbf{Z}$.

How to Extend the FS to CMDS When exploring multivariate data the basic idea underlying the FS is that of forming growing size subsets of units through a step-by-step process that starts from an "outlier-free core" and then carries on by adding one unit at a time according to the ordering on Mahalanobis distance (MD). Then, by applying a statistical method of interest, it is possible to both detect some kinds of perturbation in data and monitor their impact on analysis results (Atkinson et al. 2004). Regarding outlier detection, the FS has proved to be particularly effective in disclosing "masking" and "swamping" effects, arising typically when multiple outliers are present. Masking denotes the situation in which a unit is not recognized as outlier since many other outliers are present, thus masking its existence. Swamping occurs when a unit is wrongly considered as outlier because of the presence of a group of outliers (Atkinson et al. 2004).

In CMDS framework the FS can be applied through the following steps:
(1) The forward subsets $S^{(m)}$ have to be formed through standard application of the FS. That is, starting from an initial subset of size m_0 a total number $M = n - m_0 + 1$ of subsets $S^{(m)}$ of size m are formed by entering the remaining units one at a time ($m = m_0, \ldots, n$). This is done on the basis of squared MD, which is computed for both the units entering and not entering the subset $S^{(m)}$: $d_{im}^2 = (\mathbf{y}_i - \boldsymbol{\mu}_m)^t \boldsymbol{\Sigma}_m^{-1} (\mathbf{y}_i - \boldsymbol{\mu}_m)$, for $i \in S^{(m)}$, $i = 1, \ldots, m$, and for $i \notin S^{(m)}$, $i = m + 1, \ldots, n$, being the centroid $\boldsymbol{\mu}_m$ and the variance-covariance matrix $\boldsymbol{\Sigma}_m$ both computed on the m units in $S^{(m)}$. At the subsequent step, the $m + 1$ units with the smallest squared MD will form the subset $S^{(m+1)}$. This procedure is then iterated until all units are considered. The initial subset $S^{(m_0)}$ can be formed according to many different criteria. The standard computation method is based on "robustly centered ellipses". Alternative criteria are the methods based on either "bivariate boxplots" or a robust computation of the overall centroid and variance-covariance matrix. The initial subset could indeed be formed at random or by any arbitrary choice, since it was noticed that the starting point has no crucial influence when detecting outliers (Atkinson et al. 2004).
(2) CMDS is then applied on each dissimilarity matrix $\boldsymbol{\Delta}^{(m)}$ of order m computed from the corresponding $(m \times p)$ data matrix $\mathbf{Y}^{(m)}$, which comprises row vectors of $\mathbf{Y}$ related to units in the subset $S^{(m)}$. The starting data matrix $\mathbf{Y}$ usually involves not directly comparable variables, so that standardization is strongly recommended. As known, if the input dissimilarity measure is one of the Minkowski family of distances, standardizing variables is the same as computing weighted Minkowski distances, with weights given as functions of the reciprocal of square root of variances. On this basis, for each subset $S^{(m)}$ matrices $\mathbf{A}^{(m)}$ and $\mathbf{B}^{(m)}$ can be computed

analogously as described above. Finally, the "forward" CMDS solutions are produced by applying the SDT on matrices $\mathbf{B}^{(m)}$ and then by computing the sets of coordinates: $\mathbf{X}^{(m)} = \mathbf{\Gamma}^{(m)}\mathbf{\Lambda}^{1/2\,(m)}$ with dimension q kept fixed during the search ($m = m_0, \ldots, n$). A number M of forward CMDS solutions are thus derived.
(3) Once the forward CMDS solutions have been computed, the main results can be monitored through the forward plots. As said before, both diagnostic and comparative objectives can be pursued, especially when two different CMDS solutions are involved. For instance, dimension scores can be plotted against increasing subset size similarly as made in principal component analysis (Atkinson et al., 2004, Chap. 5). Also relative eigenvalues and the goodness-of-fit indices $\text{GOF}_1^{(m)}$ and $\text{GOF}_2^{(m)}$ can be monitored. When the input dissimilarity matrix is not Euclidean, the ratio of GOFs: $\text{GOF}_{ratio}^{(m)} = \text{GOF}_1^{(m)}/\text{GOF}_2^{(m)}$ represents a normalized measure for units in $S^{(m)}$ of how close their original space is to be perfectly reproduced in a Euclidean space.

With comparative purposes, GOF indices of different CMDS analyses can be monitored on the same graph, in order to see how far the CMDS solutions $\mathbf{X}^{(m)}$ and $\mathbf{Z}^{(m)}$ are one each other and if the discrepancy between them possibly tends to worsen as the subset size increases. A more in depth inspection relies on $R^{2\,(m)}$, that is the normalized Procrustes statistic computed on each subset $S^{(m)}$. The forward plot of $R^{2\,(m)}$ allows two main aspects to be highlighted. Firstly, it permits to disclose the influence of outlying units on the lack of matching between two configurations. Secondly, it highlights the role of some units that, though not outliers, could cause apparent mismatches between the two configurations at a certain step of the search. For instance, suppose that unit u^* joins $S^{(m)}$ to form the subset $S^{(m+1)}$. Then a value of $R^{2\,(m+1)}$ "much greater" than $R^{2\,(m)}$ denotes that unit u^* has markedly altered the geometric structure underlying $S^{(m)}$, so that it is not possible to make $\mathbf{X}^{(m+1)}$ and $\mathbf{Z}^{(m+1)}$ as similar as they were on subset $S^{(m)}$ through transformations like rotation, reflection, dilation or translation. To assess the influence of each single unit, a measure of its contribution to the normalized Procrustes statistic is required. Here we propose the following simple one:

$$\mathcal{D}^{(m+1)} = R^{2\,(m+1)} - R^{2\,(m)}, \tag{1}$$

($m = m_0, \ldots, n-1$). It is near to zero if unit u^*, when added to the subset $S^{(m)}$, does not cause notable changes in the geometric structure; it is positive whenever unit u^* contributes to worsen the lack of matching between the two configurations in the sense above described; it is negative when unit u^* contributes to improve the matching degree between the two configurations, thus becoming more similar.

3 A Case Study: Linosa Dataset

To show the potentiality in the application of the FS in CMDS we considered a subset of the data collected by the Linosa Heart Study (Lucini et al. 2006), which was addressed to study the problem of the onset of neurovegetative disorders and their relationship with the Metabolic Syndrome (MetS). The variables here considered, fifteen in all, regard some autonomic, metabolic and cardiovascular characteristics collected on 140 subjects living on Linosa island. These variables are not directly comparable, thus requiring standardization or distance weighting.

Two main applications of CMDS on these data are here illustrated: in the first, input data are made of Euclidean distances ($\boldsymbol{\Delta}_{\mathrm{E}}$); in the second, City-block distances are used ($\boldsymbol{\Delta}_{\mathrm{C}}$). Two CMDS solutions, respectively, $\mathbf{X}_{\mathrm{E}}$ and $\mathbf{X}_{\mathrm{C}}$, are obtained in $q = 3$ dimensions. In both cases the first dimension is related to the MetS condition: the more the values are highly positive, the more the subjects are healthy. The more highly negative, the more in the MetS condition. The other two dimensions are connected with, respectively, the insulin levels and the neurovegetative system. GOF indices are equal to: $\mathrm{GOF}_{1,E} = \mathrm{GOF}_{2,E} = 0.5709$ for $\boldsymbol{\Delta}_{\mathrm{E}}$ and $\mathrm{GOF}_{1,C} = 0.4099$, $\mathrm{GOF}_{2,C} = 0.5021$ for $\boldsymbol{\Delta}_{\mathrm{C}}$. As it was expected, $\boldsymbol{\Delta}_{\mathrm{E}}$ is reproduced better than $\boldsymbol{\Delta}_{\mathrm{C}}$, since this latter matrix requires a greater number of dimensions. Moreover, in this last case the ratio of GOFs is equal to 0.8164, denoting that 18.36% of the entire City-block distance structure cannot be recovered by real coordinates. Finally, straight comparisons between the two configurations $\mathbf{X}_{\mathrm{E}}$ and $\mathbf{X}_{\mathrm{C}}$ through Procrustes analysis brings to a value of $R^2 = 0.2643$, denoting that the matching between them is not so high.

Standard application of the FS on Linosa dataset has been carried out by computing the initial subset of size $m_0 = 16$ through all the criteria mentioned in Sect. 2. In case of random assignment, analyses with 10 randomly formed initial subsets have been considered. In addition, a "not-outlier-free" subset with the 16 most distant units from the overall centroid according to MD distance has been also constructed. All the analyses agree in disclosing the presence of three outlying units, labelled, respectively, with codes "1132", "1116" and "1003" (the forward plots of scaled MD are here omitted), and confirm the substantial independence of the FS from its starting point also in our case. The unit "1003" is "the most healthy" one in the dataset, getting fairly small values on the MetS risk factors. The other two units "1132" and "1116" are featured instead by the highest health hazards, being in the MetS condition with the highest values on risk factors. Whatever the input dissimilarity matrix is, these three units are then expected to influence CMDS results to a great extent.

The next figures show several results of the application of the FS to CMDS carried out as outlined in Sect. 2 and using robustly-centered-ellipses method to form the initial subset. During the search weighted Euclidean and City-block distances have been re-computed on each forward subset. Figure 1 is concerned with the application of CMDS on weighted Euclidean distance matrices. From the left-hand panel, showing the forward plot of first dimension scores, it is apparent that the three outliers "1132", "1116" and "1003" enter the subsets in the final steps of the search and that they assume the highest scores in absolute value. Their influence is however mostly confined to the variability of this dimension, in the sense that they do not markedly affect neither the magnitude or the order of the majority of units' scores, given that their trajectories are quite stable, horizontal lines. In the right-hand panel of Fig. 1 trajectories of relative eigenvalues $\tilde{\lambda}^{(1)}$ of the first three dimensions are

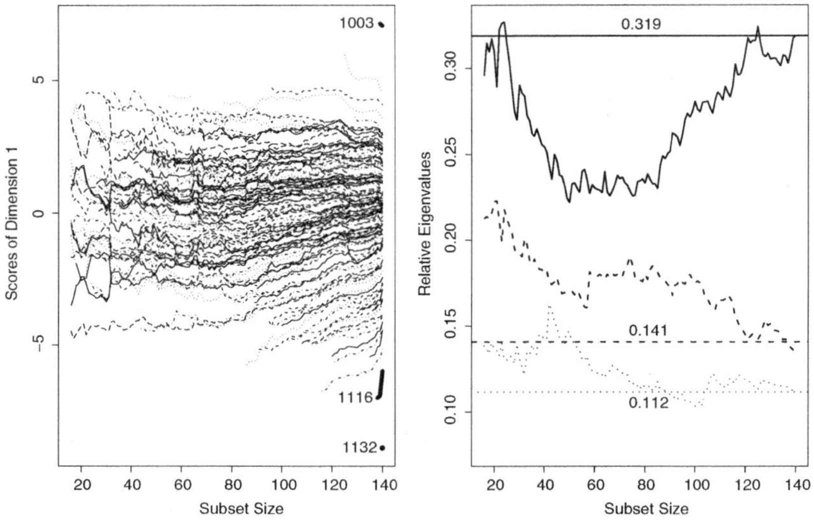

Fig. 1 CMDS on Linosa dataset: Monitoring of first dimension scores (*left panel*) and relative eigenvalues (*right panel*) starting from Euclidean distances

plotted along with their values computed on the entire dataset (horizontal lines). It is worth noting that the relative eigenvalue of dimension 1 falls slightly from nearly step 35 to step 80 of the search, reaching its minimum value of 0.2220; after that it begins to increase gradually towards its final value of 0.319 at the end of the search.

Figure 2 reports results related to comparisons among different CMDS solutions, obtained by starting from different distances. In the left-hand panel the monitoring of GOF indices for both solutions $\mathbf{X}_{\mathrm{E}}$ and $\mathbf{X}_{\mathrm{C}}$ is shown. Two remarks are worth making: first, during the search GOF indices corresponding to $\mathbf{X}_{\mathrm{C}}$ assume constantly lower values than $\mathbf{X}_{\mathrm{E}}$, with the exception of the first steps. In the last steps especially, values of GOF indices for $\mathbf{X}_{\mathrm{C}}$ seem to slightly increase since outlying units enter the search. Second, during the search the gap among the three trajectories tends to increase, becoming stable in its last steps. As regards the comparison of GOF indices, it is worth recalling that a greater number of dimensions is necessary to reproduce "at best" matrix $\boldsymbol{\Delta}$ in a Euclidean space if this is not Euclidean. Obviously, such a number is strictly related to the total unit set size, so that the trend represented in this figure can be thought as typical. In a similar fashion, the gap between $\mathrm{GOF}_{1\mathrm{C}}$ and $\mathrm{GOF}_{2\mathrm{C}}$ increases as the subset size grows, in that the number of negative eigenvalues tends to become greater if the input dissimilarity matrix is of higher order. In the right-hand panel of Fig. 2 three different CMDS solutions are compared by plotting on the same graph ratios of GOF indices. In addition to Euclidean and City-block, also input Lagrange distance is considered here. Of course, ratio of GOFs is always equal to one in the Euclidean case. In the other two cases the percentage of positive eigenvalues tends to reduce considerably as the subset size increases, especially when Lagrange distance is concerned.

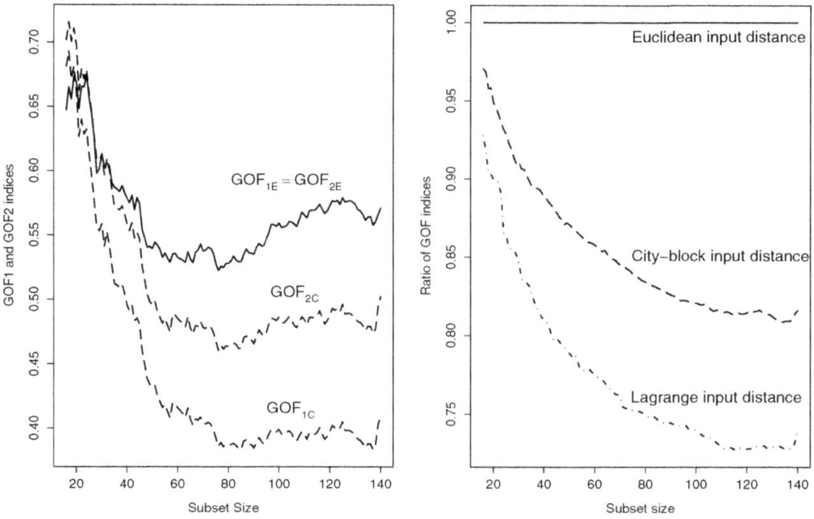

Fig. 2 CMDS on Linosa dataset: Monitoring of GOF indices in the Euclidean and City-block cases (*left panel*) and Ratio of GOFs for Euclidean, City-block and Lagrange solutions (*right panel*)

In Fig. 3 results regarding Procrustes analysis are displayed. The left-hand panel shows the monitoring of the normalized Procrustes statistic when matching $\mathbf{X}_C$ to $\mathbf{X}_E$. It is apparent that $\mathbf{X}_C$ differs substantially from $\mathbf{X}_E$ in three specific moments: the first is around the 40th step of the search; the second is nearly at the center, between steps 75 and 95; the third is in the final steps. While this last situation can be explained by the entry of outlying units into the subset, the other two seem quite unexpected, especially the first one, in that they involve units not detected as outliers. A persuasive argument could be that the units joining the subset at those steps would lead to large changes in the geometric structure and consequently the City-block configuration can hardly be considered Euclidean.

A further inspection can be carried out with the forward plot of the marginal contribution $\mathcal{D}^{(m+1)}$ to normalized Procrustes statistic [formula (1)]. The right-hand panel of Fig. 3 shows that, although in most of steps the trajectory of $\mathcal{D}^{(m+1)}$ lies almost around zero, the individual contribution of units with codes "1029" (in the subset with $m = 37$) and "1081" (in the subset with $m = 48$) is not negligible, as it can be seen from the corresponding clear-cut peaks. Nevertheless, when afterwards unit "1078" enters the subset of size $m = 38$ as well as unit "1046" enters the subset of size $m = 49$, two large reductions on Procrustes statistic can be clearly observed. It seems then that these two units contribute to restore a configuration of points that makes the City-block solution much closer to admit a Euclidean representation.

Finally, it is worth remarking that starting the search from an initial subset formed according to one of the criteria mentioned in Sect. 2 could generally lead to different Procrustes analyses. In other terms, depending on the composition of the forward subsets the units altering the geometric structure could be different over the

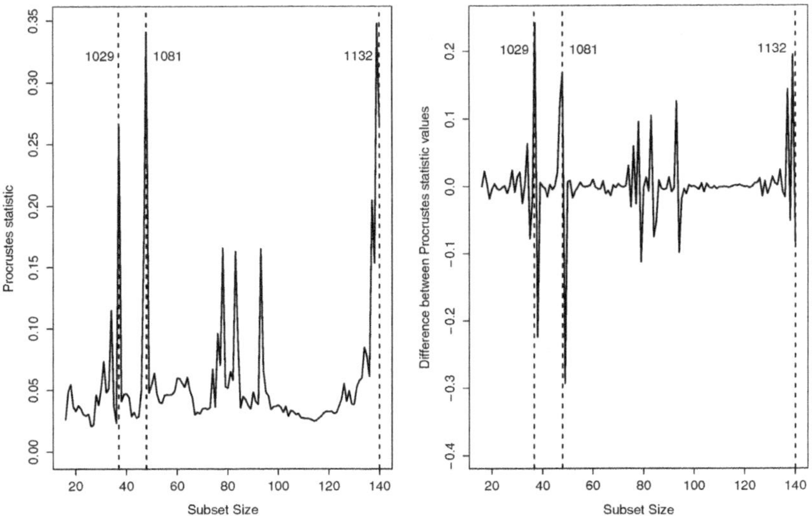

Fig. 3 CMDS on Linosa dataset: Monitoring of normalized Procrustes statistic (*left panel*) and absolute contribution of units to normalized Procrustes statistic (*right panel*)

analyses. In our study outlying units turn out to be also changing-structure units. In any case, this matter would require a more careful consideration.

4 Conclusions

The extension of the FS to CMDS has revealed that it is a very powerful tool in monitoring the influence of units on principal results derived from a single CMDS application as well as in comparisons among different CMDS solutions. However, many questions are still open so that further inspections would be necessary. For instance, the monitoring of Procrustes statistic has revealed that units can have a role other than outlier. How could these units be treated under a unified theoretical system? In addition, another crucial question should be at issue: the FS is basically carried out by employing Mahalanobis distances, requiring then a quantitative starting data matrix. Data can however have a more general nature in that the starting data matrix can be exclusively qualitative or involve both types of variables. Another question concerns the situation in which the starting data matrix is completely unknown, so that only proximity data are available. How could the FS idea be applied in such situations? Finally, in this work we have focused on the extension of the FS to a specific MDS model, that is classical MDS, thus not covering many other MDS methods, such as Sammon's non linear mapping or non-metric MDS, which would require a specific treatment. These all are very important aspects to which future research should be addressed.

References

Atkinson, A. C., Riani, M., & Cerioli, A. (2004). *Exploring multivariate data with the forward search*. New York: Springer.

Borg, I., & Groenen, P. J. F. (2005). *Modern multidimensional scaling – Theory and applications* (2nd edition). New York: Springer.

Cox, T. F., & Cox, M. A. A. (2001). *Multidimensional scaling* (2nd edition). New York: Chapman and Hall/CRC.

Lucini, D., Cusumano, G., Bellia, A., Kozakova, M., Di Fede, G., Lauro, R., Pagani, M., Lauro, R. (2006). Is reduced baroreflex gain a component of the Metabolic Syndrome? Insights from the Linosa Study. *Journal of Hypertension, 24*, 361–370.

Part IV
Multivariate Analysis and Application

Discriminant Analysis on Mixed Predictors

Rafik Abdesselam

Abstract The processing of mixed data – both quantitative and qualitative variables – cannot be carried out as explanatory variables through a discriminant analysis method. In this work, we describe a methodology of a discriminant analysis on mixed predictors. The proposed method uses simultaneously quantitative and qualitative explanatory data with a discrimination and classification aim. It's a classical discriminant analysis carried out on the principal factors of a Mixed Principal Component Analysis of explanatory mixed variables, i.e. both quantitative and transformed qualitative variables associate to the dummy variables. An example resulting from real data illustrates the results obtained with this method, which are also compared with those of a logistic regression model.

1 Introduction

The methodology of quantification qualitative variables evolved in the context of Mixed Principal Component Analysis (MPCA) (Abdesselam 2006) is used here in a discrimination and classification aim on explanatory mixed variables. Discriminant analysis in its usual version use only quantitative predictors (Fisher 1938). Since, a methodology called DISQUAL method (Saporta 1977) allows to extend the context of discriminant analysis to qualitative predictors. The proposed Mixed Discriminant Analysis (MDA) approach allows to implement a discriminant analysis with the two types of predictors, this is the main aim of this work; to extend the discriminant model context for using mixed predictors like, for example, logistic model or discriminant partial least squares (PLS) approach. The proposed approach is evaluated then compared to the logit model on the basis of real mixed data. These analyses are carry out by discrimination with two groups on principal factors procedure of

R. Abdesselam
ERIC EA 3038, University of Lyon 2, 69676 Bron, France
e-mail: rafik.abdesselam@univ-lyon2.fr

F. Palumbo et al. (eds.), *Data Analysis and Classification*,
Studies in Classification, Data Analysis, and Knowledge Organization,
DOI 10.1007/978-3-642-03739-9_13,

SPAD software for MDA and by logistic procedure of SAS software for the logistic model.

2 Mixed Discriminant Predictors

We use the following notations to explain the methodology which consists in transforming the qualitative explanatory variables on quantitative variables for the discriminant model. Let us denote:

- $Z_{(n,r)}$ the qualitative data matrix associated to $\{z^t; t = 1, r\}$, the dummy variables of the variable z with r modalities or groups that we wish to discriminate.
- $X_{(n,p)}$ the quantitative data matrix associated to the set of p discriminant variables $\{x^j; j = 1, p\}$, with n rows-individuals and p columns-variables.
- $(y_1, \ldots, y_l, \ldots, y_m)$ the set of m qualitative discriminant variables with $q = \sum_{l=1}^{m} q_l$ dummy variables $\{y_l^k; k = 1, q_l\}_{\{l=1,m\}}$.
- $Y_{l\,(n,q_l)}$ the dummy variables matrix associated to the q_l modalities of the variable y_l.
- $Y_{(n,q)} = [Y_1, \ldots, Y_l, \ldots, Y_m]$ global matrix, juxtaposition of the matrix $Y_{l\,(n,q_l)}$.
- $E_z = R^r$, $E_x = R^p$ and $E_y = \oplus\{E_{y_l}\}_{\{l=1,m\}} = R^q$ are the individual subspaces associated by duality respectively to the data matrix $Z_{(n,r)}$, $X_{(n,p)}$ and $Y_{(n,q)}$.
- $D = \frac{1}{n} I_n$ diagonal weights matrix of the n individuals and I_n the unit matrix with n order.
- $N_x = \{x_i \in E_x; i = 1, n\}$ and $N_{y_l} = \{y_i \in E_{y_l}; i = 1, n\}$ are the configurations of the individual-points associated to the rows of the matrix $X_{(n,p)}$ and $Y_{l\,(n,q_l)}$.
- $M_x = V_x^+$ and $M_{y_l} = \chi^2_{y_l}$ are the matrix of inner product, the Mahalanobis distance in E_x and the Chi-square distance in E_{y_l}.
- $V_{xy_l} = {}^t XDY_l$ the matrix of covariances.
- $P_{E_{y_l}}$ the orthogonal projection operator in subspace E_{y_l}.

The quantification of qualitative data is made with the statistical and geometrical construction of m configurations of individual-points $\hat{N}_x^{y_l} = \{P_{E_{y_l}}(x_i); x_i \in N_x\} \subset E_{y_l}$. For all $l=1$ to m, we note $\hat{X}^{y_l} = X V_x^+ V_{xy_l}$ the data matrix of order (n, q_l) associated to the project configuration of individual points $\hat{N}_x^{y_l}$; the subspace E_{y_l} is considered as an explanatory subspace on which we project the configuration of individual points N_x of quantitative data in the explain subspace E_x.

It is shown in Abdesselam (2006) the following remark and property concerning the Mixed Principal Component Analysis (MPCA).

Remark. The PCA $(\hat{X}^{y_l}; \chi^2_{y_l}; D)$ is equivalent to Multivariate ANalysis Of Variance (MANOVA) between the p quantitative variables and the q_l dummy variables associated to the levels of the explained factor y_l, which $I(\hat{N}_x^{y_l}) = trace(V_{y_l x} V_x^+ \times V_{xy_l} \chi^2_{y_l})$, the explained inertia is equal to Pillai's trace.

Property. The MPCA of the mixed data table $[\, X \mid Y \,]_{(n;p+q)}$ consists to carry out the standardized PCA of the data table $[\, X \mid \tilde{Y} \,]_{(n;p+q)}$.

Where, $\tilde{Y}_{(n,q)} = [\tilde{Y}_1, \ldots, \tilde{Y}_l, \ldots, \tilde{Y}_m]$ is a juxtaposition matrix of transformed qualitative data, with $\tilde{Y}_l = Y_l - {}^tG_l$ the quantitative data matrix of order (n, q_l) associated to the configuration of individual points $N_{\tilde{y}_l} \subset E_{y_l}$ that inertia $I(N_{\tilde{y}_l}) = q_l - 1$, where $G_l = {}^t\hat{X}^{y_l} D\, 1_n$, is the mean vector of the variables $\hat{X}^{y_l}$ and 1_n the unit vector with n order.

Note that MPCA is equivalent to Mixed Data Factorial Analysis (MDFA) (Pagés 2004). The main aim of these two methods is to research principal components, noted F^s, which maximize the following mixed criterion, proposed in square correlation terms in Saporta (1990) and geometrically in terms of square cosinus of angles in Escofier and Pagés, J. (1979):

$$\sum_{j=1}^{p} r^2(x^j, F^s) + \sum_{l=1}^{m} \eta^2(y_l, F^s) = \sum_{j=1}^{p} \cos^2 \theta_{js} + \sum_{l=1}^{m} \cos^2 \theta_{ls}$$

where, r^2 and η^2 are respectively the square of the linear correlation coefficient of quantitative variables and the correlation ratio of qualitative variables with the sth factor, and θ the angle between the correspondent vectors. These two expressions are equal in view of fact that the variables are normalized.

In a methodological point of view, the MDA appears as a chain of two procedures: a projection procedure of configurations of points corresponding to the MANOVA coordinates to quantify the qualitative variables, we take into account the correlation ratios, then a standardized PCA procedure to synthesize the linear correlations between all variables, quantitative and transformed qualitative variables.

Definition 1. The MDA $[\, X \mid Y \,]_{(n;p+q)} \longrightarrow Z_{(n;r)}$ consists to carry out a discriminant analysis on the principal factors of the MPCA of mixed data table $[\, X \mid Y \,]_{(n;p+q)}$.

So, this extension methodology of discriminant analysis on mixed variables, that we can call DISMIX method (DIScrimination on MIXed variables), is like DISQUAL method (DIScrimination on QUALitative variables), which consists to make a discriminant analysis on factors of Multiple Correspondence Analysis (MCA) of explanatory variables (Saporta 1977).

We can note that the first principal factors of MPCA (respectively MCA) are not necessary the better discriminant factors of DISMIX (respectively DISQUAL) method, but we can select only the significant discriminant factors. We obtain satisfactory discrimination results with these methods.

3 Application Example

To illustrate this approach then to compare it with logistic model, we use data of an application example taken from the library SAS System. In this study of the analgesic effects of treatments on elderly patients with neuralgia, two test treatments and a placebo are compared. This data set contains the responses of $p = 2$ explanatory quantitative variables: Age of the patients and the Duration of complaint before the treatment began and $m = 2$ explanatory qualitative variables with $q = 5$ modalities in total: Treatment (A, B, Placebo) and Sex (Female, Male) of the patients according to the response explain variable Pain with two groups: is whether the patient reported pain or not (Yes_{25}, NO_{35}).

This sample of size $n = 60$ patients is subdivided into two samples: a basic-sample or "training set" composed of $n_1 = 55$ (90%), randomly drawn from the whole data set for the discriminant rule and a test-sample or "validation set" of size $n_2 = 5$ (10%) for next evaluated the performance of this rule.

Moreover the fact to compare the two test treatments and a placebo, the aim is to bring to the fore the mixed characteristics which well differentiate the two groups of patients.

3.1 Predictor Analysis

First, we analyze and describe only the predictors using Mixed Principal Component Analysis (MPCA). This analysis extracts in total five factors $(p + q - m)$ given in Table 1.

Table 2 gives the linear correlations between mixed predictors and MPCA factors. Figure 1 shows the graphical representations of the quantitative and transformed qualitative variables, on the MPCA factorial planes which explain 90.16% of the total variability. The first axis (30.18%) opposes men to women patients, the second one (22.69%) compares treatment B and placebo. While the third axis (21.21%) summarizes the transformed variable treatment A, the fourth axis (16.07%) synthesizes and opposes the age variable to duration variable.

Table 1 MPCA eigenvalues

Number	Eigenvalue	Proportion (%)	Cumulative (%)
1	2.1129	30.18	30.18
2	1.5886	22.69	52.88
3	1.4850	21.21	74.09
4	1.1249	16.07	90.16
5	0.6885	09.84	100.00

Table 2 Correlation mixed variables – factors

Iden.	Wording variables	Factor 1	Factor 2	Factor 3	Factor 4	Factor 5
AGE	Age of the patient	−0.37	−0.03	−0.13	+0.73	+0.55
DURA	Duration	+0.14	+0.49	+0.04	−0.65	+0.57
TREA	Treatment A	+0.31	−0.01	+0.94	+0.15	+0.05
TREB	Treatment B	−0.15	+0.82	−0.49	0.14	−0.18
TREP	Treatment placebo	−0.17	−0.82	−0.45	−0.29	+0.13
FEMA	Female-sex	+0.95	−0.05	−0.26	+0.15	+0.06
MALE	Male-sex	−0.95	+0.05	+0.26	−0.15	−0.06

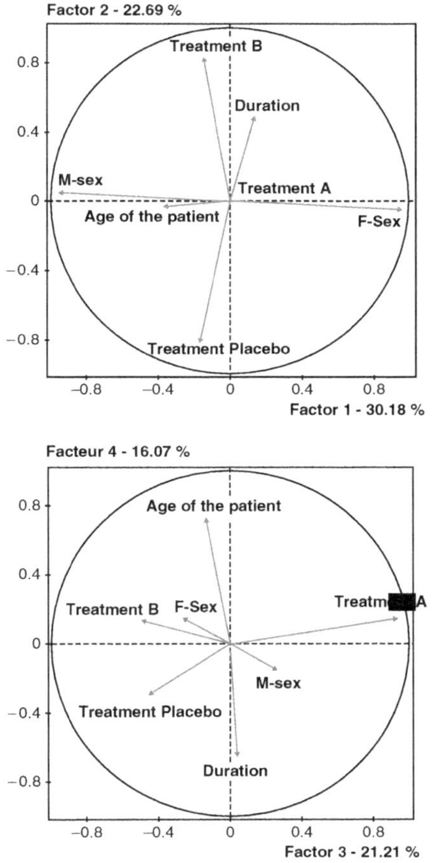

Fig. 1 Circles of correlations: mixed predictors on the first and second MPCA factorial planes

3.2 Discriminant Analysis

We use a discriminant analysis on the significant MPCA factors (corresponding to the four first components with an eigenvalue larger than unity) which explain 89.90% of the variance kept for the discrimination (see Table 1).

Table 3 presents the Fisher discriminant linear function of the MDA with two groups on MPCA factors of explanatory mixed variables. This discriminant rule is computed from the training set of 55 observations. The obtained results show that the discriminant model overall is very significant, the probability (PROBA = 0.0001) is less than the classical significance level of 5%.

So, among the four introduced mixed variables, we can note that, with a significance level less or equal to 5%, neither the duration nor the treatment A differentiate the two groups of patients (PROBA > 5%).

Indeed, the patients who did not report pain are women less elderly who had been given treatment B. However, the group patients reporting the most pain are more elderly men who had been given the placebo.

Table 4 presents some results of logistic model applied to the same training set, implement with the logistic procedure of SAS System. The estimation and the significance of the parameters estimated by the binary logistic model are presented. In this model, the reference modalities for explain variable "Pain" and explanatory variables "Treatment" and "Sex" are respectively "No pain", "Placebo" and "Male". The likelihood ratio, score and Wald tests lead all to reject the nullity hypothesize of the set of coefficients. So, with a classical error risk of 5%, only Duration and Treatment A don't have a significant marginal apport in this full model.

3.3 Comparison

In this part, we use the criterion of misclassification rates to evaluate and compare the performances of the discrimination rules of MDA and Logistic methods.

Table 3 Mixed discriminant analysis – SPAD results

FISHER'S LINEAR FUNCTION

VARIABLES NUM	IDEN	LABEL	PARAMETER FUNCTION DISC.	ESTIMATE REGRESSION	STANDARD DEVIATION (RES. TYPE REG.)	T STUDENT	PROBA
2	AGE	Age of the patient	-0.2186	-0.0646	0.0218	2.97	0.005[a]
3	DURA	Duration	0.0137	0.0041	0.0097	0.42	0.677
4	TREA	Treatment A	0.8076	0.2387	0.1547	1.54	0.129
5	TREB	Treatment B	1.1590	0.3426	0.1584	2.16	0.036[b]
6	TREP	Placebo	-1.9666	-0.5814	0.1551	3.75	0.000[a]
7	FEMA	Female patient	0.9656	0.2855	0.1111	2.57	0.013[b]
8	MALE	Male patient	-0.9656	-0.2855	0.1111	2.57	0.013[b]
		INTERCEPT	14.606855	4.248122			

R2 = 0.42246 F = 7.16850 PROBA = 0.0001
D2 = 2.89710 T2 = 38.76842 PROBA = 0.0001

[a]Significance less or equal than 1%
[b]Significance 1% - 5%

Table 4 Binary logistic model – SAS results

		Model fit statistics			
		Criterion	Intercept only	Intercept and covariates	
		AIC	76.767	57.280	
		SC	78.74	69.324	
		−2 Log L	74.767	45.280	
		Testing global null hypothesis: BETA = 0			
	Test	Chi-square	DF	Pr > ChiSq	
	Likelihood ratio	29.4864	5	<0.0001[a]	
	Score	23.2353	5	0.0003[a]	
	Wald	13.2742	5	0.0209[b]	
		Analysis of maximum likelihood estimates			
Parameter	DF	Estimate	Standard error	Wald chi-square	Pr > ChiSq
Intercept	1	17.4418	6.8320	6.5176	0.0107[b]
Treatment A	1	0.7498	0.5324	1.9836	0.1590
Treatment B	1	1.2554	0.6128	4.1970	0.0405[b]
Sex female	1	0.9682	0.4119	5.5247	0.0188[b]
Age	1	−0.2457	0.0953	6.6448	0.0099[a]
Duration	1	0.0183	0.0350	0.2737	0.6009

[a]Significance less or equal than 1%
[b]Significance 1–5%

Table 5 Comparison – number of observations (percent) well classified into group

	Reported pain groups	MDA	Logistic	Total
Basic sample (90%)	No pain	30 (93.75%)	28 (87.50%)	32
	Yes pain	15 (65.22%)	18 (78.26%)	23
	Total	45 (81.82%)	46 (83.64%)	55
Test sample (10%)	No pain	3 (100.00%)	3 (100.00%)	3
	Yes pain	1 (50.00%)	1 (50.00%)	2
	Total	4 (80.00%)	4 (80.00%)	5

Table 5 shows that the classification results obtained by these two methods on the basic and test samples, are very similar. Indeed, on the training set of 55 observations, the estimations of well classification probabilities are practically the same, namely 81.82% for MDA and 83.64% for logistic model. This corresponds with 45 and 46 observations, respectively.

When we estimate the misclassification probabilities based on the validation set that consists of the remaining five observations, we obtain the same results for MDA and Logistic model.

4 Conclusion

In this work, the methodology to extend discriminant analysis to mixed variables is presented as a methodological chain of known factorial methods. Simple in concept and easy to use, it finds interest in the context of the classification and prediction techniques, when user is confronted with analyzing objects characterized by mixed variables, as is often the case, especially in economics, financial and insurance fields.

The Mixed Discriminant Analysis proposed allows to implement a discriminant analysis on the two types of predictors. This method comes up to one of the disadvantages of discriminant analysis in relation to logistic regression. The latter being a rival if we look at it from discrimination and prediction method point of view.

Finally, it will be interesting to compare the performances of this approach with those of PLS Discriminant Analysis.

References

Abdesselam, R. (2006). Mixed principal component analysis. In M. Nadif & F. X. Jollois (Eds.), *Actes des XIIIémes Rencontres SFC-2006* (pp. 27–31). Metz, France.

Escofier, B., & Pagés, J. (1979). Traitement simultané de variables quantitatives et qualitatives en analyse factorielle. *Cahier de l'analyse des données, 4*(2), 137–146.

Fisher, R. (1938). The statistical utilization of multiple measurements. *Annals of Eugenics, VIII*, 376–386.

Geoffrey, J., & McLachlan (2005). *Discriminant analysis and data statistical pattern recognition*. New York: Wiley.

Hand, D. (1981). *Discrimination and classification*. New York: Wiley.

Hubert, M., & Van-Driessen, K. (2004). Fast and robust discriminant analysis. *Computational Statistics and Data Analysis, 45*, 301–320.

Lachenbruch, P. (1975). *Discriminant analysis*. New York: Hafner Press.

Pagés, J. (2004). Analyse factorielle de données mixtes. *Revue de Statistique Appliquée, LII*(4), 93–111.

Saporta, G. (1977). *Une méthode et un programme d'analyse discriminante sur variables qualitatives*. Journées internationales, Analyse des données et informatique, INRIA.

Saporta, G. (1990). Simultaneous analysis of qualitative and quantitative data. In *Atti XXXV Riunione Scientifica della Societa Italiana di Statistica* (pp. 63–72).

Sjöström, M., Wold, S., & Söderström, B. (1986). PLS discrimination plots. In: E. S. Gelsema & L. N. Kanals (Eds.), *Pattern recognition in practice II*. Amsterdam: Elsevier.

Tenenhaus, M. (1998). *La régression PLS: Théorie et pratique*. Paris: Technip.

Tomassone, R., Danzart, M., Daudin, J. J., & Masson, J. P. (1988). *Discrimination et classement* (172 pp.). Paris: Masson.

A Statistical Calibration Model for Affymetrix Probe Level Data

Luigi Augugliaro and Angelo M. Mineo

Abstract Gene expression microarrays allow a researcher to measure the simultaneous response of thousands of genes to external conditions. Affymetrix GeneChip® expression array technology has become a standard tool in medical research. Anyway, a preprocessing step is usually necessary in order to obtain a gene expression measure. Aim of this paper is to propose a calibration method to estimate the nominal concentration based on a nonlinear mixed model. This method is an enhancement of a method proposed in Mineo et al. (2006). The relationship between raw intensities and concentration is obtained by using the Langmuir isotherm theory.

1 Introduction

The measure of gene expression microarray intensities has become very important in life science, for example, to identify gene functions, transcriptional patterns related to drug treatment and so on. Different technologies are used, and one of the most used is the Affymetrix GeneChip® expression array technology. Affymetrix GeneChip® is characterized by short specific oligonucleotide probes that are tethered and immobilized on the surface of Affymetrix array. Target cDNA is fluorescently labeled and hybridized on the array. A 2D image is then generated, with each probe being identified by a position and an intensity. Each probe is 25 bases long, and each gene is represented by a set of 11–20 probe-pair called *probe-set*. Each probe-set comprises a perfect match probe (PM) and a mismatch probe (MM). The PM and MM have almost the same base sequences but the middle base of the MM probe is changed to the complementary of the PM probe middle base.

Affymetrix (2002) has proposed to use MM probes to quantify and remove two types of errors: *optical noise*, that is an intensity read even if biological sample is not labeled, and *non-specific binding*, that is when a single stranded DNA sequence

A.M. Mineo (✉)
Dipartimento di Scienze Statistiche e Matematiche, Università di Palermo, Viale delle Scienze, Edificio 13, 90128, Palermo, Italy
e-mail: elio.mineog@dssm.unipa.it

F. Palumbo et al. (eds.), *Data Analysis and Classification*,
Studies in Classification, Data Analysis, and Knowledge Organization,
DOI 10.1007/978-3-642-03739-9_14,

binds to a probe sequence which is not completely complementary. Then, the aim of the MM probe is to measure non-specific hybridization. Therefore, by design, the MM probe acts as a background measurement for its corresponding PM probe. *Probe-level analysis* is the process to estimate the gene expression level from PM and MM probes in the corresponding probe-set.

In microarray technology, a number of critical steps are then required to convert the raw measurements into data reliable for biologists and clinicians. This procedure is commonly referred to as preprocessing (Gentleman et al. 2005). Three steps are typically carried out in preprocessing:

- *Background correction*: in this step probe intensities are correct for optical noise and non-specific binding.
- *Normalization*: in this step probe intensities are correct in order to remove systematic bias due to technical variations, such as a different scanner setting or physical problems with arrays.
- *Summarization*: in this step raw intensities read from a probe set are summarized in a gene expression index.

Several methods have been proposed for preprocessing; some methods are reviewed in the following section. In this work we present a statistical framework for Affymetrix GeneChip® that permits to increase the accuracy of the gene expression level and integrate the preprocessing steps.

The paper is organized as follows. In Sect. 2 we give a brief review on preprocessing methods; in Sect. 3 we present our method, that is a calibration method based on a nonlinear mixed model; in Sect. 4 we compare our method with the most used ones and finally in Sect. 5 we draw some conclusions.

2 A Brief Review on preprocessing Methods

Statistical models for preprocessing can be divided in two groups; in the first group we have methods using a modular approach; in other words, in this case we have a specific model for each step of preprocessing of the raw intensity data. A drawback of splitting up the analysis of gene expression data into separate steps is that the error associated with each step is ignored in the downstream analysis (Hein et al. 2005). In the second group we have probabilistic models. Probabilistic models are developed in order to take into account the variability present at different steps of the analysis. In this case we have a unique model by which we can consider the different sources of variation that occur in a microarray experiment. Examples of models of the first group are MAS 5.0 (Affymetrix 2001), in which the background correction is based on a very simple deterministic model, or RMA (Irizarry et al. 2003); this method is based on the assumption that the PM values are sum of two components: a signal component, supposed exponentially distributed, and a background component, supposed normally distributed; the gene-specific signal is obtained as conditional expectation. GC-RMA (Wu and Irizarry 2005) is an extension of RMA by considering that the MM values also contain some information about the true signal.

Examples of models that we can include in the second group are the Bayesian Gene Expression (BGX) (Hein et al. 2005), the mgMOS model (Liu et al. 2005) and the Frequentist Gene Expression Index (FGX) (Purutçuoğlu and Wit 2007). All these models are based on the assumption that PM and MM are correlated, because they are sharing a common gene expression signal. In particular, the BGX model is a Bayesian hierarchical model in which the gene expression index is computed by the median of the posterior signal distribution. The parameters of the model are estimated by MCMC, so it is very computational demanding for large data set. In order to reduce the computational cost, the FGX model is estimated by means of the maximum likelihood method.

In order to evaluate a preprocessing model, Irizarry et al. (2006) have proposed different measures to assess the accuracy (low bias) and the precision (low variance) of a given gene expression index. In particular, to evaluate the accuracy we can use the slope of the regression of the gene expression index on the nominal $\log_2$-concentration (*Signal Detect Slope*). The ideal value of this index is 1. Because the accuracy depends on the overall expression, the authors propose to separate the *Signal Detect Slope* into three components, according to low (nominal concentration less than 4 pM), medium (nominal concentration between 4 and 32 pM) and high (nominal concentration greater than 32 pM) expressed genes (pM stands for picoMolar). We shall use these indices in Sect. 4.

3 The Proposed Calibration Method

Our model is based on the idea that a gene expression index should be closely related to the technological features of the Affymetrix GeneChip® in order to reduce the influence of each preprocessing step on gene expression measures. In this way, we can increase the accuracy of the gene expression index. To do this, we define a nonlinear regression model for the PM probe raw intensities.

The availability of the Affymetrix spike-in studies has led to a significant effort in exploring the relationship between concentration and microarray signal. Figure 2 is obtained using a free-available data set called Spike-In133. This data set consists of three technical replicates of 14 separate hybridizations of 42 spiked transcriptions in a complex human background at concentration ranging from 0 to 512 pM. Thirty of the spikes are isolated from a human cell line, four spikes are bacterial controls and eight spikes are artificially engineered sequences believed to be unique in the human genome. In Fig. 1(a), we can clearly see a sigmoidal growth curve between $\log_2$-concentration ($\log_2$pM) and microarray signal on $\log_2$ scale ($\log_2$PM); a similar result is obtained using the other free-available data sets. In Fig. 1(b) we can see that the hybridization capabilities of the PM probes are not the same due to the differences in the probe composition. This effect is called *probe effect*.

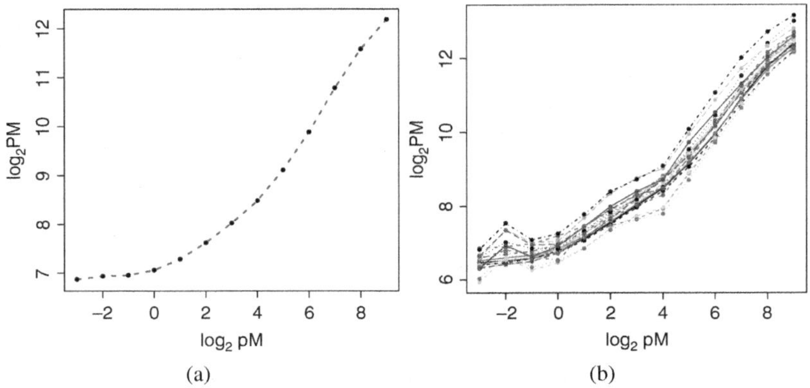

Fig. 1 Panel (**a**) shows the relationship between $\log_2$-concentration ($\log_2$pM) and microarray signal, on $\log_2$-scale ($\log_2$PM), read from the PM probes in the Spike-In133 data set. Panel (**b**) shows the relationship for different probes

Li and Wong (2001) show that, even after subtracting MM, there is a strong probe effect; Naef and Magnasco (2003) propose a simple model to describe the probe effect, by considering only the sequence composition of the probes. To quantify the probe effect they define the *affinity* ξ_{ij} of the i-th probe for the j-th gene as the sum of position-dependent base affinities:

$$\xi_{ij} = \sum_{k=1}^{25} \sum_{m \in \{A,T,C,G\}} \mu_{m,k} 1_{b_k = m} \quad \text{with} \quad \mu_{m,k} = \sum_{l=0}^{3} \beta_{m,l} k^l$$

where m is the base letter index, $k = 1, \ldots, 25$ indicates the position along the probe, b_k represents the base at position k, $1_{b_k=m}$ is an indicator function, and $\mu_{m,k}$ represents the effect of having base m in position k. The correction of the probe intensities for the probe-effect reduces the variability for different amount of concentration. Then, a possible model to estimate the gene expression level should not be based on the assumption of a linear relationship between raw intensities and amount of concentration, as in FGX (Purutçuoğlu and Wit 2007), multi-mgMOS (Liu et al. 2005) or BGX (Hein et al. 2005), for example. Following Hekstra et al. (2003), we define a calibration functional form from Langmuir adsorption model (Atkins 1994). A Langmuir adsorption isotherm is an elementary model of surface adsorption, that assumes the probe intensities are linearly dependent on the fraction of occupied probe sites. In this way, we can use some parameters to model the normalization step and other parameters to obtain a calibration of the gene expression measures. Normalization parameters are useful to remove the difference in the lower and upper intensity thresholds, due for example to limits on detection of the instrument, instrumental saturation and so on, while calibration parameters are used to reduce the bias in the gene expression measures.

Let Y_{ijk} be the $\log_2$ intensity read from the i-th probe of the j-th gene of the k-th array, we propose the following nonlinear mixed-effect model:

$$\log_2 PM_{ijk} = Y_{ijk} \sim N\left(\mu_{ijk}, \sigma^2\right)$$

with $\mu_{ijk} = \alpha_{ijk} + (\alpha_1 - \alpha_{ijk})2^{\psi_{jk}}$, $\psi_{jk} = (\eta_j - \beta_{0,k})/\beta_1$, $\alpha_{ijk} = \xi_{ij} + \alpha_{0,k}$ and $\alpha_{0,k} \sim N\left(\alpha_0, \sigma^2_{\alpha_0}\right)$, $\beta_{0,k} \sim N\left(\beta_0, \sigma^2_{\beta_0}\right)$. The advantage of this approach is that we can give a natural interpretation to the model parameters; in particular, ξ_{ij} is the affinity of the i-th probe for the j-th gene, α_0 is the mean of the non-specific signal, which consists of nonspecific hybridization, background and stray signal, α_1 is the saturation level of Affymetrix GeneChip®, that can depend on the scanner setting of GeneChip® used in the experiment, β_0 is the middle level of amount of concentration on $\log_2$ scale, β_1 is the scale parameter. In order to use replicated arrays, we assume that $\alpha_{0,k}$ and $\beta_{0,k}$ are random parameters that are used to describe the difference between arrays, while the scale parameter (β_1) and the saturation level (α_1) are fixed parameters, specific for the Affymetrix GeneChip®.

Then, a possible workflow to apply the proposed model is the following:

1. Estimate the parameters $\alpha_0, \alpha_1, \beta_0$ and β_1 by means of the calibration studies developed from the Affymetrix.
2. Use the method proposed by Naef and Magnasco (2003) to estimate the affinity ξ_{ij}.
3. Since it is not appropriate to use all the data to estimate the random parameters, we propose two different methods:

   (a) Following Hill et al. (2001), use hybridization control genes to define a subset of genes to estimate the random parameters $\alpha_{0,k}$ and $\beta_{0,k}$. For example, the GeneChip® Eukaryotic Hybridization Control Kit contains six vials that are composed of a mixture of biotin-labeled cRNA transcripts of *bioB*, *bioC*, *bioD* and *cre* prepared in staggered concentration (1.5, 5, 25 and 100 pM, respectively).

   (b) Use the *invariant method* (Li and Wong 2001) to obtain a set of genes with constant level of expression across all the arrays. In this case, we have $\psi_{jk} = \phi_{0,k} + \phi_j$, where ϕ_j is the unscaled $\log_2$-concentration. The $\log_2$-concentration can be estimated by the following relationship: $\eta_j = \beta_1 \phi_j$.

4. Finally, estimate the concentration $\hat{\eta}_j$ for each gene by means of maximum likelihood estimators.

4 A Comparison with the Most Popular Methods

In order to evaluate the proposed method we have used the free-available calibration data set called Spike-In133, described in Sect. 3. We have randomly split the data set in a training and a test set. The training set is composed by 21 arrays.

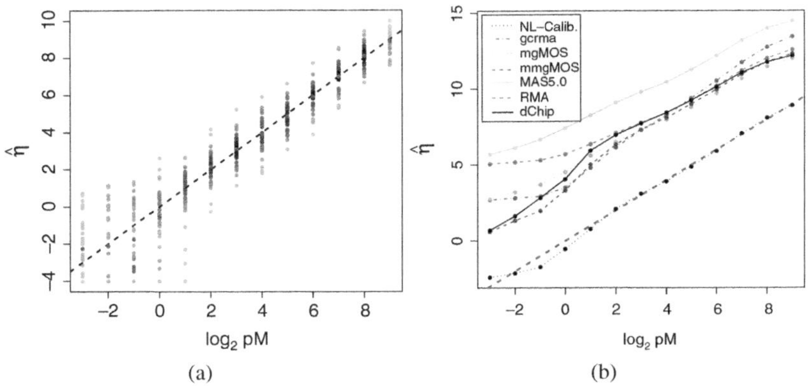

Fig. 2 Panel (**a**) shows the relationship between known levels of $\log_2$ concentration ($\log_2$pM) and estimated $\log_2$-concentration ($\hat{\eta}$). Panel (**b**) shows the comparison between the proposed model (NL-Calib.) and the most popular methods (the ideal line is the *red dashed line* with slope equal to 1)

Table 1 Signal detect slopes for the considered methods computed using the Spike-In133 data set. (The optimal value for these indices is 1, Irizarry et al. 2006)

Method	slope.all	slope.low	slope.medium	slope.high
NL-Calibration	1.002	0.907	0.859	1.003
GC-RMA	0.999	0.680	1.036	0.974
RMA	0.678	0.404	0.739	0.799
dChip	0.983	1.303	0.755	0.688
mgMOS	0.814	0.770	0.752	0.766
Multi-mgMOS	1.033	1.174	0.842	0.788
MAS 5.0	0.767	0.694	0.734	0.766

In order to compare the proposed method with the most popular ones, we have used the procedure proposed in Irizarry et al. (2006) and implemented in the R package *affycomp*. Figure 2(a) shows accuracy and precision of the proposed calibrated gene expression index. Figure 2(b) shows that by considering the Affymetrix GeneChip® features, such as the saturation level and the global scale parameter, with our method the accuracy of the gene expression level is very high and better than that of the other methods.

In Table 1 we compute the overall Signal Detect Slope and the Signal Detect Slope separated into three components to take into account low, medium and high expressed genes. Then, slope.all, slope.low, slope.medium and slope.high are the slopes obtained from regressing the calibrated $\log_2$ expression values on nominal $\log_2$-concentrations for all the genes, genes with low intensities, genes with medium intensities and genes with high intensities, respectively. We can see how the GC-RMA method seems better than the other popular methods when we consider medium and high expressed genes. Because the GC-RMA is based on the assumption that exists a linear relationship between microarray signal and amount

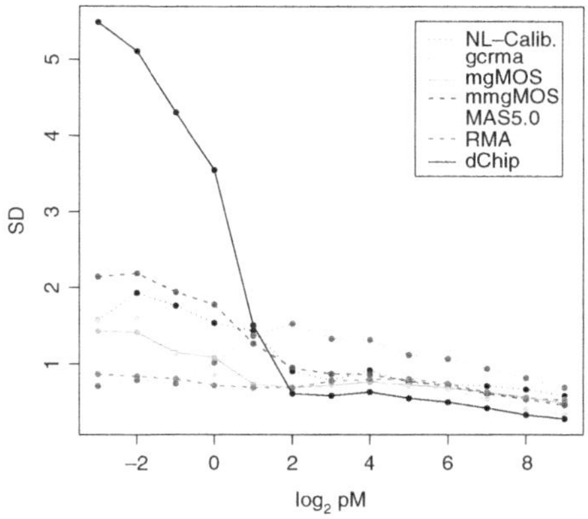

Fig. 3 Standard deviations of the estimated gene expression index computed using the proposed method (NL-Calib.) and the most popular ones

of concentration, this method shows a reduction in the accuracy when we consider genes with low expression. Differently from the most popular models, our model is based on a nonlinear relationship: in this way, we obtain a very high level of accuracy almost for each level of expression. In particular, our model has a very good performance for low level concentration in respect of the other methods.

In Fig. 3 we compare the standard deviations of the considered gene expression indices for different levels of concentration. We can see how the RMA method is characterized by a very low level of variability. Our method is characterized by a level of precision that increases with the level of expression. Anyway, from this point of view the behavior of our method is comparable with that of the other methods. In particular, the proposed method has a similar performance to multi-mgMOS (Liu et al. 2005).

5 Conclusions

In this paper we have proposed a statistical model by which we can integrate all the preprocessing steps for microarray raw intensity data in a unique statistical model, that increases the accuracy of the gene expression index by means of the Affymetrix GeneChip® features. The index is obtained by using the maximum likelihood method, that gives, as it is known, easy and fast to compute estimators. The proposed method takes into account for the experimental design and fits replicated data into a single measure. Moreover, we can use the proposed method to obtain preprocessed data for an high level analysis, for example for a statistical test for the SAM (Tusher et al. 2001) or the EBAM (Efron et al. 2001).

In conclusion, with this model we unify normalization, background correction and summarization steps. The proposed method has two main features: first of all, we have an increase in the accuracy of the gene expression measure at any level of the nominal concentration; secondly, for medium-high levels of nominal concentration we obtain a reduction of the variability of the gene expression measure, when compared with the RMA method, that is from this point of view the best considered method.

Acknowledgements The authors want to thank the University of Palermo for supporting this research.

References

Affymetrix. (2001). *Statistical algorithms reference guide*. Santa Clara, CA: Author.

Affymetrix. (2002). *GeneChip expression analysis: Data analysis fundamentals*. Santa Clara, CA: Author.

Atkins, P. (1994). *Physical chemistry* (5th edition). Oxford: Oxford University Press.

Efron, B., Tibshirani, R., Storey, J., & Tusher, V. (2001). Empirical Bayes analysis of a microarray experiment. *Journal of the American Statistical Association, 96*(456), 1151–1160.

Gentleman, R., Carey, V., Huber, W., Irizarry, R., & Dudoit, S. (2005). *Bioinformatics and computational biology solutions using R and bioconductor*. New York: Springer.

Hein, A. M., Richardson, S., Causton, H. C., Ambler, G. K., & Green, P. J. (2005). BGX: A fully Bayesian gene expression index for Affymetrix GeneChip data. *Biostatistics, 6*(3), 349–373.

Hekstra, D., Taussig, A. R., Magnasco, M., & Naef, F. (2003). Absolute mRNA concentrations from sequence-specific calibration of oligonucleotide arrays. *Nucleic Acids Research, 31*(7), 1962–1968.

Hill, A. A., Brown, E. L., Whitley, M. Z., Kellogg, G. T., Hunter, C. P., & Slonim, D. K. (2001). Evaluation of normalization procedures for oligonucleotide array data based on spike cRNA controls. *Genome Biology, 2*(12), 1–13.

Irizarry, R., Hobbs, B., Collin, F., Beazer-Barclay, Y., Antonellis, K., Scherf, U., et al. (2003). Exploration, normalization and summaries of high density oligonucleotide array probe level data. *Biostatistics, 4*(2), 249–264.

Irizarry, R. A., Wu, Z., & Jaffee, H. A. (2006). Comparison of Affymetrix GeneChip expression measures. *Bioinformatics, 22*(7), 789–794.

Li, C., & Wong, W. (2001). Model-based analysis of oligonucleotide arrays: Expression index computation and outlier detection. In *Proceedings of the National Academy of Science USA, 98*, 31–36.

Liu, X., Milo, M., Lawrence, N. D., & Rattray, M. (2005). A tractable probabilistic model for Affymetrix probe-level analysis across multiple chips. *Bioinformatics, 21*(18), 3637–3644.

Mineo, A. M., Fede, C., Augugliaro, L., & Ruggieri, M. (2006). Modelling the background correction in microarray data analysis. In *Proceedings in computational statistics, 17th COMPSTAT Symposium of the IASC* (pp. 1593–1600). Heidelberg: Physica.

Naef, F., & Magnasco, M. O. (2003). Solving the riddle of the bright mismatches: Labeling and effective binding in oligonucleotide arrays. *Phyical Review. E, Statistical, Nonlinear, and Soft Matter Physics, 68*(1 Pt 1), 011906.

Purutçuoğlu, V., & Wit, E. (2007). FGX: a frequentist gene expression index for Affymetrix arrays. *Biostatistics, 8*(2), 433–437.

Tusher, V. G., Tibshirani, R., & Chu, G. (2001). Significance analysis of microarrays applied to the ionizing radiation response. *Proceedings of the National Academy of Sciences USA, 98*(9), 5116–5121.

Wu, Z., & Irizarry, R. A. . Stochastic models inspired by hybridization theory for short oligonucleotide arrays. *Journal of Computational Biology, 12*, 882–893.

A Proposal to Fuzzify Categorical Variables in Operational Risk Management

Concetto Elvio Bonafede and Paola Cerchiello

Abstract This contribution is deemed in the view of the authors, as a methodological proposal in order to employ the well know fuzzy approach in a context of operational risk management. Even though the available data can not be considered native fuzzy, we show that modelling them according to fuzzy intervals is useful from two point of view: it allows to take into account and to exploit more information and, on the other hand, either unsupervised or supervised models applied to this kind of data present comparatively good performance. The paper shows how to obtain fuzzy data moving from a classical database and later on the application of fuzzy principal components analysis and linear regression analysis.

1 Fuzzy Approach

In the daily life we typically face with inaccurate data since in many cases it is not possible or feasible to observe and to measure a phenomenon avoiding an arbitrary degree of accuracy. This imprecision leads to difficulties in managing and constructing models especially in the case of complex problems. A fuzzy approach is a method able to simplify complexity by taking into account a reasonable amount of imprecision, vagueness, and uncertainty (Kruse et al. 1994).

With such an approach the degree of knowledge about a measure, an event or some qualitative data can be specified via a membership function ($\mu(x)$). With the membership function a classical set of value (or a fact) is fuzzified into another domain. There are several useful methods to assign a membership function (Kruse et al. 1994). With the membership function we characterize the fuzzy set which describes the data and that is defined as: $A = \{(x, \mu_A(x))|x \in X\}$. The process of creating a fuzzy set is called fuzzification and it should be done ad hoc for each situation but sometimes it is hard to look for a best procedure.

P. Cerchiello (✉)
University of Pavia, Corso Strada Nuova 65, Italy
e-mail: paola.cerchiello@unipv.it

F. Palumbo et al. (eds.), *Data Analysis and Classification*, Studies in Classification, Data Analysis, and Knowledge Organization,
DOI 10.1007/978-3-642-03739-9_15, © Springer-Verlag Berlin Heidelberg 2010

A typical class of fuzzy set is constructed on the basis of a triangular function identified by: a centre (m) that describes the maximum certainty about a situation or a measure; a left spread (L) representing the degree (or velocity) to approach to m^-; a right spread (R) representing the degree to move away from m^+. Each of this parameter will depend upon different elements. A general class of $\mu(x)$, called LR_2, is identified as

$$\mu_A(x) = \begin{cases} L(\frac{m_{1A}-x}{l_A}), & x \leq m_{1A}, (l > 0), \\ 1, & m_{1A} \leq x \leq m_{2A}, \\ M(\frac{x-m_{2A}}{r_A}), & x \geq m_{2A}, (r > 0). \end{cases}$$

For such a function the parameters describing a database are the set of centers $M = (m_{1A}, m_{2A})$, the left spreads (L) and the right spreads (R).

Starting from fuzzy data it is possible to apply clustering analysis, regression procedure and principal component analysis (Coppi et al. 2006a,b; Hoppner et al. 1999). In this contribution we employ fuzzy principal components analysis (PCAF).

PCAF models the parameters of the membership functions of a dataset in terms of components. The components are calculated on the basis of membership function parameters M, L and R by minimizing the overall distance between such parameters and the observed ones M^*, L^* and R^* (for more details see Coppi et al. 2006a).

2 The Problem

According to Basel Committee and common practice, Operational Risk can be defined as "the risk of loss resulting from inadequate or failed internal processes, people and systems or from external events". In particular we focus on IT-intensive organizations, where great relevance is given to losses coming from processes and systems intrinsically tied up to the core business of Telco enterprises. In this context there is a great demands of models able to reveal and to measure the relations existing between the noticed causes of a specific occurred problems (interruptions), the resulting loss and the customer type. The available dataset is provided by an Israeli company offering Value Added Services (VAS) to SMEs (Small Medium Enterprises), basically, a set of communications services, both voice and data. Those data deal with the IT problems risible from a communications service provider. In particular we analyze a database compiled by technicians called to solve the clients' IT-problems. Moreover we pay attention to two set of variables: "Problem description" and "Severity". The former is a qualitative variable containing the wording of the problem occurred to the client, for example: "INT", "Security", "SFW04", etc. The latter is a quantitative ordinal variable referring the impact of the loss in term of severity: 1 "high", 2 "medium", 3 "low".

In Table 1 there is a sample of the original database; "PBX.No" is the client number.

Table 1 Original database

PBX.No	177 Problem levels	31 Problem levels	5 Problem levels	Severity
10001	*Planned site visit*	*SEC04*	*Security*	3
11005	*Application problem*	*INT*	*Interface*	2
11015	*Version change on site*	*SFW04*	*Software*	1
11025	*Server problem*	*NTC*	*Network communications*	2

This structure can not be considered a classical one in the context of Operational Risk analysis, mainly for the measurement of variable "Severity" (Bonafede and Giudici 2007; Cornalba and Giudici 2004; Cruz 2002). Being it ordinal, we do not have an estimate of the monetary loss occurred either to the client or to the Telco enterprise. Besides that, the original "problem description" variable has a very high number of different levels (177) that can not be evaluated meaningfully with a number of events equal to 1,201. Thereby we decided to reduce the original variable according to this criteria: we decrease the original 177 levels to 31 following an hierarchical approach, i.e. grouping problems referring to the same area. Afterwards, starting from the obtained 31 levels, we employ once again this grouping strategy finally coming up to five levels. Of course the degrees of generality are rather different and those databases also represent our benchmarks for the proposal that we explain in the next section.

3 The Proposal

Since the analysis and the useful exploitation of a IT-Operational Risk is not straightforward, we propose to employ the fuzzy approach in order to better take advantages of the available not complete information. Our original data can not be considered vague and imprecise in the sense explained in the first section. But we believe that, representing them according to the fuzzy paradigm, we can obtain good and useful results.

Thereby our analysis is divided into two subsequent stages: (1) by means of a PCAF model we reduce the original number of variables contained in the database then (2) we employ the scores obtained during the previous phase as input variables in a linear regression analysis, (see Giudici 2003), where the target variable is represented by the Severity.

Therefore, before starting the analysis, we need a method to fuzzify data. We have started by counting for each client, identified by a PBX number, the occurrence of each variable (considering 31 different problem descriptions). Those numbers will be the centre of the membership function.

Later on we employ two kind of methods to generate left (L) and right (R) spreads:

1. First hypothesis: the centre is distributed as a binomial variable. We have used the confidence interval (95% level) with a Normal approximation for L and R spread: $\pm(1.96 \times \sqrt{n_{ij} \times (1 - \frac{n_{ij}}{N_j})})$; where n_{ij} is the counting of i-client with the j-problem and N_j is the total occurrence of the j-problem.
2. Second hypothesis: we generate 5,000 samples for each problem description, employing a Bootstrap procedure (Efron and Tibshirani 1994). Afterwards, we have calculated the variance and the standard error for each combination of client and problem. The L and R spread is the following approximated confidence interval (95% level):$\pm(1.96 \times \sqrt{v_{ij}})$, where v_{ij} is the variance for i-th client and j-th problem.

The above procedures result in several databases divisible into two main groups: one characterized by a symmetry around the centre, i. e. L and R spreads are equal, and another group in which L and R spreads are different. In particular L spread is set equal to 0 when the left side of interval is negative since in our application there is no sense for a negative occurrence.

We finally come up with five databases. A short extract of one of them is reported in Table 2.

To perform a PCAF we have used square root, triangular and parabolic membership functions, which offer different fuzziness level, with parameters L, R and M of Table 2. The general function form is

$$\mu_{ij}(x) = \begin{cases} 1 - (\frac{m_{ij}-x}{l_{ij}})^{\alpha}, & x \leq m_{ij},\ (l_{ij} > 0), \\ 1 - (\frac{x-m_{ij}}{r_{ij}})^{\alpha}, & x \geq m_{ij},\ (r_{ij} > 0), \\ 0, & \text{otherwise}, \end{cases}$$

where $\mu_{ij}(x)$ is the membership function for the j-th problem and the i-th customer. With r_{ij}, l_{ij}, m_{ij} we indicate the coefficients for each pair (*Client, Problem*).

Varying coefficient α, different membership functions are obtained:

- $\alpha = 1/2 \rightarrow$ Square root.
- $\alpha = 1 \rightarrow$ Triangular.
- $\alpha = 2 \rightarrow$ Parabolic.

Table 2 Database samples

Database[a]	PBX.No	$L.(SEC)$	$R.(SEC)$	$M.(SEC)$	$L.(INT)$	$R.(INT)$	$M.(INT)$
Bin. sim.	11005	1.9508	1.9508	1	2.7483	2.7483	2
Bin. asim.	11005	0	1.9508	1	0	2.7483	2
Boot sim. $\neg N$	11005	2.2834	2.2834	1	0.7323	0.7323	2
Boot sim. N	11005	1.1682	1.1682	0.8288	0.9419	0.9419	5.0765
Boot asim.	11005	0	2.2834	1	0.7323	0.7323	2

[a]With "Boot sim. $\neg N$" we indicate the database without normalization, on the contrary "Boot sim. N" is normalized

Their fuzziness level increases from square root to parabolic membership function, as shown in Coppi et al. (2006b).

What explained so far concerns the first part of our analysis. In fact, once evaluated the optimal number of factors to be extracted, we move towards a supervised model, i.e. linear regression analysis, to compare the performance of the different employed fuzzification methods, not only to each other, but also in comparison to our defined benchmarks (31 original variable levels and derived 5 ones with an hierarchical approach). To achieve that objective we need to apply a reasonable and useful transformation of the target variable, that is "Severity". The original scale of the above mentioned variable is ordinal with only three levels: 1 high impact, 2 medium impact, 3 low impact. In Operational Risk context, this type of measurement is atypical and not particular useful, thereby we decided to propose monetary intervals according to the three levels. Taking into account the type of analyzed company and the typology of occurred problems we suggest the following intervals:

- $High(1) \in [10000, 20000]$
- $Medium(2) \in [3000, 10000)$
- $Low(3) \in [500, 3000)$

To finalize the approach, we simulate the possible monetary losses moving from three uniform distribution, each based on the above intervals. The resulting simulated target variable is obviously continuous but it always respects the original three levels.

4 Results

We remind that the objective of our analysis is to reveal the relation existing between the problems causing a loss and that loss itself. In particular we need a predictive model able to give an estimate of the risible loss when a specific combination of problems occurs. Thereby once applied the PCAF model to the obtained five fuzzy dataset, we fix a minimum threshold concerning the quota of variability explained by the components. In particular we extract seven (see Table 3) components from each PCAF analysis, assuring a 60% level of explained inertia at least. We result in five databases containing for each observation the scores obtained from the extract components. The target variable, represented by the Severity, is simulated as explained in the above section, i.e. transforming an ordinal variable into a continuous one according to reasonable monetary intervals.

In Table 4 we report the results obtained from the linear regression analysis in terms of AIC, adjusted R^2 and R^2 computed on the validation set.

It clearly appears that, among the fuzzifized databases, the best performance is attained by the symmetric boostrap (N). Moreover, considering our benchmark databases "DB total" (31 covariates) and "DB5levels" (five covariates), the R^2 (val) obtained by our best fuzzyfied database is between them, i.e. perfectly comparable. Furthermore if we consider the well known parsimony principle stating the

Table 3 An extract of the resulting PCAF database reporting the centers (M) employing a parabolic membership function

Database	PBX.No	Severity	PCA1	PCA2	PCA3	PCA4	PCA5	PCA6	PCA7
Bin. sym.	11002	2,492	−0.6362	0.4492	−0.147	0.0417	0.1834	−0.0198	−0.0521
Bin. asym.	11002	2,492	−0.9879	0.721	−0.2878	0.1782	0.1013	0.0459	−0.0229
Boot sym.	11002	2,492	−0.6617	0.4789	−0.1773	0.0466	0.407	0.2303	−0.1877
Boot asym.	11002	2,492	−1.7707	0.3353	−0.7751	0.1909	0.6051	0.2057	−0.1858

Table 4 Results from regression and out-sample evaluation

Database	No. covariates	R^2 (val)	AIC	Adj R^2
DTotal[a]	(31)	0.8990	20,533.93	0.9261
Boot sim. N	(7)	0.7955	21,433.92	0.8404
Boot sim. $\neg N$	(7)	0.7161	22,049.75	0.7335
Bin. sim.	(7)	0.7138	22,154.11	0.7094
Bin. asim.	(7)	0.7063	22,271.97	0.6797
Boot asim.	(7)	0.7012	22,212.59	0.6949
D5Class[a]	(5)	0.6966	22,031.07	0.7374

[a] With "D5Class" and "DTotal" we indicate respectively the databases with 5 and 31 problem descriptions

preference towards models simple and with a number of variable as little as possible, we can say that our proposed approach can be profitably employed.

5 Conclusions

This paper proposes a method to fuzzify qualitative variables in the context of IT Operational Risk. Considering this specific application framework, a set of information about problems, interruption and losses occurred to IT enterprises, is available. However, because of the still new topic and of the not yet consolidated statistical methodologies, there are not standard representation and measurements for what concerns the variables of interest. Thereby we suggest how to exploit the not ideal available information making use of the fuzzy approach. The IT Operational Risk application represents only the initial motivation for the development of a methodology that can be employed every time a qualitative variable is available. Typically a qualitative variable is transformed into a binary or frequency counts one without considering further important information. We show that reporting the L-R spreads for each counts is useful. In particular we propose two different approach to calculate the spreads (i.e. the intervals of interest): binomial intervals for the counts and bootstrap intervals. Once created this dataset structure, we employ PCAF to extract a smaller number of components to be used within a predictive model. A proper transformation of the target variable ("Severity" = the occurred loss) is employed and finally several comparative linear regression models are fitted. The obtained results are interesting: among all the fuzzified databases (employing different membership functions and different methods of fuzzification) the symmetric bootstrap approach shows good performance in terms of AIC, adjusted R^2 and validation R^2 in comparison to our benchmarks databases (not fuzzified ones).

Acknowledgements The authors acknowledge financial support from national Italian grant MIUR-FIRB 2006–2009 and European grant EU-IP MUSING (contract number 027097). The authors also thank Chiara Cornalba for the construction of the hierarchical maps of "Problem description" variable.

References

Bonafede, C. E., & Giudici, P. (2007). Bayesian networks for enterprise risk assessment. *Physica A*. doi:10.1016/j.physa.2007.02.065

Coppi, R., Gil, M., & Kiers, H. (2006a). The fuzzy approach to statistical analysis. *Computational Statistics and Data Analysis, 51*(1), 1–14.

Coppi, R., Giordani, P., & D'Urso, P. (2006b). Component models for fuzzy data. *Psychometrika, 71*, 733–761.

Cornalba, C., & Giudici, P. (2004). Statistica models for operational risk management. *Physica A, 338*, 166–172.

Cruz, M. G. (2002). *Modeling, measuring and hedging operational risk*. Chichester: Willey.

Efron, B., & Tibshirani, R. (1994). *An introduction to the bootstrap*. New York: Chapman & Hall/CRC.

Giudici, P. (2003). *Applied data mining*. London: Willey.

Hoppner, F., Klawonn, F., Kruse, R., & Runkler, T. (1999). Fuzzy cluster analysis. Chichester: Willey.

Kruse, S., Gebhardt, J., & Klawonn, F. (1994). Foundations of fuzzy systems. Chichester: Willey.

Common Optimal Scaling for Customer Satisfaction Models: A Point to Cobb–Douglas' Form

Paolo Chirico

Abstract The first aim of this paper is to present a singular algorithm of ALSOS's (Alternating Least Squares with Optimal Scaling). It allows to assign the same scaling to all variables measured on the same ordinal scale in a categorical regression. The algorithm is applied to a regression model to measure and evaluate Customer Satisfaction (CS) in a sanitary case. The results seem to support the use of multiplicative models like Cobb–Douglas', to analyze how the overall CS of goods or services is shaped. According to this evidence, the second aim intend to suggest a theory about the overall CS very similar to theory about utility in Marginal Economics. After a brief introduction to the CS measurement and evaluation methods (Sect. 1), the algorithm is presented on the Sect. 2. Sections 3 and 4 concern the application and the theory about overall CS. Conclusions are reported in Sect. 5.

1 Features of a Customer Satisfaction model

In the last 20 years several statistical methods have been proposed to measure and to evaluate the satisfaction degree of a customer about goods or services, namely Customer Satisfaction (CS). A brief overview of these methods is not a target of the present paper, nevertheless it is useful to consider some features that can characterize and distinguish a method.

The first feature concern the measurement scale. The natural scale of CS is typically an ordinal scale (e.g. very dissatisfied, dissatisfied, neither satisfied nor dissatisfied, satisfied, very satisfied) but, unfortunately, this measurement doesn't always allow very meaningful analysis. The most diffused approaches to overcome this limit are:

- Adopting a Likert scale
- Determining a metric scale from a probabilistic model
- Introducing an Optimal Scaling algorithm

P. Chirico
Dipartimento di Statistica e Matematica applicata, Via Maria Vittoria 38, 10100, Torino, Italy
e-mail: paolo.chirico@unito.it

F. Palumbo et al. (eds.), *Data Analysis and Classification*,
Studies in Classification, Data Analysis, and Knowledge Organization,
DOI 10.1007/978-3-642-03739-9_16,

Table 1 Features of some popular statistical method for CS

Methods	Scaling method	Observation	Free distribution
SERVQUAL	Likert	Indirect	Yes
Categorical regression	Optimal scaling	Direct	Yes
Categorical PCA	Optimal scaling	Indirect	Yes
Rasch analysis	Probabilistic	Indirect	No
PLS path model	Likert	Indirect	Yes
LISREL	Likert	Indirect	No

The Likert scale (see Brasini et al. 2002, pp. 164–168) consists on replacing ordinal categories with their ranks. Such transformation is very easy and is adopted by several statistical methods (see moreover Table 1), but is obviously arbitrary and can be considered acceptable only if categories are conceptually equidistanced. Probabilistic approaches are the Thurstone's method and the Rasch Analysis model (see Andrich 1988), but either approach imply the choice of distributional assumptions. Optimal Scaling (OS) is instead a class of distribution free methods, that allow to assign numerical values to categorical variables in a way which optimizes an analysis model (see Boch 1960; Kruskal 1965). Conceptually Rasch Analysis can be considered like a OS method, but historically OS methods are free distribution, while Rasch Analysis is not. An another feature regards if the CS is directly observable or not. In many cases the customer can be asked for his satisfaction degree (direct observation), but this observation can be considered a effective degree of satisfaction only if we can assume the customer's rationality, in other words this means that his answer in not affected by environmental and psychological influences. Otherwise the CS had to be estimated from other observable variables by means of appropriate models (indirect observation). In the following table are reported some popular statistical method used for CS measurement and evaluation (for SERVQUAL see Parasuraman et al. 1988, for LISREL see Jöreskog and Sörbom 1996). They are compared in regard to the features discussed (see also Montinaro & Chirico 2007).

In the following sections a singular Categorical Regression model is proposed for CS evaluation. It is based on an ALSOS algorithm (Alternating Least Squares with Optimal Scaling, see Sect. 2) and allows to obtain a common scaling for all evaluation model variables measured on the same ordinal scale. This does not normally happens with the standard ALSOS programs.

2 Categorical Regression with Common Optimal Scaling

The ALSOS algorithms are OS methods that permit the optimization of a model adopting Alternating Least Squares (ALS) and Optimal Scaling (OS) principles (see Young et al. 1976 and 1981). More specifically they are based on a iterative two-steps estimation process (Fig. 1), which permits to get least squares estimations

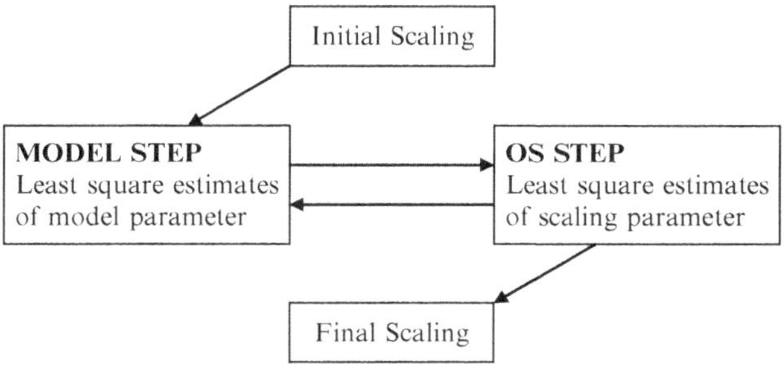

Fig. 1 The ALSOS algorithm

of scaling values and model parameters. Every algorithm starts with an exogenous scaling and terminates when the iterative solution converges. The models involved are linear models, which can be performed by an optimization (Regression, Principal Component Analysis, ...); the corresponding analysis is also named with the term "categorical".

2.1 The Pattern of the Model

Let $\tilde{Y}$ the overall satisfaction degree of a customer about a good or service and $\tilde{X}_1, \tilde{X}_2, \ldots, \tilde{X}_p$ the satisfaction degrees of some aspects of the good or service. All satisfactions are measured on a scale of k ordinal categories $c_1, c_2, \ldots, c_k$. The target is to convert a qualitative scale into a quantitative one by means of a common transformation $z(\ )$ in order to minimize the error ε of regression:

$$Y = \beta_0 + \beta_1 X_1 + \cdots + \beta_p X_p + \varepsilon, \tag{1}$$

where $Y = z(\tilde{Y})$, $X_1 = z(\tilde{X}_1), \ldots, X_p = z(\tilde{X}_p)$. Practically the transformation $z(\ )$ is defined by k ordered values $z_1 \leq z_2 \leq \cdots \leq z_k$ corresponding to the k ordered categories. Assuming data are observed on n customers, the score $\mathbf{y}, \mathbf{x}_1, \mathbf{x}_2, \ldots$ of each scaled variable can be got in the following way:

$$\begin{aligned} \mathbf{y} &= \mathbf{U_y z}, \\ \mathbf{x_j} &= \mathbf{U_j z}, \end{aligned} \tag{2}$$

where $\mathbf{U_y}, \mathbf{U_j}$ are the typical indicator matrix (the generic element $u_{i,h}$ is 1 if the i-th customer respond c_h about the corresponding variable, else 0); $\mathbf{z}$ is the vector of the scaling parameters $z_1, z_2, \ldots, z_k$.

So the model (1) can be described in the classic form:

$$\mathbf{y} = \mathbf{X}\beta + \varepsilon. \tag{3}$$

This form is useful for the model step, but not for the OS step, because does not point out the scaling parameters. For it, it needs to rewrite the classic form in the following scaling form:

$$(\mathbf{U_y} - \mathbf{B})\mathbf{z} = \beta_0 \mathbf{1} + \varepsilon, \tag{4}$$

where $\mathbf{B} = \sum_{j=1}^{p} \beta_j \mathbf{U_j}$

2.1.1 The Algorithm of the Parameters Estimation

According the approach of ALSOS (Fig. 1), the algorithm is described by the following steps:

Initialisation: an arbitrary $\mathbf{z}$ is chosen.
Model step: β is estimated by classic estimator:$\beta = (\mathbf{X}'\mathbf{X})^{-1}\mathbf{X}'\mathbf{z}$.
OS step: a new $\mathbf{z}$ is estimated by minimizing SSE in the model (4) with the constrains $z_1 = z_{min}$ and $z_k = z_{max}$.
Control step: if the absolute difference between the two last $\mathbf{z}$ is less than a suitable convergence, the *Final Results* are obtained; else it need to go back to *Model Step*.
Final results: the last $\mathbf{z}$ and β are the final results.

It is easy to note that the OS model above does not include constrains for the monotonicity of transformation: If initial scaling is monotone and customer responses are rational, they are not needed, but there are no problems to include them. Indeed the minimum and the maximum of scaling parameters are fixed. It is due to avoid the algorithm produces the dummy solution $z_1 = z_2 = \cdots = z_k$. Generally it needs to fix two constrains to define a metric intervals scale (average and standard deviation, minimum and maximum, etc.) and the constrains adopted are very suitable in a linear optimization problem. The convergence is guaranteed because the sum of squares errors (SSE) decreases at every step and round. There is one hooker: the ALSOS procedure does not guarantee convergence on the global least squares solution. Nevertheless every final scaling is better (in terms of SSE) than an initial, supposed good scaling.

3 Multiplicative Models for CS

The proposed model was applied to a survey on CS in a Piedmont ASL (Local Sanitary Firm): 525 patients were asked about their satisfaction degree on:

- Whole service (overall satisfaction)
- Some aspects of the service (waiting time, suitable environment, courtesy, professionalism, etc.)

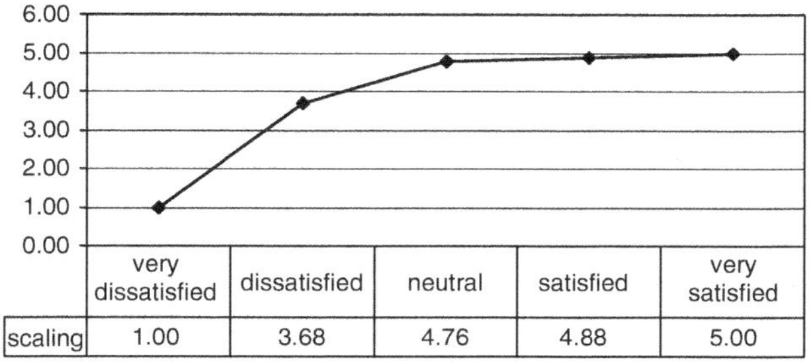

Fig. 2 The optimal scaling in a sanitary case

Responses scale was: very dissatisfied, dissatisfied, neutral, satisfied, very satisfied. Here only the final scaling is reported in the Fig. 2 (for more details see Chirico 2005).

This result contrasts with idea of conceptual equidistance among categories. Nevertheless it is possible to partially recover equidistance with a power transformation like:

$$\dot{\mathbf{z}}' = [a^{z_1}, a^{z_2}, \ldots, a^{z_k}] \tag{5}$$

with $a > 1$. It means that the scaling $\mathbf{z}$ could be viewed (see Fig. 2) as the logarithmic transformation of a more realistic scaling $\dot{\mathbf{z}}$. Then the model (1) should be the logarithmic transformation of the model:

$$a^Y = a^{\beta_0+\beta_1 X_1+\cdots+\beta_p X_p+\varepsilon} \tag{6}$$

that can be rewritten as

$$\dot{Y} = \beta_0 \times \dot{X}_1^{\beta_1} \times \cdots \times \dot{X}_k^{\beta_p} \times \varepsilon, \tag{7}$$

where variables with the point above correspond to $a^{variable}$. Now the new variables' values, in model (7), better represent the categories $c_1, c_2, \ldots, c_k$. This fact suggests that the better relation between overall satisfaction and partial satisfactions might be multiplicative, like a Cobb–Douglas function, rather than linear. The linear model, thanks to the proposed algorithm, is useful to estimate the parameters $\beta_0, \beta_1, \ldots, \beta_p$ (they do not change in the multiplicative model) and the pre-final scaling $z_1, \ldots, z_k$.

3.1 Some Observations

Final Scaling. The final scaling $\dot{z}_1, \ldots, \dot{z}_k$ could be get from $z_1, \ldots, z_k$ by means of a power transformation with basis $a > 1$:

$$\dot{z}_j = a^{z_j}. \tag{8}$$

Unlikely it is not clear which value of a is better to get the final scaling, because not every value of a determines the same effects in terms of ratio and intervals among $\dot{z}_1, \ldots, \dot{z}_k$. If a conceptual equidistance among the categories $c_1, \ldots, c_k$ is assumed, a could be chosen in order to minimise the variability of the differences: $\dot{z}_h - \dot{z}_{h-1}$ $(h = 2, \ldots, k)$. Other criteria can be adopted; each one determines different final scaling and consequently different values of position indicators like the mean, for example. Indeed the parameters $\beta_0, \beta_1, \ldots, \beta_p$ (which indicate the importance of every factor $X_0, X_1, \ldots, X_p$) not change and not their significance (see next section).

Weighting. As Least Squares methods are applied on the linear model (1), the fit of the multiplicative model (7) is worse in correspondence of greater value of $\dot{Y}$. To reduce this effect, it is possible to change the two estimation steps introducing a weighted least squares estimation method.

4 A Theory About Overall CS

According with the results underlined in the last section, the following theory about the CS is proposed:

- Every customer determines his/her own satisfaction about a good or service (Overall Customer Satisfaction: *OCS*) composing the relative evaluations of some fundamental aspects of the good or service (Partial Customer Satisfaction: *PCS*).
- The composition criterion is approximated by a multiplicative model of the Cobb–Douglas type.

$$OCS = \alpha \times PCS_1^{\beta_1} \times \cdots \times PCS_k^{\beta_k} \tag{9}$$

The first assumption is typical of the most of CS model (SERVQUAL, ACSI, ECSI). The second one shapes the Customer Satisfaction similar to the customer utility in the marginal consumer theory (for more details see Varian 2005). In fact it is easy to prove that

$$\beta_j = \frac{d(OCS)/OCS}{d(PCS_j)/PCS_j} \tag{10}$$

that means β_j is the elasticity of OCS respect to PCS_j. If customer's responses are rational, all β_j will be positive or null (negative estimates of these parameters could be obtained, but they ought to be not different from zero at the test). Generally $\alpha = 1$ and $\sum \beta_j = 1$ are expected (scale effects do not have sense!). The second assumption involves

$$0 < \beta_j < 1. \tag{11}$$

Therefore β_j indicates the importance of the j-th aspect for the CS. Another similitude to marginal consumer theory is that the marginal overall satisfaction determined by each partial satisfaction is decreasing. In fact:

$$\frac{d(OCS)}{d(PCS_j} = \beta_j \frac{OCS}{PCS_j}. \tag{12}$$

If PCS_j increases, the OCS increases less proportionally [see (9) and (11)] and consequently $d(OCS)/d(PCS_j)$ decreases. This means the improvement of one level from *satisfied* to *very satisfied* in an aspect produces a smaller increase of the overall satisfaction than the improvement of one level from *neutral* to *satisfied* in the same aspect. In other words, improvements from low quality levels are more important for customers than improvements from high quality levels. This deduction from the model (9) is consistent with the psychology of the majority of the customers. If the OCS of a good or service ought to be improved, the best strategy is not always to improve the most important aspect (that one with the biggest β_j). It could be more effective to improve another aspect with a low quality level. Each possible improvement ought to be considered and valued in regard to his marginal satisfaction and, of course, his cost (costs of needed actions to get the improvement).

5 Conclusions

The algorithm presented in this paper has the typical features of ALSOS programs: free distribution method and convergence of estimates obtained by analytic functions. It also ensures a common scaling for all data measured on same ordinal scale, whereas ALSOS programs included in the most popular statistic software do not. In fact these programs, as general approach, assign different scaling to every qualitative variable, whether it is measured on a common scale or not. However the same values should be assigned to same categories, if the scaling gives a metric significance to the measurement of qualitative data (see Chirico 2005).

The application of algorithm in a CS evaluation study has pointed out that the relation between the overall satisfaction and its factors seems to be formalized better by multiplicative models, like Cobb–Douglas ones. In other words: the overall satisfaction and its factors are conceptually comparable to overall utility and its factors in the marginal consumer theory (the Cobb–Douglas' function was originally proposed like production function, but subsequently it was also used to confirm the marginal consumer theory). This model form permits to formalized the concept of "decreasing marginal satisfaction" that involves the strategic importance of improving the low quality aspects. At present, further studies on how to get the final scaling in a multiplicative model are being carried on.

References

Andrich, D. (1988). *Rasch models for measurement*. Beverly Hills, CA: Sage.

Boch, R. D. (1960). *Methods and applications of optimal scaling* (Psychometric laboratory report, 25), University of North Carolina.

Brasini, S., et al. (2002). *Statistica aziendale e analisi di mercato*. Bologna: Il Mulino.

Chirico, P. (2005). *Un metodo di scaling comune per modelli multivariati di valutazione della customer satisfaction* (Working paper). Dipartimento di Statistica e Matematica Applicata, Universita' degli Studi di Torino.

Jöreskog, K.G., Sörbom, D. (1996), LISREL 8: User's Reference Guide, Scientific Software International.

Kruskal, J. B. (1965). Analysis of factorial experiments by estimating monotone transformations of the data. *Journal of Royal Statistical Society, Series B, 27*, 251–263.

Montinaro, M., & Chirico, P. (2007). Customer satisfaction measurement procedures: One-dimensional and multi-dimensional approach. *Statistica Applicata, 2006*(3), 277–296.

Parasuraman, A., et al. (1988). SERVQUAL: A multiple-item scale for measuring customer perceptions of service quality. *Journal of Retailing, 64*(1), 12–40.

Varian, H. R. (2005). *Microeconomic analysis*. New York: Norton.

Young, F. W. (1981). Quantitative analysis of qualitative. *Psychometrika, 46*(4), 357–388.

Young, F. W., et al. (1976). Regression with qualitative and quantitative variables: An alternating least squares method. *Psychometrika, 41*(4), 505–529.

Structural Neural Networks for Modeling Customer Satisfaction

Cristina Davino

Abstract The aim of this paper is to provide a Structural Neural Network to model Customer Satisfaction in a business-to-business framework. Neural Networks are proposed as a complementary approach to PLS path modeling, one of the most widespread approaches for modeling and measuring Customer Satisfaction. The proposed Structural Neural Network allows to overcome one of the main drawbacks of Neural Networks because they are usually considered as black boxes.

1 Introduction

Nowadays the determining factor that can help an enterprise to remain with success on the market is the level of service that it can offer to the customers. That is true in particular for business-to-business markets where the competitive advantage develops also through the offer of a *package* of services. It is obvious that Customer Satisfaction (CS) evaluation plays a very important role for the companies that, being sensible to the customers' needs and desires, can realize competitive and customized offers thus increasing the customer loyalty.

The aim of this paper is to provide a Structural Neural Network (SNN) (Lee et al. 2005) for modeling CS in a business-to-business framework. We introduce Neural Networks (Bishop 1995) as a complementary approach to PLS path modeling (Esposito Vinzi et al. 2008), one of the most widespread approaches to model CS. The proposed SNN allows to define in an objective way the structure of the network and to interpret its parameters. Moreover, the SNN plays a crucial role in case of non linear relationships.

The paper is organized as follows: CS measurement and PLS path modeling methodology are described in Sect. 2, classical NN and the proposed SNN are detailed in Sect. 3, while Sect. 4 regards the description of a SNN for modeling CS of the Cucine Lube Srl, an Italian business-to-business enterprise.

C. Davino
University of Macerata, Dipartimento di Studi sullo sviluppo economico, Italy
e-mail: cdavino@unimc.it

F. Palumbo et al. (eds.), *Data Analysis and Classification*,
Studies in Classification, Data Analysis, and Knowledge Organization,
DOI 10.1007/978-3-642-03739-9_17,

2 Customer Satisfaction and PLS Path Modeling

Customer Satisfaction is a concept that cannot be directly measured because it is a complex concept related to mental constructions. Different factors influence the CS and for each of them it is possible to find a set of indicators. A well-established model in the CS framework is the one used to describe the European CS Index (ECSI) (Tenenhaus et al. 2005) (Fig. 1). The unobservable CS concept is caused by four factors (Image, Customer Expectation, Perceived quality, Perceived value) and its level has two consequences (Loyalty and Complaints). Each factor of the model is an unobservable factor and it can be measured through subjective indicators corresponding to the customers behavior.

PLS path modeling is one of the most widespread approaches to estimate the unobservable factors of a CS model. In the PLS path modeling terminology, the seven factors in Fig. 1 are called latent variables (LVs) and the indicators measured for each of them are called manifest variables (MVs). The idea behind the model is to measure the LVs through the MVs and to describe the causal connections among the LVs (the arrows in the model). A PLS path model is made of a measurement model relating each set of MVs to the corresponding LVs and a structural model connecting the LVs in accordance with a network of causal relationships. In the measurement model there are different ways to relate the MVs to their LVs; in this paper we will refer to the formative way where each LV is a linear function of its MVs or to the reflective way where each MV reflects its LVs.

Let ξ_j be one of the J LVs, it is measured by a set of x_{jh} $(h = 1, \ldots, H_j)$ MVs and by a set of ξ_{ji} $(i = 1, \ldots, I_j)$ LVs. In the measurement model in the case of both a formative and reflective scheme, each LV is a linear function of its MVs:

$$\xi_j = \sum_{h=1}^{H_j} w_{jh} x_{jh}. \tag{1}$$

In the structural model, a set of linear equations allows to relate LVs:

$$\xi_j = \beta_{j0} + \sum_{i=1}^{I_j} \beta_{ji} \xi_i. \tag{2}$$

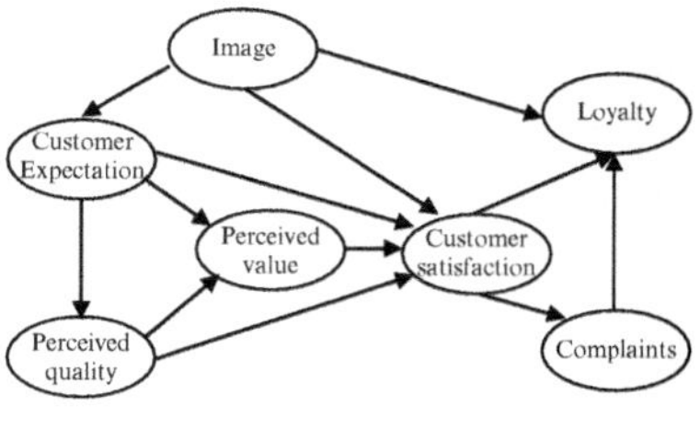

Fig. 1 The structure for modeling CS

The PLS algorithm is based on an iterative process. According to the original Wold's PLS approach (Wold 1982), starting from an arbitrary vector of weights, they are normalized and used for the external estimation of LVs with unitary variance. These LVs are then updated by considering the relationships with adjacent latent variables in the causal network so as to yield internal estimates. Upon convergence of the iterative procedure, the next step carries on the estimation of the structural equations by individual OLS multiple regressions or PLS regressions in case of strong multicollinearity between the estimated LVs. This option is a brand new option available only in the PLSPM module XLSTAT-PLSPM software (XLSTAT 2008).

3 Customer Satisfaction and Neural Networks

Neural Networks can represent a complementary approach to PLS path modeling, mainly if the relations in Fig. 1 are not linear.

Inspired by biological systems, NN can be defined as computational models represented by a collection of simple units, the neurons, interlinked by a system of connections, the weights (Bishop 1995). Usually, the neurons are organized in layers or hidden layers, being the latter not directly connected to the external stimulus. The single neuron elaborates a certain number of operations in input in order to produce one single value as output. The inputs can be given by either external stimulus or stimulus induced by other neurons of the network; the outputs represent either the final results of the network or the inputs for other neurons. A generic neuron can be represented graphically by a circle whereas the connection between a couple of neurons by a direct arrow showing the direction of the information flow.

Let j be the generic neuron of a network, it receives a set of inputs $X = x_1, x_2, \ldots, x_n$ playing a different role as measured by the intensity of the connections: $W = w_{1j}, w_{2j}, \ldots, w_{nj}$. The input of each neuron also known as *activation state* or *potential* P_j is usually given by the weighted sum of the input values:

$$P_j = \sum_{i=1}^{n} x_i w_{ij}. \tag{3}$$

The output of the neuron results from the application of the *transfer function* to the potential:

$$y_j = f(P_j) = f\left(\sum_{i=1}^{n} x_i w_{ij}\right). \tag{4}$$

The transfer function can be of any type, a good function could be a sigmoidal function because of its capacity to be a universal approximator (White 1992).

In constructing a NN the most important phase is the "learning" through suitable algorithms (Rumelhart et al. 1986) that are used to update the weights which are the parameters to estimate in the NN. Once the weights are identified the NN is built up and it can be used for prediction of unseen cases. The wide success of NN can be attributed to some key features: they do not impose distributional hypotheses, they are able to analyze complex problems characterized by non-linear relationships and finally they are universal function approximators, as they can approximate any continuous function to any desired accuracy (Davino et al. 1997).

In this paper the well established methodological structure of the PLS path modeling is borrowed and it is adapted to be used in a NN context. The result is a NN, namely a Structural Neural Network, where the number of hidden layers and neurons is defined by the model and it is not user dependent. The SNN allows to overcome one of the main drawbacks of NN: they are usually considered as black boxes because their internal structure (hidden layers and neurons) is not data driven but it is arbitrarily defined. In the proposed SNN it is not possible to distinguish input, hidden and output layers because each neuron and each layer can be connected to external stimulus. Such a kind of network allows to precisely mimic the CS model in Fig. 1, the only difference being the jargon: the LVs are hidden layers with just one neuron and the MVs are the inputs of each of them.

The output of each hidden layer represents the measurement model:

$$\xi_j = \varphi_1 \left(\sum_{h=1}^{H_j} w_{jh} x_{jh} \right), \tag{5}$$

where φ_1 can be any function, even non linear.

The final outputs are multivariate functions obtained through the structural part of the model by connecting the hidden neurons through the weights $\mathbf{z}$, according to the model structure:

$$y = \varphi_2 \left(\sum_{i=1}^{I_j} z_{ji} \xi_{ji} \right) = \varphi_2(\varphi_1(\mathbf{x}; \mathbf{w}); \mathbf{z}). \tag{6}$$

In the SNN the coefficients of both the measurement model and the structural one are estimated at the same time. Thus, the approach does not require an alternating procedure as in the PLS algorithm.

4 A Structural Neural Network for Modeling CS

A SNN has been defined and trained for Modeling CS of an Italian company named Lube. The Lube company is today ranked as one of the top three Italian kitchen producing companies and it has more than 1,500 sales points in Italy. In order to measure the CS of the Lube sales points, a questionnaire regarding the following factors of their satisfaction has been submitted to a random sample of 600 responsible persons of the sales points:

- Image: notoriety of the brand, coherence of the image, width of the products range, capacity to renew the product range.
- Customer Expectation (Technical/functional features): timely in delivering the goods, delivery precision, promptness in replacing products, availability of the kitchen planning software.
- Customer Expectation (Sell-out support): easiness of the catalogue, merchandising, showroom support, offer of training for sellers.
- Customer Expectation (Relational features): easiness of contacts, kindness of the staff, availability of technical assistance.
- Perceived quality (Technical/functional features): timely in delivering the goods, delivery precision, promptness in replacing products, availability of the kitchen planning software.
- Perceived quality (Sell-out support): easiness of the catalogue, merchandising, showroom support, offer of training for sellers.
- Perceived quality (Relational features): easiness of contacts, kindness of the staff, availability of technical assistance.
- Perceived value: quality/price ratio, value with respect to competitors.
- Customer satisfaction: overall satisfaction, fulfillment of expectations, satisfaction with respect to the ideal service.
- Complaints: care about customer complaints.
- Loyalty: long-last relation with the company, enlargement of the showroom, valorisation of the brand.

The LV *Complaints* has not been considered. The defined SNN is a feed-forward network where the direction of the information is from the inputs to the outputs. For this reason each input variable (MV) is connected to its corresponding hidden neuron (LV) in a formative way. As the output variables are both the MVs of the Loyalty LV and those of the Satisfaction LV, it is necessary to divide the learning phase into two steps: in the first step the network outputs are the Mvs of the Satisfaction LV, while the second one, starting from the estimated Image and Satisfaction LV, aims to reconstruct the MVs of Loyalty. All the MVs are scaled from 1 to 10, where scale 1 expresses a very negative point of view on the service while scale 10 a very positive opinion. Before training the network, the MVs are standardized even if they are expressed in the same scale because this transformation allows to have a faster learning phase and it avoids local minima.

Following the PLS path modeling approach, the Perceived quality and the Customer Expectation are considered second order LVs because no direct overall MV is observed. The approach commonly followed (hierarchical PLS path model) is to create two super block LVs generated by all the MVs related to the Perceived quality and to the Customer Expectation. In this case, the scores of the LVs of the first order (Technical/functional features, Sell-out support, Relational features) are considered as partial scores while the scores of the super block LVs are considered as global scores.

Using linear transfer functions for each neuron in the hidden layers, the SNN is able to estimate LVs which are for the most part highly correlated to the ones deriving from a PLS path model. In the following, the correlation coefficients between SNN and PLS LVs:

– Image: 0.96
– Cust.E xp. (Technical/functional features): 0.18
– Cust. Exp. (Sell-out support): 0.39
– Cust.Exp. (Relational features): 0.98
– Customer Expectation: 0.85
– Perc. Qual. (Technical/functional features): 0.16
– Perc. Qual. (Sell-out support): 0.72
– Perc. Qual. (Relational features): 0.94
– Perceived quality: 0.90
– Perceived value: 0.98
– Customer satisfaction: 0.99
– Loyalty: 0.99

It can be noticed that estimates of first order latent variables related to Customer Expectation and Perceived Quality are less concordant across the two methods. In spite of that, the corresponding second order LVs still show a strong correlation. Moreover, it is worthy of notice (Fig. 2) that LV Loyalty is mainly explained by its Mvs and not by the LVs Image and Satisfaction.

In order to evaluate the generalization capability of the model and following the classical approach of the NN framework, the sample has been divided into a training set (67.5%) and a test set (32.5%). The values of a performance measure (Akaike

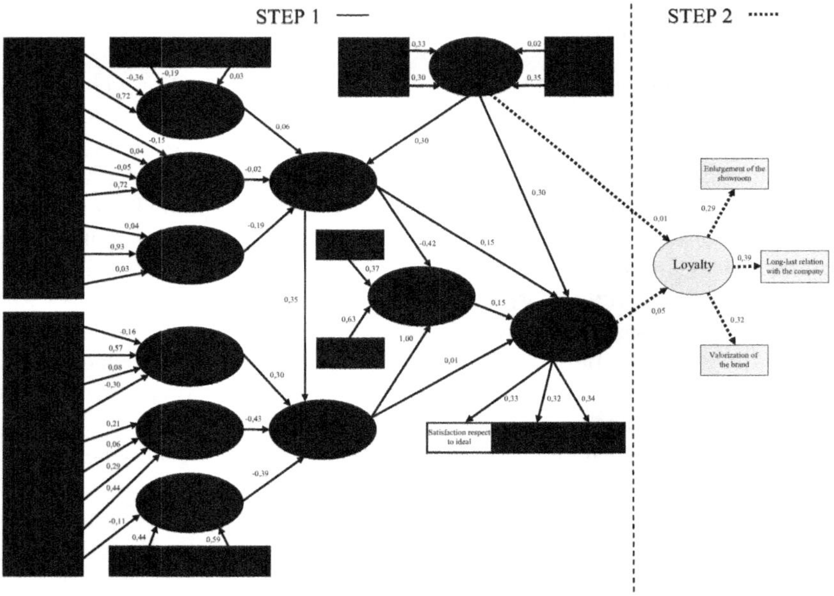

Fig. 2 The SNN for modeling Lube CS

Table 1 Performance measures of the SNN

	Training	Test
Step 1	AIC = −2,877.6	AIC = −1,210.8
Step 2	AIC = −1,620.5	AIC = −632.1

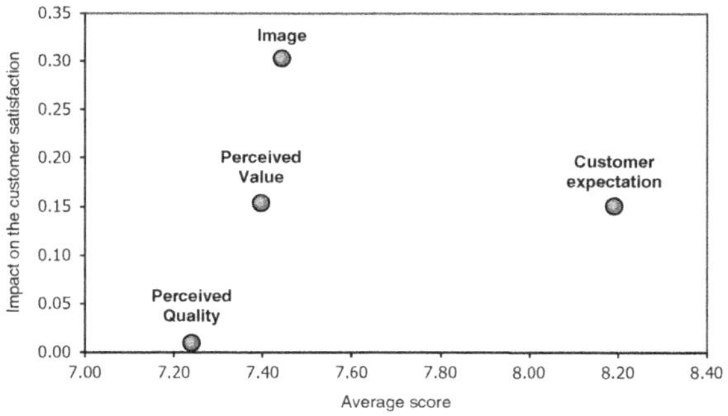

Fig. 3 Strategic levers for the improvement

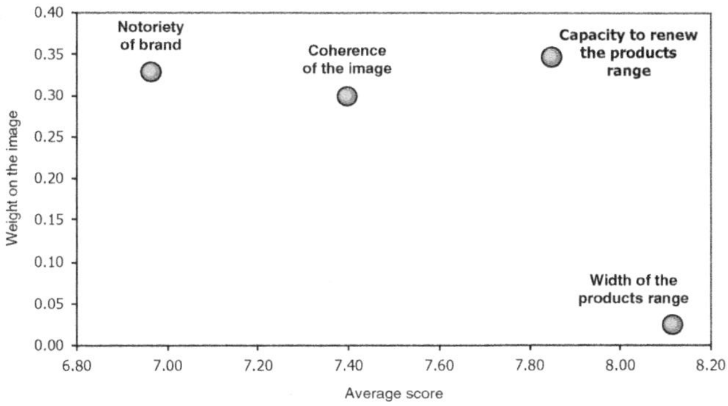

Fig. 4 Manifest variables of image latent variable

Information Criterion) derived both form step 1 and from step 2 of the model are satisfactory for both the training and the test data (Table 1).

In order to identify the strategic levers for the improvements, it is necessary to interpret some of the estimated weights, namely the coefficients of the model. In Fig. 3 the LVs impacting on the CS are plotted on the basis of their average scores and of their impacts.

The image LV is the one with the highest impact on the CS (the highest coefficient) but a low average score if compared to Customer expectation. A strategic lever could thus be an improvement of the image that can be realized reflecting on the coefficients of its MVs (Fig. 4).

If the coefficients of the indicators of Image are analyzed, it results that the factor that has the highest weight on the image is the capacity to renew the products range (coefficient equal to 0.347). This factor receives also a good score (mean value equal to 7.849) so it represents a consolidated key point in the image characterization. On the other hand, notoriety of the brand is the true critical point as its weight on the image is high as well (coefficient equal to 0.328) but the average score is below 7. It could be advisable to invest on the notoriety of the brand in order to improve the customer satisfaction because improving the satisfaction on that factor has a great impact on the overall satisfaction. Regarding the width of the products range, it has a very high average score but its weight is almost zero; such a variable is uninteresting as a lever for image.

5 Concluding Remarks

The proposed SNN suggests the strategic levers to improve CS. For example, LUBE company should invest on the image and in particular on the notoriety of the brand to gain a good impact on CS.

From a methodological point of view, the proposed approach has to be considered as a complementary approach to PLS path modeling that could be particularly valuable in case of non linear relationships among MVs and LVs. From the NN point of view, the PLS methodological framework is exploited to overcome one of the main drawbacks of NN: they are usually considered as black boxes while the structure of the proposed NN is suggested by the model and it is also possible to interpret the weights. From the PLS point of view, the proposed approach allows to obtain an overall estimate of the weights while in the PLS approach there are two phases, one for the structural part of the model and one for the measurement one.

Finally, the presence of an optimization function in the learning phase of a NN represents an element in favor of the proposed SNN but also an evidence of the flexibility of PLS because it has provided LVs quite correlated to SNN results but using a completely different criterion.

References

Bishop, C. M. (1995). *Neural network for pattern recognition*. Oxford: Clarendon.

Davino, C., Mola, F., Siciliano, R., & Vistocco, V. (1997). A statistical approach to neural networks. In K. Fernandez-Aguirre & A. Morineau (Eds.), *Analyses Multidimensionnelles des Donnees* (pp. 37–51). Saint-Mande: CISIA Ceresta.

Esposito Vinzi, V., Chin, W. W., Henseler, J., & Wang, H. (Eds.) (2008). *Handbook of partial least squares: Concepts, methods and applications*. Berlin: Springer.

Lee, C., Rey, T., Mentele, J., & Garver, M. (2005). Structured neural network techniques for modeling loyalty and profitability. In *Proceedings of SAS User Group International (SUGI 30)*.

Rumelhart, D. E., Hinton, G. E., & Williams, R. J. (1986). Learning internal representation by error propagation. In *Parallel distributed processing: Explorations in macrostructure of cognition* (Vol. 1, pp. 318–362). Cambridge, MA: Badford Books.

Tenenhaus, M., Esposito Vinzi, V., Chatelin, Y.-M., & Lauro, C. (2005). PLS path modeling. *Computational Statistics and Data Analysis, 48*, 159–205.
XLSTAT (2008). Addinsoft. Paris, France. Retrieved from http://www.xlstat.com.
White, H. (1992). *Artificial neural networks*. New York: Springer.
Wold, H. (1982). Soft modeling. The basic design and some extensions. In Jöreskog Wold (Eds.), *Systems under indirect observation* (Vol. II). Amsterdam: North-Holland.

Dimensionality of Scores Obtained with a Paired-Comparison Tournament System of Questionnaire Items

Luigi Fabbris

Abstract In this paper we deal with the dimensionality of preferences expressed by a sample of respondents on a set of items. We analyze the data obtained from a sample of entrepreneurs to whom a set of questionnaire items was administered through the single scoring method and some variants of the paired-comparison tournament method. Cardinality and pattern of the preference data are analyzed through multivariate methods.

1 Preference Data Collection

In the following, we refer to methods of preference data collection suitable for computer-assisted data collection surveys and remote interviewing systems. We want to determine the relative importance of a set of p items about which a sample of n respondents are asked to express their preferences. We are concerned with two basic methods. The first one is based on an incomplete series of hierarchical comparisons between distinct pair of items. This method, as suggested by Fabbris and Fabris (2003), consists in:

- Ordering the set of p items (for the sake of simplicity p is supposed to be even) according to a specified criterion
- Submitting, in a first round of choices, the $p/2$ couples of adjacent items to the sample of respondents and obtaining their preferences for one of the two items
- Administering, in a hierarchical fashion (see Fig. 1 for $p = 8$), couples (or triplets if p is not a power of 2) of first-round-choices till the last round where the most preferred item is sorted out
- Summarizing the individual choices either in a frequency distribution or a dominance matrix (Sect. 2), and finally estimating items' scores (Sect. 3)

L. Fabbris
Statistics Department, University of Padua, Via C. Battisti 241, 35121 Padova, Italy
e-mail: luigi.fabbris@unipd.it

F. Palumbo et al. (eds.), *Data Analysis and Classification*, Studies in Classification, Data Analysis, and Knowledge Organization, DOI 10.1007/978-3-642-03739-9_18,

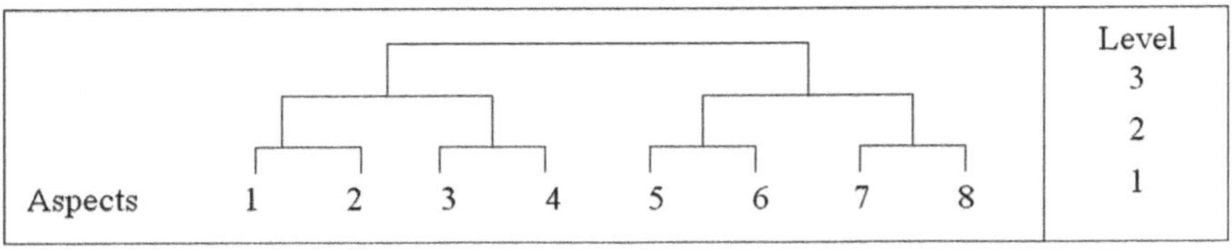

Fig. 1

This method, named *tournament* for its similarity to a soccer championship, is a reduced version of the well-known pair-comparison method, where all distinct couples of items are submitted for choice to respondents and, at each pair-comparison, a preferred item is detected (this method can be named "choose $1/2$"). Either methods can be juxtaposed to other methods popular in preference elicitation: (1) the single stimulus scoring, where each item of a set is rated on a ordinal or interval scale with reference to a common dimension, (2) that of sorting one or few items ("pick h/p"), and (3) that of ranking either a subset or all items ("order h/p").

The tournament method showed to be empirically preferable to the single scoring and direct ranking methods because the latter is not feasible for telephone interviews with more than three or four items, and the single scoring method is subject to social desirability effects and lacks discriminatory power.

The tournament method is superior to the full pair-comparison one as the number of items diverges: in fact, the number of questions to be administered to each respondent, if the latter method applies, is $p(p-1)/2$; instead, the tournament criterion requires the administration of a number of questions even lower than p, the number of items. In fact, suppose $p = 2^L$, where L is the number of levels (rounds), the number of questions is $p - 1$. The main drawback of the tournament method is that the sequence of administration of the couples may influence the intermediate order of preferences.

In Sect. 4, we analyze the data obtained from a sample of 394 entrepreneurs surveyed in 2004 in the Venetian District. The preferences concern their motivation to work and were obtained by administering a set of questionnaire items through the single scoring method and some variants of the paired-comparison tournament method. Entrepreneurs were partitioned into four independent groups and randomly assigned to experimental modes: one was the 1–10 cardinal scale rating and the other three concerned the tournament method and were labelled according to first-level ordering of items: (1) random order, (2) maximum distance, and (3) minimum distance between items after implicit item ordering.

2 Preference Data

The paired-comparison data may be organized in a frequency distribution. For item j at unit h, preferences may be measured by y_{hj} $(h = 1, \ldots, n;\ j = 1, \ldots, p)$, which may be either the number of times item j was preferred by the unit to the other items in L comparisons, or the final dichotomous preference of the unit for item j ($y_{hj} = 1$) or for another item ($y_{hj} = 0$). We can organize, too, the relationships between all pairs of items in a dominance, or preference, matrix:

$$\boldsymbol{P} = \left\{ p_{ij} = \frac{n_{ij}}{n_{ij} + n_{ji}} \right\} \quad (i \neq j = 1, \ldots, p), \tag{1}$$

where n_{ij} is the number of times item i was preferred to item j by n respondents. If ties are absent, $n_{ij} = n - n_{ji}$ (say, $p_{ij} + p_{ji} = 1$). Hence, matrix **P** is antisymmetric, and irreducible (all cells are positive but the diagonal which is null) mathematical relations between items may be *transitive*, i.e. if item i was preferred to j and the latter was preferred to k ($i, j, k = 1, \ldots, p$) by sample unit h ($h = 1, \ldots, n$), then i is to be preferred to k even if they did not match directly at that unit. The transition rule is relevant for completing the matrix **P** (because cells of the non-matched pairs are empty). Hence we can apply the *weak transition rule* (Coombs 1976):

$$\text{if } (p_{ij} > 0.5 \cap p_{jk} > 0.5) \rightarrow p_{ik} > 0.5. \tag{2}$$

3 Scoring Algorithms

The item scoring may be based on alternative algorithms. The relevance score of item j ($j = 1, \ldots, p$) may be a proportion of the number of times it top-ranked:

$$\overline{y}_j = \frac{1}{n} \sum_{h=1}^{n} y_{hj}, \tag{3}$$

where y_{hj} is a dichotomous variable equalling 1 if item j was preferred at the top level by respondent h and 0 otherwise. Estimates (3) vary between 0 and 1. Bradley and Terry (1952) showed this estimator is maximum likelihood. Another estimator may be based at all comparison levels on formula (3), where y_{hj} is the number of times item j was preferred by respondent h. If standardized with its maximum value, L, the latter average varies between 0 and 1. It may be shown that the same statistical properties of estimator (3) apply to estimator (4) (Fabbris and Fabris 2003).

A *naïve* unstandardized score for item i may be estimated with the sum of row preferences of matrix **P**,

$$p_{i+} = \sum_{i} p_{ij}. \tag{4}$$

A more refined estimator is that based on the sum of row balances between preference data in symmetric positions,

$$d_{i+} = \sum_{j} d_{ij}, \tag{5}$$

where $d_{ij} = p_{ij} - p_{ji} = 2p_{ji} - 1$. The rationale of this estimator is that symmetric preferences conflict to each other and should be balanced. For instance, if we want to detect the dominance of i over j, we should compensate the frequency that $i > j$ with that of $j > i$. If they balance out perfectly, $n_{ij} = n_{ji}$, there is no dominance of either item: $d_{ij} = d_{ji} = 0$. Both estimators (4) and (5) may be standardized with their maximum value:

$$d_i^* = \frac{d_{i+}}{Max(d_{i+})} = \frac{d_{i+}}{p-1}, \tag{6}$$

so that the former varies between 0 and 1 and the latter between -1 and 1. The maximum value occurs for both criteria if item i was preferred to all others by all respondents and the minimum if it was uniformly dominated by the other $p - 1$ items. If matrix $\mathbf{P}$ is irreducible a large eigenvalue, λ_1, exists to which a vector with positive values $\mathbf{w} = w_1, \ldots, w_p;\ w_i > 0$ can be associated.

According to the Perron–Frobenius theorem for positive matrices, the following equality holds[1] (Saaty 1977):

$$\mathbf{Pw} = p\mathbf{w}, \quad \text{with constraint} \quad \mathbf{w}'\mathbf{w} = 1. \tag{7}$$

If a one-dimensional pattern underlying the conjoint preferences can be hypothesized, items can be projected on a real axis according to the intensity of saturation of the "importance" dimension, that is: $p_{ij} \propto \delta_{ij}$, where δ_{ij} is the Euclidean distance between points i and j. The more consistent the pairwise preferences, the further apart the stimulus points should be in the real axis (Brunk 1960). If $\mathbf{P}$ is multidimensional, scores may be estimated through factor analysis, eventually after manipulation for satisfying triangular properties and/or rank transformation of dominances.[2] The multivariate analysis can give "ordered metric" solutions, i.e. real values that best reproduce the order relations among the stimuli. Besides, a multivariate solution can be considered a failure in applying preference analysis because it does not deliver a unique linear "mental" dimension of respondents, while they express their preferences, but a nonlinear, or even a more complex function that may be very difficult to perceive from a factorial solution.

4 Dimensions in Preference Data

We analyse the answers on six motivating aspects obtained in a survey on entrepreneurs' motivation to work: income, professional accomplishment, social prestige, social relationship making, frequent travelling, and availability of spare time. For the sake of simplification, we will compare the data on maximum logi-

[1] The $\mathbf{w}$-vector is also the limiting solution of normalized row sums of $\mathbf{P^a}$ for diverging a.

[2] If the data in the upper triangle of matrix $\mathbf{P}$ do not match the triangular property: $p_{ij} + p_{jk} \geq p_{ik}$ they may be rank-transformed before a multidimensional scaling method application.

Table 1 Estimates of aspects' scores of Venetian entrepreneurs' motivations by random (A) and maximum distance (B) pairing with tournament method and single rating of aspects (C)

	A		B		C
	Top choice	All comparisons	Top choice	All comparisons	
Income	7.3	13.0	6.1	9.5	6.42
Professional accomplishment	55.1	35.9	51.5	34.1	7.76
Social prestige	2.9	9.8	3.0	17.0	6.33
Social relationship making	21.7	22.1	28.8	24.2	7.49
Frequent travelling	8.7	10.1	4.5	4.9	4.53
Spare time availability	4.3	9.1	6.1	10.2	5.07
Total	100.0	100.0	100.0	100.0	=
(n)	(69)		(66)		(192)

cal distance between aspects ($n = 66$) and those on random coupling at the first level ($n = 69$) with those obtained with the single scoring method ($n = 192$). The tournament design implies that respondents firstly choose within the three initial pairs of aspects and finally among the three aspects chosen at the first level. The distributions obtained with the tournament and single item scoring methods, as described in Table 1, show that professional accomplishment is uniformly the most relevant motivation for work and social relationship-making is the second. All other motivations rank differently according to estimation and data collection approaches.

Estimates may be refined by eliminating the inter-item correlation which may influence estimate scores by introducing redundancy on preferences. The correlation coefficients obtained with a tournament approach, other than random-start, show (Table 2) that the aspects coupled at the first round are systematically negatively correlated. It can be shown that the level of negative correlation between two aspects is a function of the dominance of an aspect upon the other (the larger the dominance, the closer the estimate of the correlation coefficient to 1, its maximum) and of the differential chance an aspect is chosen at the last comparison instead than just to the first level. We conclude that, for unbiased inter-item correlation estimation, we should start a tournament of comparisons by randomly coupling items at the initial stage of the tournament. In any case, the correlation coefficients between the preference data with the random start tournament method differ from the analogous correlations computed on single scoring data. Moreover, the tournament random-start data are mildly less correlated than the single stimulus method, but with an almost constant shift, and so the two methods show a similar general pattern. If we apply a factor analysis on the correlation matrices considered in Table 2, we may put forward the following considerations:

(a) The number of dimensions emerging from the analysis of single scoring and random pairing data is one,[3] while it is two or more on preferences expressed

[3] Factor loadings are 0.578, 0.738, 0.680, 0.634, 0.517, 0.466 from the analysis of single scoring data, and 0.527, 0.766, 0.455, −0.543, −0.724, −0.431 for random pairing tournament data. Neither the rankings of the two solutions converge, nor the rankings converge with the estimates in Table 1.

Table 2 Correlation matrix between motivations for Venetian entrepreneurs by random (A) and maximum distance (B) pairing with tournament method and single rating of aspects (C)

		X_2	X_3	X_4	X_5	X_6 = spare time availability
X_1: Income	A	0.112	−0.131	−0.404	−0.273	−0.288
	B	−0.073	−0.002	−0.791	0.106	0.033
	C	0.440	0.267	0.044	0.170	0.188
X_2: Professional accomplishment	A		0.087	−0.444	−0.494	−0.293
	B		0.064	−0.300	−0.736	−0.200
	C		0.396	0.391	0.108	0.168
X_3: Social prestige	A			−0.177	−0.412	−0.226
	B			0.013	−0.107	−0.879
	C			0.352	0.190	0.142
X_4: Social relationship making	A				0.038	−0.179
	B				−0.062	−0.103
	C				0.304	0.147
X_5: Frequent travelling	A					0.114
	B					0.135
	C					0.288

with the other tournament approaches.[4] It is to be noticed that the rank-orders of the aspects in the unique dimension drawn from single scoring data and random – start tournament data differ because of the position of the aspects' scores in Table 1.

(b)The circular shape of the graphical factorial solution from the maximum distance design brings back the contrasts between initial pairings. Hence, similarly to correlation coefficients, the factorial solution suffers overdimensionality because of the fixed pairing at the first comparison level. As a matter of fact, it is difficult to put the aspects in a plausible sequence and assign them definite scores. The opportunity not to adopt factor analysis for data reduction with tournament preference data is evident.

Hence, we applied the one-dimensional estimation method (7) by extracting the main eigenvalue and the correspondent eigenvector of the preference matrix. For any two items that did not match directly, to avoid a sparse matrix, a stronger rule rather than formula (2) was adopted (Coombs 1976):

[4] Eigenvalues are: 2.08, 1.29, 1.11, 0.79, 0.74, 0.00 on random pairing data analysis; 2.08, 1.83, 1.55, 0.38, 0.16, 0.00 on maximum distance and 2.23, 1.07, 0.98, 0.70, 0.63, 0.39 on single scoring data. At least one eigenvalue from the analysis of comparison data is nil because of a linear dependence.

$$if\ (p_{ij} > 0.5 \cap p_{jk} > 0.5) \rightarrow p_{ik} = \max(p_{ij}, p_{jk}). \quad (8)$$

The results of extensive estimation attempts, which involved also the application of the transition rule (8), are described in Table 3. The analysis highlights what follows:

(a) Preferences expressed with a tournament approach are one-dimensional. In other words, whatever the coupling of the stimuli at the first comparison round, respondents' preferences are one-dimensional.
(b) The transition rule generated data with the same dimensionality as the collected preference data. The matrix composed of just transitive data is one-dimensional with approximately the same scores as the really collected data. It is to be underlined that transitive data are the "mirror image" of the relationships among preferences expressed by the interviewees and the transitive preferences are internally consistent provided the real preferences are consistent to each other.
(c) The minimum-distance approach for initially coupling the items is the least consistent of the three tournament approaches. If we compare the estimates derived from all approaches, through an absolute distance indicator, the minimum-distance approach gives worst results even than the data inputted through the transitivity rule.
(d) The solution closest to the one drawn from the analysis of fully – expressed preferences (Column E in Table 3) pertains to data obtained with random coupling of the items (Column B).
(e) Whatever the pairing approach, the largest estimate score with tournament data pertains to professional accomplishment. This is definitely the most relevant stimulus for entrepreneurship. The second is the opportunity of establishing social relationships. The least relevant is the chance of achieving social prestige. In between, income expectation and the opportunities of disposing of own

Table 3 One-dimensional analysis of preference and transitivity data matrices obtained from Veneto entrepreneurs with tournament method, by analytical approach (A: Random pairing, without transitivity data; B: Random pairing with transitivity data; C: Maximum distance: D: Minimum distance; E: Any type of pairing; F: Just transitivity imputed data)

First eigenvector	*A*	*B*	*C*	*D*	*E*	*F*
Income	0.363	0.339	0.327	0.323	0.327	0.307
Professional accomplishment	0.623	0.635	0.637	0.634	0.632	0.648
Social prestige	0.274	0.246	0.230	0.121	0.231	0.129
Social relationship making	0.500	0.499	0.537	0.454	0.498	0.502
Frequent travelling	0.302	0.317	0.271	0.334	0.312	0.335
Spare time availability	0.253	0.267	0.269	0.401	0.307	0.324
Eigenvalue: 1	3.18	3.15	3.08	3.07	3.16	3.02
Eigenvalue: 2	0.74	0.82	0.89	0.74	0.82	0.88
Eigenvalue: 3	0.53	0.51	0.51	0.59	0.51	0.53
Sample size × choices	68 × 5	68 × 7	66 × 7	65 × 7	199 × 7	199 × 2
Absolute distance from E column data	0.158	0.076	0.124	0.276	=	0.179

spare time and travelling for work. First and second stimuli are the same as the frequency-based estimates.[5]

5 Conclusions

The tournament method, our suggested variant of the paired comparison method, showed to be applicable to elicit preferences with computer assisted interviewing systems. The estimation of the stimulus scores, which involved also the statistical analysis of the data inputted according to a transition rule, highlighted what follows:

- Different methods for the collection of preference data give frequency distributions, correlation matrices and multivariate patterns that are different but the extremely preferred stimuli. That is, the top item is the same whatever the preference elicitation method the bottom one is frequently the same; scores and ranks of the intermediate items fluctuate.
- A one-dimensional solution can be given for granted either if we apply factor analysis on single scoring data or the singular decomposition on a preference matrix whatever, the initial pairing for a tournament of comparisons. This is a basic requirement for multivariate preference elicitation.
- Scores and ranks of the various solutions are just similar. Differences depend both on structural constraints and the psychological meaning of choices expressed by respondents according to the data collection design. Hence, which is the best preference scoring method remains a matter for discussion.

Acknowledgements The author wishes to thank Prof. Michael Greenacre, Dr. Giovanna Boccuzzo and Mr. Lorenzo Maragoni for helping him in refining a first draft of the paper.

References

Bradley, R. A., & Terry, M. E. (1952). Rank analysis of incomplete block designs: I. The method of paired comparisons. *Biometrika, 39*(3/4), 324–345.

Brunk, H. D. (1960). Mathematical models for ranking from paired comparisons. *Journal of the American Statistical Association, 55*, 503–520.

Coombs, C. H. (1976). *A theory of data*. Ann Arbor, MI: Mathesis Press.

Fabbris, L., & Fabris, G. (2003). Sistema di quesiti a torneo per rilevare l'importanza di fattori di customer satisfaction mediante un sistema CATI. In L. Fabbris (Ed.), *LAID-OUT: Scoprire i rischi con lanalisi di segmentazione* (pp. 299–322). Padova: Cleup.

Saaty, T. L. (1977). A scaling method for priorities in hierarchical structures. *Journal of Mathematical Psychology, 15*, 234–281.

[5] Scores obtained from factor analysis are forced to be different from those obtained with singular value decomposition since factor loadings vary between -1 and 1, whilst the $\mathbf{w}$-vector elements are positive and their squared values add up to 1.

Using Rasch Measurement to Assess the Role of the Traditional Family in Italy

Domenica Fioredistella Iezzi and Marco Grisoli

Abstract In the last two decades, Italian family has undergone profound changes. The role of traditional family has become weaker and new living arrangements have gained importance. The ISTAT multipurpose survey on "Households and social aspects" collected the Italianopinion about traditional role of the family. The aim of this study is to assess "Italian" "opinion" about traditional family pattern. Construct and content validity are supported using classical test theory and Rasch modeling in order to test an eight-item Likert scale.

1 Introduction

The gradual decline in mortality and fertility rates in European countries has induced deep changes within the structure of traditional family (Nazio and Blossfeld 2003). It is difficult to find a widely accepted definition of family. Originally the UN/ECE definition of the family unit was based on the "conjugal family concept". Subsequent revisions should be considered as the response to the growing magnitude of living arrangements, in particular for unmarried cohabitation, and extramarital births. "These developments raise questions about the hegemony of legal marriage as the basis of family life and many of the assumptions on which public policies are built" (Kiernan 2002). In Italy, deep changes have been registered in household composition and family structure with a gradual increase of the amount of family households. The Italian National Institute of Statistics (ISTAT) has estimated an increase of Italian families by 2,500,000 from 1998 to 2003 and a decrease of the mean family size, due to 6.5% more singles and 2.1% more couple without children households. As a result of a changing society, from 1995 to 2005 have been registered a 13.5% of decrease in marriages; as a consequence, the number of couples living together doubled in ten years. Similar trends were reported from birth data: the incidence of children born outside a marriage grew from the 8% of total birth

D.F. Iezzi (✉)
Università degli Studi di Roma "Tor Vergata", Italy
e-mail: stella.iezzi@uniroma2.it

F. Palumbo et al. (eds.), *Data Analysis and Classification*,
Studies in Classification, Data Analysis, and Knowledge Organization,
DOI 10.1007/978-3-642-03739-9_19,

in 1995 to 15% in 2005. Moreover, the number of divorces and separations from 1995 to 2005 registered an increase of 74% and 54.7% respectively. The paper aims to assess Italians' opinions on traditional family pattern. We apply a Rating Scale Rasch model (Smith and Smith 2004) to build a Traditional Family index (TFI) and multilevel regression (Snijders and Bosker 2000) to study the possible relationships between socioeconomic and demographic variables.

2 Data and Descriptive Analysis

We studied ISTAT multipurpose survey on "Households and social aspects", produced in 2003 and replicated every five years on a representative sample of 60,000 Italian people. Data collection survey is based on a two-stage sample (ISTAT 2007). This design is employed by selecting towns at the first stage of sampling, followed by the selection of either clusters of families within town at the second stage. The interview involves some Paper and Pencil technique (PAPI). The eight-items set here analysed is a part of the orange self-administered questionnaire, one of the three of the survey, in which adults were asked to answer questions connected to social aspects like study and work history, occupational condition, parents occupation, weekly activities, leaving home parents, daily life. The item set includes the following questions: I_1 marriage is an outdated institution, I_2 a couple can live together without having intention of getting married, I_3 a woman can decide to have a child as a single parent, I_4 children, aged between 18 and 20, should leave the parents home, I_5 it is right that a couple in unhappy marriage decide to divorce even if there are children, I_6 if parents divorce, children should stay with their mothers and not their fathers, I_7 if parents need care, it is the daughters who should be responsible for that, I_8 for a woman being an housewife is better than having a job out-side. For each item, respondents had to express the degree of agreement on a five-category Likert-type scale: strongly agree (1), agree (2), neither agree nor disagree (3), disagree (4), strongly disagree (5). Only 2.1–2.5% of total answers are missing. Figure 1 shows that 80% of Italian people thinks that marriage is not an outdated institution, while more than 60% believes that a couple can live together without having intention of getting married; only 50% agrees that a single parent woman could have a child. Eighty-three percent of Italian people feels that an unhappily married couple should ask for separation or divorce. The majority of subjects (84%) thinks that children should be entrusted to the mother after the divorce.

Moreover, 50% thinks that daughters have to help parents and a woman is better as housewife than having a job outside. The questions from I_1 to I_5 measure the attitude towards a new family model, while the items I_6 to I_8 evaluate the inclination to traditional family. The questions that compose the item set can have one of two different directions. We gave the same polarity to all the items, from strongly agreement (1) to strongly disagreement (5) as to the attitude towards the traditional family. We use Spearman rank correlation coefficient to measure the degree of agreement among subjects as regards their opinion about traditional family. All items present a significant correlation at the 0.01 level (2-tailed).

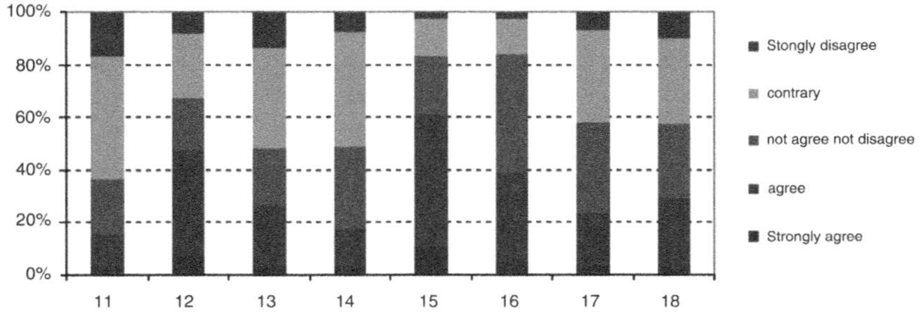

Fig. 1 Descriptive analysis of the eight-items set

3 Method

We used item analysis because it provides a quick way of controlling some important properties of items, as their reliability and unidimensionality. Item analysis is a big box containing a lot of methods to analyse a test, following two different approaches: Classical Test Theory (CTT) and Item Response Theory (IRT). Item Response Theory (IRT) does not require assumptions about sampling or normal distributions, which makes it ideal for performance assessment with different item structures. It also does not require that measurement error be considered the same for all persons taking a test. IRT allows users to create interval-scale scores for both the difficulty of items and the ability of persons tested. These scores are reported in units called logits and are typically placed on a vertical ruler called a logistic ruler. The Rasch model uses mathematical formulas to calculate the probability that a person will choose a given category. When the expected values calculated with the equation of the model are very different from the observed values (chosen categories), that means mathematical model does not fit data structure. This kind of control is performed through the use of fit statistics. In the following analysis we applied Classical test theory (CTT) to select the best items for inclusion in a TFI, to identify poorly written test items and areas of weakness for individual people. We calculated New Family Pattern Index (NFPI) Standard Deviation (SD), Discrimination Index (DI) and Discrimination Coefficient (DC) and reliability (Haladyna 2004). NFPI is a measure of the behaviour to traditional family pattern (1).

$$NFPI = \overline{x}/x_{max}, \tag{1}$$

where $\overline{x}$ is the mean credit obtained by all users attempting the item, and x_{max} is the maximum credit achievable for that item. A test has higher or lower open mind to new family patterns when the results ranging from 0.2 to 1. Standard Deviation (SD) measures the spread of answers in the response population. Discrimination Index (DI) measures how performance on one item correlates to test as a whole. There should always be some correlation between item and test performance, however it is expected that discrimination will fall in a range between 0.2 and 1.0. It provides a rough indicator of the performance of each item to separate high affection to traditional family vs. less-affection (2). the mean credit obtained by all users attempting the item, and x_{max} is the maximum credit achievable for that item. A test has higher or lower open mind to new family patterns when the results ranging from 0.2 to 1. Standard Deviation (SD) measures the spread of answers in the response population. Discrimination Index (DI) measures how performance on one item correlates to test as a whole. There should always be some correlation between item and test performance, however it is expected that discrimination will fall in a range between 0.2 and 1.0. It provides a rough indicator of the performance of each item to separate high affection to traditional family vs. less-affection (2).

$$NFPI = \frac{x_{top} - x_{bottom}}{n}, \tag{2}$$

where x_{top} is the sum of the fractional credit (achieved/maximum) obtained at this item by the one-third of users having that highest grades in the whole test (i.e. number of correct responses in this group), x_{bottom} is the analogue sum for users with the lower one-third grades for the whole test and n is analogue sum for users with the lower one-third grades for the whole test and n is the number of responses given to this question sx is the standard deviation of fractional scores for this question. This parameter can take values between +1 and −1. If the index goes below 0.0 it means that more of the weaker learners got the item right than the stronger learners. Such items should be discarded as worthless. In fact, they reduce the accuracy of the overall score for the test. The correlation between on the observations on an item (or by a person) and the person raw scores (or item marginal scores) are crucial for evaluating whether the coding scheme and person responses accord with the requirement that "higher observations correspond to more of the latent variable" (and viceversa). The coefficient r_{pb} indicates the strength of relationship (correlation) between how individuals answer an item and their score total (3).

$$r_{pb} = \frac{\sum_{i=1}^{n} (x_i - \bar{x})(y_i - \bar{y})}{n s_x s_y}, \tag{3}$$

where$\sum_{i=1}^{n} (x_i - \bar{x})(y_i - \bar{y})$is the sum of the products of deviations for item scores and overall test scores, n is the total sample size, s_x is the standard deviation of fractional scores for item and, s_y is the standard deviation of scores at the test as a whole. Reliability are calculated by Cronbach's alpha (α). It measures how well a set of items measures a single unidimensional latent construct. When data have a multidimensional structure, Cronbach's alpha will usually be low. Cronbach's alpha can be written as a function of the number of test items and the average inter-correlation among the items. We show the formula for the standardized:

$$\alpha = \frac{n\bar{r}}{1 + \bar{r}(n-1)}, \tag{4}$$

where n is the number of items and $\bar{r}$ is the average inter-item correlation among the items. CTT employs relative simple mathematical procedures, but it has several important limitations, e.g. FI and DI are both sample dependent. We used also Rasch analysis to measure Italian opinions about traditional family and to transform ordinal-scale measures into interval-scale measures that provides good precision (reliability). The TFI was obtained by applying Rating Scale model (Smith and Smith 2004). This model defines the probability, π_{nix}, of person n of location β_n on the latent variable continuum being observed in category x of item i with location δ_i as

$$\pi_{nix} = \frac{\exp \sum_{j=0}^{x} \left[\beta_n \left(\delta_i + \tau_j\right)\right]}{\sum_{k=0}^{m} \exp \sum_{j=0}^{x} \left[\beta_n \left(\delta_i + \tau_j\right)\right]}, \tag{5}$$

where the categories are ordered from 0 to m, and the coefficients τ_j are the rating scale structure parameters. Multilevel model was applied to study the TFI relationships with predictor variables (socioeconomic status, gender, level of education, profession, civil state and age), within regions of Italy (Snijders and Bosker 2000).

4 Discussion

The eight items of the original dataset were subject to item analysis, using standard procedures of CCT. First of all, the r_{pb} is lower than 0.20 for I_4, I_6, I_7, and I_8, (Table 1). DI shows that I_5, has a low discrimination for traditional family affection. The selection of items based on CCT shows that it is necessary to reduce the number of items in the TFI. Cronbach's alpha increases if I_4, I_6, I_7, and I_8 are deleted.

The Rasch model underlying the item set is a rating scale model, where the category coefficient patterns are the same across all items. WINSTEPS software 2006 program was used to conduct Rasch Analysis (Linacre 2006). This technique enables calibration of items to construct a scale on which linear measures underlying the observations are defined. Rasch analysis showed that the original rating scale had not adequate separation level for person (0.45), as well as adequate reliabilities for person (1.02) and item (1.00) (Table 2).

Items included in the analysis are connected to concepts like marriage, divorce, cohabitation and single-parent family. Items excluded are connected to the role of children in the family, responsibility in caring parents and the necessity for a woman to work outside home. By choosing the four-item model, the item-person map (Fig. 2) is the better way for evaluating the conjoint contribution of individual abilities and item difficulties. On the right side, items are quite well separated in the

Table 1 Summary of measures

Items	FI	SD	DI	r_{pb}	α If item deleted
I_1	0.722	0.986	67.600	0.411	0.535
I_2	0.562	1.082	44.799	0.591	0.465
I_3	0.672	1.057	59.355	0.474	0.511
I_4	0.680	0.909	43.867	0.167	0.606
I_5	0.486	0.899	21.801	0.362	0.552
I_6	0.547	0.828	33.179	0.069	0.626
I_7	0.645	0.939	33.834	0.170	0.606
I_8	0.640	1.026	53.508	0.166	0.611

Table 2 Reliability Estimates by item

Item set		Item			
	Reliability	Infit	Outfit	α	Item deleted
8	1.00	1.000	1.010	0.514	–
7	1.00	1.000	1.010	0.588	I_8
6	1.00	0.990	1.020	0.643	I_7
5	1.00	0.990	1.020	0.704	I_6
4	1.00	0.990	1.000	0.739	I_4
		Person			
	Reliability	Infit	Outfit	α	Item deleted
8	0.450	1.010	1.010	0.514	–
7	0.530	1.020	1.010	0.588	I_8
6	0.590	1.030	1.020	0.643	I_7
5	0.650	1.030	1.020	0.704	I_6
4	0.690	1.000	0.990	0.739	I_4

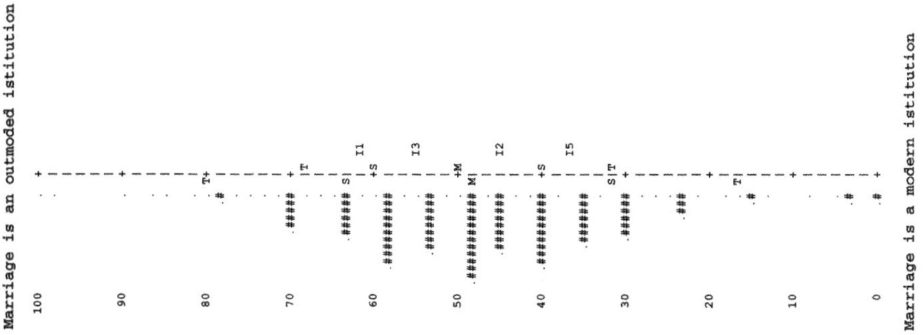

Fig. 2 Item-person map

middle of the axis; as a result, aspects like updating of marriage as institution and single-parent households without stable relation are quite difficult to be accepted; on the contrary, civil union and divorce seem to be more shared between people. On the left side, it results that persons with highest position on the scale are the most open-minded about new family models.

However, the information-weighted mean-square (outfit) statistics were 1.01 and 1.01, respectively, suggesting that some variables were noisy and not contributing to measure the opinions about the traditional family. In the second calibration, which four items were deleted (Table 1). Cronbach α confirms the necessity to exclude four items. It results that the four-item model is just satisfactory to accept the residual item set as a scale. Items included in the analysis are connected to concepts like marriage, divorce, cohabitation and single-parent family. Items excluded are connected to the role of children in the family, responsibility in caring parents and the necessity for a woman to work outside home. By choosing the four-item model, the item-person map (Fig. 2) is the better way for evaluating the conjoint contribution of individual abilities and item difficulties. On the right side, items are quite well separated in the middle of the axis; as a result, aspects like updating of marriage as institution and single-parent households without stable relation are quite difficult to be accepted; on the contrary, civil union and divorce seem to be more shared between people. On the left side, it results that persons with highest position on the scale are the most open-minded about new family models. The analysis showed a hierarchical structure of the data, that we urged to apply a multilevel regression model in which the first level units were 60,000 Italian persons, whereas the second level units were 20 regions. The effects of regions is high for age ($p = 0.000$), gender ($p = 0.000$) and level of education ($p = 0.003$). Women and young people are more open to new lifestyles, above all in Sardegna and in Molise, that are the more conservative regions. The multipurpose survey on "Households and social aspects" didn't collect data on actually family topics, i.e. homosexual couples, relationship duration, monogamy vs. promiscuity, number of children being raised, rates of intimate partner violence, the sharp drop in the birth-rate, the reduced marriage rate, and the increase in immigration. The TFI is a measure of the Italians' attitude towards traditional family values.

References

Haladyna, T. (2004). *Developing and validating multiple-choice test items* (3rd edition). Mahwah, NJ: Lawrence Erlbaum.

ISTAT. (2007). *Il matrimonio in Italia: un'istituzione in mutamento. Anni 2005–2005, Rome, Note per la stampa, 12 febbraio 2007*. Rome: Author.

Kiernan, K. (2002). The state of European unions: An analysis of partnership formation and dissolution. In M. Macura & G. Beets (Eds.), *Dynamics of fertility and partnership in Europe. Insights and lessons from comparative research* (Vol. I). New York: United Nations.

Linacre, J. M. (2006) *User's guide to WINSTEPS Rasch-model, computer programs*. Chicago: MESA Press. Retrieved from http://www.winsteps.com.

Nazio, T., & Blossfeld, H. P. (2003). The diffusion of cohabitation among young women in West Germany, East Germany and Italy. *European Journal of Population, 19*(2), 47–82.

Smith, E. V., & Smith, R. M. (2004). *Introduction to Rasch measurement: Theory, models and applications*. Maple Grove, MN: JAM Press.

Snijders, T., & Bosker, R. (2000). *Multilvel analysis*. London: Sage.

Preserving the Clustering Structure by a Projection Pursuit Approach

Giovanna Menardi and Nicola Torelli

Abstract A projection pursuit technique to reduce the dimensionality of a data set preserving the clustering structure is proposed. It is based on Silverman's (J R Stat Soc B 43:97–99, 1981) critical bandwidth. We show that critical bandwidth is scale equivariant and this property allows us to keep affine invariance of the projection pursuit solution.

1 Introduction

In the last decades, advances in technology have led to the opportunity of collecting and storing enormous amount of data. High-dimensional data sets present many opportunities because of the increase of information but several problems occur when the number of dimensions becomes high: data are difficult to explore, the computational effort required to run any technique increases, the solutions are affected by the curse of dimensionality, the interpretation of results becomes more and more tangled. Reducing the dimension of the original data, prior to any model application, can be useful to overcome all these problems.

In mathematical terms, the problem can be formalized as follows: given the d-dimensional variable $x = (x_1, \ldots, x_d)'$, the goal is to find a lower dimensional mapping of it, $z = \varphi(x)$, with $z = (z_1, \ldots, z_p)'$ and $p \leq d$, that captures the information of the original data, according to some criterion. In general, the optimal mapping $z = \varphi(x)$ will be a non-linear function. However, there is not a systematic way to generate non-linear transformations, and the problem is strongly data dependent. For this reason, most of dimension reduction techniques produce linear representations of the original, with each of the p components of z being a linear combination of the original variables: $z_i = a_{1i}x_1 + \cdots + a_{di}x_d$, for $i = 1, \ldots, p$; that is $z = Ax$, where $A_{p \times d}$ is the linear transformation weight matrix.

G. Menardi (✉)
Department of Economics and Statistics, P.le Europa,
1 Trieste, Italy
e-mail: Giovanna.Menardi@econ.units.it

F. Palumbo et al. (eds.), *Data Analysis and Classification*,
Studies in Classification, Data Analysis, and Knowledge Organization,
DOI 10.1007/978-3-642-03739-9_20,

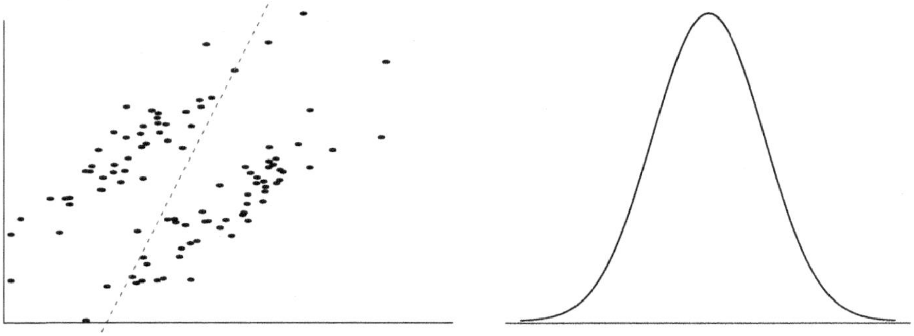

Fig. 1 *To the left:* direction of the first PC (*dashed line*) calculated from 100 data points drawn from a bimodal distribution. The *right panel* displays the probability function of the projected data

In this paper we investigate the opportunity of reducing the dimensionality of a set of data points while preserving its clustering structure and we propose a technique aimed at this goal. Principal component analysis (PCA), traditionally used to reduce the dimensionality of a multidimensional space, is not adequate in this context because it may completely fail in keeping the original structure of groups (see Fig. 1). We consider, instead, the more flexible *projection pursuit* methods.

Projection pursuit methods (see Huber 1985) seek to find "interesting" low-dimensional linear projections of multivariate data by numerically optimizing an objective function called *projection index*. The projection index has to be chosen in such a way that it takes large values when the projected data are interesting.

Literature about projection pursuit usually considers interesting a projection which exhibits departure from normality and uses as projection indexes any statistic sensitive to departure from normality (see, for example, Huber 1985; Friedman 1987; Hall 1989).

In fact, we are interested in projections which enhance the clustering structure, but the departure from normality does not entail that the data are grouped. Hence, we need to identify a projection index that takes large values when the reduced data keep the clustering structure unaltered.

In order to guarantee the affine invariance of the projection pursuit solution, a further requirement is the location and scale invariance of the projection index (Huber 1985), that is $I(sz + m) = I(z)$, where I is the projection index calculated on the transformed data z and $s, m \in \mathbb{R}$. Otherwise, the invariance may be obtained by *sphering* the data before running the projection pursuit (Friedman 1987).

The paper is organized as follows: Sect. 2 shows how the Silverman's (1981) critical bandwidth, originally introduced to test the hypothesis of multimodality, can be suitably adjusted to obtain an affine invariant projection index. In Sect. 3 the new technique is compared with other standard methods of dimensionality reduction and some numerical evidence based on a simulation study and a few real data applications is presented. In Sect. 4 there are some concluding remarks.

2 Projection Pursuit for Preserving the Clustering Structure

In a cluster analysis framework, the more evident the groups, the more interesting the projection. In Hartigan (1975), clusters are defined as *regions of high density separated from other such regions by regions of low density*. Following this approach, it is natural to consider projection indexes which reveal the structure of the modes of the density function underlying data. Therefore, any statistic to test unimodality is a reasonable candidate.

Montanari and Lizzani (1998) investigated into the use of some indexes to preserve the multimodality of the data: the critical bandwidth (Silverman 1981), the *dip* statistic (Hartigan and Hartigan 1985), the *excess mass* (Müller and Sawitzki 1992) and compared their performances on some simulated and real data sets. Krause and Liebscher (2005) used the dip statistic as a projection index. They showed that this statistic has desirable properties of continuity and differentiability when the projection varies.

In this work we consider the opportunity of using the non-parametric critical bandwidth (Silverman 1981) as a projection index.

In the sequel, $\mathcal{X} = (\mathbf{x}_1, \ldots, \mathbf{x}_n)'$, $\mathbf{x}_i \in \mathbb{R}^d$, $i = 1, \ldots, n$, denotes the matrix of the observations. The problem of univariate projection pursuit using an index I can be formalized as follow:

$$\hat{Z} = \mathcal{X}\hat{\mathbf{a}}, \qquad \hat{\mathbf{a}} = \operatorname{argmax}_{\mathbf{a}'\mathbf{a}=1} I(\mathcal{X}\mathbf{a}).$$

2.1 *The Critical Bandwidth to Test Multimodality*

Silverman's (1981) approach for investigating into the number of the modes in the density underlying the data is based on the observation of the behaviour of the kernel density estimate, keeping the data points fixed but allowing the window width to vary. Let $\mathcal{Y} = (y_1, \ldots, y_n)'$ be a sample of univariate *i.i.d* observations drawn from an unknown density function f.The kernel density estimator of f is: $\hat{f}_h(y; \mathcal{Y}) = \sum_{i=1}^{n} \frac{1}{h} K\left(\frac{y-y_i}{h}\right)$. Here K is a kernel function satisfying $\int K(y)dy = 1$, and h is the bandwidth or window width. The bandwidth determines the amount of smoothing of the estimator and, hence, the number of the modes in the estimate. Silverman shows that the number of local maxima of the estimated density is monotone decreasing in h for the normal kernel. As a consequence, there exists a critical value h_{crit} of h defined as follow:

$$h_{crit} = \inf\{h : \hat{f}_h(\cdot, \mathcal{Y}) \text{ has at most one mode}\}. \tag{1}$$

Silverman shows that, for n large, h_{crit} approaches to zero under the null hypothesis of unimodality but it remains bounded away from zero otherwise. This behaviour occurs because, when data come from a multimodal distribution, a considerable amount of smoothing is necessary to get an unimodal density estimate.

The problem of using h_{crit} as a projection index is that it is not affine invariant. However, sphering the data is not recommended in this framework, because the clustering structure may be altered (Cook et al. 1995). To overcome this problem, Montanari and Guglielmi (1994) estimated the relationship between h_{crit}, n and the variability σ_Z of the projected data under the hypothesis of unimodality and proposed to use an approximately scale invariant index. We suggest an alternative adjustment of the critical bandwidth based on the following:

Theorem 1. *The critical bandwidth is location invariant and scale equivariant.*

Proof. Let $\hat{f}_h(\cdot, \mathcal{Y})$ be the kernel density estimate based on the data $\mathcal{Y} = (y_1, \ldots, y_n)'$, $y_i \in \mathbb{R}, i = 1, \ldots, n$ and built using a kernel function $K_h = \frac{1}{h}K(\frac{x}{h})$. We aim to show that there is a biunivocal correspondence between the local maxima of $\hat{f}_h(\cdot, \mathcal{Y})$ and the local maxima of:

1. $\hat{f}_h(\cdot; \mathcal{Y} + \alpha)$
2. $\hat{f}_{\alpha h}(\cdot; \alpha\mathcal{Y})$,

where $\mathcal{Y} + \alpha = (y_1 + \alpha, \ldots, y_n + \alpha)'$ and $\alpha\mathcal{Y} = (\alpha y_1, \ldots, \alpha y_n)'$.

1. Let choose $y \in \mathbb{R}$ arbitrarily. Then

$$\begin{aligned}\hat{f}_h(y + \alpha; \mathcal{Y} + \alpha) &= \frac{1}{n}\sum_{i=1}^{n} K_h\left((y + \alpha) - (y_i + \alpha)\right)\\ &= \frac{1}{n}\sum_{i=1}^{n} K_h(y - y_i) = \hat{f}_h(y; \mathcal{Y}).\end{aligned}$$

   It follows from the arbitrariness of y that, if $\tilde{y}$ is a local maximum of $\hat{f}_h(\cdot; \mathcal{Y})$, then $\tilde{y} + \alpha$ is a local maximum of $\hat{f}_h(\cdot; \mathcal{Y} + \alpha)$.
2. Moreover

$$\begin{aligned}\hat{f}_{\alpha h}(\alpha y; \alpha\mathcal{Y}) &= \frac{1}{n}\sum_{i=1}^{n} K_{\alpha h}(\alpha y - \alpha y_i)\\ &= \frac{1}{n}\sum_{i=1}^{n} \frac{1}{\alpha h}K(\frac{\alpha y - \alpha y_i}{\alpha h})\\ &= \frac{1}{n}\sum_{i=1}^{n} \frac{1}{\alpha h}K(\frac{y - y_i}{h})\\ &= \frac{1}{n}\sum_{i=1}^{n} \frac{1}{\alpha}K_h(y - y_i).\end{aligned}$$

   Therefore, if $\hat{f}_h'(\cdot; \mathcal{Y})$ is the first derivative of $\hat{f}_h(\cdot; \mathcal{Y})$ with respect to y,

$$\hat{f}'_{\alpha h}(\alpha y;\alpha\mathcal{Y}) = \frac{1}{n}\sum_{i=1}^{n}\frac{1}{\alpha}K'_h(y-y_i) = \frac{1}{\alpha}\hat{f}'_h(y;\mathcal{Y}).$$

In a similar way one can show that $\hat{f}''_{\alpha h}(\alpha y;\alpha\mathcal{Y}) = \frac{1}{\alpha^2}\hat{f}''_h(y;\mathcal{Y})$. Therefore, if $\hat{f}_h(\cdot;\mathcal{Y})$ has a local maximum at $\tilde{y}$,then $\hat{f}_h(\cdot;\alpha\mathcal{Y})$ has a local maximum at $\alpha\tilde{y}$.

It follows that, if h_{crit} is the critical bandwidth of $\mathcal{Y}$, it is also the critical bandwidth of $\mathcal{Y}+\alpha$, and αh_{crit} is the critical bandwidth of $\alpha\mathcal{Y}$. □

2.2 Projection Pursuit Using the Adjusted Critical Bandwidth

As immediate corollary of the above theorem, the critical bandwidth of a linear projection of the data is proportional to the standard deviation of the projected data.

For this reason, we propose the use of the *adjusted critical bandwidth* as a projection index, that is defined as follows:

$$I_h(Z) = \frac{\inf\{h : \hat{f}_h(\cdot; Z) \text{ is unimodal}\}}{\sigma_Z} = \inf\{h : \hat{f}_h(\cdot; \frac{Z}{\sigma_Z})\text{is unimodal}\}, \quad (2)$$

with σ_Z is the standard deviation of Z. It is worthwhile to note that using the (2) as a projection index prevents us from rescaling the data before searching the projections.

A possible way to generalize the described procedure for moving from a d-dimensional space to a p-dimensional space ($p \leq d$), consists in finding subsequent orthogonal univariate projections.

3 Numerical Results

The illustrated technique has been evaluated both on simulated and real data.

3.1 A Simulation Study

A simulation study has been conducted for evaluating the ability of the I_h index in reducing the dimensionality by preserving the original composition of the clusters, evaluating the efficiency of the procedure when p varies and comparing the performance of the I_h index with the principal components.

To these aims we have generated a large number of samples from several multimodal distributions (mixtures of gaussian densities) defined on $\mathbb{R}^d$, when d varies. Here, we show for $d = 5, 7, 10$ the results concerning the simulation from one density function. For each sample, from two to five principal components and from

Table 1 Percentiles of the empirical distribution of the ARI obtained by running the three clustering algorithms on two to five projections maximizing I_h and two to five principal components. Data come from a five-dimensional distribution

		k-Means					Ward					AT method				
		5%	25%	50%	75%	95%	5%	25%	50%	75%	95%	5%	25%	50%	75%	95%
2	PP	0.33	0.42	0.51	0.56	0.63	0.34	0.45	0.51	0.58	0.62	0.62	0.67	0.69	0.83	0.90
	PC	0.07	0.13	0.18	0.23	0.31	0.05	0.09	0.13	0.19	0.29	0.43	0.46	0.50	0.57	0.63
3	PP	0.33	0.46	0.50	0.56	0.61	0.27	0.36	0.42	0.50	0.57	0.66	0.68	0.71	0.88	0.97
	PC	0.08	0.15	0.25	0.31	0.39	0.08	0.15	0.18	0.25	0.37	0.30	0.55	0.59	0.61	0.64
4	PP	0.69	0.74	0.82	0.85	0.89	0.37	0.44	0.48	0.53	0.61	0.88	0.93	0.95	0.96	0.97
	PC	0.00	0.01	0.11	0.30	0.43	0.10	0.20	0.26	0.31	0.41	0.71	0.86	0.89	0.91	0.94
5	PP	0.34	0.38	0.40	0.51	0.55	0.39	0.47	0.53	0.57	0.63	0.68	0.92	0.95	0.96	0.98
	PC	0.00	0.07	0.21	0.43	0.48	0.14	0.23	0.32	0.38	0.45	0.54	0.66	0.81	0.90	0.95

two to five projections maximizing I_h have been obtained and three clustering procedures have been applied in order to reconstruct the original clustering structure: one hierarchical (the Ward method), one partitional (the k-means method), one density based (the AT method, Azzalini and Torelli 2007). The number of clusters has been fixed to the actual number of clusters except for the AT method which automatically detects the modes of the densities underlying data. We have compared the detected clusters with the real ones in terms of Adjusted Rand Index (ARI, Hubert and Arabie 1985). The ARI derives from the Rand Index which evaluates the agreement between two partitions as the proportion of couples of data in the same class according to both the partitions. Its expected value is equal to zero and, like the Rand Index, the ARI takes value 1 when the two partitions overlap. Results are in Tables 1, 2, and 3. Although the projection pursuit procedure does not produce uniformly better results than PCA, it avoids the blurring of the original clusters. When the groups lie parallel to the direction of the maximum variability of the data, the principal components are not able to catch the clustering structure. Moreover, in most of the considered situations, the Adjusted Rand Index calculated on data reduced by maximizing I_h is larger than the corresponding index calculated on the data reduced by PCA. In the remaining situations the difference is not appreciable. From the simulation study it arises that bivariate or trivariate projections of the data usually reveal the clusters while augmenting the dimensionality may result in confounding the structure if there are several dimensions which are not relevant. This result is remarkable because it increases the usefulness of the proposed technique which allows us to take advantage of the graphical tools to explore the data.

3.2 Real Data Applications

We have run the projection pursuit procedure on two real data sets, that are typical examples used to run the supervised classification techniques: the *iris* data (Fisher 1936) and the *olive oil* data (Forina et al. 1983). The *iris* data set gives the

Table 2 See Table 1. Data have been generated from a seven-dimensional distribution

	k-Means					Ward					AT method				
	5%	25%	50%	75%	95%	5%	25%	50%	75%	95%	5%	25%	50%	75%	95%
2 PP	0.71	0.74	0.78	0.80	0.84	0.75	0.85	0.90	0.94	0.99	0.44	0.80	0.84	0.86	0.89
PC	0.82	0.85	0.87	0.89	0.91	0.81	0.88	0.91	0.94	0.96	0.83	0.86	0.88	0.90	0.93
3 PP	0.70	0.84	0.86	0.88	0.92	0.84	0.90	0.94	0.96	0.98	0.53	0.77	0.81	0.86	0.87
PC	0.82	0.85	0.88	0.89	0.92	0.86	0.89	0.91	0.93	0.95	0.69	0.80	0.85	0.87	0.91
4 PP	0.61	0.63	0.70	0.94	0.96	0.58	0.67	0.75	0.86	0.88	0.00	0.41	0.51	0.76	0.87
PC	0.62	0.65	0.68	0.79	0.82	0.44	0.59	0.61	0.79	0.85	0.00	0.32	0.48	0.63	0.84
5 PP	0.64	0.68	0.70	0.81	0.84	0.71	0.73	0.75	0.87	0.89	0.00	0.00	0.37	0.50	0.82
PC	0.33	0.56	0.68	0.80	0.83	0.16	0.39	0.52	0.84	0.86	0.00	0.00	0.00	0.32	0.52

Table 3 See Table 1. Data have been generated from a ten-dimensional distribution

	k-Means					Ward					AT method				
	5%	25%	50%	75%	95%	5%	25%	50%	75%	95%	5%	25%	50%	75%	95%
2 PP	0.54	0.78	0.83	0.86	0.89	0.52	0.68	0.72	0.74	0.78	0.63	0.64	0.66	0.67	0.69
PC	0.54	0.62	0.76	0.81	0.88	0.69	0.73	0.78	0.81	0.90	0.69	0.70	0.71	0.73	0.75
3 PP	0.65	0.72	0.76	0.79	0.82	0.69	0.72	0.76	0.78	0.81	0.92	0.94	0.96	0.97	0.98
PC	0.59	0.62	0.65	0.67	0.68	0.58	0.62	0.65	0.67	0.70	0.66	0.70	0.72	0.73	0.82
4 PP	0.60	0.64	0.85	0.97	0.99	0.80	0.86	0.93	0.97	0.99	0.92	0.96	0.97	0.98	0.99
PC	0.60	0.63	0.66	0.68	0.69	0.59	0.64	0.65	0.69	0.70	0.28	0.68	0.71	0.73	0.93
5 PP	0.60	0.62	0.64	0.91	0.99	0.81	0.90	0.94	0.97	0.99	0.94	0.98	0.99	0.99	1.00
PC	0.60	0.63	0.65	0.67	0.69	0.60	0.63	0.65	0.67	0.70	0.38	0.64	0.76	0.99	1.00

Table 4 ARI obtained by applying the clustering methods on *iris* and *olive oil* data

Iris data	k-Means	Ward	AT
2 Projections	0.86	0.82	0.65
3 Projections	0.65	0.75	0.90
4 Projections	0.73	0.76	0.75

Olive oil data	k-Means	Ward	AT
2 Projections	0.36	0.40	0.70
3 Projections	0.28	0.25	0.75
4 Projections	0.46	0.37	0.81
5 Projections	0.54	0.80	0.81

measurements of the variables sepal length and width and petal length and width, respectively, for 50 flowers from each of three species of iris. The species are Iris setosa, versicolor, and virginica. The *olive oil* data consists of the percentage composition of eight fatty acids (palmitic, palmitoleic, stearic, oleic, linoleic, eicosanoic, linolenic, eicosenoic) found in the lipid fraction of 572 Italian olive oils. For each oil the geographical origin of the olives is known (North of Italy, South of Italy, Sardinia). We have applied some clustering algorithms on the data reduced by the projection pursuit in order to reconstruct the label class of each data. Results are in Table 4. The real data applications are strongly dependent on the used clustering technique. With regard to the distance based algorithms, the clusters of the *iris* data are already revealed in the bidimensional space, while five dimensions catch the groups in the *olive oil* data. However, results deriving from the application of AT procedure to both data suggest that the projection pursuit algorithm based on I_h is able to preserve the clustering structure in two or three dimensions keeping the structure of high density regions. A further remarkable consideration concerns the apparent ability of the proposed index in emphasizing the clustering structure as well as preserving it. In fact, the AT procedure cannot separate two of the three groups in the original data, but it does it easily in the reduced data.

4 Concluding Remarks

In this paper we showed a useful property of the critical bandwidth and proposed an adjustment aimed at using it as an affine invariant projection index when reducing the dimensionality by projection pursuit methods. Results from simulation studies and real data applications have proved that the use of the proposed technique is effective in preserving the clustering structure while reducing dimensionality. When sample size is large and the number of variables is larger than 15, the use of standard optimization algorithms implies a computational burden that makes difficult the application of the proposed technique. There is room for improvement and looking for more effective algorithms will be one of the focuses of future research.

References

Azzalini, A., & Torelli, N. (2007). Clustering via nonparametric density estimation. *Statistical Computing, 17*, 71–80.

Cook, D., Buja, A., Cabrera, J., & Hurley, C. (1995). Gran tour and projection pursuit. *Journal of Computational and Graphical Statistics, 4*, 155–172.

Fisher, R. A. (1936). The use of multiple measurements in taxonomic problems. *Annals of Eugenics, 7*, 179–188.

Forina, M., Armanino, C., Lanteri, S., & Tiscornia, E. (1983). Classication of olive oils from their fatty acid composition. In H. Martens & H. Russwurm Jr. (Eds.), *Food research and data analysis* (pp. 189–214). London: Applied Science.

Friedman, J. H. (1987). Exploratory projection pursuit. *Journal of American Statistical Association, 82*, 249–266.

Hall, P. (1989). Polynomial projection pursuit. *Annals of Statistics, 17*, 589–605.

Hartigan, J. A. (1975). Clustering algorithms. New York: Wiley.

Hartigan, J. A., & Hartigan, P. M. (1985). The dip test of unimodality. *Annals of Statistics, 13*, 70–84.

Huber, P. J. (1985). Projection pursuit. *Annals of Statistics, 13*, 435–475.

Hubert, L., & Arabie, P. (1985). Comparing partitions. *Journal of Classification, 2*, 193–218.

Krause, A., & Liebscher, V. (2005). Multimodal projection pursuit using the dip statistic, Preprint-Reihe Mathematik, 13.

Montanari, A., & Guglielmi, N. (1994). Exploratory projection pursuit maximizing departure from unimodality. In *Proc. XXXVII Riun. Scient. Soc. Ital. Stat.* (pp. 245–251).

Montanari, A., & Lizzani, L. (1998). Projection pursuit and departure from unimodality. *Metron, 56*, 139–153.

Müller, D. W., & Sawitzki, G. (1992). Excess mass estimates and test for multimodality. *Journal of American Statistical Association, 86*, 738–746.

Silverman, B. W. (1981). Using kernel density estimates to investigate multimodality. *Journal of the Royal Statistical Society: Series B, 43*, 97–99.

Association Rule Mining of Multimedia Content

Adalbert F.X. Wilhelm, Arne Jacobs, and Thorsten Hermes

Abstract The analysis of video sequences is of primary concern in the field of mass communication. One particular topic is the study of collective visual memories and neglections as they emerged in various cultures, with trans-cultural and global elements (Ludes P., Multimedia und Multi-Moderne: Schlüsselbilder, Fernsehnachrichten und World Wide Web – Medienzivilisierung in der Europäischen Währungsunion. Westdeutscher Verlag, Opladen 2001). The vast amount of visual data from television and web offerings make comparative studies on visual material rather complex and very expensive. A standard task in this realm is to find images that are similar to each other. Similarity is typically aimed at a conceptual level comprising both syntactic as well as semantic similarity. The use of semi-automatic picture retrieval techniques would facilitate this task. An important aspect is to combine the syntactical analysis that is usually performed automatically with the semantic level obtained from annotations or the analysis of captions or closely related text. Association rules are in particular suited to extract implicit knowledge from the data base and to make this knowledge accessible for further quantitative analysis.

1 Introduction

In mass communication research visual memories are systematized and decoded by means of the concept of "key visuals", which encompasses visual images and stereotypes. Researchers in this field typically code the visual material manually and then further explore the material using content analysis techniques. While it seems unreasonable to strive for an automatic routine that retrieves all relevant pictures out of a heterogeneous data base, it seems reasonable to automatize specific tasks within a more homogeneous data archive and with a clearly defined target in mind.

A.F.X. Wilhelm (✉)
Jacobs University Bremen, P.O. Box 75 05 61, D-28725 Bremen, Germany
e-mail: a.wilhelm@jacobs-university.de

F. Palumbo et al. (eds.), *Data Analysis and Classification*,

Studies in Classification, Data Analysis, and Knowledge Organization,
DOI 10.1007/978-3-642-03739-9_21,

Person recognition is such a specific task that is of relevance in the context of key visuals. Research in computer graphics on automated person recognition in video sequences focuses on movements and biometrical features, see for example Yam et al. (2004). For the mass communication application the focus is not on identifying a particular person, but one aims in classifying the person(s) presented in the video sequence according to their role in society, e.g. to extract all video sequences from a given set of news reports that show a political leader or a sports person.

The main challenge is to develop models that intelligently combine syntactic information as used in the automatic process of picture recognition and the semantic information as provided by the manual coding of the material. Decision trees, neural networks and association rules are potential vehicles that can be used to learn the expert knowledge, presented in the form of semantic coding and descriptions in the data base (Perner 2004). Then the combined models might be used to reduce the amount of manual coding while still keeping a high rate of successful recognition. The quality of the models for the semantic information now depends on the ability to connect best to the syntactic information extracted.

2 Syntactic Analysis of Video Data

Video sequences constitute a massive data set and a variety of different techniques are available to reduce the amount of storage. The common approach of data reduction for storage purposes is to compress video streams using a codec. For data analysis purposes this reduction is not sufficient, because it still leaves to much material to be analyzed. Independent of the compression that is used, a further reduction of the data is typically possible by extracting representative images for a given video scene. A *scene* is typically the smallest semantic unit of a video sequence. People usually interpret a video as a sequence of scenes and for the human mind it is fairly easy to determine scene boundaries in a video. Typically, such a human analysis also takes into account the associated audio track. Automatic detection of scene boundaries is much more complicated. So instead of scene detection, one does a shot detection and tries to find the key frames of a scene. Hence, the scene is segmented in different shots and only a few still images are used as a representation of the scene.

There are a variety of different approaches to shot boundary detection, see Lienhart (1998), Yusoff et al. (1998) and the proceedings of the annual TRECVID competition (http://www-nlpir.nist.gov/projects/tvpubs/tv.pubs.org.html). The principle methodology is always the same. For every nth frame of a video sequence the differences for consecutive frames are calculated and compared to a given threshold. The algorithms differ in the kind and number of features that enter the calculation of the differences and in the thresholds used. In our research, we use a system of the Technologie Zentrum Informatik (TZI) in Bremen, which is based on RGB histograms, see Jacobs et al. (2004). Color histograms are used to detect candidates for a hard cut or a gradual cut. Additionally, texture, position and shape features, are used for automatic image retrieval in the system *PictureFinder* (Hermes et al. 2004).

The color analysis of the *PictureFinder* system performs a color segmentation of the image that groups together pixels of similar colors to regions. The differences of two neighboring pixels in hue, lightness and saturation are assessed and constitute the basis for the creation of regions. Pixels belong to the same region as long as the assessed difference does not exceed pre-specified thresholds. The algorithm used is an extension of the blob-coloring algorithm presented in Ballard and Brown (1982). This algorithm delivers as output a segmentation of the image in different color regions, defined by their bounding box, their centers of gravity, their colors, and statistical characteristics (mean, standard deviation, range, minimum and maximum) of the hue, lightness, and saturation values of all pixels in the region. The texture analysis is derived by region-based and edge-based methods, see Hermes et al. (2000).

Once an image is divided by color and texture analysis into regions, a shape analysis is performed. The input for the shape analysis are object images which are black and white images where all pixels of the object have the value one, all other pixels are set to the value zero. This object image is also represented as contour line, polyline and convex hull. These different representations are used to extract various features, such as the position of the region, its relative size, the number of holes (lakes) in the object, the number of edges and vertices, the mean length of lines, the perimeter, the vertical and horizontal balance, the convexity factor, the lake factor and the main bay direction.

To bring the various things together data mining methods can be used to automatically create object classifiers. Using the standard approach of splitting the sample into a training and test set, the user has to manually select some regions of the training samples and assign classifiers to them. Decision trees, concept descriptions or neural nets are common choices for classifiers to be learned in the training phase. In the testing phase, the regions will be assigned the class label that has the highest confidence given the algorithms used.

During the training and the testing phase segmentation and feature extraction can be based on any method discussed above or on a combination of these methods. The choice of the method should be done with care and depends also on the domain from which the images originate.

When segmenting according to color for example, the actual frame is compared with the five neighboring frames by calculating the sum of the five difference values

$$\Delta_{RGB}(n)_k = |R_n - R_{n+k}| + |G_n - G_{n+k}| + |B_n - B_{n+k}|,$$

where R_n, G_n, and B_n denote the average R, G, and B value of the n-th frame and k runs from 1 to 5, and the squared differences of consecutive value pairs

$$\begin{aligned} Sqr\_Dif\_Sum(n) = & (\Delta_{RGB}(n)_1 - \Delta_{RGB}(n)_2)^2 \\ & + (\Delta_{RGB}(n)_2 - \Delta_{RGB}(n)_3)^2 \\ & + (\Delta_{RGB}(n)_3 - \Delta_{RGB}(n)_4)^2 \\ & + (\Delta_{RGB}(n)_4 - \Delta_{RGB}(n)_5)^2 \\ & + (\Delta_{RGB}(n)_5 - \Delta_{RGB}(n)_1)^2 . \end{aligned}$$

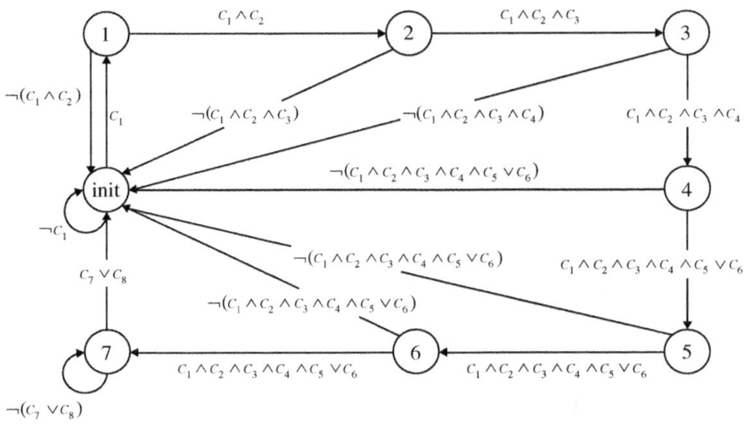

Fig. 1 Schematic representation of the finite state machine for detecting gradual transitions

A hard cut candidate is detected if the difference values for the first and fifth frame comparison exceed pre-specified thresholds, i.e.

$$\Delta_{RGB}(n)_1 \geq th_{RGB} \vee \Delta_{RGB}(n)_5 \geq th_{RGB}.$$

Decreasing the threshold th_{RGB} results in more candidates and hence increases the recall, but at the same time decreases the precision of detection.

Further hard cut candidates and gradual transitions are detected by a finite state machine which is illustrated in Fig. 1.

Block motion analysis is used to either confirm or reject hard cut candidates. Many hard cut candidates are based on the use of flash light. Hence a step of flash light detection is included to filter false alarms of shot boundaries that are only based on the appearance of a flash light.

3 Semantic Analysis

The last decades have been marked by an extensive growth of stored data. Besides text this includes also images and video sequences. To handle large amount of visual data, as in video or image archives, image retrieval technologies have been developed to support the user. To enhance queries and search methods, it is necessary to enhance the visual data with content-based annotations. Since a complete manual annotation is very costly, partial automation of the annotation process is desirable. For this purpose, one uses automatic procedures to identify objects in images, to extract features, and to deduce classifiers.

Research in mass communication requires that the video material is on the one hand summarized and on the other hand enhanced by background information.

Typically, this is done by manual annotation of the video scenes. According to a specified coding scheme, all video scenes will be watched and enriched by specific questions on the domains who, when, where, what and why? Coding schemes might comprise 200 or more features, such as "Is the main actor in the scene a statesmen, head of government, a sports champion, a celebrity?", etc. From a statistical point of view, these coding schemes can be seen as a set of binary variables indicating the presence or absence of a specific feature in a particular scene. Coding of the video material can be done with pre-specified boundaries of the scene or in an open fashion such that the coders can specify the scene boundaries as they see them. One of the first questions for analysis from the semantic point of view is whether certain features occur together in a scene and whether this information can be used to group scenes together. A task that is closely related to association rules.

4 Association Rules

Association Rules are a typical data mining procedure and aim to describe relationships between items that occur together. They have been proposed by Agrawal et al. (1993) in the context of market basket analysis to provide an automated process, which could find connections among items, that were not known before. In market basket analysis the *database* D is a set of *transactions*, every transaction T being a subset of the set of possible items $\mathcal{I} = \{i_1, i_2, \ldots, i_p\}$. An *association rule* is an implication of the form $X \rightarrow Y$, where X and Y are mutually exclusive itemsets (i.e. $X, Y \subset \mathcal{I}$ and $X \cap Y = \emptyset$).

Instead of concentrating on each transaction and the items bought in it, we prefer the statistical approach and concentrate on the items themselves. We identify every item with a binary variable which takes the value "1" if the item occurs in the transaction and "0" if it doesn't. Hence, an association rule $X \rightarrow Y$ can be simply described by a contingency table representing two dummy variables, one for the body (or antecedent) of a rule and one for the head (or consequence). The cross-classification of those dummy variables yields a table as follows with cell entries being the corresponding counts.

	Y	$\neg Y$	
X	$n_{X \wedge Y}$	$n_{X \wedge \neg Y}$	n_X
$\neg X$	$n_{\neg X \wedge Y}$	$n_{\neg X \wedge \neg Y}$	$n_{\neg X}$
	n_Y	$n_{\neg Y}$	n

The standard measures of association rules are support $s = \frac{n_{X \wedge Y}}{n}$ and confidence $c = \frac{n_{X \wedge Y}}{n_X}$ and usually association rules are ordered according to confidence ensuring that a minimum amount of support is maintained.

An association rule $X \rightarrow Y$ holds with *confidence* $c = c(X \rightarrow Y)$, if $c\%$ of transactions in D that contain X also contain Y. The rule $X \rightarrow Y$ has *support* s in the database D, if $s\%$ of transactions in D contain $X \cup Y$. Discovery of association rules is based on the frequent item set approach. Typically some variation of the a priori algorithm (Agrawal et al. 1993) is used with the aim of generating all association rules that pass some user-specified thresholds for support ($minsup$) and confidence ($minconf$). The problem is, that depending on the specified thresholds for confidence and support a vast amount of rules may be generated (Agrawal and Srikant 1994).

Association rule methods can now be used on the annotated material of video sequences and result in specifications of features that are related to the same video scene. Some attempts to use association rules for multi media data go back to Ordonez and Omiecinski (1999) and Ding et al. (2002) focusing on the pixel level of the images. An alternative, perceptual approach requiring a visual thesaurus is presented in Tešiç (2004).

For the actual association rule learning we have taken annotated video material from the SFB 240 in Siegen and we used the Apriori-algorithm as implemented in the software Pissarro (Keller et al. 2002). We generated all rules comprising between four and six items. The minimum support was set to be 1%, the minimum confidence was set to be 20%. Starting with about 200 binary features and a total of 3,152 scenes, we obtained 2,452 rules satisfying the conditions. Trying out various pruning methods as implemented in Pissarro, see also Bruzzese and Davino (2001), helped in reducing the number of rules to a manageable size. The most important rules where built upon the features "Thematic Field: Politics", "Main actor: Member of Government", "Thematic Field: Economy", "Main actor: Labor Unions", "Presentation Form: image", "Thematic Field: Society", "Region: West Germany", "Region: US", and "Region: GDR". An evaluation of the importance of these rules depends on the combination of the syntactic and semantic level which is yet to be performed.

5 Synergy Effects

In the previous sections we have described the different ingredients of video analysis on a syntactic and semantic level. The challenge is now to combine the two levels in order to enhance the two individual analysis stages. A general procedure is not yet in reach. But specific domain knowledge can be used to make significant advances in that direction. For this purpose, we focus on the task of extracting all video sequences for a given set of news reports that show persons with a particular feature, e.g. political leaders, sport champions, head of states. In an explorative manner, first steps to combine the two levels of analysis have been done. The results of the syntactic segmentation can be straightforwardly used in the coding process. Before manual annotation is started, the video sequences are automatically segmented in shots and a representing key frame is created. Now, the coder has all levels available. On one hand the original video sequence but also the shot boundaries and the representing key frame. Hence, a lot of guidance for the semantic annotation is given. On the other hand the coder can manually correct wrongly determined scene boundaries or insufficient key frames. An open question is how these corrections could be fed back to the automatic detection procedure.

For the selected video sequences we have been analyzing, as indicated in Sect. 4 only a few features as described in the coding scheme show up in the important association rules. Hence, some features are (at least for the task we have been looking for) redundant and could be eliminated to reduce the labor intensive work. Moreover, for video sequences that are manually annotated we can extract association rules and use the results to select those features which are commonly occurring together. As next steps, we intend to do a correlation analysis to correlate the frequent features together with the features derived in the syntactic analysis such as color histograms, texture and shape. The challenge is that so far there is no integrated system to detect features. The context information coming from the semantic analysis provides guidance which can be used by a human, but it is still difficult to formalize it for automatic detection.

The main challenge, however, is to create a similarity measure that combines both semantic and syntactic information in order to search video data bases for similar scenes. The next step will be to train classifiers that include both features from the semantic and syntactic level. The choice of semantic features will be determined from the results of the association rule method, but also a decision tree approach will be used. On the syntactic level the feature extraction is done by a decision tree. The work is ongoing in that direction.

References

Agrawal, R., Imielinski, T., & Swami, A. (1993). Mining associations between sets of items in massive databases. In *Proceedings of the ACM-SIGMOD 1993 International Conference on Management of Data* (pp. 207–216).

Agrawal, R., & Srikant, R. (1994). *Fast algorithms for mining association rules* (IBM Research Report RJ9839).

Ballard, D., & Brown, C. (1982). *Computer vision*. Englewoods Cliff: Prentice-Hall.

Bruzzese, D., & Davino, C. (2001). Statistical pruning of discovered association rules. *Computational Statistics, 16*, 387–398.

Ding, Q., Ding, Q., & Perrizo, W. (2002). Association rule mining on remotely sensed images using p-trees. In: M.-S. Cheng, P. S. Yu, & B. Liu (Eds.), *PAKDD Volume 2336 of Lecture notes in computer science* (66–79). Berlin: Springer.

Hermes, T., Miene, A., & Kreyenhop, P. (2000). On textures: A sketch of a texture-based image segmentation approach. In: R. Decker & W. Gaul (Eds.), *Classification and information processing at the turn of the millenium* (pp. 219–226). Berlin: Springer.

Hermes, T., Miene, A., & Herzog, O. (2005). Graphical search for images by PictureFinder. *International Journal of Multimedia Tools and Applications. Special Issue on Multimedia Retrieval Algorithmics, 27*, 229–250.

Jacobs, A., Miene, A, Ioannidis, G., & Herzog, O. (2004). Automatic shot boundary detection combining color, edge, and motion features of adjacent frames. In *TREC Video Retrieval Evaluation Online Proceedings*.

Keller, R., Schlögel, A., Unwin, A., & Wilhelm, A. (2002). *PISSARRO*. Retrieved from http://stats.math.uni-augsburg.de/Pissarro.

Lienhart, R. (1998). Comparison of automatic shot boundary detection algorithms. In: M. M. Yeoung, B.-L. Yeo, & C. A. Bouman (Eds.), *Proc. SPIE, Storage and Retrieval for Image and Video Databases VII* (Vol. 3656, pp. 290–301).

Ordonez, C., & Omiecinski, E. (1999). Discovering association rules based on image content. In *ADL* (pp. 38–49).

Perner, P. (Ed.) (2004). *Advances in data mining, applications in image mining, medicine and biotechnology, management and environmental control, and telecommunications, 4th industrial conference on Data Mining, ICDM 2004, Leipzig, Germany, July 4–7, 2004, Revised Selected Papers*. Berlin: Springer.

Tešić, J. (2004). *Managing large-scale multimedia repositories*. Ph.D. Thesis, University of California, Santa Barbara.

Yam, C.-Y., Nixon, M. S., & Carter, J. N. (2004). Automated person recognition by walking and running via model-based approaches. *Pattern Recognition, 37*, 1057–1072.

Yusoff, Y., Christmas, W.-J., & Kittler, J. (1998). A study on automatic shot change detection. In: D. Hutchison & R. Schäfer, (Eds.), *ECMAST, volume 1425 of Lecture Notes in Computer Science* (pp. 177–189). Berlin: Springer.

Part V
Classification and Classification Tree

Automatic Dictionary- and Rule-Based Systems for Extracting Information from Text

Sergio Bolasco and Pasquale Pavone

Abstract The paper offers a general introduction to the use of meta-information in a text mining perspective. The aim is to build a meta-dictionary as an *available linguistic resource* useful for different applications. The procedure is based on the use of a hybrid system. The suggested algorithm employs, conjointly and in a recursive way, dictionaries and rules, the latter both lexical and textual. An application on a corpus of diaries from the Time Use Survey (TUS) by Istat is illustrated.

1 Introduction

The importance of meta-data for the automatic extraction of information from texts is undoubted and unanimously agreed upon (Basili and Moschitti 2005; Poibeau 2003). Generally, in the field of natural language processing, the meta-data consist of annotations and categorizations of lexical and textual units (Bolasco 2005). In the present work, a procedure based on a hybrid system is proposed in order to construct linguistic resources that can be used – in a perspective of text mining – for the extraction of entities from more than one corpus of textual data. To this purpose, the integration between the levels of lexical analysis and textual analysis is crucial.

In the phase of lexical analysis, the object of study is the lexicon of a corpus of texts. The unit of analysis of the text is the "word" as a *type*. Each word-token is both uniform – since it is a lexia, that is, an elementary unit of meaning which is not decomposable further – and mixed, since it can consist of an inflected form,

*The present research was funded by MIUR 2004 – C26A042095. This paper is the result of joint work by two authors, Bolasco and Pavone. Sections 1, 2 were written by Bolasco and Sect. 3 by Pavone.

S. Bolasco (✉)
Dipartimento di Studi Geoeconomici, Linguistici, Statistici, Storici per l'Analisi Regionale, Sapienza, University of Rome, Via del Castro Laurenziano 9, Roma
e-mail: sergio.bolasco@uniroma1.it

F. Palumbo et al. (eds.), *Data Analysis and Classification*,
Studies in Classification, Data Analysis, and Knowledge Organization,
DOI 10.1007/978-3-642-03739-9_22,

a multiword or a lemma. The annotations are performed on a "vocabulary" table, which results from the numerical coding of the corpus types (parsing).[1]

Lexical annotations of *types* (meta-data) can be of various kinds: (a) linguistic, (b) semantic, (c) quantitative, (d) statistic. These annotations are produced at different steps of the processing: text normalization, grammatical and/or semantic tagging, calculus of the TFIDF index. This makes it possible to extract and select significant parts of the vocabulary in order to describe the lexical characteristics of the corpus: for example, significant elements of each part of speech (a), such as verb, noun, pronoun, adjective, adverb. This selection of elements can then be used for the interpretation of the factorial maps.

Usually, for the purposes of text mining, meaningful parts of the vocabulary are selected: (b) a *peculiar* language (over/under-used with respect to the expected use according to a frequence dictionary of reference), (c) a *relevant* language (extracted through the TFIDF index (Salton 1989) which discriminate between the documents), and (d) a *specific* language (characteristic of some partitions of the corpus).

In the phase of textual analysis, the object of study is the corpus analyzed as a collection of documents (fragments) to be "categorized". The unit of analysis of context is the fragment, which can be anything from a single phrase to the whole document. Textual analysis is characterized by the selection or extraction of significant elements from the investigated corpus. Depending on the specific case, single types, classes/categories of a specific type or even named entities (names, toponyms, companies) are searched for. Relations between types, classes, categories and entities are searched for as well, by establishing searching criteria and rules based on logic operators.

The annotations for the textual analysis are carried out on a "fragments" table that contains both a priori variables (the categories of partition of the corpus) and variables that are the result of the textual analysis. In this case too, the annotations can be of one of four kinds: *linguistic* (individuation of structures or syntagms with variable elements: Balbi et al. 2002), *semantic* (from concepts up to more complex structures such as ontologies: Pazienza 2003), *quantitative* (relevance of fragment established by using TFIDF with respect to a query), or *statistic* (probability of different meanings of the same word).

These annotations are the result of a process of Extraction, Transformation and Loading (ETL) able to search non-structured information within a text and to transform it into structured information in the fragments table. The latter is useful for subsequent work phases, as it remains readily available. The annotation can be done in several ways: information presence (yes/no), number of times of appearance, record of what follows the entity searched in the text. Each information extracted from a text is an entity of interest in itself. The entity search is performed by writing a query as a regular expression, which is a typical operative function of text mining.

[1] Sometimes this is repeated due to further re-coding (re-parsing) of the types caused by lexicalisation, disambiguation, lemmatisation and/or stemming.

Query execution produces a list of the entities found together with their frequency within the corpus and their distribution in each fragment.[2]

2 A Model for Creating a Meta-dictionary by Means of a Hybrid System

The meta-data are obtained via models and it is possible to re-use them again through resources. The differences between a model and a resource are the following.

Within the field of automatic text analysis, a model is a set of "open" instructions which express one or more rules. The model, when applied to corpora different from those it is made for, produces new but also unexpected results. A lexical query such as **nipot**, for example, extracts terms such as *nipote/i/ino*, or *pronipote*, from a corpus concerning the description of daily activities. In a different corpus, say a collection of press articles, the query finds the same terms plus additional ones, such as *nipotastro*, *pronipotino*, *bisnipote*, *arcipronipote*, and includes false positives (presence of noise), e.g., *plenipotenziario*, *inipotizzabile*. A model also gives the opportunity to retrieve *false negatives* (reduction of silence), since it recognizes spelling mistakes compatible with the query (*nipotiini*, *nipotiva*).

A resource is instead a set of "closed" instructions defined in a list (dictionary). Each time it is applied, at most it reproduces itself. Therefore, it does not discover new elements, nor does it introduce false positives (absence of noise). On the other hand, a resource does not allow for the discovery of false negatives (it cannot reduce the silence).

A *hybrid system* is an algorithm for the extraction of information from a text, characterized by the combined and iterated use of dictionaries (DIZ) and rules (REG). A hybrid system produces as a final result a list of entities (meta-dictionary).

A *dictionary* consists of a list of predefined lexias. When these lexias are multiwords, a new entry in the vocabulary of the corpus is produced upon their recognition (*lexicalization*).

A *rule* defines a condition for an entity search in the text. Often, it allows one to identify entities through a correlation between one or more categories and/or types. The application of the same rule to different corpora results in both predictable and unexpected entities: in the latter case, new elements are discovered which are permissible under that rule. However, some entities can be *false positives*, because they are not pertinent with respect to the information being sought. Therefore, they must be eliminated from the list. Examples of lexical rules are queries for the search of lexemes, of infixes and of morphemes in the dictionary of the corpus. Examples of textual rules are queries written by means of regular expressions that combine

[2] This function is available in computer programmes for the analysis of texts, such as, for instance, TaLTaC2 (http://www.taltac.it).

classes of types obtained from the application of dictionaries via boolean operators (AND, OR, AND NOT).

The application of a dictionary and/or of a lexical rule allows for the annotation with a label of both the types of the dictionary and the corresponding tokens in the corpus. The elements that have the same label constitute a *class* and are equivalent to each other, like "synonyms". A *meta-dictionary* is the result of the application of several dictionaries and rules which constitute the model based on a hybrid system. Once controlled and cleaned up to eliminate the false positives, it constitutes the resource to be re-applied to textual *corpora* of the same type.

As is well-known, every *model* is created in three stages. A first phase, of construction, is required for empirically determining the basic components of the structure of the model (*training*); that is, the single entries of a dictionary or the *operanda* of each rule. These are put to test many times on the dictionary and/or the text, until a definitive choice is made. A second phase consists of the formalization of the model by means of the creation of the meta-list and the meta-query (see below). The third phase is the application of the theoretical model: it applies the model to the corpus being studied or to other *corpora* of the same type.

An algorithm organizes *dictionaries* and *rules* (also in a recursive way) into processes that are *explorative* – first lexical (see step (A) below), then textual (B) – and subsequently, after the model formalization step (C), applicative, textual (D) and lexical (E). It is articulated in the following steps:

(A) Predispose classes of types *at the lexical level* by means of uni-label (lists) or multi-label (tables) *dictionaries*, and/or *lexical queries* (uni-label dynamic dictionaries produced from elementary rules on single lexias: prefixes, lexical morphemes, infixes or suffixes).[3] This phase allows one to explore and define the constituent parts of the structure of the model.

(B) Look for relevant entities through *textual queries* by applying regular expressions *f(x)* that localize sequences of words in the corpus. Each *f(x)* combines two or more of the classes realized at step A, producing a list of individuated sequences, both as vocabulary of entities and in terms of positioning of tokens.

(C) Perform the *model formalization as a set of rules*. Once the dictionaries, the lexical queries and the single *f(x)s* have been validated, in order to repeat with a single action the annotations in the vocabulary of the corpus, a *meta-list* and a single textual *meta-query* (subsuming all individual *f(x)s*, so obtaining the model in its total structure) are defined.

(D) Proceed to the *application* of the meta-query in order to make the *model* up-to-date (*final list*). This vocabulary of individuated entities[4] supplies redundant occurrences, because each *f(x)* puts into action an automaton with finite states that scans the text byte by byte and counts all the entities individuated by each single

[3] Such dictionaries and lexical queries feed with an equal amount of labels the CATSEM field in TaLTaC2.

[4] This list of the entities contains a lot of "compatible trash" (analogous in English to things like "me house", "on the table", "in the bed", "in a rooms") and consequently "grasps" the phenomenon fully, beyond spelling and grammar.

query. Therefore, shorter entities, e.g., *<house>*, are included in longer ones, such as *<my house>*, *<my mother's house>*, *<relatives' house>* and so on.

(E) Re-apply this dictionary of entities, depurated of false positives and assumed as *meta-dictionary* (*available resource*), for a semantic tagging aimed to lexicalize the entities found. With such an operation, the occurrences of every entity (as lexias of the corpus vocabulary) are made exact: that is, in the above example, the tokens of *<house>* do not include those of *<my house>*, etc.

3 Application to the Istat TUS Survey

In what follows, an application of the hybrid system is described, which has been carried out on the corpus of 50,000 diaries of the Time Use Survey (henceforth, TUS) performed by *Istat* 2002–2003. In TUS, each diary is written in free text, and describes the activity performed by a person in the course of the day, according to intervals of 10 min (minimum). Contextually, the place and/or means of transport in which the activity takes place are annotated. The corpus amounts to approximately 9 million occurrences (Bolasco et al. 2007). The construction of the model has the objective of characterizing the thousands of locutions used in order to describe the places of the daily activities. These have as their basic linguistic structure a *prepositional syntagm* composed, in Italian, as follows:

$$PREPOSITION+(ADJECTIVE)+SUBSTANTIVE+(ADJECTIVE). \tag{1}$$

The adjectives are placed between parentheses because their presence is optional. For example, as regards the elementary locution "a casa" ("at home"), the model recognizes sequences of the type: "a casa mia" ("in my house"), "nella mia seconda casa" ("in my second house") and similar ones. The prepositional syntagm can be found, even several times, in a single sentence, with adjectival function with respect to the main substantive (e.g., "on the seat — of the car"). In the diaries, contractions such as "vicino casa" (standing for "vicino a casa") can also be found. Table 1 illustrates the typology of space locutions relative to the entity "means of transport".

In the exploratory stage, the basic constituents of the model were defined, preparing dictionaries (see Table 2) composed of: the list of prepositions; a multi-label table of adjectives, distinguishing between possessive and qualificative; lexical queries regarding the substantives.

The construction of these elements was performed according to various criteria: the prepositions were categorized in different ways on the basis of their position in the structure;[5] substantives and adjectives individuated by applying lexical queries based on a reduction to lexical or grammatical morphemes (for example,

[5] The list PREP1 contains the main (simple, articulated and improper) prepositions compatible with the sense of the prepositional structure. The list PREP2, contains instead only the simple and articulated form of the preposition "of".

Table 1 Some examples of the structure of prepositional syntagms

PREP1	ADJ	SUBST	PREP2	ADJ	SUBST
in		auto / macchina / treno / automobile / autobus			
dentro / fuori / presso la / davanti / vicino alla		macchina			
nella	mia	automobile			
sul	nuovo	autobus			
in		macchina		mia / sua / loro	
sull'		auto	di un		vicino
verso la		fermata	dell'		autobus

Table 2 Some elements of dictionaries

PREP1	POSS	ADJ	SUBST	PREP2	POSS	ADJ	SUBST
da un	loro	altro	auto/autovettura	di	mia	cara/o/i	amica/i/o/he
dal/dalla/ dall'	mia	nuovo	autobus	dei	mio	nuovo/a	azienda
in	sua	nuova	autocarro	del	sua		collega
in un/una			automobile	dell'	suo		ditta
nel/nella			autostrada	della/e			figlia/o/e
dentro il/l			macchina	degli			mamma
su un/una/un			moto/motocicletta/ motorino	di un			nipote
sul/sulla/...			tram				nonni/o/e
			metropolitana				
			treno				

in English: auto*, moto*, *ary, *ation). With these queries, unpredictable entities were obtained, by adding both elements compatible with the rule (e.g., from auto*: *autobus*, *autocar*, *automotive*), and false positives (*autobiographic*, *autogestion*, *autonomy*).

The model was completed by means of the repeated application of textual queries written with regular expressions. The aim was to reconstruct specific parts of the structure of the graph. For instance, using sub-lists of prepositions for some substantives, locutions of place are only individuated when supported by those prepositions (e.g., *<to the pub>*, *<to the doctor>*, so inhibiting sequences, such as *<of the pub>* or *<by the doctor>*, that are not locutions of place).

The graph in Fig. 1 formalizes the definitive model expressed in formula (1) above. In the second stage, on the basis of this graph, a single meta-list (Table 3) and a single textual meta-query were reconstructed. In more detail, the query was composed by a regular expression consisting of 39 elements (sequences) in "OR" *(e.g., "PREP1 SUBST" OR "PREP1 ADJ SUBST" OR "PREP1 POSS ADJ SUBST" OR ... OR "PREP1 ADJ SUBST PREP2 ADJ SUBST" ...).*

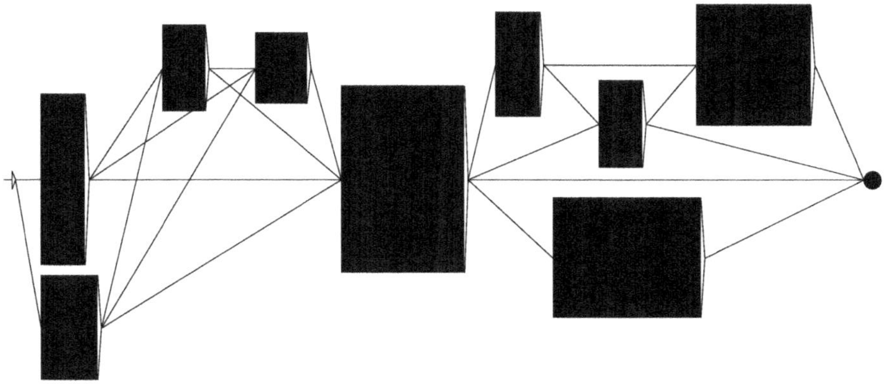

Fig. 1 The formalization of the model

Table 3 Sample of the meta-list

Type	Label	Type	Label
a	prep1	casa	sost
alla	prep1	auto	sost
da	prep1	...	
dal	prep1	di	prep2
...		del	prep2
mia	poss	...	
sua	poss	futura	agg
...		nuova	agg

Only then has one moved to the third phase, that is, to the application of the meta-query. This application individuated 6,388 entities, for a total of 1,731,630 "gross" (redundant[6]) occurrences. The entities were cleaned up to get rid of the false positives, obtaining as a final result 5,404 locutions of place. These will constitute a reference point (meta-dictionary) for any future survey. By applying this resource for a semantic tagging, the 5,404 entities of the TUS corpus were lexicalized (Table 4).

In general, the results of the above-mentioned queries, as pointed out in Sect. 1, produce new variables that are inserted in the matrix of the fragments (individual diaries). These variables constitute a representation of the "concepts" or relations among concepts that are to be correlated to the a priori information (e.g., the structural variables of the individuals).

In this case, it is possible to emphasise the correlations between the locutions and individual characteristics via factorial analysis. The latter allows one to reconstruct

[6] See step D in the Sect. 2.

Table 4 Some examples of locutions of place

Locution	Occurrences	Locution	Occurrences	Locution	Occurrences
da casa mia	377,866	sul divano	7,344	nella mia cameretta	90
a piedi	72,428	in ufficio	5,481	su una panchina	88
in macchina	43,712	in spiaggia	3,347	nel cortile della scuola	64
a letto	38,113	in giro	2,161	ad una festa di compleanno	48
in cucina	18,766	nell' orto	2,145	in mezzo alla natura	35
al bar	15,169	presso la propria abitazione	320	vicino al caminetto	32
a scuola	14,880	alla fermata dell'autobus	290	sulla sedia a rotelle	24
in bagno	14,684	dal giornalaio	233	fuori dal mio paese	15
al lavoro	11,244	davanti alla tv	202	verso il centro commmerciale	11
per strada	10,094	sotto l'ombrellone	186	tra i negozi dell'ipermercato	2

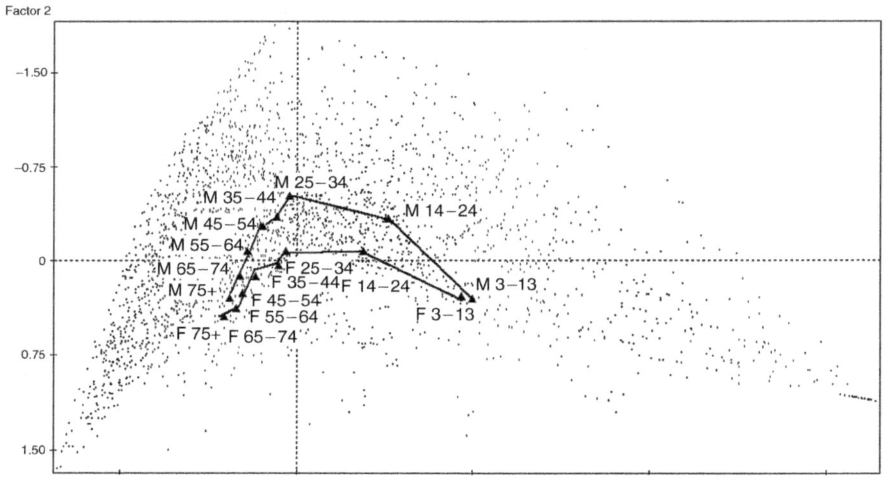

Fig. 2 Factorial analysis of the locutions of place by age × sex groups – TUS 2002–2003

in detail the relationship between the various kinds of locutions and the individuals, by partitioning the corpus of the diaries according to age × sex.

From the overall analysis of all the locutions (a matrix $n \times p$, where $n = 5,404$ locutions and $p = 16$ classes age × sex), such strong relationships emerge that the resulting map – shown in Fig. 2, where each point individuates a locution and the barycentres of the age × sex classes are connected by a line – can be described according to the slogan "Each age has its places".

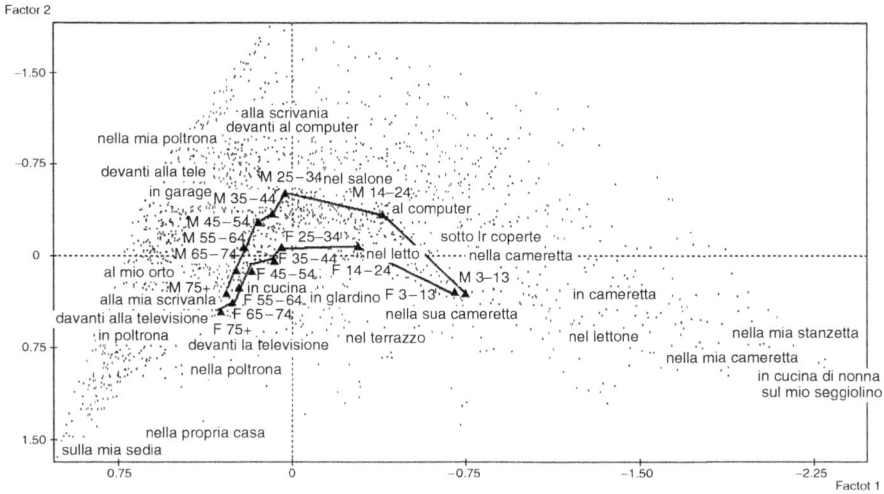

Fig. 3 Factorial analysis of the locutions of place with reference to "places inside one's house" according to age × sex groups – TUS 2002–2003

As can be observed in the factorial plane, in the young age there is a marked variability of places; the latter increases as age increases (the maximum is reached around the age of 20–25, in proximity to the origin of the factors), and then decreases as the old age approaches.

In more detail, let us consider the thematic list with reference to "places inside one's house" (Fig. 3). The maximum variety of places in a day exists for the age groups "in-between" (*alla scrivania*, *davanti al computer*, *nel salone*, *in giardino*, ..., *nel terrazzo*), while as the years go by mobility (which begins in the early years: *sul mio seggiolino*, *nella mia cameretta*) gets more and more limited (*davanti alla televisione*, *in poltrona*, *davanti al camino*) and eventually disappears (*sulla mia sedia*).

It is interesting to note how the differences between the sexes gradually increase around the intermediate age (see Fig. 3, M: *in garage*; F: *in cucina*), then tend to disappear in the older age (*nella propria casa*). Furthermore, the barycentre of each sub-class of women is slightly more to the left, that is, towards the older age. This is consistent with the greater life expectancy of women.

The 5,000 expressions of place, although so many, can not represent all places. If we consider the place "sea" (*mare*), for example, the TUS corpus provides the Italian correlates of expressions such as <at sea>, <over the sea>, <close to the sea>, <on the seashore>, <in a village on the sea>, etc., but it does not contain an expected expression such as <at the bottom of the sea>. On the other hand, the selected items do constitute an exhaustive list concerning where everyday activities take place. Therefore, the meta-dictionary produced by the application considered in this paper is indeed a re-usable resource, *in primis* for the next Time Use Survey which Istat planned for 2007–2008.

References

Balbi, S., Bolasco, S., & Verde, R. (2002). Text mining on elementary forms in complex lexical structures. In A. Morin & P. Sébillot (Eds.), *JADT 2002* (pp. 89–100), St. Malo, March 13–15, IRISA-INRIA, Rennes.

Basili, R., & Moschitti, A. (2005). *Automatic text categorization. From information retrieval to support vector learning*. Rome: Aracne.

Bolasco, S. (2005). Statistica testuale e text mining: Alcuni paradigmi applicativi. *Quaderni di Statistica, 7*, 17–53.

Bolasco, S., D'Avino, E., & Pavone, P. (2007). Analisi dei diari giornalieri con strumenti di statistica testuale e text mining. In M. C. Romano (Ed.), *I tempi della vita quotidiana. Un approccio multidisciplinare all'analisi dell'uso del tempo* (pp. 309–340). Rome: ISTAT.

Pazienza, M. T. (Ed.) (2003). *Information extraction in the Web era. Natural language communication for knowledge acquisition and intelligent information agents, Lecture Notes in Computer Science* (Vol. 2700). Berlin: Springer.

Poibeau, T. (2003). *Extraction automatique d'information*. Paris: Hermes Lavoisier.

Salton, G. (1989). *Automatic text processing: The transformation, analysis and retrieval of information by computer*. Reading, MA: Addison-Wesley.

Several Computational Studies About Variable Selection for Probabilistic Bayesian Classifiers

Adriana Brogini and Debora Slanzi

Abstract The Bayesian network can be considered as a probabilistic classifier with the ability of giving a clear insight into the structural relationships in the domain under investigation. In this paper we use some methodologies of feature subset selection in order to determine the relevant variables which are then used for constructing the Bayesian network. To test how the selected methods of feature selection affect the classification, we consider several Bayesian classifiers: Naïve Bayes, Tree Augmented Naïve Bayes and the general Bayesian network, which is used as benchmark for the comparison.

1 Introduction

Classification is one of the basic tasks in data analysis that requires the identification of information from a database (set of cases or instances). Numerous approaches to this problem have been proposed which often ignore the relationships among the variables. The performance of the classification can be increased by taking into account the dependencies between the variables. We will evaluate the power and usefulness of the Bayesian network as probabilistic classifier with the ability of giving a clear insight into the structural relationships in the domain under investigation. Most of the learning algorithms identify the complete model but especially those constructed from large databases tend to have a high number of variables and in/dependence relationships resulting in increased structural complexity; therefore part of the structure may not be relevant for classification (Kohavi and George 1997). In this study we address the problem of efficiently identifying a small subset of variables (also called features) from a large number, upon which to focus the attention in building the classification model. Our aim is to adopt feature selection techniques, since they select a subset of variables, preserving their original

A. Brogini (✉)
Department of Statistics, University of Padova, via Cesare Battisti 241, 35121, Padova, Italy
e-mail: brogini@stat.unipd.it

F. Palumbo et al. (eds.), *Data Analysis and Classification*, Studies in Classification, Data Analysis, and Knowledge Organization, DOI 10.1007/978-3-642-03739-9_23,

semantics and offering the advantage of interpretability by a domain expert. In the context of classification, feature selection techniques can be organized into three categories: filter, wrapper and embedded methods; for a review see Saeys et al. (2007). In this paper we report the result of some computation studies evaluating how accurately the methods of feature selection find relevant subsets of variables for learning the Bayesian network which will be compared with the performance of other Bayesian classifiers. In the literature, several simplified Bayesian structures for classification have been proposed; these include Naïve Bayes (Langley 1992), Tree Augmented Naïve Bayes (Friedman et al. 1997) and BN Augmented Naïve Bayes (Cheng and Greiner 1999). We concentrate on problems involving complete databases, i.e. without missing cases, for a set of discrete variables. The remainder of the paper reviews basic concepts of Bayesian network (Sect. 2) and feature subset selection (Sect. 3), we describe the methods of variable selection which are involved in the learning process of the Bayesian network used as a probabilistic classifier, then we present the experimental results over a set of learning problems and the conclusions (Sect. 4).

2 Bayesian Networks and Classification

A Bayesian network, BN for short, is a graphical probabilistic model consisting of:

- A finite set of random variables $\mathbf{X} = \{X_1, \ldots, X_n\}$.
- A direct acyclic graph G consisting of nodes and edges. All nodes of G correspond one to one to members of $\mathbf{X}$, whereas the edges indicate direct dependencies between the variables.
- A joint probability distribution P over the variable set $\mathbf{X}$.

The graph G encodes in/dependence relationships of the domain, that can be read from G by means of the d-separation criterion (Lauritzen 1996). G and P are connected by the Markov condition (Pearl 1988): each variable is probabilistically independent of its non descendants given its parents in G. P is expressed through the concise factorization $P(\mathbf{X}) = \prod_i P(X_i | Pa_{X_i})$, where Pa_{X_i} denotes the set of direct parents of X_i in G (the nodes pointing to X_i in the graph). A BN is faithful to a joint probability distribution P over the random variable set $\mathbf{X}$ if all dependencies entailed by G and the Markov condition are also present in P (Spirtes et al. 2000). For any variable $X_i \in \mathbf{X}$, the Markov Blanket $MB(X_i) \subseteq \mathbf{X}$ is the set of variables such that for any $X_j \in \mathbf{X} \setminus \{MB(X_i) \cup \{X_i\}\}$, X_i is independent from X_j given $MB(X_i)$. In any faithful BN on variables $\mathbf{X}$, $MB(X_i)$ is the set of parents, children and parents of children of X_i which d-separates X_i from any other variables in $\mathbf{X}$ (Neapolitan 1990); for every X_i, $MB(X_i)$ is unique (Tsamardinos and Aliferis 2003). In this paper we focus on discrete BNs, typically used in applications of machine learning and data modelling, and on faithful probability distributions which are a very large class as proven in Meek (1997).

2.1 Learning Bayesian Networks

Learning a BN from data is a process which is divided in two phases: finding the structure G of the network and estimating the conditional probability distribution on $P(\mathbf{X})$ (the parameters of the BN), given the structure G (see Heckerman 1999 for a tutorial on the subject). Methods for automatic induction of BN models fall into two main different categories. The first considers network construction as a constraint satisfaction problem by computing conditional independence statistics (Spirtes et al. 1993). The second considers network construction as an optimization problem searching among candidate network structures for the optimum (Cooper and Herskovits 1992; Heckerman et al. 1995). In this paper we consider the latter approach, introducing a scoring function, also called metric, that evaluates each network with respect to the data, and searching for the optimal network according to the score. One method for deriving a score is based on Bayesian consideration; the K2 and the BDe metrics are the most common choices (see Cooper and Herskovits 1992; Heckerman et al. 1995 for a complete description). In learning a BN, no distinction is made between the classification node and other nodes, even if a BN can be used for classification (Friedman et al. 1997; Cheng and Greiner 1999). In Madden (2003) it is proven that BNs constructed by the Bayesian approach perform well in classification on benchmark databases, so we adopt this procedure.

2.2 Classifiers Based on Bayesian Networks

Probabilistic classification is the process of mapping an assignment of values of the variables set $\mathbf{X}$ into a probability distribution for a distinguished variable T (target or class). As usual, we assume that all data instances are drawn (i.i.d.) from a certain probabilistic distribution. Although a BN may be used for classification task, the classification node, i.e. the target, is not explicitly identified and the structure may have an increased complexity when databases with many variables are considered. Several simplified Bayesian structures, intended specifically for classification tasks, have been proposed: Naïve Bayes (NB for short) (Langley 1992), Tree Augmented Naïve Bayes (TAN) (Friedman et al. 1997), and BN Augmented Naïve Bayes (BAN) (Cheng and Greiner 1999). In all of these structures it is assumed that the classification variable is the root node and it cannot depend on other variables.

3 Feature Subset Selection for Classification

Feature Selection, FS for short, aims to identify the minimal subset of $\mathbf{X}$ that is relevant for probabilistic classification. FS is important for various reasons: improving prediction performance, running time requirements and interpretational issues imposed by the problem itself. FS are divided into three categories, depending on how the feature selection search is combined in machine learning with the construction of the classification model: filter, wrapper and embedded methods.

Filter methods are performed as a pre-processing step to learning and they assess the relevance of features by the properties of the data and by a relevance score, removing low scoring features. These methods are computationally simple, fast and independent from the classification algorithm. We focus on the Information Gain criteria, IG for short, as univariate method where each feature is considered separately, and on Markov Blanket discovery, MB for short, as multivariate approach which considers features dependencies. Several algorithms have been developed or proposed for identifying the Markov Blanket (Margaritis and Thrun 1999; Frey et al. 2003; Tsamardinos et al. 2003); we focus on HITON algorithm (Aliferis et al. 2003) as it is developed to improve the performance of other Markov Blanket discovery algorithms in literature.

Wrapper methods employ a search through the space of all possible feature subsets using the estimated accuracy from a classification algorithm. These methods are computationally intensive and dependent on the classification algorithm, but they take into account feature dependencies. We focus on the use of a simple Genetic algorithm, GA for short, as search procedure: it evolves good feature subsets by using random perturbations of a current list of candidate subsets (Goldberg 1989).

In *Embedded* methods, the optimal subset of features is built into the classifier construction. The interaction with the classification model improves the computational complexity and takes into account feature dependencies. We focus on Decision Tree, DT for short, that is a mapping of instances about an item to conclusion. We use C4.5 algorithm (Quinlan 1993) because it has been shown that it provides a good classification accuracy and it is the fastest among the compared main-memory algorithms for machine learning and data mining.

4 Experimental Results and Conclusions

In this section we report the methodology used for comparing the classifiers and the experimental results.

IG evaluates variables by measuring their gain ratio with respect to the data. We choose $\alpha = 0.05$ as threshold to select the variable subset. A way to enlarge the number of variables could be done by decreasing the threshold, for instances to 0.01, but with the drawback of missing the variable space simplification.

We apply HITON with a G^2 statistical independence test, also called Maximum Likelihood Statistic, with a significance level set to 0.05. Preliminary experimental runs show that there are small discrepancies in the results with different value of α.

We use GA as search method through the space of subsets of variables. We set to 0.6 the probability of crossover and to 0.033 the probability of mutation. We fix to 20 the population size, i.e. the number of individuals, or variable sets, in the population, and this is evaluated for 20 generations. The parameter setting is comparable to the typical values mentioned in the literature (Mitchell 1996).

Both pruned and unpruned DT are constructed using five-fold cross-validation to evaluate the performance, by fixing to 10 the minimum number of instances per leaf.

To test how the selected methods of feature selection affected the classification, we consider the NB, the TAN and the general BN which is used as benchmark for the comparison and it is learned as described in Sect. 2.1. These are learned both to the original databases and to the databases filtered through the feature selection algorithms. We select seven databases from the UCI repository of machine learning database (http://www.ics.uci.edu/~mlearn/MLRepository.html) and from the Department of Statistics, University of Munich (http://www.stat.uni-muenchen.de/service/datenar-chiv/welcome_e.html); the databases differ for the number of variables and the number of cases. Usually in the literature, to determine the performance of a classifier each database is randomly divided into two-thirds for training and one-third for testing and the accuracy of each classifier on the testing data is measured. This is repeated 10 times, except for particularly small databases for which the number of repetitions increases to 50 in order to reduce the variability. Further experimental runs show that there are small discrepancies in the results when using the 10-fold cross validation approach.

Table 1 shows the main characteristics of each database used in the analysis. The databases for which the number of split repetitions is 50 are marked with symbol $*$. Table 2 shows the number of the variables used in the learning phase, the classification accuracy, in terms of percentage of correctly classified instances, and the standard deviation of the three compared Bayesian classifiers. Following usual conventions, for each database and for each subset of selected features the classifier with best accuracy (if one) is highlighted in boldface. Where two classifiers have statistically indistinguishable performance, based on a corrected resample t-test (Nadeau and Bengio 2000), and they outperform the other algorithm, they are both highlighted in bold. All the experiments are performed with WEKA software (WEKA 2004).

From the experimental results, we can see that, most of the time, when a classifier performs well with the original database, it also performs well with the variable subset databases. This confirms the results in Madden (2003). When the number of variables in the original database is relatively small, using the NB leads to decreasing the percentage of corrected classified instances. Furthermore, when the

Table 1 Databases characteristics

Database name	No. of variables (without T)	No. of instances
Auto	45	793
KRvsKP	36	3,196
Spect$*$	22	80
Credit	20	1,000
Lympho$*$	18	148
TicTacToe	9	958
Nursery	8	12,960

Table 2 Performance of the Bayesian classifiers with respect to the number of variables, percentage of correctly classified instances and standard deviation

Auto					
	Original 45	IG 4	Hiton 10	GA 18	DT 18
BN	36.34 ± 1.81	36.75 ± 0.10	36.97 ± 0.93	36.68 ± 0.97	37.01 ± 0.73
TAN	36.72 ± 1.36	33.02 ± 0.10	36.01 ± 2.47	35.16 ± 2.87	34.95 ± 2.62
NB	35.24 ± 3.15	34.79 ± 3.31	38.08 ± 2.29	34.53 ± 2.00	36.09 ± 2.48

KRvsKP					
	Original 36	IG 3	Hiton 20	GA 27	DT 22
BN	**97.01 ± 0.82**	90.34 ± 0.55	**97.01 ± 0.47**	**94.23 ± 0.72**	**96.78 ± 0.57**
TAN	92.84 ± 0.82	90.34 ± 0.55	94.36 ± 0.82	**94.16 ± 0.62**	87.37 ± 1.23
NB	87.95 ± 0.85	90.34 ± 0.55	92.61 ± 0.65	63.21 ± 1.59	92.05 ± 1.56

Spect					
	Original 22	IG 11	Hiton 3	GA 10	DT 22
BN	66.46 ± 10.62	70.08 ± 8.08	74.32 ± 8.35	67.84 ± 6.82	75.00 ± 7.87
TAN	70.18 ± 7.25	73.39 ± 5.89	76.22 ± 7.39	71.58 ± 7.71	77.58 ± 7.57
NB	72.45 ± 6.94	78.84 ± 6.92	78.73 ± 7.36	**78.16 ± 6.35**	78.38 ± 7.80

Credit					
	Original 20	IG 1	Hiton 13	GA 10	DT 11
BN	71.97 ± 1.87	69.32 ± 2.08	73.44 ± 1.54	69.06 ± 3.37	71.51 ± 1.51
TAN	74.76 ± 2.67	–	75.18 ± 2.00	71.50 ± 2.10	**75.06 ± 1.95**
NB	76.38 ± 2.38	68.32 ± 2.08	76.56 ± 2.48	73.06 ± 2.20	**74.62 ± 1.74**

Lympho					
	Original 18	IG 15	Hiton 14	GA 7	DT 10
BN	81.21 ± 4.58	80.33 ± 5.01	80.68 ± 4.89	74.94 ± 5.29	82.55 ± 5.15
TAN	84.87 ± 4.00	83.28 ± 3.84	82.96 ± 3.86	77.15 ± 4.40	84.90 ± 4.59
NB	84.27 ± 3.84	85.38 ± 3.68	84.26 ± 3.74	78.70 ± 4.78	84.63 ± 3.87

(continued)

Table 2 (continued)

TicTacToc					
	Original 9	IG 1	Hiton 5	GA 8	DT 9
BN	**76.40 ± 4.84**	70.76 ± 2.68	78.84 ± 4.25	70.50 ± 4.86	76.40 ± 4.84
TAN	**75.42 ± 1.97**	–	72.38 ± 2.74	75.12 ± 2.18	75.42 ± 1.97
NB	69.74 ± 2.59	70.76 ± 2.68	72.88 ± 2.61	70.08 ± 2.82	69.74 ± 2.59

Nursery					
	Original 8	IG 3	Hiton 6	GA 8	DT 8
BN	**93.26 ± 0.73**	**89.34 ± 0.51**	**91.58 ± 0.91**	**93.26 ± 0.73**	**93.26 ± 0.73**
TAN	**94.03 ± 0.59**	**89.34 ± 0.51**	**91.84 ± 0.72**	**94.03 ± 0.59**	**94.03 ± 0.59**
NB	90.50 ± 0.40	87.82 ± 0.59	89.64 ± 0.43	90.50 ± 0.40	90.50 ± 0.40

Table 3 Auto database: number of selected variables by using the GA approach with different parameter setting. The numbers refer to population sizes of 50 and 20 respectively

	Prob. of Crossover				
Prob. of mutation	1.0	0.7	0.5	0.25	0.0
0.05	19,16	19,18	17,17	18,17	17,18
0.01	19,17	17,18	18,19	18,19	19,20
0.001	19,17	17,19	19,19	19,19	21,22

number of instances of the databases is small, the results are affected by a degree of variability which makes it difficult to statistically compare the classifiers. Using filter univariate FS methods, as IG, leads to select a small number of relevant variables, and often the performance of the classifier decreases. When the HITON algorithm is used to select the relevant variable for the target, the average performance of each classifier increases with respect to complete database, especially when the original variable set is large. The genetic algorithm seems to lead to no particularly significant results. This could be due to the choices of the parameters used in the search. More researches have been done in order to investigate the choice of optimal parameters. Table 3 shows the number of selected variables by using the GA approach with different parameter setting. For lack of space, we report the results only for one database. We set the population size to 50 and to 20, the probability of crossover to 1.0, 0.7, 0.5, 0.25, 0.0, and the probability of mutation to 0.05, 0.01, 0.001. Considering this wide range of values, there are no significant differences with respect to the results obtained by fixing the typical values mentioned in the literature. With respect to the previous methods, DT selects a higher number of variables which, used for the construction of classifiers, provides a classification accuracy similar to the complete databases.

Finally, the objective of this paper is to compare some computational results which combine the use of feature subset selection methods with the Bayesian classifiers based on BN structure. Whereas we determine the choice of the feature selection algorithms by motivations of good performance in literature, we select the classifiers which are generally used in the literature of the Bayesian probabilistic classifiers and which are simple to construct with the current available software.

References

Aliferis, C. F., Tsamardinos, I., & Statnikov, A. (2003). HITON: A novel Markov blanket algorithm for optimal variable selection. In *Proceedings of the 2003 American Medical Informatics Association (AMIA) Annual Symposium* (pp. 21–25).

Cheng, J., & Greiner, R. (1999). Comparing Bayesian network classifiers. In *Proceedings UAI-99*.

Cooper, G. F., & Herskovits, E. (1992). A Bayesian method for the induction of probabilistic networks from data. *Machine Learning, 9*(4), 309–348.

Frey, L., Fisher, D., Tsamardinos, I., Aliferis, C. F., & Statnikov, A. (2003). Identifying Markov blankets with decision tree induction. In *Proceedings of third IEEE International Conference on Data Mining (ICDM)* (pp. 59–66).

Friedman, N., Geiger, D., & Goldszmidt, M. (1997). Bayesian network classifiers. *Machine Learning, 29*, 131–161.

Goldberg, D. E. (1989). *Genetic algorithms in search, optimization and machine learning*. Reading, MA: Addison-Wesley.

Heckerman, D. (1999). A tutorial on learning Bayesian networks. In *Learning graphical models*. Cambridge, MA: MIT Press.

Heckerman, D., Geiger, D., & Chickering, D. M. (1995). Learning Bayesian networks: The combinations of knowledge and statistical data. *Machine Learning, 20*, 197–243.

Kohavi, R., & George, H. J. (1997). Wrappers for feature subset selection. *Artificial Intelligence, 1*(2), 273–324.

Langley, P., Iba, W., & Thompson, K. (1992). An analysis of Bayesian classifiers. In *Proceedings of AAAI-92* (pp. 223–228).

Lauritzen, S. L. (1996). *Graphical models*. Oxford: Clarendon Press.

Madden, M. G. (2003). The performance of Bayesian network classifiers constructed using different techniques. In *Working notes of the ECML PkDD-03 workshop* (pp. 59–70).

Margaritis, D., & Thrun, S. (1999). Bayesian network induction via local neighborhoods. In *Proceedings of conference on Neural Information Processing Systems (NIPS-12)*, MIT Press.

Meek, C. (1997). *Graphical models: Selecting causal and statistical models*. Ph.D. Thesis, Carnegie Mellon University.

Mitchell, M. (1996). *An introduction to genetic algorithms*. Cambridge, MA: MIT Press.

Nadeau, C., & Bengio, Y. (2000). Inference for the generalization error. *Advances in Neural Information Processing Systems, 12*, 293–281.

Neapolitan, R. E. (1990). *Probabilistic reasoning in expert systems: Theory and algorithms*. New York: Wiley.

Pearl, J. (1988). *Probabilistic reasoning in intelligence systems*. Los Altos, CA: Morgan Kaufmann.

Quinlan, J. R. (1993). *C4.5: Programs for machine learning*. Los Altos, CA: Morgan Kaufmann.

Saeys, Y., Inza I., & Larrañaga, P. (2007). A review of feature selection techniques in bioinformatics. *Bioinformatics, 23*(19), 2507–2517.

Spirtes, S., Glymour, C., & Scheines, R. (1993). *Causation, prediction and search*. Berlin: Springer.

Spirtes, P., Glymour, C., & Scheines, R. (2000). *Causation, prediction, and search*. New York: MIT Press.

Tsamardinos, I., & Aliferis, C. F. (2003). Towards principled feature selection: Relevancy, filters and wrappers. In *Proceedings of the ninth international workshop on Artificial Intelligence and Statistics*.

Tsamardinos, I., Aliferis, C., & Statnikov, A. (2003). Algorithms for large scale Markov blanket discovery. In *Proceeding of the sixteenth international FLAIRS conference*.

WEKA. (2004). *On-line documentation*. Waikato University, New Zeland. Retrieved from http// www.cs.waikato.ac.nz/ml/weka/.

Semantic Classification and Co-occurrences: A Method for the Rules Production for the Information Extraction from Textual Data

Alessio Canzonetti

Abstract Information extraction is a field of computer science research which explores the problem of detecting and retrieving desired information from textual data. This paper proposes a two-steps method that enables the detection of relevant information within a corpus of textual data. The first phase consists of observing the most recurrent structures through the study of textual co-occurrences and collocations, while the following phase consists of deriving rules from these structures which make it possible to create an inventory of all the expressions that identify a particular concept of interest, that is, the desired information.

1 Introduction

The field of Information extraction explores the problem of detecting and retrieving desired information from textual data. By "desired information" we mean any entity of interest appearing within a text. Such an entity generally constitutes a particular concept of interest, and this may simply consist of a single word, but, more often, the interest focuses on observing the presence of a particular sequence of words or, in more general terms, of particular structures.

The purposes of information extraction may be multiple, but the two principal aims are, on the one hand the inventory of the forms in which a concept of interest can be expressed, and on the other, document classification through the attribution, to the documents themselves, of metadata extracted from the text (Poibeau 2005). In this work a method will be introduced that enables the detection of pertinent information inside a corpus of textual data, composed of a set of economic-financial news. This will occur through two stages: the first stage recognizes the most recurrent structures, in absolute and relative terms, while the second stage turns these structures into rules thus obtaining a list of all the expressions that identify a particular concept of interest, that is, the desired information.

A. Canzonetti (✉)
Dipartimento Studi Geoeconomici, Linguistici, Statistici, Storici per l'Analisi regionale - Facolta' di Economia - Sapienza Universita' di Roma, Via del Castro Laurenziano 9, Roma
e-mail: alessio.canzonetti@uniroma1.it

F. Palumbo et al. (eds.), *Data Analysis and Classification*, Studies in Classification, Data Analysis, and Knowledge Organization, DOI 10.1007/978-3-642-03739-9_24,

2 The Analyzed Corpus and Semantic Classification

The corpus is composed of 1,332 economic-financial news articles published between the end of 1996 and 2000. This kind of text is characterized by linear language, in which the "fact" is the principal happening ("Fiat si accorda con General Motors"), around which may appear other connected situations ("il titolo Fiat compie un balzo di + 3.3%").

In order to extract these facts and situations, that constitute the desired information, it is necessary to build the syntactic structures in which they tend to appear, in other words the "sense molecules" containing them. In the case of the former above-mentioned example, the structure could be synthesized as such: <enterprise> <verb> <enterprise>. Therefore, first of all it is necessary to identify the "atoms" which compose the unities of sense meaning. This was carried out through a vocabulary analysis and compared with a reference dictionary,[1] with the aim of identifying the peculiar language of the text, that is the subset of the vocabulary over-represented in comparison with an average reference.

This comparison highlighted the presence of a large variety of proper names and enterprises. Besides these sets of words which identify the information subjects, several verbs were also detected. that we can define terminological (<to foresee>, <to earn>, <to lose>, <to rise>, etc.) representing the principal existing relationships among the aforesaid subjects, Furthermore, a strong presence of numbers, places and temporal references were recorded.

Each of these sets of forms represent, in our model, a particular semantic class, so we have the classes, *enterprise*, *names*, *verbterm*, *places*, *time*, *num* and others. These classes are the "sense atoms" and the associations between them detect the desired information.

Resorting to a semantic classification of a vocabulary subset is a strategic choice for the purpose of our aim: considering sets of words rather than single forms as pivots of the desired information, enables the identification of highly generic and hence more exhaustive rules.

3 Rules Production Using Co-occurrences and Collocations

The detection of typical structures of desired information was carried out through the analysis of the co-occurrences and the collocations of the words belonging to the semantic classes planned in the preceding phase.

A co-occurrence happens when two words appear in the text within the same context. According to the situation, the definition of this context can change. In our

[1] The reference dictionary used is REP90 (Bolasco and Canzonetti 2005), was built of articles over a 10 year period from the Italian daily newspaper "La Repubblica". This dictionary is a statistic-linguistic resource of the software used for the analysis TaLTaC2 (http//www.taltac.it).

Table 1 Co-occurrences matrix (extract)

	Num	Verb	Verbterm	Enterprise	Place	Stock exchange
Num	16,130	6,199	6,028	2,003	1,388	1,065
Verb	6,199	4,062	5,333	2,512	1,460	943
Verbterm	6,028	5,333	3,408	2,382	1,306	1,007
Enterprise	2,003	2,512	2,382	2,068	959	236
Place	1,388	1,460	1,306	959	1,192	403
Stock exchange	1,065	943	1,007	236	403	358

case, a context is defined as a boundary having a maximum width of n words, and it cannot extend over the limits defined by the punctuation inside the text. For our analysis we used $n = 20$.

The main result of the co-occurrence calculation is a square matrix of the type *semantic class x semantic class*, an extract of which is shown in Table 1.

This table shows the co-occurrences between the respective row classes and column classes.

The results provide evidence of a strong relationship between the semantic classes. However, some reflections need to be pointed out.

Firstly, the co-occurrences counted between two classes depend on the dimensions, in terms of occurrences, of the classes themselves: the class *num* appears to be associated with the class *enterprise* double the amount compared to the class *stock exchange*. However, we have to consider that *enterprise* has, 7,183 occurrences against *stock exchange* with 2,344 occurrences.

Secondly, some relationships are of a functional nature, so to speak. The two classes *verbterm* and *verb* (terminological verbs and other frequent verbs) have high values due to the fact that verbs are fundamental for building sentences, and their correct use is dependent on close "proximity" to the subjects and objects they set in relationship.

Finally, the co-occurrence matrix doesn't show any information about the relative position of the forms in the text, when any single co-occurrence is verified.

Thus, the criteria for selecting the most interesting relationships has to be set taking into account the reflections above. According to the first two of the three points, it must be concluded that the amount in absolute terms of co-occurrences is not always a sufficient criterion for selection. Relationships based only on syntactic reasons have to be avoided. At worst, however, it is better to choose classes belonging to the lexical domain of the text in analysis (in our case *verbterm* rather than *verb*), since most likely they will concern interesting and not trivial relationships.

Even in cases where two classes meet the requirement above, the value of co-occurrences in absolute terms may still not be a good indicator of an interesting relationship. From the first reflection, it follows that we must also take into account the occurrences of the two classes under consideration. A value of a few tens of co-occurrences could highlight a close association if occurrences in the text of the two classes were also in the same order of magnitude.

Therefore, to properly evaluate these cases, the concept of relatives co-occurrences is introduced. This index makes it possible to obtain for each class, a different ranking of the main associations in which the class is involved. The vector of the co-occurrences of a class, which takes in this context the role of pivot, must relate to the value of occurrences of the co-occurring class. In this way, the value obtained records the degree of exclusivity of the relationship: a value close to 1 indicates that the co-occurring class is almost always near the class pivot, even if the absolute value of the co-occurrences is very low. Conversely, a high absolute value of co-occurrences will be reduced if the class co-occurring has a high number of occurrences.

Therefore, we must distinguish two types of relationships: high frequency relationships and high exclusivity. A selection of high frequency main relationships is shown in Table 2.

After having appraised the most interesting associations, bearing in mind the previous reflections, we can analyze the existing collocations between a particular class and all the others.

The collocations make it possible to obtain the positional distribution of the co-occurrences of a particular semantic class, in other words the number of times that a co-occurring class appears in preceding or successive n positions with regards to a pivot class.

As can be noted, the classes *num* and *verb* strongly associate with class *enterprise*, and these associations tend to fall when the distance from the pivot rises (that is, going towards positions −8 and 8 of Table 3[2]). Furthermore, an auto-association

Table 2 High frequency co-occurrences (extract)

Class1	Class2	Co-occurrences
Impresa (enterprise)	Impresa (enterprise)	2,052
Impresa (enterprise)	Num	2,011
Borsa (stock exchange)	Num	1,055
Impresa (enterprise)	Luogo (place)	953
Indice (index)	Num	438
Borsa (stock exchange)	Luogo (place)	400
Crescita (grow)	Num	393
Fatturato (sales)	Num	388
Gruppo (group)	Impresa (enterprise)	371
Perdita (loss)	Num	326
Capitale (capital)	Num	324
Azione (share)	Num	293
Borsa (stock exchange)	Indice (index)	242
Borsa (stock exchange)	Impresa (enterprise)	233
Impresa (enterprise)	New	217

[2] Table 3 does not show the collocations over a distance +8/−8. However, the total number of co-occurrences (column *Co-occ Tot* in table) concerns a context of 20 words (+10/−10).

Table 3 Collocations of the pivot class <enterprise> in absolute terms (extract)

Positions ->	−8	−7	−6	−5	−4	−3	−2	−1	1	2	3	4	5	6	7	8	Co-occ Tot	Occ Tot
Num	39	51	48	82	76	106	167	49	64	201	191	195	165	144	116	90	2,011	22,860
Verb	134	142	186	186	116	124	72	14	652	232	138	87	59	53	47	33	2,593	18,473
Verbterm	79	97	119	136	161	119	187	93	327	283	153	111	82	73	68	65	2,409	14,973
Impresa	70	63	107	87	98	96	280	129	129	280	96	98	87	107	63	70	2,052	8,620

Table 4 Main recurring structures with at least four poles (first pole is the class enterprise)

Sequence	Poly-cooccurrences
Enterprise LAG verb LAG terminological verb LAG num	43
Enterprise LAG terminological verb LAG num LAG currency	32
Enterprise LAG verb LAG num LAG currency	32
Enterprise LAG terminological verb LAG num LAG num	29
Enterprise LAG verb LAG num LAG num	16
Enterprise LAG terminological verb LAG num LAG enterprise	14
Enterprise LAG terminological verb LAG num LAG num LAG currency	14
Enterprise LAG verb LAG terminological verb LAG enterprise	14
Enterprise LAG verb LAG terminological verb LAG terminology	13
Enterprise LAG num LAG num LAG currency	12

exists also inside the class *enterprise* itself, in the immediate proximities and then at distance 6. Analyzing these profiles we can conclude about the existence of a structure of the type: *<enterprise> LAG <terminological verb> LAG <enterprise>* ("Fiat si accorda con General Motors"), where the LAG labels point out a number of words, variable and possible, that can be interposed among the poles of the structure, and its optimum quantification is inferable from the collocations analysis.

We observe also that the num tend to place themselves to the right of the enterprise and they show not negligible quantities to high distances. Therefore, the above-mentioned structure can be extended as it follows: *<enterprise> LAG <terminological verb> LAG <enterprise> LAG <num>* ("Mediobanca detiene il 9.91% delle Generali rispetto al precedente 11.73%", etc.).

It is worth noting that the co-occurrences and collocations are not able to identify the highly complex structures mentioned above. The resulting relationships can be set between a maximum of two classes.

These more complex structures can be verified in a better way by analyzing the poly-cooccurrences (Martinez 2003), that is the co-occurrences between three or more forms/classes, and not between couples of classes (see Table 4).

The output consists of an inventory, with frequencies, of all class sequences observed in the context considered. Unlike the matrix of co-occurrences, the order of presentation of the classes actually reflects that found in the text. Moreover, this inventory is redundant, as an inventory of repeated segments (Salem 1987).

Sorting and/or filtering this inventory allows the detection of the most interesting, or the most desired relationships. The following table shows the main recurring structures with at least four poles, where the first pole is the class enterprises.

In the example, the structure *<enterprise> LAG < terminological verb > LAG <num> LAG <enterprise>* capture the following phrases:

Vodafone Airtouch conta di arrivare a controllare oltre il 50% di Mannesmann
Olivetti incorpora un premio del 25% su Tecnost
Telecom si e' detta disposta a pagare per arrivare al 29.9% di Pagine Gialle
Fininvest possiede il 48.3% di Mediaset
Fininvest acquisira' il 10% delle societa' del gruppo Benetton
Compart controllava il 36.1% del capitale con diritto di voto della Montedison
Mediobanca scende sotto il 10% nelle Generali
Mediobanca detiene il 9.91% delle Generali
Consob Mediobanca detiene il 9.91% delle Generali
Erg rileva 230 stazioni di servizio Shell
Bz Gruppe detiene il 16.3% dei diritti di voto Roche

These phrases concern mainly shares held by a company against another, or with the intention to alter its stake.

The structures we have detected at this point using co-occurrences in absolute terms represent the "mass" of the information inside the text. We have noticed, in fact, that this result is also due to the dimensions of the classes *num* and of the *verb/verbterm*, and this can lead to the extraction of information somehow banal or expected. With the aim of capturing a kind of less evident information in absolute terms, but which is no doubt pertinent and interesting, we have decided to consider the ratio: frequencies of the collocations/frequency in the vocabulary of the co-occurring class. This involves a clean rearrangement between the co-occurring classes.

Rearranging the collocations of the class *enterprise* on the basis of co-occurrences in relative terms (see Table 5), *opa* now appears to be the class with

Table 5 Collocations of pivot class <enterprise> rearranged on relative co-occurrences (extract)

Positions ->	−8	−7	−6	−5	−4	−3	−2	−1	1	2	3	4	5	6	7	8	Relative co-occ	Occ Tot
Opa	3	5	7	14	30	14	41	14	0	4	5	8	10	3	5	6	0.63	288
Vendita	4	3	1	12	5	4	11	0	0	0	2	4	2	0	4	1	0.48	122
Integrazione	0	2	1	1	5	7	14	2	0	0	2	1	0	1	1	1	0.40	112
Advertisement	5	5	5	2	15	3	2	0	0	0	3	0	0	4	2	1	0.38	136
Controllata	1	2	1	0	5	3	19	19	0	1	1	3	3	0	0	1	0.38	166
Accordo	9	3	7	8	17	16	45	5	0	1	3	6	9	4	6	2	0.37	425
Fusione	1	3	7	8	19	8	38	0	0	0	1	5	4	3	4	3	0.35	311
Amministratore	0	0	3	1	6	46	7	0	0	1	1	0	0	0	0	1	0.32	214
Intesa	2	2	7	3	5	3	9	0	0	0	2	1	3	1	2	1	0.31	144

the most number of associations (only 288 occurrences in the corpus). An in-depth analysis has made it possible to identify the structure: *<enterprise> LAG <opa (takeover bid)> LAG <enterprise>*. The following sentences provide examples of this structure:

Consob ha dato l'ok al prospetto dell'Opa depositato da Compart
Consob non ha ancora ricevuto la bozza di prospetto dell'Opa di Compart
Compart ritiene di non avere obbligo di Opa a cascata in quanto gia' controlla oltre 30% di Montedison
Vodafone segue il successo dell' Opa su Mannesmann
Compart ha vincolato l'Opa su Montedison
Compart ha deliberato il lancio di una Opa sul totale del capitale di Montedison
Consob non potra' far altro che dare semaforo verde all'Opa di Mediobanca Compart lancia infatti un' Opa sulla totalit del capitale di Montedison
Consob dara' l'ok all'Opa di Compart

These sentences detect the takeover bid launched by Compart against Montedison and form a summary of events.

Another interesting relationship which can be inferred from the analysis of relative co-occurrences is *<enterprise> LAG <agreement> LAG <enterprise>*, which produces:

Mannesmann ha approvato oggi l'accordo di fusione con la Vodafone
Pirelli in netto rialzo in vista del closing dell'accordo con Cisco
Generali su Ina e' stato raggiunto un accordo per la cessione della quota Ina in Banco San Paolo
Pirelli ha chiuso l'accordo con Cisco
Pirelli e' poi in dirittura d'arrivo la formalizzazione dell'annunciato accordo con Cisco
Mannesmann ha raggiunto un accordo con Vodafone
Nissan Motor e 4 banche creditrici hanno raggiunto un accordo di principio sul piano di salvataggio di Nissan
San Paolo Imi che ha fatto un accordo con Nokia e Wind
Nissan Motor e quattro banche creditrici hanno raggiunto un accordo di principio sul piano di salvataggio di Nissan

These are some examples for all agreement, acquisition, assignment operations that it is possible to find within the text.

4 Future Developments and Improvements

Unfortunately, this document base did not have additional information associated with the single documents. The date and the time of publication, in fact, would have been very useful for analyzing the temporal evolution of the extracted information. Moreover, a more structured application of the poly-cooccurrences together with an analysis of the specificities (Lafon 1984) would perhaps imply a significant improvement. Finally, beyond the semantic class and by the same standards, grammatical categories could be worth considering.

Acknowledgement The present research was funded by MIUR 2004 – C26F040421.

References

Bolasco, S., & Canzonetti, A. (2005). Some insights into the evolution of 1990s standard Italian using text mining and automatic categorization. In M. Vichi, P. Monari, S. Mignani, & A. Montanari (Eds.), *New development in classification and data analysis* (pp. 293–302). Berlin: Springer.

Lafon, P. (1984). *Dépouillements et statistiques en lexicométrie*. Paris: Slatkine-Champion.

Martinez, W. (2003). *Contribution à une méthodologie de l'analyse des cooccurrences lexicales multiples dans les corpus textuels*. Thèse de Doctorat enSciences du Langage, Université de la Sorbonne nouvelle – Paris 3, Paris.

Poibeau, T. (2005). Una metodologia per l'annotazione semantica e l'estrazione di informazione. In S. Bolasco, A. Canzonetti, & F. M. Capo (Eds.), *Text mining – Uno strumento strategico per imprese e istituzioni* (pp. 37–44). Rome: CISU.

Salem, A. (1987). *Pratique des segments répétés*, Publications de L'InaLF, collection "Saint-Cloud", Klimcksieck, Paris.

The Effectiveness of University Education: A Structural Equation Model

Bruno Chiandotto, Bruno Bertaccini, and Roberta Varriale

Abstract The evaluation of the effectiveness of higher education is a crucial aspect of competitiveness of modern economies. In this contribution we investigate the quality and effectiveness of higher education in Italy using a structural equation model; in particular, we evaluate the performance of the university system from the users' point of view, both immediately following (internal effectiveness), and one year after (external effectiveness), the completion of the degree. The model allows the construction of synthetic indexes and hence the ranking of study programs.

1 Introduction

Higher education is a crucial aspect for the competitiveness of modern economies and this justifies the prominent interest of government institutions on this topic. In recent years, many authors have been interested in the evaluation of public education systems (Chiandotto 2004; Draper and Gittoes 2004). The evaluation of the global performance of a university and, generally, of a public activity can be divided into two phases: the first deals with how resources are spent to reach particular objectives (efficiency analysis), the second deals with the adherence of results to the planned objectives (effectiveness analysis). Both phases can be analyzed from an internal or external perspective.

Quality and effectiveness of higher education can also be investigated from at least three different points of view: the student, the university institution and the society in general. As shown in Table 1, we suggest to modify the scheme proposed by Lockheed and Hanushek (1994) and evaluate the performance of a university system from the users' point of view, both immediately following (internal effectiveness), and one year after (external effectiveness), the completion of the degree. We propose a Structural Equation Model (SEM) which allows the simultaneous construction of synthetic indexes of satisfaction measuring the internal and external

R. Varriale (✉)
Università degli Studi di Firenze, Dip.to di Statistica 'G. Parenti', Italy
e-mail: roberta.varriale@ds.unifi.it

F. Palumbo et al. (eds.), *Data Analysis and Classification*, Studies in Classification, Data Analysis, and Knowledge Organization, DOI 10.1007/978-3-642-03739-9_25,

Table 1 Concepts of educational efficiency and effectiveness (Chiandotto 2004)

	Internal	External
Physical aspects	Internal effectiveness *(effect of university or study program on the learning capacity of the student)*	External effectiveness *(effect of university or study program on the skills of the graduate)*
Satisfaction	Satisfaction of the student with respect to the study program	Satisfaction of the graduate with respect to the occupational condition
Monetary aspects	Internal efficiency *(costs/returns analysis of the investments)*	External efficiency *(economic return due to the study program attended)*
Satisfaction	Satisfaction of the student with respect to the employed resources	Satisfaction of the graduate with respect to the economic condition

effectiveness of the universities; these indexes can be used to rank the different study programs offered by the university institutions.

2 The Model

The university administrative data are merged with the data collected by the ALMALAUREA surveys on graduates profile and on their employment status one year after the degree. This dataset includes variables on individual characteristics, objective measurements of performance, subjective measurements of satisfaction with the university experience and variables relating to the working conditions after the degree. Many of these variables are a consequence of the university education and are a direct or indirect expression of the latent variable "satisfaction". We can hence adopt the approaches proposed in literature on customer satisfaction based on SEM models (Eskildsen et al. 2000; Martensen et al. 2000; Chiandotto et al. 2007). These models were originally developed in the economic context and later they were also used to measure satisfaction with public services (O'Loughlin and Coenders 2002). While Partial Least Squares method (Wold 1985) was the technique initially proposed for the estimation of the latent variables in Costumer Satisfaction Indexes models (Fornell 1992; Bayol et al. 2000), in this work we use the SEM approach. The reasons for such a choice are the flexibility in the specification of the model parameters and the possibility of testing the significance of omitted parameters, such as error co-variances and loadings on more than one latent variable; moreover, they are general since they can include, for example, observed variables with an ordinal measurement level, latent and observed categorical variables and they also allow the implementation of a two-level data structure and handling of missing data (Chiandotto et al. 2007).

Following the standard LISREL notation (Bollen 1989), the measurement model is

$$\mathbf{y} = \Lambda_y \boldsymbol{\eta} + \boldsymbol{\delta}_y,$$
$$\mathbf{x} = \Lambda_x \boldsymbol{\xi} + \boldsymbol{\delta}_x,$$

where the vector of indicators $\mathbf{y}$ of dimension p is related to an underlying m-vector $(m < p)$ of endogenous latent variables $\boldsymbol{\eta}$ through the $p \times m$ factor loading matrix Λ_y. Similarly, the vector of indicators $\mathbf{x}$ of dimension q is related to an underlying n-vector $(n < q)$ of exogenous latent variables $\boldsymbol{\xi}$ through the $q \times n$ factor loading matrix Λ_x. The vectors $\boldsymbol{\delta}_y$ and $\boldsymbol{\delta}_y$ are the error terms with dimension, respectively, p and q. The structural (latent) model is

$$\boldsymbol{\eta} = B\boldsymbol{\eta} + \Gamma\boldsymbol{\xi} + \boldsymbol{\varsigma}, \quad (1)$$

where the $m \times m$ matrix B describes the relationships among the latent variables $\boldsymbol{\eta}$ and the $m \times n$ matrix quantifies the influence of $\boldsymbol{\xi}$ on $\boldsymbol{\eta}$. Common assumptions in SEM are: $\boldsymbol{\xi} \sim N(\boldsymbol{\mu}_\xi, \Omega_\xi)$, $\boldsymbol{\varsigma} \sim N(\mathbf{0}, \Omega_\varsigma)$, $\boldsymbol{\delta}_y \sim N(\mathbf{0}, \Sigma_y)$, $\boldsymbol{\delta}_x \sim N(\mathbf{0}, \Sigma_x)$, $\boldsymbol{\delta}' = (\boldsymbol{\delta}'_y, \boldsymbol{\delta}'_x)$, $cov(\boldsymbol{\xi}, \boldsymbol{\delta}') = 0$, $cov(\boldsymbol{\varsigma}, \boldsymbol{\delta}') = 0$, $cov(\boldsymbol{\xi}, \boldsymbol{\varsigma}) = 0$, $(I - B)$ is non singular. Some constraints need to be placed on Λ_y and Λ_x for identifiability.

The observed variables considered as potential satisfaction outcomes are:

1. Reasons to enroll at the university choosing a particular study program
2. Grades achieved and time to get the degree
3. Evaluation of the relationships with teaching and non-teaching staff and with fellow students
4. Evaluation of the university facilities (classrooms, laboratories, libraries, cafeterias, etc.)
5. Intention to proceed with studies
6. Overall job satisfaction and satisfaction with respect to specific aspects of working conditions
7. Hypothesis of re-enrollment at the university

The model we propose is an adaptation of the ECSI model (ECSI Technical Committee 1998) to the available data that allows to analyse the users' perception of the quality of the university both at the completion of, and one year after, the degree.

Figure 1 shows the structure of the proposed model: ellipses and squares represent respectively the endogenous and the exogenous latent variables, while the boxes indicate the observed indicators. The model assumes that the overall satisfaction (SATI2), represented by the satisfaction one year after the degree, is a function of the following latent variables:

1. Perceived Quality: it refers to an assessment of students on the characteristics of the university facilities (Quality of facilities – QUAFA) and the quality of

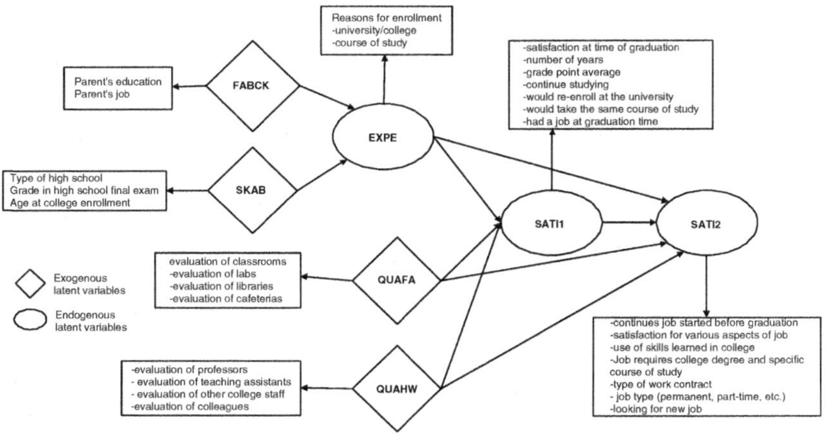

Fig. 1 Diagram of the hypothesised model

the relationships with the academic staff (perceived quality of humanware – QUAHW). Both factors are assumed to exert a direct and positive effect on the overall satisfaction.

2. Satisfaction at the completion of the degree (SATI1): it represents the value of the perceived quality at the end of the academic experience; we assume that SATI1 is a function of the Perceived Quality and positively contribute to SATI2.
3. Expectations (EXPE): it indicates the level of service quality that the user expects to receive. It is assumed to be an exogenous latent factor that has a positive influence on both SATI1 and SATI2 and it is a function of the two exogenous latent variables family background (FABCK) and pre-enrollment individual skills and abilities (SKAB).

The relationships between the factors FABCK, SKAB, QUAFA, QUAHW that are assumed to influence, directly or indirectly, both the satisfaction at degree time (SATI1) and the job satisfaction (SATI2) are described by the arrows in the diagram.

In our analysis we used data on 13,428 students who graduated during the solar year 2005 at the Italian Universities participating to the ALMALAUREA consortium, that are working one year after the degree.

Some of the observed indicators are continuous, some are ordinal and some are dichotomous; we measure each latent variable with indicators of the same type to simplify the interpretation of the model.

The model is estimated using Mplus, version 5 (Muthén and Muthén 1998–2007). Given the presence of some ordinal indicators, we adopt a robust weighted least squares estimator based on tethracoric or polychoric correlations (Muthén and Muthén 1998–2007). For ordinal indicators regressed on latent factors, probit regressions with proportional odds are estimated.

3 Main Results

As suggested in the literature (Bollen 1989), a confirmative factor analysis (CFA) should be initially used to validate the proposed structural equation model in order to evaluate the quality of the indicators employed as a gauge of the latent components and, at the same time, eliminate aspects that include sources of variability other than those contemplated.

The values of two goodness-of-fit indexes, TLI = 0.947 and RMSEA = 0.055,[1] suggest to use a model with 31 indicators and five factors, instead of the seven factor model initially hypothesized:

- BCKG: family background and pre-enrollment experiences
- QUAFA: facilities quality
- QUAHW: perceived quality of humanware
- SATI1: satisfaction on university experience at the degree time
- SATI2: job satisfaction one year after the degree

The latent variable BCKG is obtained merging the three latent variables EXPE, FABCK and SKAB. The correlations between BCKG – QUAFA and BCKG – QUAHW are set to be zero.

The CFA model was therefore transformed into a ECSI-SEM model and the regression equations between the latent components were re-specified. The proposed ECSI-SEM model did not converge within an acceptable number of iterations; taking into account the values of the modification indexes and the a-priori knowledge on the phenomenon under study, some links between factors were eliminated.

The final model is represented in Fig. 2. The values of the estimated coefficients statistically differing from zero are shown on the arrows representing the causal links. All the effects go in the expected direction. The hypothesized links between factors that did not result to be statistically significant (BCKG, QUAFA and QUAHW on SATI2) were eliminated. The values of the TLI (0.960) and RMSEA (0.044) indexes indicate the good fitting of the model.

For lack of space, we limit our comments to the structural part of the model. The quality of the humanware has the highest influence on the satisfaction for the university system: its effect is twice the effect of the facilities. Furthermore, the higher is the background level, the lower is the satisfaction. For example, students with high educated parents and a good performance in high school are probably more demanding and, hence, less satisfied than the others. As expected, job satisfaction is positively affected by the satisfaction at the degree.

[1] TLI and RMSEA are fit indexes both varying in [0, 1]. Values greater than 0.95 for the first one and value less than 0.06 for the second one are evidence of a good fitting to the data. For a brief review of the fit indexes see Hox and Bechger (1998).

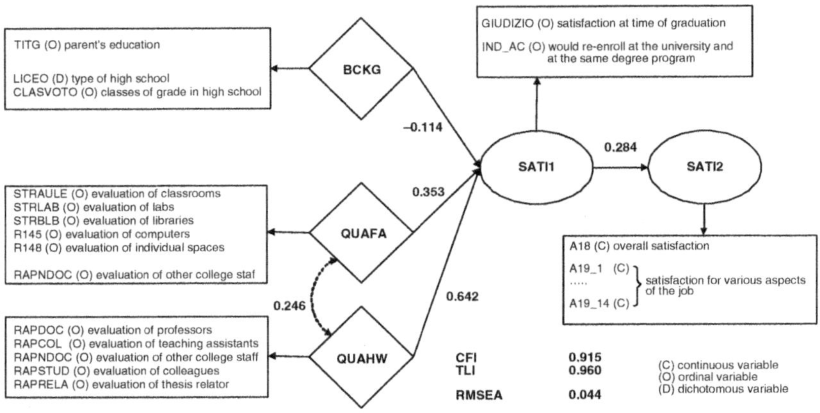

Fig. 2 Significant indicators and coefficients of the structural part of the model

In structural equation models the main aim is to analyse the latent variables after the manifest variables have been observed (Bartholomew and Knott 1999); this information is derived from the conditional density $h(\eta|y) = h(\eta)g(y|\eta)/f(y)$. From the point of view of social behavioural scientists, this means locating units on the dimensions of the latent space (finding the *factor scores*); units with the same response pattern will be assigned the same factor score.

Since the aim of this work is the evaluation of the performance of the university system, the analysis focused on the two dependent latent variables SATI1 and SATI2. To obtain a measure of the effectiveness of course programs we need to aggregate the individual factor scores (satisfaction indexes). Even if non completely satisfactory, the simplest method[2] is to compute the mean of the factor scores for SATI1 and SATI2 for each course program, or for each university.

Figure 3 shows how groups of study programs are located with respect to the two dimensions of the satisfaction. In the first (third) square there are the programs with high (low) level of both SATI1 and SATI2. In the second (fourth) square are located programs with high (low) level of satisfaction at the degree time but a low level (high) of job satisfaction.

The Medical group has the highest levels of satisfaction, followed by Chemistry, Engineering and Education. On the contrary, Psychology, Law and Political Science are the worse. In the fourth square there are only two groups (Physical Education and Architecture) whit very low level of SATI1 but medium level of SATI2. The

[2] Since the available data have a hierarchical structure (students at the first level are nested in study programs or in universities), multilevel techniques could be used to take into account the hierarchical structure of the data and obtain the latent factor scores of the second level units. In this work, multilevel techniques for structural equation models (Skrondal and Rabe-Hesketh 2004) are not feasible with the available software because of the high number of latent variables involved in the model.

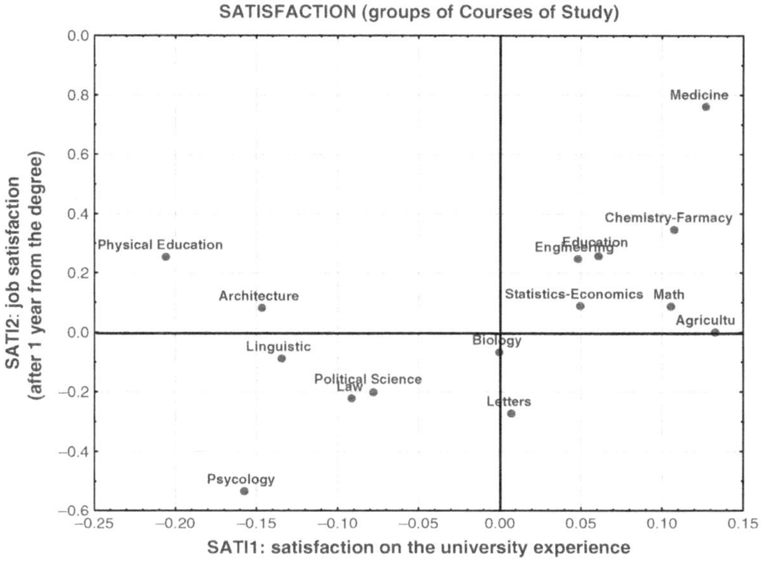

Fig. 3 Rank of groups of study programs with respect of the two analysed satisfaction dimensions

same analysis was also conducted for every university to benchmark the same study programs belonging to different universities.

4 Conclusions

This work investigated the effectiveness of higher education from the students' point of view. The evaluation of the performance (effectiveness) of the Italian university system was conducted through the definition and the estimation of a SEM based on the European Customer Satisfaction Index.

The model was built attributing causal meaning to the links between factors. This traditional way of interpreting SEM as causal models is unsatisfying. In our opinion, different statistical models, such as graphical models (Pearl 2000; Spirtes et al. 2000), should be used. Future research will re-specify the model following this approach.

Moreover, the initial model was simplified aggregating some factors and deleting some links also because converge problems raised during the estimation process. In order to deal with computational problems and in order to compare the proposed analysis with other available techniques more oriented to latent variable scores prediction we will estimate and evaluate the hypothesised model also through the PLS path modeling algorithm (Tenenhaus et al. 2005).

References

Bartholomew, D. J., & Knott, M. (1999). *Latent variable models and factor analysis*. London: Arnold.

Bayol, M. P., de la Foye, A., Tellier, C., & Tenenhaus, M. (2000). Use of PLS path modelling to estimate the European consumer satisfaction index (ECSI) model. *Statistica Applicata, 12*, 361–375.

Bollen, K. A. (1989). *Structural equations with latent variables*. New York: Wiley.

Chiandotto, B. (2004). Sulla misura della qualitá della formazione universitaria. *Studi e note di Economia, 3*, 27–61.

Chiandotto, B., Bini, M., & Bertaccini, B. (2007). Quality assessment of the University Educational Process: An application of the ECSI model. In *Effectiveness of the University Education in Italy*. Heidelberg: Physica.

Draper, D., & Gittoes, M. (2004). Statistical analysis of performance indicators in UK higher education. *Journal of the Royal Statistical Society A, 167*(3), 449–474.

ECSI Technical Committee. (1998). *European customer satisfaction index: Foundation and structure for harmonized national pilot projects*. Report prepared for the ECSI Steering Committee.

Eskildsen, J. K., Martensen, A., Gronholdt, L., & Kristensen, K. (2000). Benchmarking student satisfaction in higher education based on the ECSI methodology. *Sinergie Rapporti di Ricerca, 9*, 385–400.

Fornell, C. (1992). A national customer satisfaction barometer: The Swedish experience. *Journal of Marketing, 56*, 6–22.

Hox, J. J., & Bechger, T. M. (1998). An introduction to structural equation modeling. *Family Science Review, 11*, 354–373.

Lockheed, M. E., & Hanushek, E. R. (1994). Concepts of educational efficiency and effectiveness. In T. Husen & T. N. Postlethwaite (Eds.), *International encyclopedia of education* (pp. 1779–1784). Oxford: Pergamon.

Martensen, A., Gronholdt, L., Eskildsen, J. K., & Kristensen, K. (2000). Measuring student oriented quality in higher education: Application of the ECSI methodology. *Sinergie Rapporti di Ricerca, 9*, 371–383.

Muthén, L. K., & Muthén, B. O. (1998–2007). *Mplus users guide* (5th edition). Los Angeles, CA: Muthén and Muthén.

O'Loughlin C., & Coenders G. (2002). *Application of the European customer satisfaction index to postal services. Structural equation models versus partial least squares*. Departament d'Economia, Universitat de Girona.

Pearl, J. (2000). *Causality: Models, reasoning, and inference*. New York: Cambridge University Press.

Skrondal, A., & Rabe-Hesketh, S. (2004). *Generalized latent variable modeling Anders*. Boca Raton: Chapman and Hall.

Spirtes, P., Glymour, C., & Scheines, R. (2000). *Causation, prediction and search* (2nd edition). Massachusetts, CA: The MIT press.

Tenenhaus, M., Vinzi, V. E., Chatelin, Y. M., & Lauro, C. (2005). PLS path modeling. *Computational Statistics and Data Analysis, 48*, 159–205.

Wold, H. (1985). Partial least squares. In S. Kotz & N. L. Johnson (Eds.), *Encyclopedia of statistical sciences* (Vol. 6, pp. 581–591). New York: Wiley.

Simultaneous Threshold Interaction Detection in Binary Classification

Claudio Conversano and Elise Dusseldorp

Abstract Classification Trunk Approach (CTA) is a method for the automatic selection of threshold interactions in generalized linear modelling (GLM). It comes out from the integration of classification trees and GLM. Interactions between predictors are expressed as "threshold interactions" instead of traditional cross-products. Unlike classification trees, CTA is based on a different splitting criterion and it is framed in a new algorithm – STIMA – that can be used to estimate threshold interactions effects in classification and regression models. This paper specifically focuses on the binary response case, and presents the results of an application on the Liver Disorders dataset to give insight into the advantages deriving from the use of CTA with respect to other model-based or decision tree-based approaches. Performances of the different methods are compared focusing on prediction accuracy and model complexity.

1 Introduction

In statistical modeling, a-priori hypotheses about the distribution of data and theoretical considerations among the relationships existing between predictors allow the analyst to specify model interaction terms. One impractical and time-consuming possibility is testing all possible interactions and retain the most important ones.

The Regression Trunk Approach (RTA) has been proposed in Dusseldorp and Meulman (2004) to overcome this problem. RTA is restricted to prediction problems involving a continuous response. Its strength lies in the ability to automatically detect a regression model with multiple main effects and a parsimonious amount of higher order interaction effects. Dusseldorp et al. (2009) improved RTA in terms of computational efficiency and estimating capabilities, and carefully investigated the features of the pruning step of the algorithm. RTA has been used successfully

C. Conversano (✉)
Department of Economics, University of Cagliari, Viale Fra Ignazio 17, I-09123, Cagliari, Italy
e-mail: conversa@unica.it

F. Palumbo et al. (eds.), *Data Analysis and Classification*,
Studies in Classification, Data Analysis, and Knowledge Organization,
DOI 10.1007/978-3-642-03739-9_26,

in a psychological study involving the choice of the appropriate treatment for panic disorder patients (Dusseldorp et al. 2007).

This paper extends threshold interaction detection in regression analysis by introducing a new model that integrates two existing methods: generalized linear modeling (GLM) (McCullagh and Nelder 1989) and tree-based modeling (Breiman et al. 1984). It estimates the main effects part of the model on the basis of the standard Iteratively Reweighted Least Squares algorithm characterizing GLM, and then it looks for the presence of interaction terms to be added sequentially to the main effects model in each iteration of the estimation algorithm. The possible identification of such interaction terms derives from a recursive binary partitioning algorithm based on a suitable (model-based) splitting criterion. The tree deriving from such a splitting procedure usually requires two to seven splits to capture the interactive structure in data. Because of its reduced size, it is called "trunk". As it usually happens in GLM, depending on the nature of the response variable the model can be used for many types of data: if the response is continuous(binary), the estimated model is similar to a linear(logistic) regression model with interaction terms and it can be used to predict (classify) new observations. Since the framework is quite general, the same approach can also be used to model interaction effects in other cases such as, for example, in polychotomous or Poisson regression. The core element of the proposed approach is a different representation of interaction effects, that is, as a threshold interaction (see Sect. 2), as opposed to the commonly used cross-product representation.

In the following, Sect. 2 summarizes the cross-product vs. the threshold representation of interaction terms in regression models and the alternative approaches used for their estimation. Section 3 introduces the trunk model and the estimation algorithm. In Sect. 4 the focus is restricted to binary classification and a comparison of different methods, including the Classification Trunk Approach (CTA), is presented by analyzing a benchmark dataset. Concluding remarks are reported in Sect. 5.

2 Modeling Interaction Effects in Regression Analysis

In regression analysis, interaction between two or more predictors occurs if their separate effects do not combine additively (de Gonzalez and Cox 2007) or, equivalently, when over and above any additive combination of their separate effects, they have a joint effect (Cohen et al. 2003). Dealing with huge datasets with many predictors, as in data mining applications, requires a "manual" search of all the possible interaction terms. This is even more complicated when interactions involve multi-attribute predictors, as well as when higher order interactions are detected. In both cases, the estimated model can lack of parsimony and its results are not easily interpretable.

Tree-based models, such as CART (Breiman et al. 1984), allow a different representation of interaction terms. In a binary tree, interaction effects occur when the effect of a predictor on the response outcomes is different for cases who score above a certain threshold value on another (splitting) predictors compared to cases who score below that threshold value. As a result, tree-based modeling points out threshold interaction terms that are automatically identified within the model estimation procedure. Notwithstanding the appeal of its self-selection mechanism of the interaction terms, the recursive partitioning algorithm tends to be biased towards selecting predictor variables that afford more splits, in this way penalizing categorical predictors. Moreover, the main effect of some continuous predictors is not captured by the model and, particularly for large-sized trees, it is difficult to distinguish the joint effect of some predictor from their separate ones when interpreting the results. Beyond all these limitations, another characteristic of trees is that they assign the same predicted value to all the subjects falling in a terminal node: in the simple case of a tree with only two splits (three terminal nodes), the threshold values defining an interaction term (that is, the split points) lead to a model that might lack of fit since it simply produces three alternative estimated values.

The proposed trunk approach takes benefit from the possibility of trees to automatically detect threshold interactions but it also accounts for the lack of fit because the predicted value for all the subjects falling in a terminal node of the trunk may differ, since the regression coefficients for that set of observations are estimated separately by the procedure.

Figure 1 provides a graphical interpretation of the different ways of detecting interactions with standard regression (Panel a), tree-based regression (Panel b) and the trunk approach (Panel c). In the first case, interaction is defined by the

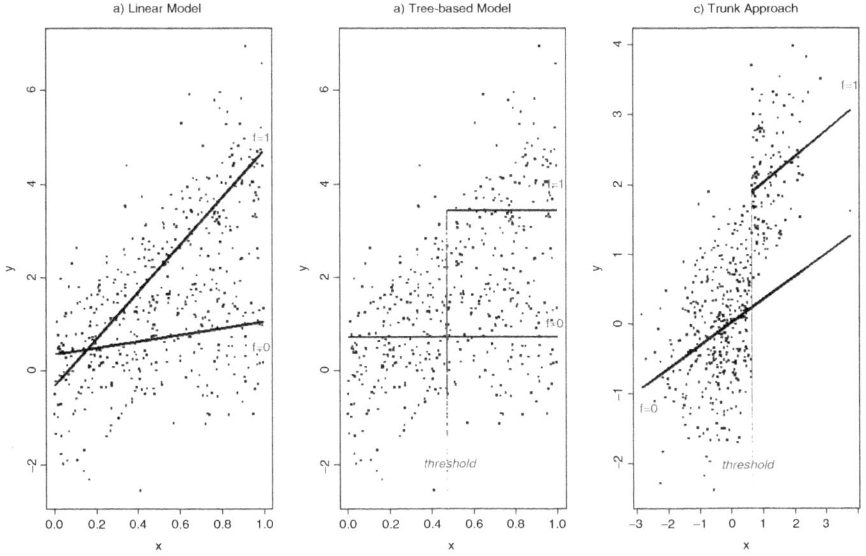

Fig. 1 An example of interaction effect in linear regression (**a**), tree-based regression (**b**) and the trunk approach (**c**) with respect to an interaction term between a numeric variable x (x-axis) and a categorical variable f, having two values (0 or 1)

cross-product between x and a 0–1 factor f, whereas in the other cases it depends on a threshold value of x identified by a suitable splitting of data. In particular, in the trunk approach this splitting criterion takes into account of the possibility of improving the overall goodness of fit of the regression model when introducing the interaction term by estimating two separate regression coefficients for each of the two child nodes identified by the split point.

3 The Trunk Model

Let y be a univariate response variable to be explained by a vector $x' = (x_1, \ldots, x_J)$ of numerical or categorical predictors and let us assume that y follows an exponential family density $\rho_y(y;\theta;\phi)$ with a natural parameter θ and a scale parameter ϕ (see Fahrmeir and Tutz 2001 p. 18); we introduce a model where the mean $\mu = E(y|x_1, \ldots, x_J)$ is linked to the x_j's via

$$g(\mu) = \underbrace{\beta_0 + \sum_{j=1}^{J} \beta_j x_j}_{\text{main effect}} + \underbrace{\sum_{m=1}^{M-1} \beta_{J+m} I\left\{(x_1 \leq s_1 \cap \ldots \cap x_j \leq s_j) \in R_m\right\}}_{\text{interaction effect}}. \quad (1)$$

The first term relates to the main effects model estimated via GLM and the second to the interaction effect estimated using recursive partitioning. Main idea is to fit a trunk over and above the linear main effects of predictors to identify interaction effects. Since the trunk corresponds to a reduced-size tree, the overall estimation method is named classification or regression trunk approach (CTA or RTA) depending on the distribution of y: if y follows a standard normal distribution, $g(\cdot)$ is the identity function and (1) reduces to the regression trunk model. Whereas, if y follows a binomial (or multinomial) distribution we obtain a logistic (or multinomial logistic) regression model with threshold interaction terms. In (1), M is the number of terminal nodes of the trunk; $\beta_0 + \beta_{J+m}$ is the intercept of the regression line fitted for observations falling into terminal node R_m ($i = 1, \ldots, M$). The indicator function $I(\cdot)$ assigns observations to one of the terminal nodes based on the splitting values s_j of the splitting predictors x_j ($j = 1, \ldots, J$). The total number of indicator variables $I(\cdot)$ included in the model equals $M - 1$, since one of the terminal nodes of the trunk serves as reference group. As a result, $M - 1$ threshold interaction terms are automatically identified by the trunk.

The estimation algorithm for both CTA and RTA is named STIMA (*Simultaneous Threshold Interaction Modeling Algorithm*) and consists of a tree growing step and a pruning step. In the ℓ-th iteration of the tree growing process ($\ell = 1, \ldots, L$), the ℓ-th interaction term entering the model is the one maximizing the effect size $f^{(\ell)}$, i.e., the relative decrease in the residual deviance when passing from the model with $\ell - 1$ terms to the one with ℓ terms. In practice, for each possible combination of splitting variable x_j, split point s_j and splitting node $R_m^{(\ell-1)}$ (i.e., a terminal node after the split $\ell - 1$), the best split is chosen according to the combination, say (x_j^*, s_j^*, R_m^*), maximizing the effect size $f^{(\ell)}$. The highest effect size determines the highest relative decrease in deviance when moving from a more parsimonious model to a less parsimonious one. Tree growing proceeds until the user-defined maximum number of splits L is reached.

Once the tree growing is complete, pruning is carried out using CART-like V-fold cross-validation. The "best" size of the trunk corresponds to the one minimizing the cross-validated prediction accuracy as well as its standard error. Likewise in CART, a "$c \cdot SE$" rule is used, where c is a constant. Simulation studies reported in Dusseldorp et al. (2009) suggest that a reasonable value for c is between 0.50 and 0.80. Pruning is a fundamental step of the algorithm since the number of terminal nodes of the trunk and their relative split points, as well as the splitting predictors, determine the number, order and type of threshold interactions terms to be included in the classification trunk model. Taking advantage of the extreme flexibility of the recursive partitioning algorithm, RTA and CTA are applicable to all types of predictors and can be used to model all types of interactions.

4 Empirical Evidence

In the following, we focus on binary classification problems and we describe the effectiveness of CTA and its advantages compared to some alternative approaches by analyzing the "Liver Disorders" data (UCI Machine Learning Repository) as a benchmark dataset. The goal is to learn how to classify 345 individuals that were already classified as high alcohol consumers or regular drinkers on the basis of the number of half-pint equivalents of alcoholic beverages drunk per day (*drinks*), and blood tests measurements which are thought to be sensitive to liver disorders that might arise from excessive alcohol consumption: the mean corpuscular volume (*mcv*), the alkaline phosphotase (*alkphos*), the alamine aminotransferase (*sgpt*), the aspartate aminotransferase (*sgot*) and the gamma-glutamyl transpeptidase (*gammagt*).

CTA is performed in order to find threshold interactions among predictors. Figure 2 summarizes the output of the classification trunk model. Figure 2a relates to the pruning step of the STIMA algorithm: a "$0.50 \cdot SE$" rule is used to select the final trunk. The maximum number of interaction terms specified by the user (nsplit) is 9, and 10-fold cross validation suggests the best trunk has three terminal nodes (nsplit = 2), since its CV-error (REcv) of 0.252 is comparable with its maximum (using the "$0.50 \cdot SE$" rule, $0.252 + 0.50 \times 0.023$). The selected classification trunk model is depicted in Fig. 2b: the first split of the trunk is made on *sgpt* and the second on *gammagt*. The terminal nodes of the trunk highlight three different regions (R_1, R_2 and R_3): R_1 identifies regular drinkers as those presenting a level of *sgpt* lower than 21.5, whereas subjects presenting a level of *sgpt* higher than 21.5 are identified as regular drinkers if their reported level of *gammagt* is higher than 22.5 (R_2) or as non-regular drinkers if this value is lower than 22.5 (R_3). In

nsplit	dev	RE	SE	REcv	SEcv
1	.000	.296	.025	.304	.025
2	.027	.272	.024	.299	.025
3	.064	.246	.023	.252	.023
4	.026	.246	.023	.264	.024
5	.020	.241	.023	.287	.024
6	.025	.229	.023	.293	.024
7	.021	.212	.022	.293	.024
8	.010	.206	.022	.293	.024
9	.010	.214	.022	.299	.025

a)

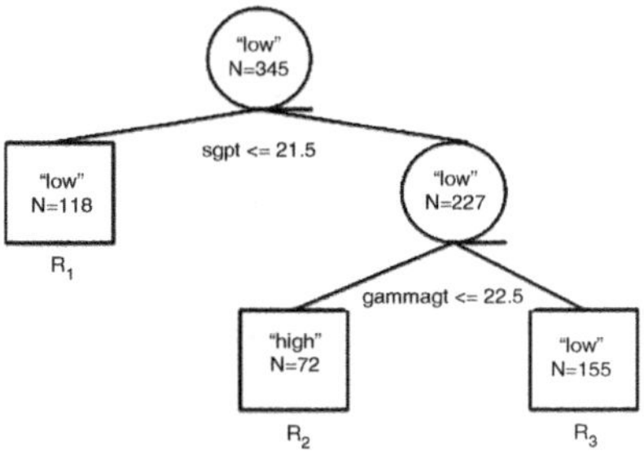

b)

	Beta	Std. Error	Std. Beta	Exp (Beta)	Z	p-value
(Intercept)	5.74	2.91	−0.40	0.00	1.97	0.05
mcv	0.07	0.03	0.33	1.08	2.32	0.02
alkphos	0.02	0.01	0.42	1.02	3.19	0.00
sgot	−0.12	0.03	−1.23	0.89	−4.87	0.00
gammagt	−0.01	0.01	−0.34	0.99	−1.64	0.10
drinks	0.09	0.04	0.30	1.09	−2.10	0.04
sgpt	0.05	0.01	1.03	1.05	4.17	0.00
R_1	−2.05	0.39	−0.97	0.13	−5.25	0.00
R_3	−1.96	0.41	−0.98	0.14	−4.81	0.00

c)

Fig. 2 Output of the STIMA algorithm estimating the classification trunk model for the Liver Disorder data: results of the cross-validated classification trunks (**a**); selected trunk (**b**); summary of the classification trunk model (**c**)

addition, Fig. 2c reports the model summary with the estimated regression coefficients, their standard errors and their odds ratios (Exp(Beta)). R_1 and R_3 are added to the model as dummy variables identifying threshold interaction terms, whereas R_2 is used as reference category. The joint effect of *sgpt* and *gammagt* on y is explained through R_1 and R_3. In particular, the significance of the coefficients associated with R_1 and R_3 and the values of their odds ratio (0.13 and 0.14) seem to emphasize the major role jointly played by *gammagt* and *sgpt* in reducing the alcohol consumption levels. In particular, the probability of classifying a new subject as high alcohol consumer decreases if *sgpt* decreases or if *gammagt* decreases once that *sgpt* is higher than 21.5.

Finally, the effectiveness of CTA is assessed performing a comparative analysis of its performance with respect to that of other alternative approaches such as: GLM, GLM with stepwise variable selection, GAM (Hastie and Tibshirani 1990), GAM with stepwise variable selection, CART, Multivariate Adaptive Regression Splines (Friedman 1991) (MARS), Support Vector Machines (Vapnik 1998) (SVM), Random Forest (Breiman 2001), Bagging (Breiman 1996) and (AdaBoost) Boosting

Table 1 Benchmarking the classification trunk model

Model	*CV error*	*Standard error*	*False positive*	*False negative*	*No. of parameters*
CTA	**0.249**	0.024	0.150	0.386	8
GLM	0.301	0.025	0.190	0.455	7
Stepwise GLM	0.287	0.024	0.195	0.414	11
GAM	0.255	**0.023**	0.185	**0.352**	21
Stepwise GAM	0.432	0.027	0.415	0.455	7
CART	0.287	0.024	0.125	0.510	**2**
MARS	0.307	0.025	0.240	0.400	9
SVM	0.293	0.024	0.210	0.401	10
Random forest	0.261*		0.155*	0.407*	
Bagging	0.336*		0.255*	0.483*	
Ada boost	0.278*		**0.006***	0.579*	

*Indicates out-of-bootstrap estimates

(Freund and Schapire 1997). For each method cross-validation is used to estimate prediction accuracy, expect for the last three methods that are typically based on the bootstrap and provide a most optimistic prediction accuracy. Results of the comparative analysis are summarized in Table 1. The performance of the model presented in Fig. 1 is in line with that of its competitors: it is worth noting that CTA provides the lowest CV Error and results as the second best in terms of the False positive rate after Boosting, whose accuracy is measured in terms of out-of-bootstrap estimate. The same ranking is obtained also with respect to the False negative rate: CTA is the second best after GAM, which is the model requiring the highest number of parameters.

5 Concluding Remarks

The classification trunk model based on CTA appears as a valuable alternative in the estimation of GLM with threshold interaction terms for binary classification problems. This particular representation of interactions distinguishes CTA from classical GLM. Nevertheless, CTA differs also from a classification tree mainly with respect to three aspects: (a) the partitioning criterion takes into account both separate effects (also called main effects) of predictors and interaction effects; (b) the partitioning criterion is based on the global fit of the GLM model.

Future research will be addressed to the assessment of a suitable pruning rule for the classification trunk carrying out a large scale simulation study as it has been done in Dusseldorp et al. (2009) for the regression trunk. Moreover, the possibility of extending this approach to the multi-class response classification problem will also be considered.

References

Breiman, L. (1996). Bagging predictors. *Machine Learning, 24*, 123–140.

Breiman, L. (2001). Random forests. *Machine Learning, 45*, 5–32.

Breiman, L., Friedman, J. H., Olshen, R. A., & Stone, C. J. (1984). *Classification and regression trees*. Belmont, CA: Wadsworth.

Cohen, J., Cohen, P., West, S. G., & Aiken, L. S. (2003). *Applied multiple regression/correlation analysis for the behavioral sciences* (3rd edition). Mahwah NJ: Lawrence Erlbaum.

de Gonzalez, A. B., & Cox, D. R. (2007). Interpretation of interaction: A review. *Annals of Applied Statistics, 1*(2), 371–375.

Dusseldorp, E., & Meulman, J. (2004). The regression trunk approach to discover treatment covariate interactions. *Psychometrika, 69*, 355–374.

Dusseldorp, E., Spinhoven, P., Bakker, A., Van Dyck, R., & Van Balkom, A. J. L. M. (2007). Which panic disorder patients benefit from which treatment: Cognitive therapy or antidepressants? *Psychotherapy and Psychosomatics, 76*, 154–161.

Dusseldorp, E., Conversano, C., & Van Os, B. J. (2009). *Combining an Additive and tree-based regression model simulatenously: STIMA*, Journal of Computational and Graphical Statistics, to appear.

Fahrmeir, L., & Tutz, G. (2001). *Multivariate statistical modelling based on generalized linear models* (2nd edition). New York: Springer.

Freund, Y., & Schapire, R. (1997). A decision-theoretic generalization of on-line learning and an application to Boosting. *Journal of Computer and System Sciences, 55*(1), 119–139.

Friedman, J. H. (1991). Multivariate adaptive regression splines (with discussion). *Annals of Statistics, 19*, 1–141.

Hastie, T. J., & Tibshirani, R. J. (1990). *Generalized additive models*. London: Chapman & Hall.

McCullagh, P., & Nelder, J. A. (1989). *Generalized linear models* (2nd edition). London: Chapman & Hall.

Vapnik, V. (1998). *Statistical learning theory*. New York: Wiley.

Detecting Subset of Classifiers for Multi-attribute Response Prediction

Claudio Conversano and Francesco Mola

Abstract An algorithm detecting a classification model in the presence of a multi-class response is introduced. It is called *Sequential Automatic Search of a Subset of Classifiers* (SASSC) because it adaptively and sequentially aggregates subsets of instances related to a proper aggregation of a subset of the response classes, that is, to a super-class. In each step of the algorithm, aggregations are based on the search of the subset of instances whose response classes generate a classifier presenting the lowest generalization error compared to other alternative aggregations. Cross-validation is used to estimate such generalization errors. The user can choose a final number of subsets of the response classes (super-classes) obtaining a final tree-based classification model presenting an high level of accuracy without neglecting parsimony. Results obtained analyzing a real dataset highlights the effectiveness of the proposed method.

1 Introduction

Tree-based classifiers are typically used to predict the membership of some objects in the classes of a (categorical) response. This goal is obtained by means of their measurements on one or more inputs. Basically, they can be applied on rough data stored in large databases, that are not originally collected for statistical purposes. Because of such extreme flexibility, they are among the main statistical tools commonly used to solve data mining problems. In such situations, the response used in classification tree modelling rarely presents a number of attributes that allow to apply the recursive partitioning algorithm in the most accurate manner.

It is well known that: (a) a multi-class response, namely a factor with several attributes, usually causes prediction inaccuracy; (b) multi-class and numeric inputs play often the role of splitting variables in the tree growing process in disadvantage of two-classes inputs, causing selection bias. To overcome the above-mentioned

C. Conversano (✉)
Department of Economics, University of Cagliari, Viale Fra Ignazio 17, I-09123, Cagliari, Italy
e-mail: conversa@unica.it

F. Palumbo et al. (eds.), *Data Analysis and Classification*,
Studies in Classification, Data Analysis, and Knowledge Organization,
DOI 10.1007/978-3-642-03739-9_27,

problems, ensemble of classifiers (Dietterich 2000) and hybrid methods (Gama 2004) have been recently introduced. The first ones are based on aggregation of predictions coming from the application of the recursive partitioning algorithm on several (dependent or independent) samples drawn from the original dataset, whereas the second ones usually derive from the incorporation of (parametric or nonparametric) models into the recursive partitioning algorithm. Both these approaches usually over-perform standard classification tree algorithms, such as CART (Breiman et al. 1984). But, particularly for ensemble methods, their "black-box" mechanism produces a loss of interpretability of the classification model.

To account for the problems deriving from the prediction inaccuracy of tree-based classifiers grown for multi-class response, as well as to reduce the drawback of the loss of interpretability induced by ensemble methods in these situations, an algorithm based on a *Sequential Automatic Search of a Subset of Classifiers* (SASSC) is hereby introduced. It produces a partition of the set of the response classes into a reduced number of disjoint subgroups and introduces a parameter in the final classification model that improves its prediction accuracy, since it allows to assign each new object to the most appropriate classifier in a previously-identified reduced set of classifiers. It uses cross-validated classification trees as a tool to induce the set of classifiers in the final classification model.

The remainder of the paper is as follows. Section 2 describes the SACCS algorithm and introduces the related classification model. The results from both the application of such a model on the Letter Recognition dataset and the comparison of the performance of SACCS with respect to alternative approaches are summarized in Sect. 3. Section 4 discusses the possible advantages connected to the use of the proposed method.

2 The SASSC Algorithm

SACCS produces a partition of the set of the response classes into a reduced number of super-classes. It is applicable to a dataset $\mathbf{D}$ composed of N instances characterized by a set of J (numerical or categorical) inputs X_j $(j = 1, \dots, J)$ and a response Y presenting K classes. Such response classes identify the initial set of classes $C^{(0)} = (c_1, c_2, \dots, c_K)$. Partitioning $\mathbf{D}$ with respect to $C^{(0)}$ allows to identify K disjoint subsets $\mathbf{D}_k^{(0)}$, such that: $\mathbf{D}_k^{(0)} = \{\mathbf{x}_s : \mathbf{y}_s \in c_k\}$, with $s = 1, \dots, N$. In practice, $\mathbf{D}_k^{(0)}$ is the set of instances presenting the k-th class of Y. The algorithm works by aggregating the K classes in pairs and learns a classifier to each subset of corresponding instances. The "best" aggregation (*super-class*) is chosen as the one minimizing the generalization error estimated using V-fold cross-validation.

Suppose that, in the ℓ-th iteration of the algorithm, such a best aggregation is found for the pair of classes c_{i*} and c_{j*} (with $i^* \neq j^*$ and $i^*, j^* \in (1, \dots, K)$) that allow to aggregate subsets $\mathbf{D}_{i*}$ and $\mathbf{D}_{j*}$. Denoting with $T_{(i*,j*)}$ the classifier

Table 1 The SACCS algorithm

Input:	$C = \{c_1, \ldots, c_K\}_{c_i \cup c_j = \emptyset; i \neq j; i,j \in (1,\ldots,K)}$
Set:	$C^{(0)} = C;\quad K^{(0)} = K;\ \theta_{cv}^{(0)} = 0;$
	$\mathbf{D}_k^{(0)} = \{\mathbf{x}_s : y_s \in c_k\}_{s=1,\ldots,N; k=1,\ldots,K}$
For: ℓ in 1 to K	
	$c^{(\ell)} = \{c_{i^*} \cup c_{j^*}\} : \theta_{cv}^{(\ell)}(T_{(i^*,j^*)} \vert \mathbf{D}_{i^*} \cup \mathbf{D}_{j^*}) = \min$
	$K^{(\ell)} = K^{(\ell-1)} - 1$
	$C^{(\ell)} = \{c_1, \ldots, c_{K^{(\ell)}-2+1} = c^{(\ell)}\}$
	$\mathbf{D}_k^{(\ell)} = \{\mathbf{x}_s : y_s \in c_k\}_{k=1,\ldots,K^{(\ell)}-1}$
end For	
Output:	$C^{(1)}, \ldots, C^{(K-1)};\quad T_{(1)}, \ldots, T_{(K-1)};\quad \Theta_{cv}^{(1)}, \ldots, \Theta_{cv}^{(K-1)}$

minimizing the cross-validated generalization error $\theta_{cv}^{(\ell)}$, the criteria for selecting the best classifier can be formalized as follows:

$$(i^*, j^*) = \arg\min_{(i,j)} \left\{\theta_{cv}^{(\ell)} \left(T_{(i,j)} \vert \mathbf{D}_i \cup \mathbf{D}_j\right)\right\} \tag{1}$$

The SACCS algorithm is analytically described in Table 1. It proceeds by learning all the possible classifiers obtainable by joining in pairs the K subgroups, retaining the one satisfying the selection criteria introduced in (1). After the ℓ-th aggregation, the number of subgroups is reduced to $K^{(\ell)} = K^{(\ell-1)} - 1$, since the subgroups of instances presenting the response classes c_{i^*} and c_{j^*} are discarded from the original partition and replaced by the subset $\mathbf{D}_{(i^*,j^*)}^{(\ell)} = (\mathbf{D}_{i^*} \cup \mathbf{D}_{j^*})$ identified by the super-class $c^{(\ell)} = (c_{i^*} \cup c_{j^*})$. The initial set of classes C is replaced by $C^{(\ell)}$, the latter being composed of a reduced number of classes since some of the original classes form the super-classes coming out from the ℓ aggregations. Likewise, also $\mathbf{D}_k^{(\ell)}$ is formed by a lower number of subsets as a consequence of the ℓ aggregations.

The algorithm proceeds sequentially in the iteration $\ell + 1$ by searching for the most accurate classifier over all the possible ones obtainable by joining in pairs the $K^{(\ell)}$ subgroups. The sequential search is repeated until the number of subgroups reduces to one in the K-th iteration. The classifier learned on the last subgroup corresponds to the one obtainable applying the recursive partitioning algorithm on the original dataset.

The output of the procedure is a sequence of sets of response classes $C^{(1)}, \ldots, C^{(K-1)}$ with the associated sets of classifiers $T_{(1)}, \ldots, T_{(K-1)}$. The latter are derived by learning $K - k$ classifiers ($k = 1, \ldots, K - 1$) on disjoint subgroups of instances whose response classes complete the initial set of classes $C^{(0)}$: these response classes identify the super-classes relating to the sets of classifiers $T_{(k)}$. An overall generalization error is associated to each $T_{(k)}$: such an error is also based on V-

fold cross-validation and it is computed as a weighted average of the generalization errors obtained from each of the $K - k$ classifiers composing the set. In accordance to the previously introduced notation, the overall generalization errors can be denoted as $\Theta_{cv}^{(1)}, \ldots, \Theta_{cv}^{(k)}, \ldots, \Theta_{cv}^{(K-1)}$. Of course, decreasing the number of classifiers composing a sequence $T_{(k)}$ (i.e., when moving k from 1 to $K - 1$) increases the corresponding $\Theta_{cv}^{(k)}$ since the number of super-classes associated to $T_{(k)}$ is also decreasing. This means that a lower number of classifiers are learned on more heterogeneous subsets of instances, since each of those subsets pertain to a relatively large number of response classes.

Taking this inverse relationship into account, the analyst can be aware of the overall prediction accuracy of the final model on the basis of the relative increase in $\Theta_{cv}^{(k)}$ when moving from 1 to $K-1$. In this respect, he can select the suitable number of classifiers to be included in the final classification model accordingly. Supposing that a final subset of g classifiers has been selected ($g \ll K - 1$), the estimated classification model can be represented as

$$\hat{f}(X) = \sum_{i=1}^{g-1} \sum_{m_i=1}^{M_i} \hat{\tau}_i \hat{c}_{k,i} I((X_1, \ldots, X_p) \in R_{m_i}). \tag{2}$$

The parameter τ is called "vehicle parameter". It allows to assign a new instance to the most suitable classifier in the subset g. It is defined by a set of $g - 1$ dummy variables. Each of them equals 1 if the object belongs to the i-th classifier ($i = 1, \ldots, g - 1$) and zero otherwise. The M_i regions, corresponding to the number of terminal nodes of the classifier i, are created by splits on inputs $(X_1, \ldots, X_p)$. The classification tree i assigns a new object to the class $\hat{c}_{k,i}$ of Y according to the region R_{m_i}. $I(\cdot)$ is an indicator function with value 1 if an instance belongs to R_m and value 0 if not. R_{m_i} is defined by the inputs used in the splits leading to that terminal node. The modal class of the instances in a region R_{m_i} (also called the mth terminal node of the i-th classifier) is usually taken as estimate for $\hat{c}_{k,i}$. This notation is consistent with that used in (Hastie et al. 2001).

The estimation of τ_i is based on the prediction accuracy of each classifier in the final subset g. A new object is slipped into each of the g classifiers. The assigned class $\hat{\tau}_i$ is found with respect to the tree whose terminal node better classifies the new object. In other words, a new object is assigned to the purest terminal node among all the g classifiers.

Another option of the algorithm is the possibility to learn tree classifiers to select the suitable pair of response classes satisfying (1) using alternative splitting criteria. As for CART, in the application presented in Sect. 3 we refer to both the Gini index and Twoing as alternative splitting rules. It is known that, unlike Gini rule, Twoing searches for two classes that make up together more than 50% of the data and allows us to build more balanced trees even if the resulting recursive partitioning algorithm works slower. As an example, if the total number of classes is equal to K, Twoing uses 2^{K-1} possible splits. Since it has been proved (Breiman et al. 1984, p. 95) that the tree classifier is insensitive to the choice of the splitting rule, it can be interesting to see how it works in a framework characterized by the search of the most accurate classifiers like the one introduced in SASSC.

3 Analyzing the *Letter Recognition* Dataset

In the following, SASSC is applied on the "Letter Recognition" dataset from the UCI Machine Learning Repository (Asuncion and Newman 2007). This dataset is originally analyzed in Frey and Slate (1991), who not achieve a good performance since the correct classified objects did never exceed 85%. Later on, the same dataset is analyzed in Fogarty (1992) using nearest neighbors classification. Obtained results give over 95.4% accuracy compared to the best result of 82.7% reached in Frey and Slate (1991). Nevertheless, no information about the interpretability of the nearest neighbor classification model is provided and the computational inefficiency of such a procedure is deliberately admitted by the authors.

In the Letter Recognition analysis, the task is to classify 20,000 black-and-white rectangular pixel displays into one of the 26 letters in the English alphabet. The character images are based on 20 different fonts and each letter within these 20 fonts was randomly distorted to produce a file of 20,000 unique stimuli. Each stimulus was converted into 16 numerical attributes that have to be submitted to a classifier.

Dealing with $K = 26$ response classes, SASSC provides 25 sequential aggregations. Classification trees aggregated at each single step were chosen according to 10-fold cross validation. A tree was aggregated to the sequence if it provided the lowest cross validated generalization error with respect to the other trees obtainable from different aggregations of (subgroups of) response classes.

The results of the SASSC algorithm are summarized in Fig. 1. It compares the performance of the SASSC model formed by $g = 2$ up to $g = 6$ subsets of the response classes with that of the CART algorithm using in both cases either Gini or Twoing as splitting rules. Bagging (Breiman 1996) and Random Forest (Breiman 2001) are used as benchmarking methods as well. Computations have been carried out using the R software for statistical computing (R development core Team 2009)

The SASSC model using two superclasses consistently improves the results of CART using the Gini (Twoing) splitting rule since the generalization error reduces to 0.49 (0.34) from 0.52 (0.49). As expected, the choice of the splitting rule (Gini or Twoing) is relevant when the number of superclasses g is relatively small ($2 \leq g \leq 4$), whereas it becomes negligible for higher values of g (results for $g \geq 5$ are almost identical). Focusing on Gini splitting criterion, the SASSC's generalization error further reduces to 0.11 when the number of subsets increases to 6. For comparative purposes, Bagging and Random Forest have been trained using 6 and 10 classifiers respectively and, in these cases, obtained generalization errors are worse than those deriving from SASSC with $g = 6$. As for Bagging and Random Forest, increasing the number of trees used to classify each subset

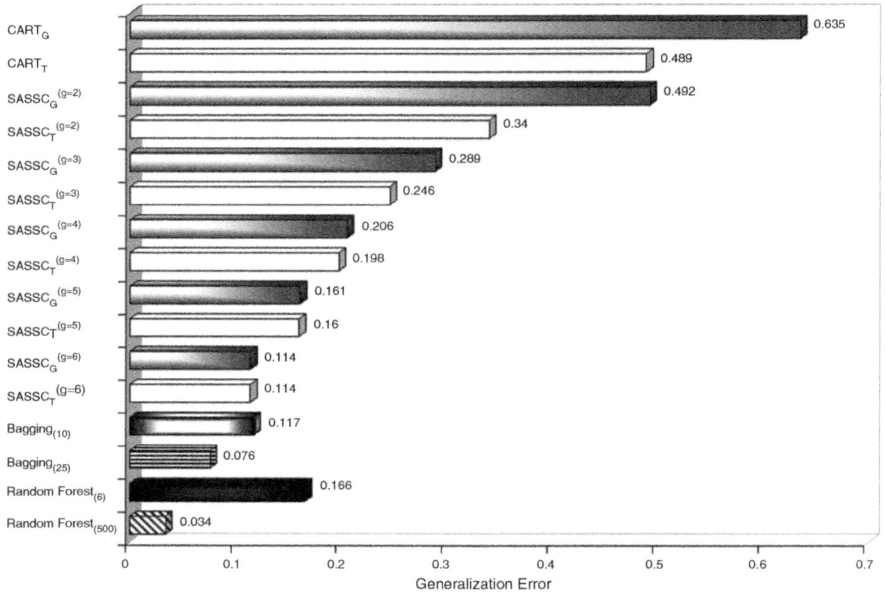

Fig. 1 The generalization errors for the Letter Recognition dataset provided by alternative approaches: as for SASSC, subscript G (T) indicates the Gini (Twoing) splitting rule, whereas apex g indicates the number of superclasses (i.e., classifiers) identified by the algorithm. The subscript for Bagging and Random Forest indicates the number of classifiers used to obtain the classification by majority voting

of randomly drawn objects improves the performance of these two methods in terms of prediction accuracy. The reason is that their predictions derive form ("in-sample") independent bootstrap replications. Instead, cross-validation predictions in SASSC derives from aggregations of classifications made on "out-of-sample" observations that are excluded from the tree growing procedure. Thus, it is natural to expect that cross-validation predictions are more inaccurate than bagged ones. Of course, increasing the number of subsets of the response classes in SASSC reduces the cross-validated generalization error but, at the same time, increases the complexity of the final classification model. In spite of a relatively lower accuracy, interpretability of the results in SASSC with $g = 6$ is strictly preserved. Figure 2 shows the classifiers obtained for $g = 6$; the user can easily understand the more influential features and their relative split points for each classifier as in standard classification trees. Whereas, the same kind of interpretation is not easily achievable in the case of Bagging (Random Forest) with 25 (500) bootstrap replications.

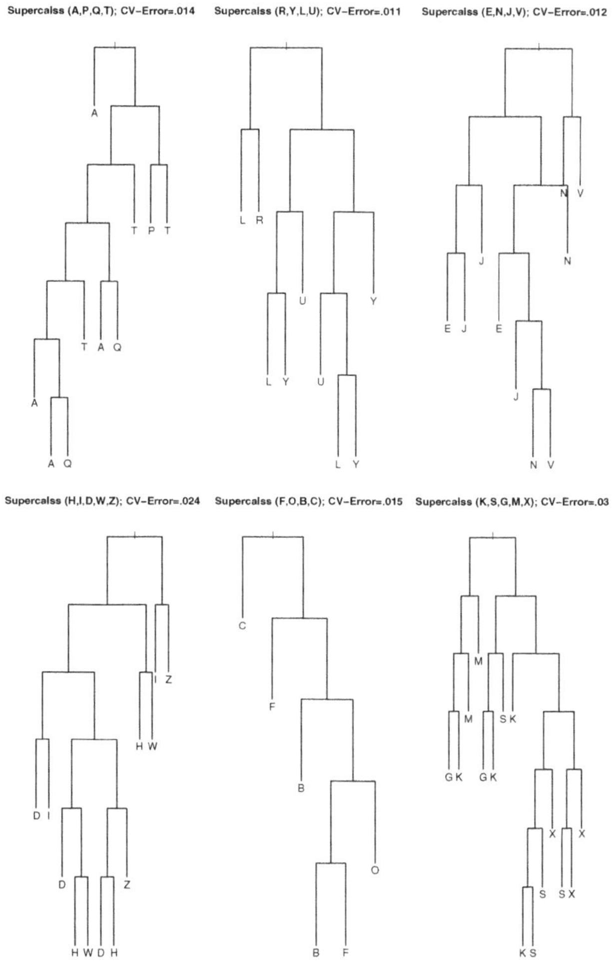

Fig. 2 The six classifiers obtained from the SASSC algorithm with $g = 6$ superclasses using the Gini splitting rule and 10-fold cross validation

4 Concluding Remarks

The motivation underlying the formalization of the SASSC algorithm derives from the following intuition: basically, since standard classification trees unavoidably lead to prediction inaccuracy in the presence of multi-class response, it would be favourable to look for a relatively reduced number of classifiers each one relating to a subset of classes of the response variable, the so called super-classes. Reducing the number of response classes for each of those classifiers naturally leads to improve the overall prediction accuracy. To further enforce this guess, an appropriate criterion to derive the correct number of super-classes and the most parsimonious tree structure for each of them has to be found. In this respect, a sequential approach that automatically proceeds through subsequent aggregations of the response classes might be a natural starting point.

The analysis of the Letter Recognition dataset demonstrated that the SASSC algorithm can be applied pursuing two complementary goals: (1) a *content-related* goal, resulting in the specification of a classification model that provides a good interpretation of the results without disregarding accuracy; (2) a *performance-related* goal, dealing with the development of a model resulting effective in terms of predictive accuracy without neglecting interpretability. Taking these considerations into account, SASSC appears as a valuable alternative to evaluate whether a restricted number of independent classifiers improves the generalization error of a classification model.

References

Asuncion, A., & Newman, D. J. (2007). *UCI machine learning repository*. School of Information and Computer Sciences, University of California, Irvine. Retrieved December 21, 2007 from http://mlearn.ics.uci.edu/MLRepository.html.

Breiman, L. (1996). Bagging predictors. *Machine Learning, 24*, 123–140.

Breiman, L. (2001). Random forests. *Machine Learning, 45*, 5–32.

Breiman, L., Friedman, J. H., Olshen, R. A., & Stone, C. J. (1984). *Classification and regression trees*. Belmont, CA: Wadsworth.

Dietterich, T. G. (2000). Ensemble methods in machine learning. In J. Kittler & F. Roli (Eds.) *Multiple classifier system, Proceedings of the First International Workshop MCS 2000* (pp. 1–15). New York: Springer.

Fogarty, T. (1992). First nearest neighbor classification on Frey and Slate's letter recognition problem (technical note). *Machine Learning, 9*, 387–388.

Frey, P. W., & Slate, D. J. (1991). Letter recognition using Holland-style adaptive classifiers. *Machine Learning, 6*, 161–182.

Gama, J. (2004). Functional trees. *Machine Learning, 55*, 219–250.

Hastie, T. J., Friedman, J., & Tibshirani, R. J. (2001). *The elements of statistical learning*. New York: Springer.

R Development Core Team. (2009). R: A language and environment for statistical computing. Vienna, Austria: R Foundation for Statistical Computing. Retrieved October 21, 2009 from http://www.R-project.org.

Clustering Textual Data by Latent Dirichlet Allocation: Applications and Extensions to Hierarchical Data

Matteo Dimai and Nicola Torelli

Abstract Latent Dirichlet Allocation is a generative probabilistic model that can be used to describe and analyse textual data. We extend the basic LDA model to search and classify a large set of administrative documents taking into account the structure of the textual data that show a clear hierarchy. This can be considered as a general approach to the analysis of short texts semantically linked to larger texts. Some preliminary empirical evidence that support the proposed model is presented.

1 Introduction

Analyzing and modeling textual information is often complicated by the size of the data collection. In fact, text, such as product descriptions, is contained in large databases. A reduction of textual data preserving essential relationships among objects is needed in order to achieve tasks as classification, clustering, or to give relevance judgement. Also, the structure of the data at hand should be properly taken into account. Statisticians and researchers in the field of information retrieval made significant progress on these topics.

We will consider in the sequel the following applied problem: we had to implement a system capable of ordering by relevance and cluster by content short sentences (10–20 words), descriptions of administrative procedures. The goal was to select a subset of procedures close by meaning (i.e. authorizations for loading/unloading zones), but with different wording and different procedural itinerary (one could be a formal declaration, another a request for authorization, etc.), and to rank them based on their cost (in terms of time, number of needed documents, fees, etc.), effectively finding the most efficient one. This tool, useful for administrative simplification, named UNISPA, was developed as a joint project with an engineering

[1] We thank WEGO S.r.l. for developing and funding the UNISPA project.

M. Dimai (✉)
Dept. of Economics and Statistics, University of Trieste, P.le Europa 1, 34127 Trieste, Italy
e-mail: matteo.dimai@gmail.com

F. Palumbo et al. (eds.), *Data Analysis and Classification*,
Studies in Classification, Data Analysis, and Knowledge Organization,
DOI 10.1007/978-3-642-03739-9_28,

company that offers business solutions and consulting to the public administration and enterprises. We will start by considering the problem of relevancy search in a large corpus of textual data, then the problem of clustering will be addressed.

For efficient relevancy search, a numeric representation of text is needed: usually, the syntax is ignored ("bag of words" approach) and the text is represented as a vector of term frequencies. This approach, although simple, is plagued with polysemy and synonyms: therefore, a way to represent the meaning of the text is needed. The importance of recognizing synonyms increases with short documents, as in our applied problem, where there is little chance that the same concept gets expressed with two synonyms. It's therefore clear that, due to the nature of the data, ordering by relevance and clustering by content of short sentences are tasks that cannot be accomplished at all by boolean search methods nor by usual relevancy search methods working at word level. Meaning in textual data can be represented through latent variable models. The basic idea is that a latent semantic structure is present in the text itself, hidden under the term frequencies: this structure is identified using probabilistic *generative models* that treat text as the result of a random process. Based on the texts analyzed, a suitable model is estimated, generally including one or more *latent*, document-specific variables. The idea of a latent semantic structure has been first developed by Deerwester et al. (1990) with the Latent Semantic Analysis (LSA) and first put on probabilistic grounds by Hofmann (1999) with the Probabilistic Latent Semantic Analysis (pLSA). The pLSA model has been further worked on and we present an application and an extension of the most successful model based on the pLSA, the Latent Dirichlet Allocation (LDA) (Blei et al. 2003) (the pLSA model can be seen as a particular case of the LSA model, as noted in Girolami and Kaban 2003).

The data analyzed are by themselves insufficient for the LDA model, but they can be linked to large, adequate documents. Procedures are defined by one or more specific laws and this allows one to establish a formal and syntactical laws-procedures hierarchy. We argue that the basic model proposed by Blei et al. (2003) could be extended to embody this situation. The paper is organized as follows. Section 2 introduces the basic LDA model and declines the procedure for inference on its parameters. In Sect. 3, an extension of the model to cope with hierarchy in document description is presented. This approach is further extended in Sect. 4.

2 The LDA Model

We define the basic elements of the model as follows:

- A *word* is an item from a set of V items, where V is the dimension of the vocabulary.
- A *document* is a sequence of N words denoted by $\mathbf{w} = (w_1, w_2, \ldots, w_n, \ldots, w_N)$, where w_n is the n-th word in the sequence. The ordering of the words, as will be noted later, is irrelevant.
- A *corpus* is a set of M documents, denoted by $\mathcal{D}$.
- A *topic* is an item from a set of k items.

The LDA model is a probabilistic generative model of a corpus. The idea behind it is that documents are represented as random mixtures over latent topics: this random mixture is then used by generate a sequence of topics and words, where each topic is characterized by a probability distribution over words. It assumes the following generative process for a document **w**, given k latent topics and a vocabulary of V words:

1. Choose $N \in \text{Poisson}(\xi)$
2. Choose $\theta \in \text{Dir}(\alpha)$
3. For each of the N words w_n:
   (a) Choose a topic $z_n \in \text{Multinomial}(\theta)$
   (b) Choose a word w_n from $p(w_n|z_n, \beta)$, a multinomial probability conditioned on the topic z_n

The documents are generated independently one of each other. Of the variables involved, only the words w_n are observed: topics z_n and topic mixtures θ are latent variables. The dimensionality k and the parameter α of the Dirichlet distribution are assumed known and fixed and the $k \times V$ β matrix is assumed fixed, but unknown and to be estimated. The assumption on α can be relaxed as a modified Newton–Raphson algorithm to estimate α is available. The number of topics k is chosen by the researcher. The i-th row of the β matrix is the multinomial distribution on words conditioned on the i-th topic. N is an ancillary variable and its randomness can be ignored. Various inferential strategies can be adopted, ranging from empirical to a fuller Bayesian approach (by assigning appropriate priors to β and α).

Given the parameters α and β, the joint distribution of a topic mixture θ, a set of N topics **z** and a set of N words **w** is given by

$$p(\theta, \mathbf{z}, \mathbf{w}|\alpha, \beta) = p(\theta|\alpha) \prod_{n=1}^{N} p(z_n|\theta) p(w_n|z_n, \beta). \tag{1}$$

The θ variables are sampled once per document and have a clear semantic interpretation (a vector of proportions of single topics in text, where topics are described by different multinomial distributions on words); these will be the numeric representations of documents we will use.

For inference we would need to evaluate the distribution of latent variables θ and **z**, $p(\theta, \mathbf{z}|\mathbf{w}, \alpha, \beta)$. Unfortunately, problems arise with $p(\mathbf{w}|\alpha, \beta)$, which is, written in terms of the model parameters:

$$p(\mathbf{w}|\alpha, \beta) = \frac{\Gamma(\sum_i \alpha_i)}{\prod_i \Gamma(\alpha_i)} \int \left(\prod_{i=1}^{k} \theta_i^{\alpha_i - 1} \right) \left(\prod_{n=1}^{N} \sum_{i=1}^{k} \prod_{j=1}^{V} (\theta_i \beta_{ij})^{w_n^j} \right) d\theta. \tag{2}$$

This distribution is intractable for exact inference, as shown by Dickey (1983): therefore, approximating methods are needed. We adopt the variational inference approach by Blei et al. (2003): variational methods try to approximate intractable functions by finding the tightest upper (or, as in this case, lower) bound on the function within a family of simpler functions, indexed by variational parameters. The appropriate values for the variational parameters (and thus the tightest lower bound on the log-likelihood) are found by minimizing the Kullback–Leibler divergence between the posterior distribution and the simpler variational distribution. These variational parameters (γ and ϕ) have a distinct semantic interpretation: k-dimensional document-specific γ variables represent the number of words of the document generated by each of the k topics (plus α), while ϕ variables represent $p(z|w)$, the probability of the topic given the word. ϕ variables are used to estimate β; on the other hand, subtracting α (which is symmetrical and has simply a smoothing effect) from the γ and normalizing we obtain estimates of θ.

It should be noted that the LDA model has an implicit justification of the "bag of words" approach as the likelihoods are the same up to a multiplicative constant regardless if the document is viewed as a sequence or a bag of words (Buntine and Jakulin 2006).

3 Considering Documents Structure: An Extension of the LDA Model

The LDA model works reasonably well with corpuses that cover a set of topics that is not too large and with an average document length that is sufficient for a clear definition of conditioned word probabilities in the estimation phase. The corpus we considered for the UNISPA project consisted of administrative documents that satisfied the first condition, but not the second: each document had an average length of 10–15 words and there was a massive presence of words without a distinct semantic meaning ("authorization", "declaration", "activity", etc.), no room was left for a clear topic estimation.

Therefore, we incorporated the laws in a two-level LDA model: laws are large documents (well over 1,000 words on average) that cover a set of topics that is not too sparse and show a sufficient use of synonyms. As they are administrative documents, they are also plagued with words that cannot be assigned to a topic: "law", "decree", "commission" and so on, but they can be filtered out without loss of informative power as would happen for descriptions of administrative procedures. The rationale behind this model is that descriptions of procedures regulated by laws will necessarily be semantically correlated to laws themselves – they're talking about the same thing, after all – and can be thought as of being "generated" by the laws they're related to.

We estimate parameters θ and β with the vocabulary (of size V) being the union of the vocabularies of laws and descriptions. The relationship between laws and procedure is in general a many-to-many one, so we assign to each description the mean $\overline{\theta}$ of the θ's of the laws related to it and "enrich" the Dirichlet prior. A new LDA model is estimated: the β matrix obtained by the first LDA model is used to initialize the model for descriptions. Not fixing the β matrix to the values estimated during the first phase allows adjustments for lexical differences as well as a more proper inclusion of words that appear in descriptions only. The topic mixtures for descriptions – ω – are estimated using a Dir($\alpha + \overline{\theta}$) as prior distribution.

More schematically the procedure can be described as follows:

1. A LDA model using laws only is estimated taking as vocabulary (of size V) the union of the vocabularies of the laws and of the procedures, filtering non-informative words (i.e. "law" or "commission') and assigning random, non-zero starting values to the β matrix.
2. Estimates of laws' θ and provisional estimates of β are obtained.
3. For each procedure, obtain the appropriate $\overline{\theta}$ vector, mean of the θ topic mixtures of the laws that regulate the procedure itself, and assign it to the procedure.
4. Estimate a LDA model for the procedures, using the provisional β as starting values and using for each procedure a $Dir(\alpha + \overline{\theta})$ prior on its topic mixture ω.
5. Obtain, by using the same variational algorithm, estimates of ω for the procedures and the final β matrix.

Procedure descriptions are therefore a priori grouped based on the semantic similarity in the laws that generate them – which is a reasonable assumption and has proved effective in practice. This approach is nothing strange for Dirichlet distributions: they are conjugates to the Multinomial and, given a vector μ of multinomial results and a Dir(α) prior on the multinomial distribution, the posterior distribution is Dir($\alpha + \mu$). Our approach mimics the multinomial updating, up to a multiplying constant: in fact, the Variational Bayes (VB) algorithm produces document-specific variational parameters γ that can be thought as parameters of posterior document-specific Dirichlets. Using $\overline{\theta}$ (mean of normalized γ) ensures that the effect of higher-level data remains of the same magnitude of α, as it's common practice in these models to set $0<\alpha\leq 1$ to ensure that topics are properly separated. Setting $\alpha = 0.5$ resulted as an appropriate solution in this context. It is sensible, after all, that factors that influence α are of its same magnitude and data should always have a fair chance to influence the posterior distribution. Still, what we call the "multiplier problem" (using normalized means of γ, unnormalized ones or something in between) must be dealt with on a case-by-case basis: it depends on the nature of the specific data sets and the relationship between them.

The UNISPA project is focused on querying and clustering the dataset, providing results that should be semantically sensible in the first place and that are therefore difficult to evaluate numerically. We present some preliminary results, aiming at demonstrating the LDA and LDA-HD capability to identify separate semantic areas. Three hundred descriptions from the dataset used for the UNISPA project have been manually classified in two classes: 117 descriptions were related to construction procedures, 183 to commerce. Using as clustering variables the parameters estimates produced by the models, two clusters of documents are identified by using the k-means algorithm: the results of the clustering were compared with the "true" classification. For both models the clustering has been repeated 100 times with random initial assignments and the most frequent configuration (the one we assume as the most stable) was chosen. With regards to the LDA algorithm, 261 descriptions have been assigned to the correct cluster, for an 87% accuracy. For the LDA-HD model, 286 descriptions have been assigned to the correct cluster, for a 95.3% accuracy.

4 Generalizing the Prior Enrichment: Collapsed Variational Bayes

The idea that, given a specific corpus to be queried and clustered by content, related documents can be included in the estimation phase, is a general approach that can be applied to a broader set of algorithms, i.e. to the Collapsed Variational Bayes algorithm (CVB, Teh et al. 2007) for the LDA model. The CVB algorithm is based on the same rationale of the VB algorithm in Blei et al. (2003), but the family of approximating distributions is simpler because θ and β parameters are marginalized out and only the distribution of $\mathbf{z}$ vectors is approximated. In this setting, β is stochastic, the same as the β matrix with a symmetric Dir(η) prior outlined in Blei et al. (2003).

Based on the assumption that latent topic tokens $\mathbf{z}$ are mutually independent, the approximating family of distributions for the vectors $\mathbf{z}$ is then a product of multinomial distributions: the variational parameters are therefore of the type $\gamma_{ijk} = q(z_{ij} = k)$, the probability that the i-th word in document j belongs to topic k (q denotes the approximating probability distribution, as the real one is unknown). From these quantities estimates of other probabilities can be found, as well as estimates of θ vectors and of the β matrix.

The variational parameters can be estimated exactly, but a less computationally expensive Gaussian approximation is available:

$$\hat{\gamma}_{ijk} \propto \left(\alpha + E_{\hat{q}}[n_{jk\cdot}^{\neg ij}]\right)\left(\eta + E_{\hat{q}}[n_{\cdot kx_{ij}}^{\neg ij}]\right)\left(W\eta + E_{\hat{q}}[n_{\cdot k\cdot}^{\neg ij}]\right)^{-1}$$
$$\times \exp\left(-\frac{Var_{\hat{q}}(n_{jk\cdot}^{\neg ij})}{2(\alpha + E_{\hat{q}}[n_{jk\cdot}^{\neg ij}])^2} - \frac{Var_{\hat{q}}(n_{\cdot kx_{ij}}^{\neg ij})}{2(\eta + E_{\hat{q}}[n_{\cdot kx_{ij}}^{\neg ij}])^2} + \frac{Var_{\hat{q}}(n_{\cdot k\cdot}^{\neg ij})}{2(W\eta + E_{\hat{q}}[n_{\cdot k\cdot}^{\neg ij}])^2}\right). \tag{3}$$

The quantities $n_{jk\cdot}^{\neg ij}$, $n_{\cdot kx_{ij}}^{\neg ij}$ and $n_{\cdot k\cdot}^{\neg ij}$ can be obtained by summing over the appropriate γ variational parameters.

The same approach of LDA-HD can be applied to CVB: the α hyperparameter can be considered as the a priori number of words in the current document assigned to each topic and the η hyperparameter can be viewed as the a priori number of times a certain word has been assigned to a certain topic. Finally, the $W\eta$ can be regarded as the a priori number of words assigned to a single topic in the whole corpus. These quantities can be enriched by adding additional terms: we can obtain estimates of a priori numbers of words assigned to a certain topic in a document from higher-level linked topics as well as estimates of topic-word probabilities ($p(z_k|x_{ij})$) and topic probabilities ($p(z_k)$).

We have tested the CVB-HD model with a set of longer documents, synopses of verdicts of administrative tribunals about the Code of Public Contracts. These documents are longer (about 50 words on average) and cover an even narrower set of topics. It is, however, a highly specialized set and there are more multitopic documents, so single topics are more difficult to detect. The documents are classified according to the article of the Code the verdict mainly refers to: we have tested the CVB and CVB-HD models by using a training set of 672 verdict summaries (very unevenly spread across Code articles) and a test set of another 45 verdict summaries (also unevenly spread across articles). These were then classified using the k-nearest-neighbor algorithm, with $k = 10$. Tests were performed for 210 various combinations of hyperparameters and number of topics ($\alpha = 0.05, 0.1, 0.3, 0.5, 0.8$, $\eta = 0, 05, 0.1$ and $k = 50, 55, \ldots, 145, 150$), recording both the percentages of documents for which the correct class is the most frequent neighbor and documents for which the correct class has been identified among those of the neighbor documents. Both quantities were necessary as the 672 training documents were spread across 258 theoretical classes, corresponding to the number of articles in the Code of Public Contracts (in practice not empty classes were about 150). Moreover, many classes only had a single item in them, so classification by simply choosing the most popular class of the neighbors, as usual in the k-nearest-neighbor, wasn't feasible.

Three different extensions of the CVB model have been considered: in each case the articles of the Code of Public Contracts were selected as higher-level documents. In the first CVB-HD extension, variational parameters for higher-level documents were estimated together with those for lower-level documents, with an updated prior enrichment at each iteration. This resembles a model for linked documents. The second CVB-HD extension dealt with higher-level and lower-level documents separately, estimating variational parameters in two separate steps. The third alternative simply consisted in enriching the training set for a simple CVB with the higher-level documents, without enforcing a prior enrichment link.

Results were promising. Of 45 test documents, with the best combinations of hyperparameters and numbers of topics, 42.2% of documents were correctly identified by the k-nn with CVB, 53.3% with CVB-HD with one-step estimation, 48.8% with CVB-HD with two-steps estimation and 48.8% with CVB-HD with no link between higher-level and lower-level documents. On the other hand, at least a neighbor of the correct class has been found with the best combination of hyperparameters and dimensionality in 80% of cases with CVB, in 82.2% of cases with CVB-HD with one-step estimation, in 64.4% of cases with CVB-HD with two-steps estimation and in 84.4% of cases with CVB-HD with no link. Note that the best combinations of hyperparameters and dimensionality with respect to k-nn classification and with respect to identification of at least one neighbor of the correct class are generally different. It should be noted that the "multiplier problem" arises in CVB-HD as well as in LDA-HD: we haven't found a conclusive solution, but in our experiments better results are generally achieved with a multiplier set to one (as in the LDA-HD).

5 Discussion

We have based our work on Latent Dirichlet Allocation, a generative model for textual corpuses that acts as a dimensionality reduction technique – but that, unlike other established techniques, has firm probabilistic grounds. Inference is done using the variational inference approximations, but other well known strategies can be used (see Griffiths and Steyvers 2002 for an application of Markov Chain Monte Carlo).

The LDA model is an extensible one – each part of the model supports viable alternatives that suit different uses of the model itself. In our case, we enriched the Dirichlet prior on descriptions by estimating an LDA model for documents that show a hierarchical relationship to descriptions – laws. This has improved the quality of the model, suggesting that it may be sensible to use information obtained from a corpus to estimate a related one whenever a relationship between documents of the two corpuses is known. The extension proposed can be useful when dealing with short texts that would otherwise be insufficient for LDA parameter estimation.

Hyperparameters in the LDA model are very difficult to evaluate, especially if we have two hyperparameters to set by putting a Dirichlet prior on the single rows of the β matrix, the direction where the latest models are pointing at. Although in Blei et al. (2003) an analytic method for determining α has been proposed, in our experiments we have found this approach to give rather inconsistent results, with a heavy dependence on the starting values. So in practice the "best" values are often found by examining a pre-determined set of alternatives or, roughly speaking, by trial and error. The influence of these arbitrary decisions can be reduced by considering the link between documents and thus give an appropriate prior distribution.

References

Blei, D., Ng, A., & Jordan, M. (2003). Latent dirichlet allocation. *Journal of Machine Learning Research, 3*, 71–80.

Buntine, W., & Jakulin, A. (2006). Discrete principal component analysis. In C. Saunders, M. Grobelnik, S. Gunn, & J. Shawe-Taylor (Eds.), *Subspace, latent structure and feature selection techniques*. Amsterdam: Springer.

Deerwester, S., Dumais, S., Landauer, T., Furnas, G., & Harshman, R. (1990). Indexing by latent semantic analysis. *Journal of the American Society for Information Science and Technology, 41*(6), 391–407.

Dickey, J. (1983). Multiple hypergeometric functions: Probabilistic interpretations and statistical uses. *Journal of the American Statistical Association, 78*, 628–637.

Girolami, M., & Kaban, A. (2003). On an equivalence between PLSI and LDA. In *Proceedings of SIGIR*.

Griffiths, T. L., & Steyvers, M. (2002). A probabilistic approach to semantic representation. In *Proceedings of the 24th annual conference of the Cognitive Science Society*.

Hofmann, T. (1999). Probabilistic latent semantic indexing. In *Proceedings of the twenty-second annual international SIGIR conference*.

Teh, Y. W., Newman, D., & Welling, M. (2007). A collapsed variational Bayesian inference algorithm for latent Dirichlet allocation. *Advances in Neural Information Processing Systems, 19*, 1353–1360.

Multilevel Latent Class Models for Evaluation of Long-term Care Facilities

Giorgio E. Montanari, M. Giovanna Ranalli, and Paolo Eusebi

Abstract The Region Umbria has conducted a survey on elderly patients living in long-term care facilities since the year 2000. Repeated measurements of items on health conditions are taken for classifying care facilities to allocate resources (RUG system). We wish to evaluate the performance of the nursing homes in terms of some aspects of health condition and quality of life of patients. To this end, a Multilevel Latent Class Model with covariates is employed. It allows for modeling a latent trait hidden behind a set of items, also in the presence of a nested error data structure coming from repeated measurements. Eleven items, surveying cognitive status, activities of daily living, behavior and decubitus ulcers, are used to measure the latent variable related to health condition and quality of life. The probability of belonging to ordinal latent classes is modeled in terms of available covariates.

1 Introduction and Data

Evaluation of long-term care facilities is a challeging issue. For example Kane et al. (2003) propose quality-of-life measures using an index-scale obtained through Factor Analysis; the Center for Medicare and Medicaid Services of the United States (Medicare & Medicaid Quality Measures 2004) is engaged in building quality measures in nursing homes using simple descriptive statistics as indicators.

In this paper we deal with this issue using data coming from a survey conducted in a region of central Italy – Umbria – since the year 2000 on elderly patients living in long-term Nursing Homes (NHs) following the classification system of RUGs (Resource Utilization Groups). The aim of the survey is to classify care facilities to allocate resources. For each individual, a nursing assistant quarterly fills up a questionnaire with a wide range of items referring to indicators of health conditions. We are interested in evaluating the performance of 14 NHs in two sanitary districts of Umbria by assessing the health condition and quality of life of the patients. The

G.E. Montanari (✉)
Dipartimento di Economia, Finanza e Statistica; Università degli Studi di Perugia, Italy
e-mail: giorgio@stat.unipg.it

F. Palumbo et al. (eds.), *Data Analysis and Classification*,
Studies in Classification, Data Analysis, and Knowledge Organization,
DOI 10.1007/978-3-642-03739-9_29,

term *performance* in the present application has a different meaning from that used when evaluating Hospitals. In that context, in fact, a person enters the hospital to be healed and then is usually discharged. Performance is defined in terms of mortality rates and length and types of treatments. In the present context, a patient enters a NH without the perspective of being healed or being discharged, so that performance is more referred to being able to preserve a status without deteriorating the health condition, both from a physical and mental point of view.

We apply a Multilevel Latent Class Model (Vermunt 2003) using data coming from 2008 questionnaires on 709 patients. Eleven polytomous items are used among all those available. Items are ordinal with categories increasing with increasing difficulty in accomplishing tasks or with increasing severity of the condition. Only items related to health status – both physical and mental – are considered, while those related to health treatments are discarded. This choice is made to disentangle health conditions from treatments, and, therefore, to classify patients according to their health status and not to the treatments they receive. This allows for NHs to be evaluated according to the health condition and quality of life of their patients, without being penalized for the treatments they provide. Note that the data used here are collected with the aim of budget allocation, so that this purposive item selection was forced to allow its use for assessing the health condition and quality of life of the patients.

Table 1 reports the RUG codes of the items used, their description and categorization. The first large group of items (BCE-items) deals with the psychological status of the patient, surveying memory issues, communication, mood and behavior. The second group (GM-items), on the other hand, deals with more physical issues as the capability of fulfilling tasks related to the Activities of Daily Living (ADL) or with the condition of the skin. We believe that these items hide a latent trait related to the health status and the quality of life of the patients. A multilevel structure arises from repeated measurements of patients within NHs. For each patient we wish to include covariates in the model. In particular, we want to consider the time since the beginning of the stay in the NH and the amount of care required at charging (Nursing home, Medical and Ancillary assistance; NMA). This last covariate, in particular, should be able to account for the patient case-mix at charging.

2 The Model

The basic idea of a Latent Class Model (LCM; Goodman 1974) is that subjects belong to one of several latent classes, and that the response of a subject to a set of items is generated by class-specific probabilities. Indeed, we want to be able to extract from the items surveyed, the latent variable that measures the health condition and quality of life of the patients. However, at the same time, we wish to model such latent variable using covariates and, when covariates are inserted, to introduce and test for a NH effect. In addition, in our case, the measurement on the patients is repeated over different time points so that, other than covariates,

Table 1 RUG code of the items used in the analysis, brief description and categorization

RUG code	Item description	Categorization
B Items: cognitive status		
B2	Short-term memory of the patient	0 – Remembers; 1 – does not remember after 5 min
B4	Ability to take decisions in everyday tasks	0 – Independent; 1 – some difficulty shown in new situations; 2 – supervision needed to take decisions; 3 – the patient never takes decisions independently
C Items: communication		
C4	Ability to communicate and being understood	0 – No problem; 1 – some difficulty, but usually understood; 2 – seldom understood; 3 – never understood / cannot communicate
E Items: mood and behavior		
E1	Synthetic indicator of depression, sad mood, anxiety in the last 30 days	0 – No problems; 1,2,3,4 – score denoting increasing issues
E4	Offensive or aggressive behavior	0 – No problem; 1 – issues in the past 7 days
G Items: ADL		
G1AA	Self mobility in bed	0 – Independent; 1 – supervised; 2 – limited assistance; 3 – intense assistance; 4 – totally dependent
G1BA	Mobility between two points	As for G1AA
G1HA	Eating	As for G1AA
G1IA	Use of toilet	As for G1AA
M Items: skin condition		
M1	Most severe stage of ulcers (any cause)	0 – No ulcers; 1 – rashes; 2 – blisters; 3 – sores; 4 – deep ulcers
M2	Most severe stage of decubitus ulcers	As for M1

we also have a nested data structure. Therefore, a Multilevel LCM as in Vermunt (2003) can be employed. We will first assume that latent classes are ordered along the latent variable and then test whether such assumption is confirmed by the data. Such assumption provides a classification of the patients that allows a comparison of NH performances.

We denote with x_{ij} the value taken by the underlying ordinal latent variable on patient i at occasion j and with k a particular latent class such that $1 \leq x_{ij} = k \leq K$. Let us consider a set of T items. A particular response on item t and the vector of T responses for measurement j on patient i are denoted by y_{ijt} and $\mathbf{y}_{ij}$, respectively. The category of a particular response variable t is denoted by m_t with $m_t = 0, 1, \ldots, M_t$ and a generic response pattern is denoted by $\mathbf{m}$.

We want the probability of belonging to a latent class to account for the nested structure and to depend on covariates. We insert the time of stay (measured as the interval of time since charging), the initial evaluation of the patients in terms of care needed (NMA) in order to adjust for the case-mix, and a NH effect to evaluate effectiveness of caregivers. These three covariates are, respectively, denoted by z_{1ij} and z_{2ij} (continuous), and $\mathbf{z}_{3ij}$ (categorical with 14 levels/dummies). The formulation of a Multilevel ordinal LCM with a random effect to account for repeated measures on each patient and covariates is the following:

$$P(\mathbf{y}_{ij} = \mathbf{m}) = \sum_{k=1}^{K} P(x_{ij} = k | z_{1ij}, z_{2ij}, \mathbf{z}_{3ij}) \prod_{t=1}^{T} P(y_{ijt} = m_t | x_{ij} = k), \quad (1)$$

with

$$\log \frac{P(x_{ij} = k | z_{1ij}, z_{2ij}, \mathbf{z}_{3ij})}{P(x_{ij} = k - 1 | z_{1ij}, z_{2ij}, \mathbf{z}_{3ij})} = \gamma_{0ki} + \gamma_1 z_{1ij} + \gamma_2 z_{2ij} + \boldsymbol{\gamma}'_3 \mathbf{z}_{3ij}, \quad (2)$$

$\gamma_{0ki} = \gamma_{0k} + \tau u_i$, for $k = 2, \ldots, K$ and $\gamma_{01} = -\sum_{k=2}^{K} \gamma_{0k}$, where u_i is a normally distributed random effect with mean 0 and variance 1 and τ is the factor loading. Model (2) is an adjacent-category ordinal logit model with a latent response variable and a random intercept to account for extra patient variability. Estimation of the models is conducted using Latent Gold 4.0 (Vermunt and Magidson 2005). This software assumes that equal distance scores along the latent trait ω^* between 0 and 1 are assigned to the K ordered classes of the latent variable. With $K = 3$, for example, the scores would be $\omega_1^* = 0$, $\omega_2^* = 0.5$ and $\omega_3^* = 1$, for latent classes 1, 2, and 3, respectively. The class size is freely estimated.

3 The Results

The first thing we note running the analysis is that items have different discrimination power in extracting the latent variable, and in particular item E4 – surveying offensive and aggressive behavior – is not significant and is therefore discarded from the analysis. Using the remaining 10 items, we estimate the Multilevel ordinal LCM with covariates with an increasing number of latent classes, $K = 2, \ldots, 8$ and note that the decrease in BIC and AIC3 stabilizes after using four classes (see Fig. 1).

Results in terms of latent classes' profiles show that the latent variable increases with better health conditions and quality of life. Conditional probabilities are summarized in Fig. 2 that shows a profile plot of the latent classes. For each item the expected category is plotted for the four classes. The expected category of polytomous items has been rescaled to the interval [0–1]. For each class the class size is also reported in the legend. The ordinal latent variable can be read from the top line to the bottom one. Two clear patterns arise: decreasing problems with G-items surveying ADL and with BC-items surveying the cognitive status and communi-

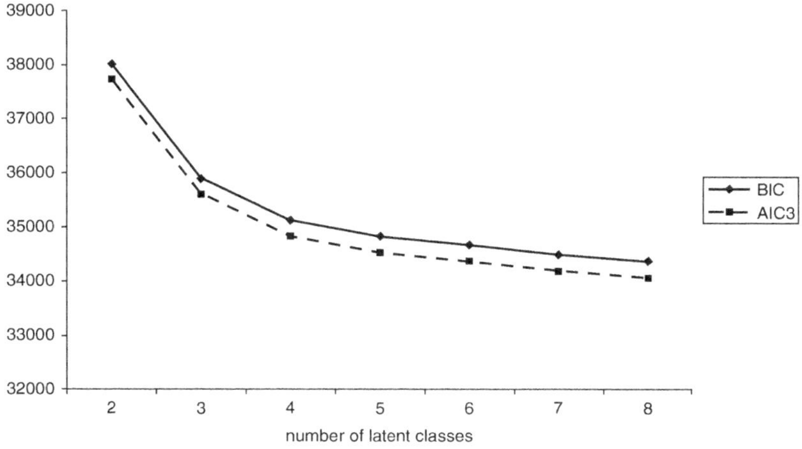

Fig. 1 BIC and AIC3 for models with an increasing number of latent classes

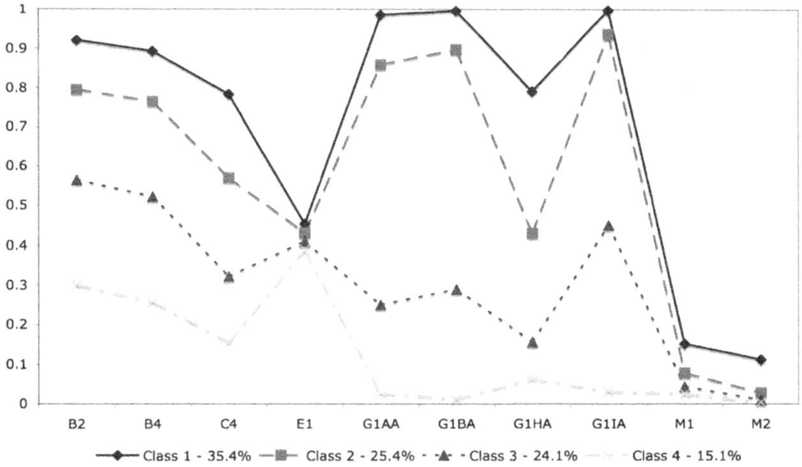

Fig. 2 Profile plot of the Multilevel ordinal LCM with four Classes. Expected categories for polytomous items were rescaled to [0, 1]

cation skills of the patients. Note that, even for the last Class issues are declared for the cognitive status and communication skills. This is consistent with the type of patients residing in the NHs. E and M-items surveying behavior and skin problems, respectively, are found significant. However, they have a much more limited discrimination power compared to the previous items.

Information criteria are also used to test whether more latent factors were present, or whether the latent classes were not ordered. Table 2 reports a summary of the values found. A model with two latent factors and with two levels for each factor – making four classes overall – shows values of the information criteria comparable

Table 2 Summary of model selection information criteria for three models based on four classes, but with different assumptions on the latent trait: one ordered variable, two ordered variables and an unordered nominal variable

Model	Log-likelihood	BIC	AIC3	# of parameters
1 Latent factor with 4 levels	−17,316.91	35,120.52	34,761.81	64
2 Latent factors with two levels each	−17,249.01	35,174.85	34,676.01	89
4 Unordered latent classes	−16,838.72	34,559.61	33,909.44	116

with the one with one latent factor – smaller AIC3, but larger BIC. Given the ambiguity of the results, we inspect the two factors and find that they had a strong association. Therefore, they are not extracting two different types of health condition; in fact, such a strong association is a sign that mental and physical issues determine together the health condition of a patient, and that there is not a clear separation of the two in the data under analysis. As of a model based on unordered latent classes, both BIC and AIC3 decrease, although not sensibly. By looking at conditional probabilities, we find that classes are not completely ordered; however, they have a very similar pattern to those reported in Fig. 2. The only difference could be summarized in terms of a profile plot with a profile of BC-items switched between classes 2 and 3. Note that in this case, model (2) becomes a multinomial model with a different set of γ-parameters for each latent class. In addition a latent trait that is not ordered does not allow the NH to be evaluated and compared in terms of an overall measure of the health condition of the patients. We do not feel that the little difference in the definition of the classes and the magnitude of the decrease in information criteria is sufficient to justify an increase in complexity of the model and a significant loss in interpretability of the parameters of the model.

As of parameter estimates of model (2), Wald tests show that they are all significant. Table 3 reports the values of the estimates, together with their standard error and z-values. Both parameters for time and NMA have a negative sign. Recall that the latent trait increases with good health. This means that as time goes by it is less likely to find a patient in good health and that patients that required more care at charging still have a worse health condition than the others, other things being equal. The factor loading τ is also significant, showing a significant correlation among observations coming from the same patient.

The estimates of the NH effect, on the other hand, have an important interpretation. They can provide, in fact, a measure of effectiveness of the care provided by the care facility other things – patient's effect, time of care, NMA – being equal. Positive parameter estimate are associated with better performing NHs (larger values of the health condition, others things being equal), while negative ones with less effective ones. Those with a non significant parameter do not have a different performance from the average. Figure 3 shows the NHs ordered according to their parameter estimate from the least performing to the best performing of all, together with an approximate 95% confidence interval that allows for direct comparisons

Table 3 Parameter estimates with associated standard errors and test statistic from model (2).

Parameter		Estimate	Standard error	z-Value
Intercept	γ_{01}	−47.532	3.669	−12.95
	γ_{02}	−8.209	0.730	−11.24
	γ_{03}	20.113	1.548	13.00
	γ_{04}	35.629	2.832	12.58
Factor loading (random effect)	τ	−38.080	2.868	−13.28
Time	γ_1	−0.012	0.003	−3.70
NMA	γ_2	−1.080	0.083	−13.07
NHs	$\gamma_{3,NH1}$	0.028	1.372	0.02
	$\gamma_{3,NH2}$	8.684	3.491	2.49
	$\gamma_{3,NH3}$	−20.367	1.873	−10.87
	$\gamma_{3,NH4}$	3.770	2.090	1.80
	$\gamma_{3,NH5}$	1.926	1.639	1.18
	$\gamma_{3,NH6}$	−6.635	2.193	−3.03
	$\gamma_{3,NH7}$	28.551	7.858	3.63
	$\gamma_{3,NH8}$	−4.619	2.273	−2.03
	$\gamma_{3,NH9}$	5.354	1.425	3.76
	$\gamma_{3,NH10}$	−8.539	4.057	−2.10
	$\gamma_{3,NH11}$	−4.535	2.822	−1.61
	$\gamma_{3,NH12}$	2.071	1.788	1.16
	$\gamma_{3,NH13}$	−6.654	2.957	−2.25
	$\gamma_{3,NH14}$	0.966	4.465	0.22

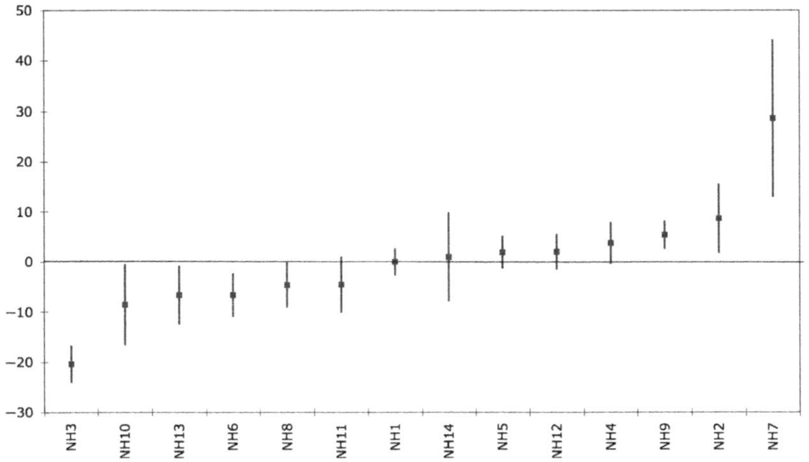

Fig. 3 NHs ordered according to their parameter estimates, together with approximate 95% confidence bounds

between NHs. We can see that the first five NHs have a performance below the average, with NH3 being the worst, and only three above the average, with NH7 being the best performing of all. We find that such classification based on parameter estimates is very robust over different models with an increasing number of levels of the latent trait.

4 Conclusions

In this work we used a Multilevel ordered latent class model to evaluate the performance of 14 Nursing Homes in two Health districts on the central Region Umbria in Italy. Data comes from a survey conducted on elderly patients living in long-term care facilities since the year 2000. Measurements of items on health conditions are taken quarterly by nursing assistants. A latent trait describing an overall measure of health condition is found, that involves both mental and physical conditions. Such latent trait is the response variable in an adjacent-category ordinal logit model with a random intercept to account for extra patient variability due to repeated measures. Time since charging and the initial evaluation of the patients in terms of care needed (NMA) are included as covariates in this model in order to adjust for the case-mix. A Nursing Home effect is also introduced to evaluate effectiveness of caregivers, other things being equal. Four levels of such a latent trait are found sufficient by looking at information criteria. The four classes are ordered according to an increasing level of health condition: a smaller level of the latent trait, in fact, is associated with larger probabilities of surveying items with the largest category, and therefore with the largest degree of difficulty or severity. All covariates in the model are found significant. The nursing home effect can be estimated by looking at their parameter estimate. Five NHs are found to perform below the average, and three above the average. Standard errors associated with parameter estimates provide confidence bounds that allow comparisons between nursing homes.

Acknowledgements The present research is financially supported by the Region of Umbria.

References

Goodman, L. A. (1974). Exploratory latent structure analysis using both identifiable and unidentifiable models. *Biometrika, 61*, 215–231.

Kane, R. A., Kling, K. C., Bershadsky, B., Kane, R. L., Giles, K., Degenholtz, H. B., et al. (2003). Quality of life measures for nursing home residents. *The Journals of Gerontology Series A, 58*, 240–248.

Medicare & Medicaid Quality Measures. (2004). *National nursing home quality measures user's manual*.

Vermunt, J. K. (2003). Multilevel latent class models. *Sociological Methodology, 33*, 213–239.

Vermunt, J. K., & Magidson, J. (2005). *Latent Gold 4.0 user's guide*. Belmont, MA: Statistical Innovations Inc.

Author–Coauthor Social Networks and Emerging Scientific Subfields

Yasmin H. Said, Edward J. Wegman, and Walid K. Sharabati

Abstract In this paper, we suggest a model of preferential attachment in coauthorship social networks. The process of one actor attaching to another actor (author) and strengthening the tie over time is a stochastic random process based on the distributions of tie-strength and clique size among actors. We will use empirical data to obtain the distributions. The proposed model will be utilized to predict emerging scientific subfields by observing the evolution of the coauthorship network over time. Further, we will examine the distribution of tie-strength of some prominent scholars to investigate the style of coauthorship. Finally, we present an example of a simulated coauthorship network generated randomly to compare with a real-world network.

1 Introduction

In this paper, we focus on demonstrating scale-free author–coauthor social networks. A common property of many large networks is that the vertex connectivities follow a scale-free power-law distribution. This feature was found to be a consequence of two generic mechanisms: (1) networks expand continuously by the addition of new vertices (growth), and (2) new vertices attach preferentially to sites that are already well connected (preferential attachment). A model based on these two ingredients reproduces the observed stationary scale-free distributions, which indicates that the development of large networks is governed by robust self-organizing phenomena that go beyond the particulars of the individual systems. Growth means that the number of vertices (actors) increases with time. Preferential attachment means that the more connected a vertex is, the more likely it is to

Y. H. Said (✉)
Isaac Newton Institute for Mathematical Sciences, Cambridge University, Cambridge, CB3 0EH UK
e-mail: ysaid99@hotmail.com
and
Department of Computational and Data Sciences, George Mason University MS 6A2, Fairfax, VA 22030, USA

F. Palumbo et al. (eds.), *Data Analysis and Classification*, Studies in Classification, Data Analysis, and Knowledge Organization, DOI 10.1007/978-3-642-03739-9_30,

acquire new edges. Intuitively, preferential attachment can be understood if we think in terms of social networks connecting people. Here an edge from actor A to actor B means that actor A "knows" or "is acquainted with" actor B. Vertices with many edges represent well-known people with lots of relations. When a new actor enters the community, he or she is more likely to become acquainted with one of those more visible actors rather than with a relative unknown. Models that satisfy these two principles are known as Barabási–Albert models (Barabási and Albert 1999). In this paper, we seek to demonstrate that author–coauthor networks in the statistical literature satisfy these two criteria.

There has been work on author–coauthor networks and the emergence of global brain in Borner et al. (2005), preferential attachment in Roth (2005), and implications for peer review in Said et al. (2008). Coauthorship relationships can be treated as a two-mode networks in which there are two types of nodes; the author nodes and paper nodes, and one relationship type; "person A authored/coauthored paper P". This two-mode social network is expressed in the PCANS model (Krackhardt and Carley 1998; Carley 2002). The PCANS model is represented in the table below:

	Person	Resource	Task
Person	N	S	A
	1-mode	2-mode	2-mode
Resource			C
			2-mode
Task			P
			1-mode

There is a one-to-one correspondence between graphs and matrices; a graph can be fully represented using a matrix. Moreover, matrix algebra is well-defined. Therefore, we will use matrix operations to obtain new socio-matrices having new properties.

Consider the two-mode "author-by-paper" binary social network AP, then

$$AP \times AP^T = AP \times PA = AA,$$

is the one-mode network of authors related through papers. Similarly,

$$AP^T \times AP = PA \times AP = PP,$$

is the one-mode network of papers related through authors.

The author-by-author socio-matrix AA is one of interest because it exhibits relationships among authors, in other words, the author-by-author matrix tells "who-wrote-with-whom".

Data on statisticians and statistics subfields were collected from the online Current Index to Statistics (CIS) database. The procedure used to harvest data involved two stages. First, we queried the database using names of well-established statisticians affiliated with prominent US universities. These data were used to build a

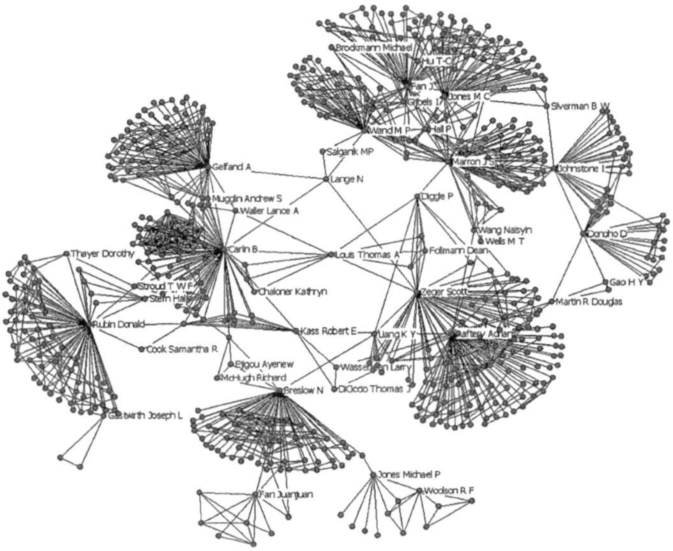

Fig. 1 Coauthorship social network of prominent statisticians

social network of coauthors and to derive the distribution of tie-strength "frequency of coauthorship" among coauthors. A different dataset was used to derive the distribution of clique size. In the second stage, we used the biopharmaceutical subfield as a keyword to query the database, the dataset was used to discover the emergence of that scientific subfield by exploring the evolution of the coauthorship social network over time as a time series.

2 Distribution of Tie Strength

In weighted coauthorship social networks, strength of a tie indicates the frequency of coauthored papers between two actors; in other words, it is a measure of how close two actors are and how much they trust each other. Therefore, studying tie-strength is a subject of interest in coauthorship social networks. We developed a MATLAB program to build the one-mode proximity matrix of the data collected from the CIS database on contributing scientists in the field of statistics. This adjacency weighted matrix was later manipulated to construct the distribution of tie-strength. The statisticians dataset contained 1,767 published papers that had 874 unique author(s)/coauthor(s), the one-mode network of coauthors is shown in Fig. 1. The distribution of tie-strength is shown in Fig. 2.

Figure 2 suggests a power law distribution (Cioffi-Revilla 2005). Because the density curve is close to linear in log-log space the distribution is power law, the next step would be computing the exponent α of the power law. This can be done either by finding the slope of the least-squares regression line in log-log space or by

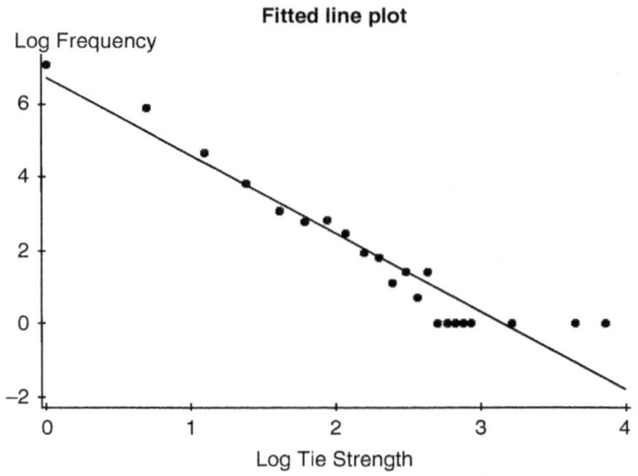

Fig. 2 Examining the attribute tie-strength: linear regression on tie-strength in log-log space

using the following aggregation method for calculating the exponent α.

$$\alpha = 1 + n \left[\sum_{i=1}^{n} \ln \frac{x_i}{x_{min}} \right]^{-1},$$

where x_i is the observed tie-strength, x_{min} is the minimum observed tie-strength (one in our problem), and n is the size of the vector.

An implementation of the aggregation method in MATLAB produced an α-value of 2.1716, the least squares regression model confirmed this result with an α-value of 2.13 and $r^2 = 0.915$. Therefore, we can observe that the distribution of tie-strength is power law with exponent value of 2.17.

Looking into the low-level processes that produced the many-some-few power law pattern, we conjecture that this behavior can be generated in view of the following reasons. First of all, there are higher chances to find two coauthors who simply published together few times. Many of these statisticians are professors who may have a number of graduates working with on a project or paper at a given time period. Upon graduation, many of these students prefer a career in the industry, therefore, they lose contact with their professors leaving behind one or two published papers with that professor. On the other hand, some scientists find themselves in the research area, as a result, the likelihood that two already coauthored individuals publish again rises. If you coauthored a good quality paper with someone and you liked him/her, chances you are going to publish with him/her again if there is mutual agreement increase. And finally, there are those authors who favor only very few coauthors; a colleague or a fellow student who maintains good contacts and relations with that author, to publish with the most.

We further investigated the distribution of tie-strength of individual authors. Figure 3 shows a typical distribution of tie-strengths. We investigated three additional authors. Surprisingly, the distribution is again power-law with exponent α

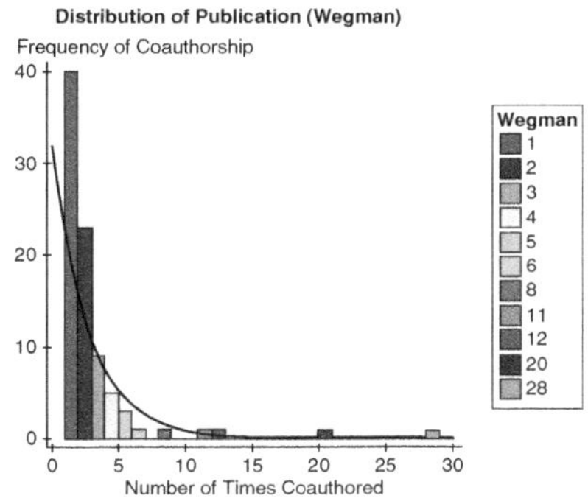

Fig. 3 Distribution of tie-strength among authors: E. Wegman

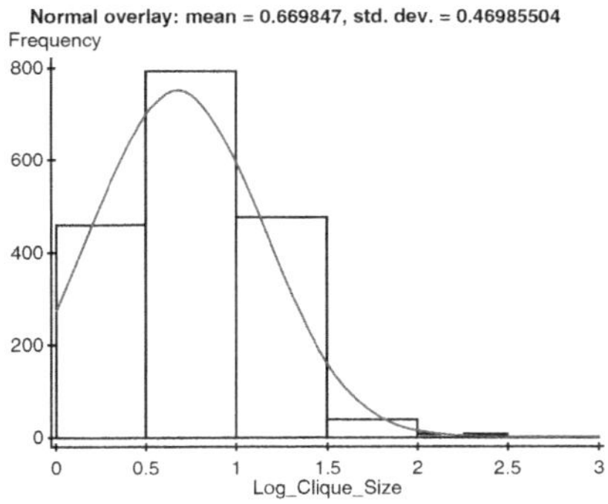

Fig. 4 Clique size in log scale

ranging 1.5–1.85. Because $\alpha < 2$ both the mean and the variance of the distribution of the power-law are not defined and hence the power-law is said to be not well-behaved. For the mean and variance of a power-law to be well-behaved α has to be greater than 3, if $2 < \alpha < 3$ only the mean is finite. We also note that the distribution of tie-strength is a self-similar power-law distribution for coauthorship social networks.

3 Distribution of Clique Size

An important factor in preferential attachment is the clique size; the number of people coauthored a single paper. Note that a paper with sole author or two coauthors is technically not considered a clique. A clique in a graph must have at least three fully connected nodes "complete graph/subgraph" Wasserman and Faust (1994). The statisticians dataset was used to construct the distribution of clique size to obtain a better understanding of how coauthors interact. Figure 4 shows the distribution of clique size. The distribution of clique size is approximately lognormal with mean $\mu = 1.954$ and standard deviation $\sigma = 1.6$.

4 Random Graph Model for Preferential Attachment

The model is based on stochastic "random" processes, in which nodes are generated randomly at each time step. At each time step, a new paper gets published and one of three things could happen:

1. New actor(s) try to attach to existing actors.
2. Already existing non-attached actor(s) attempt to make an attachment(s).
3. Already attached actor(s) strengthen their ties.

And each node has the attributes:

1. Name
2. Age
3. Weight
4. Preference
5. Status
6. Field
7. Active flag

These attributes uniquely identify actors, some of which change rapidly/slowly over time while other attributes remain the same over time. For example, the attributes "name" and "field" do not change. The evolution of "weight" and "status" attributes can be viewed as a time series because they change faster than any other attributes. "Age" changes linearly over time. Meanwhile, the "active" flag operates as a switch initially set to "on" but later could change to "off", once it is changed to "off" it remains in that state forever. Certain actors might change the attribute "preference".

The model was implemented in MATLAB and consists of approximately 350 lines of code, it exploits the distributions of tie-strength and clique-size to build the coauthorship network. Figure 5 is a two-mode author-by-paper simulated network.

0	101	102	103	104	105	106	107	108	109	110
1	1	0	0	1	0	1	0	0	0	0
2	0	1	0	0	1	1	0	0	0	0
3	0	0	1	0	0	0	0	0	0	1
4	0	0	0	1	0	0	0	1	0	0
5	0	0	0	0	0	0	1	0	0	0
6	0	0	0	0	0	0	1	0	0	0
7	0	0	0	0	0	0	0	1	0	0
8	0	0	0	0	0	0	0	0	1	0

Fig. 5 A simulated social network: two-mode author-by-paper social network

0	1	2	3	4	5	6	7	8
1	3	1	0	1	0	0	0	0
2	1	3	0	0	0	0	0	0
3	0	0	2	0	0	0	0	0
4	1	0	0	2	0	0	1	0
5	0	0	0	0	1	1	0	0
6	0	0	0	0	1	1	0	0
7	0	0	0	1	0	0	1	0
8	0	0	0	0	0	0	0	1

Fig. 6 A simulated social network: one-mode author-by-author social network

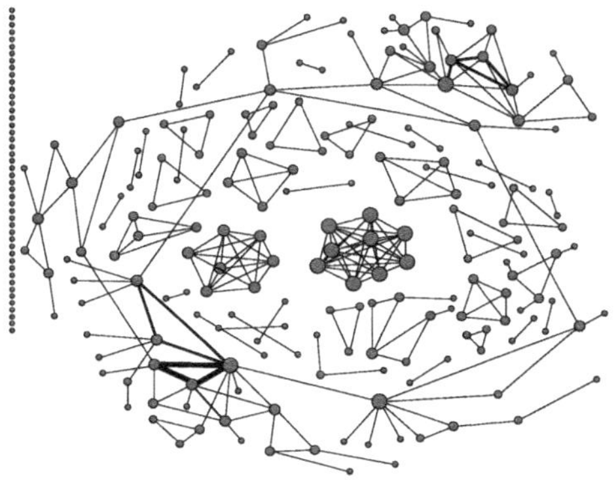

Fig. 7 A simulated coauthorship network

Note that a new publication surfaces at each time step. Figure 6 shows the one-mode coauthorship network corresponding to the matrix in Fig. 5.

Figure 7 shows a simulated coauthorship social network, the program ran for 100 iterations. The simulated network is similar to the network obtained from empirical data, see Sect. 5.

5 The Emergence of Scientific Subfields

Here we explore the social network of biopharmaceutical statisticians over time to inspect the emergence of this subfield. The data include papers published between the years 1977 and 2003. There are 157 published papers with 260 unique author(s)/coauthor(s). The structure of this network is somewhat different from the statisticians network, cliques are more evident in the biopharmaceutical social network.

Figures 8, 9, 10, and 11 show the evolution of the network over time.

In 2000, very few statisticians started writing about biopharmaceutical statistics, the graph in Fig. 8 shows an isolated authors with two cliques of size three and two dyads. In Fig. 9, we start seeing more cliques, more groups are publishing in the biopharmaceutical subfield. In Fig. 10, the network is growing tremendously with more

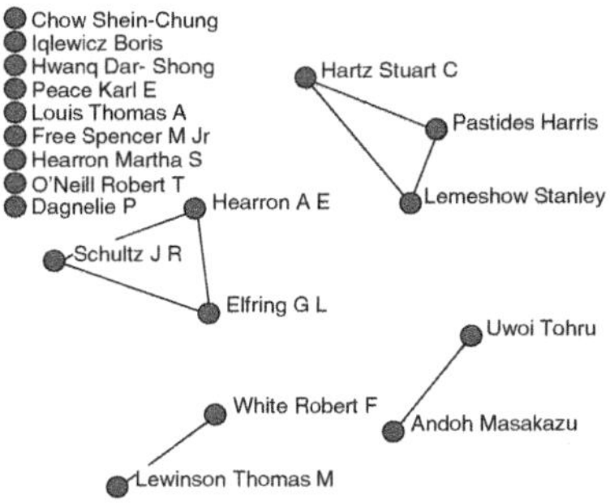

Fig. 8 The evolution of the biopharmaceutical statistical coauthorship social network: the network in 2000

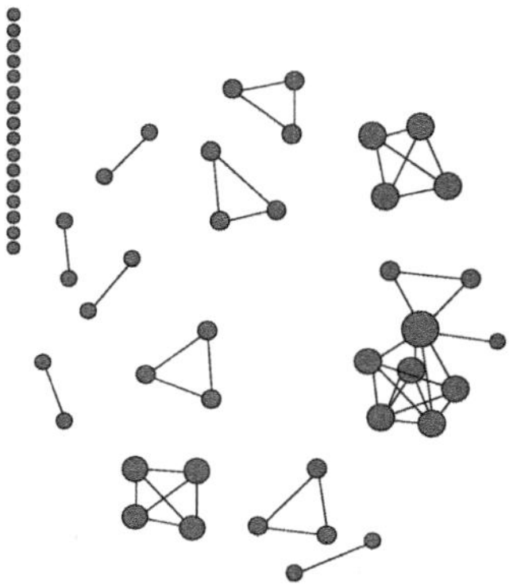

Fig. 9 The evolution of the biopharmaceutical statistical coauthorship social network: the network in 2001

individuals publishing, it seems like H. James and W. Jane are leading coauthors in the new field. Finally, in 2003, the subfield is well-established with several independent and mutually exclusive groups working simultaneously, the leading figures are still H. James and W. Jane.

Two main factors controlled the evolution of this new field. First, small groups and isolated scientists started researching the field, and then over time more scholars and larger groups are becoming more involved and interested in the subject. The second factor resides with the fact that certain coauthors became the key figures in the field, this is evident from the high number of publications they coauthored in the subfield.

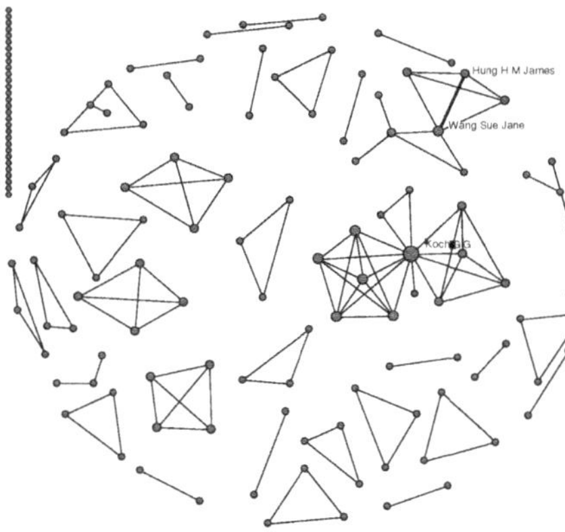

Fig. 10 The evolution of the biopharmaceutical statistical coauthorship social network: the network in 2002

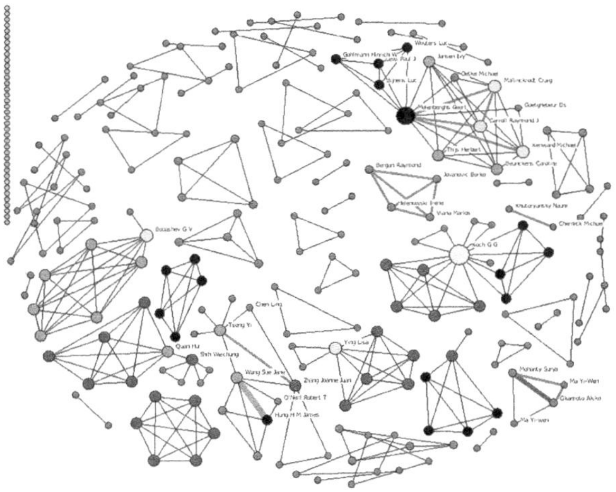

Fig. 11 The evolution of the biopharmaceutical statistical coauthorship social network: the network in 2003

6 The Network of Well-Established Scholars

Figure 1 presents the social network of prominent statisticians affiliated with US universities. In this section, we will use the method of deleting weak ties and pendants (nodes with degree = 1) to expose the important actors in the network. In coauthor social networks, weak ties and hanging nodes do not impose great impact on the status of the network, however, in other types of social networks weak ties could be crucial to the status and performance of the network. What is worth knowing in social networks is who maintains strong ties with who and who is connected

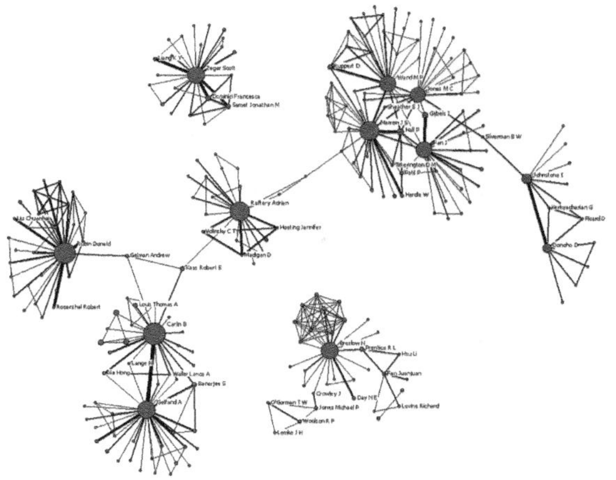

Fig. 12 Statistics social network without nodes with degree and frequency = 1

to the most actors, such authors resemble the heart of the network and their strong ties is the blood that keeps it alive and active.

To begin with, brokerage roles are evident in this network. For example, the node "Lange N" in Fig. 1 can be in the cut-point set, this author is connected to four key player scholars in the network, namely, "Gelfand A", "Carlin B", "Wand M" and "Zeger S". While maintaining good relations with prominent authors in the field of statistics, this author also connects structurally different parts of the network and styles of coauthorship. In addition, "Louis T" can also be considered in the cut-point author set, he is in contact with two mutually exclusive subgroups of authors in which none of the members of each subgroup publishes with member(s) of the other subgroup. "Hall P", "Diggle P" and "Gijbels I" are not cut-point authors but yet connected to key figures in the network, they are publishing with authors most of which are affiliated with different universities and geographically located in different continents. Further investigation reveals that some of these authors although they are not geographically in the same place, but they went to the same school, majored in the same field and spoke the same language and thus maintained good relations.

We proceed by first removing pendant authors (nodes with degree = 1) and then removing ties with weight = 1, Fig. 12 depicts the altered network. Thick edges indicate higher weight, the thicker the link is the higher the number of publications. Big nodes indicate higher degree, the bigger the node is the higher the number of coauthors that particular author has. The network is not centric, in fact, it is more like a chain-network with network diameter = 12. It contains three separate components. In this layout, "Donoho" and "Gelfand" are far away from each other. However, "Zeger" and "Breslow" form two independent subnetworks. Finally, the author "Marron", "Hall", "Fan", "Gijbels", "Wand" and "Jones" are very close and similar authors, they form inbred subnetwork.

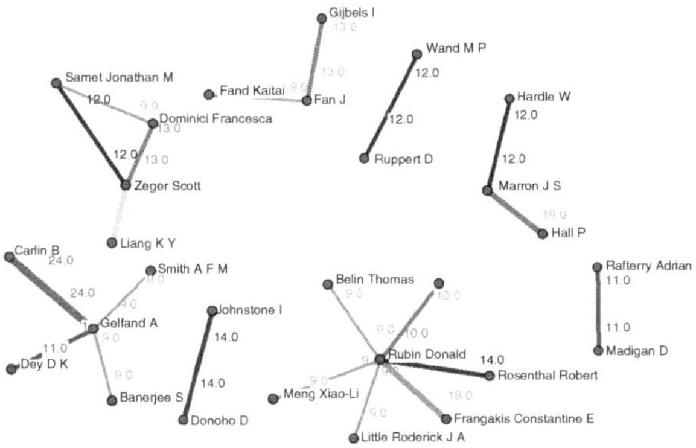

Fig. 13 Statistics social network showing authors having tie strength 7 or higher

Figure 13 shows the network of authors having tie strength of seven or higher. Clearly, there are components of the original network consist of authors with high coauthored papers, members of each component form an elite group of well-trusted authors and coauthors.

7 Conclusions

This work contained two parts; in part one, we used empirical data to investigate the distributions of tie-strength and clique-size in coauthorship social networks. The distribution of tie-strength among authors is a well-behaved power law; however, the distribution of clique size is lognormal. In the second part, we developed a program to generate coauthorship networks based on the distributions of tie-strength and clique size. The model takes into account the fact that authors/nodes status and attributes change over time. The resulting artificial network looked similar to a real-world social network in the Biopharmaceutical subfield.

Acknowledgements The work of Dr. Said is supported in part by Grant Number F32AA015876 from the National Institute on Alcohol Abuse and Alcoholism. The work of Dr. Wegman is supported in part by the Army Research Office under contract W911NF-04-1-0447. Both were also supported in part by the Army Research Laboratory under contract W911NF-07-1-0059. The content is solely the responsibility of the authors and does not necessarily represent the official views of the National Institute on Alcohol Abuse and Alcoholism or the National Institutes of Health.

References

Barabási, A., & Albert, R. (1999). Emergence of scaling in random networks. *Science, 286*(5439), 509–512. doi:10.1126/science.286.5439.509.

Borner, K., Dallasta, L., Ke, W., & Vespignani, A. (2005). *Studying the emerging global brain: Analyzing and visualizing the impact of co-authorship teams*. Bloomington IN: Indiana University.

Carley, K. (2002). Smart agents and organizations of the future. In L. Lievrouw & S. Livingstone (Eds.), *The handbook of new media* (Chap. 12, pp. 206–220). Thousand Oaks, CA: Sage.

Cioffi-Revilla, C. (2005). *Power laws in the social sciences: Discovering complexity and non-equilibrium dynamics in the social universe*. Fairfax, VA: George Mason University.

Krackhardt, D., & Carley, K. (1998). PCANS model of structure in organizations. In *Proceedings of the 1998 international symposium on Command and Control Research and Technology* (pp. 113–119), Monterey, CA. Vienna, VA: Evidence Based Research.

Roth, C. (2005). Generalized preferential attachment: Towards realistic social network models. In *ISWC 4th intl Semantic Web Conference, Workshop on Semantic Network Analysis*, Galway, Ireland.

Said, Y., Wegman, E., Sharabati, W., & Rigsby, J. (2008). Social networks of author–coauthor relationships. *Computational Statistics and Data Analysis, 52*, 2177–2184. doi:10.1016/j.csda.2007.07.021.

Wasserman, S., & Faust, K. (1994). *Social network analysis: Methods and applications*. New York: Cambridge University Press.

Part VI
Statistical Models

A Hierarchical Model for Time Dependent Multivariate Longitudinal Data

Marco Alfò and Antonello Maruotti

Abstract Recently, the use of finite mixture models to cluster three-way data sets has become popular. A natural extension of mixture models to model time dependent data is represented by Hidden Markov models (HMMs) (Cappé et al. 2005); thus, a direct generalization in the finite mixture context for solving the problem of mixing in the time dimension may be given adapting HMMs to three way data clustering. We discuss the issue of longitudinal multivariate data allowing for both time and local dependence.

1 Introduction

Clustering methods generally aim at partitioning objects into meaningful classes (also called clusters), maximizing the homogeneity (or similarity) within a group as well as the difference between groups (Everitt 1993). Standard clustering approaches have been considerably improved, allowing for solutions to some practical issues such as the choice of the number of clusters, the allocation to clusters and the clustering algorithm adopted.

Model based clustering approaches deal with these issues assuming that the objects under study are drawn from a known probabilistic model with the aim at recovering the parameters of such a process. Estimation is usually obtained through maximum likelihood, with an overfitting penalty.

Standard finite fixture approaches (see, e.g. McLachlan and Peel 2000a) have been mainly developed with multivariate normal component-specific distributions (see, e.g. McLachlan and Basford 1988); a notable exception is represented by the work on t-mixture factor analyzers of McLachlan and Peel (2000b).

Recently, three-way data sets have become popular, containing for example attributes (variables) measured on objects (statistical units) in several conditions

M. Alfò (✉)
Dipartimento di Statistica, Probabilità e Statistiche Applicate, Piazzale Aldo Moro, 5 - 00185 Roma
e-mail: marco.alfo@uniroma1.it

F. Palumbo et al. (eds.), *Data Analysis and Classification*,
Studies in Classification, Data Analysis, and Knowledge Organization,
DOI 10.1007/978-3-642-03739-9_31,

(occasions, time points, environments, etc.). Basford and McLachlan (1985) have proposed a finite mixture model for the analysis of such data, where the aim is to cluster objects by explicitly taking simultaneously into account the information on variables and occasions. This approach has been extended in the last few years (see, e.g. Hunt and Basford 2001; Meulders et al. 2002). Vermunt (2007) proposes an extension of this approach assuming that objects may be in a different latent class depending on the situation or, more specifically, objects are clustered with respect to the probability of being in a particular latent class at a given situation.

A natural extension of mixture models to model time dependent data is represented by Hidden Markov models (HMMs) (Cappé et al. 2005); thus, a direct generalization in the finite mixture context to solve the problem of mixing in the time dimension may be given adapting HMMs to three way data clustering. We introduce an extension of the finite mixture model proposed by (Basford and McLachlan (1985)), specializing the proposal of Vermunt (2007). In particular we discuss the issue of longitudinal multivariate data allowing for both time and local dependence.

The plan of the paper is as follows. In the next section model based approaches to cluster three-way data are presented and, in Sect. 3, a possible solution to take into account time dependence in clustering three-way data is proposed. Computational details and a brief simulation study are provided in Sects. 4 and 5, respectively

2 Model-Based Approach to Three-Way Data Clustering

Three-way datasets are often produced as the result of the observation of multivariate-multioccasion phenomena, characterized by various attributes measured on a set of observational units in different situations; in particular, we will refer to such data as three-mode three-way data, where a mode is defined as in Carroll and Arabie (1980).

Let $y_{i,1:P,0:T}$, $i = 1, \ldots, n$, be a PT-dimensional vector corresponding to the i-th unit. Under the mixture model proposed by Basford and McLachlan (1985) and extended by Hunt and Basford (2001), $y_{i,1:P,0:T}$ is assumed to be drawn from a finite mixture of Gaussian distribution:

$$f(y_{i,1:P,0:T}) = \sum_{g=1}^{G} \pi_g \prod_{t=1}^{T} f_g(y_{i,1:P,t}; \mu_{gt}, \Sigma_g), \tag{1}$$

where individuals belong to one of G possible groups in proportions $\pi_1, \ldots, \pi_G$, with $\sum_{g=1}^{G} \pi_g = 1$ and $\pi_g > 0$, $g = 1, \ldots, G$; μ_{gt} is the cluster-time specific mean vector and Σ_g is the cluster specific covariance matrix. As pointed out by Vermunt (2007), such modeling approach implicity assumes that the responses are conditionally (on the cluster) independent and does not take into account the possibility that individuals may move across clusters.

Starting from this consideration and developing the multilevel latent class model, Vermunt (2007) relaxes the assumption of occasion invariant clustering. In details, $y_{i,1:P,0:T}$ is drawn from one of G second level component (clusters) and a *new* element is that conditional on belonging to g, in occasion t cases are assumed to belong to one of $l = \{1, \ldots, L\}$ groups. In the following, we will assume that, conditional on the second level, the i-th response in situation t in the l-th group follows a multivariate normal distribution $y_{i,1:P,t} \sim MVN(\mu_{lt}, \Sigma_l)$:

$$f_l(y_{i,1:P,t} \mid \mu_{lt}, \Sigma_l) = \frac{1}{\sqrt{(2\pi)^P}} |\Sigma_l|^{-\frac{1}{2}} exp \left\{ -\frac{1}{2}(y_{i,1:P,t} - \mu_{lt})^T \times \Sigma_l^{-1}(y_{i,1:P,t} - \mu_{lt}) \right\}, \tag{2}$$

where the within-class covariance matrix, Σ_l, is occasion independent. The hierarchical mixture model has the following form:

$$f(y_{i,1:P,0:T}) = \sum_{g=1}^{G} \pi_g \prod_{t=1}^{T} \sum_{l=1}^{L} \pi^*_{l|g} f_l(y_{i,1:P,t} \mid \mu_{lt}, \Sigma_l), \tag{3}$$

where $\pi_g < 0$, $\sum_{g=1}^{G} \pi_g = 1$, is the prior probability that the i-th observation belongs to the g-th cluster ($g = 1, \ldots, G$), $\pi_{l|g} = \Pr(i \in l \mid i \in g)$, $\sum_{l=1}^{L} \pi_{l|g} = 1$, is the conditional probability that the i-th observation in situation t belongs to the l-th component within the g-th cluster ($l = 1, \ldots, L; g = 1, \ldots, G$). In other words the second level cluster control for potential heterogeneity across statistical units with respect to occasion specific clusters.

It should be noted that this model is equivalent to model described in (1) if $L = G$ and if $\pi_{l|g} = 1$ for $l = g$ and 0 for $l \neq g$; that is, if cases belong to the same class in each occasion. This shows that the hierarchical model extends the standard model by allowing cases to be in a different latent class in each occasion.

Due to the high dimensionality of the estimation problem, a standard EM algorithm cannot be applied; rather the upward–downward algorithm (Pearl 1988) can be used in the implementation of the E-step.

3 Multivariate Hidden Markov Model for Three-Way Data Clustering

The proposed model aims at extending the mixture model to cluster three-mode three-way data to the longitudinal setting, where situations correspond to times and observations for each unit are likely correlated. We adopt a HMM (Cappé et al.

2005) to handle time dependence assuming the hidden dynamics of the stochastic process are governed by a Markov chain. The extension is defined not only to account for individual dynamics; since units may be heterogeneous, we adopt a finite mixture model where components represent clusters and show different transition matrices for the HMMs.

We remark that the applicability of multivariate HMMs is quite wide: it applies to any multivariate time series whose dependency structure is thought to change over time. Important examples include, among others, environmental data, typically multivariate but often not exhaustively measured, and financial times series, where, e.g. the state of a national economy is a powerful qualitative mechanism that determines changes in the correlation structure among the analyzed variables.

Starting from the usual framework for multivariate Gaussian HMM (Giudici et al. see, e.g. 2000) we will focus on empirical situations where three-mode three-way data are analyzed in a hierarchical framework where one of the mode indexes time.

Let us consider a sequence $\{S_{it}\}$, $i = 1, \ldots, n$, $t = 0 \ldots, T$ of random variables whose values are in a finite and enumerable set $\mathcal{S} = \{1, \ldots, m\}$ and let $\{S_{it}\}$, $i = 1, \ldots, N$, $t = 0 \ldots, T$ be described by a homogeneous Markov chain:

$$\Pr(S_{it} = j \mid S_{i0}, \ldots, S_{it-1}) = \Pr(S_{it} = j \mid S_{it-1}), \quad j \in \mathcal{S}.$$

Developing Vermunt (2007), we model time dependence in first level clusters using a HMM framework. In detail, the hierarchical mixture (3) can be rewritten as follows:

$$f(y_{i,1:P,0:T}) = \sum_{g=1}^{G} \pi_g \sum_{\mathcal{S}^T} \left\{ \delta_{s_i 0}^{(g)} \prod_{t=1}^{T} q_{s_{it-1}, s_{it}}^{(g)} \prod_{t=0}^{T} f_{s_{it}}(y_{i,1:P,t} \mid \mu_{s_{it}}, \Sigma_{s_{it}}) \right\}, \tag{4}$$

where $\delta_{s_{i0}}^{(g)} = \Pr(S_{i0} = s_{i0} \mid g)$; $q_{s_{it-1}, s_{it}}^{(g)} = \Pr(S_{it} = s_{it} \mid S_{it-1} = s_{it-1}, g)$ and $f_{s_{it}}$ is defined as in (2).

As can be easily noted, we introduce a further assumption to accommodate time dependence in the proposed model: we do not drop the assumption that the Markov chain is time-homogeneous, but we assume that the HMM is inhomogeneous in the sense that the Markov process can be modelled as depending on the second level classification (or time-homogeneous conditional on g, $g = 1, \ldots, G$). Thus, different clusters have different propensities to be in a given state, as well as different transitions from one state to another one.

Further, the model can be viewed as a specific case of the hierarchical latent Markov model proposed by Rijmen et al. (2008) and represents an adaption of the proposal of Vermunt et al. (2008) to the framework of time-dependent data clustering.

4 Computational Details

In this section, we discuss a modified EM algorithm for MLE of the hierarchical model parameters. Each unit can be thought of as drawn from a finite mixture of G HMM. To introduce the algorithms, let us denote with

$$\gamma_{jtg} = \Pr(S_{it} = j \mid g, y_{i,1:P,0:T}) \tag{5}$$

the posterior probability, given the individual sequence and the g-th component, of being in state j at time t and with

$$\xi_{jktg} = \Pr(S_{it-1} = j, S_{it} = k \mid g, y_{i,1:P,0:T}) \tag{6}$$

the posterior probability that the unobserved sequence visited state j at time $t-1$ and made a transition to state k at time t, given the g-th component and the individual sequence.

The posterior probability that the i-th unit comes from the g-th component of the mixture is

$$\eta_{ig} = Pr(g \mid y_{i,1:P,0:T}) = \frac{\pi_g f(y_{i,1:P,0:T} \mid \theta_g)}{\sum_g \pi_g f(y_{i,1:P,0:T} \mid \theta_g)}, \tag{7}$$

where θ_g are the g-th component density parameters.

Give (3), the expected log-likelihood function has the following form:

$$\begin{aligned} E[\log L_C(\phi)] &= \sum_{i=1}^{n}\sum_{g=1}^{G} \eta_{ig} \log \pi_g \\ &+ \sum_{i=1}^{n}\sum_{g=1}^{G}\sum_{j=1}^{m} \eta_{ig}\gamma_{j1g} \log \delta_j^{(g)} \\ &+ \sum_{i=1}^{n}\sum_{g=1}^{G}\sum_{j\in S^T}\sum_{k\in S^T}\sum_{t=1}^{T} \eta_{ig}\xi_{jktg} \log q_{jk}^{(g)} \\ &+ \sum_{i=1}^{n}\sum_{g=1}^{G}\sum_{j\in S^T}\sum_{t=0}^{T} \eta_{ig}\gamma_{jtg} \log f_j(y_{it} \mid \theta_j). \end{aligned} \tag{8}$$

Our goal is to update current parameters by using old parameters and the data. Thus, the maximum value of $\hat{\delta}_j^{(g)}$ is reached at

$$\hat{\delta}_j^{(g)} = \frac{\sum_{i=1}^{N} \eta_{ig}\gamma_{j0g}}{\sum_{i=1}^{n} \eta_{ig}}. \tag{9}$$

Similarly, we obtain ML estimates for the transition matrix $Q^{(g)}$ and for the weights of the second level clusters, π_g:

$$\hat{q}_{jk}^{(g)} = \frac{\sum_{i=1}^{n}\sum_{t=1}^{T} \eta_{ig}\xi_{jktg}}{\sum_{i=1}^{n}\sum_{t=0}^{T-1} \eta_{ig}\gamma_{jtg}}, \qquad \hat{\pi_g} = \frac{\sum_{i=1}^{n} \eta_{ig}}{n}. \tag{10}$$

If we consider a specific state density of the form shown in (2); i.e. $\theta_j = \{\mu_j, \Sigma_j\}$, the second level cluster parameters estimates are

$$\hat{\mu}_j = \frac{\sum_{i=1}^{N}\sum_{g=1}^{G}\sum_{t=0}^{T} \eta_{ig}\gamma_{jtg} y_{it}}{\sum_{i=1}^{N}\sum_{g=1}^{G}\sum_{t=0}^{T} \eta_{ig}\gamma_{jtg}}, \quad \hat{\Sigma}_j = \frac{\sum_{i=1}^{N}\sum_{g=1}^{G}\sum_{t=0}^{T} \eta_{ig}\gamma_{jtg}[y_{it}-\mu_j][y_{it}-\mu_j]'}{\sum_{i=1}^{N}\sum_{g=1}^{G}\sum_{t=0}^{T} \eta_{ig}\gamma_{jtg}}. \tag{11}$$

5 Simulation Results

To investigate the empirical behavior of the proposed model when clustering multivariate three-way (time dependent) data, we have defined the following simulation study. We generate $R = 250$ samples of size $n = 100, 500, 1000$ and $T = 10$ occasions. In detail, we focus on bivariate hierarchical mixtures of HMMs according to the following scheme:

$$(y_{i,1:2,t} \mid S_{it} = j) \sim MVN(\mu_j, \Sigma_j), \quad j = 1, 2,$$

where j indexes states of the chains; while

$$\mu_1 = \begin{bmatrix} 0.2 \\ 0.7 \end{bmatrix}, \qquad \mu_2 = \begin{bmatrix} 0.5 \\ 0.4 \end{bmatrix}$$

and the covariance matrices are defined as follows

$$\Sigma_1 = \begin{bmatrix} \sigma_{111} & \sigma_{121} \\ \sigma_{211} & \sigma_{221} \end{bmatrix} = \begin{bmatrix} 1 & 0.5 \\ 0.5 & 1 \end{bmatrix}, \qquad \Sigma_2 = \begin{bmatrix} \sigma_{112} & \sigma_{122} \\ \sigma_{212} & \sigma_{222} \end{bmatrix} = \begin{bmatrix} 0.5 & 0.15 \\ 0.15 & 0.5 \end{bmatrix}.$$

Further, according to model described in Sect. 3, we consider the following *true* values for the parameter vectors, assuming $g = 1, 2$:

$$\pi = \begin{bmatrix} \pi_1 \\ \pi_2 \end{bmatrix} = \begin{bmatrix} 0.5 \\ 0.5 \end{bmatrix}, \quad \delta^{(1)} = \begin{bmatrix} \delta_1^{(1)} \\ \delta_2^{(1)} \end{bmatrix} = \begin{bmatrix} 0.8 \\ 0.2 \end{bmatrix}, \qquad \delta^{(2)} = \begin{bmatrix} \delta_1^{(2)} \\ \delta_2^{(2)} \end{bmatrix} = \begin{bmatrix} 0.4 \\ 0.6 \end{bmatrix},$$

$$Q^{(1)} = \begin{bmatrix} q_{11}^{(1)} & q_{12}^{(1)} \\ q_{21}^{(1)} & q_{22}^{(1)} \end{bmatrix} = \begin{bmatrix} 0.8 & 0.2 \\ 0.2 & 0.8 \end{bmatrix},$$

$$Q^{(2)} = \begin{bmatrix} q_{11}^{(2)} & q_{12}^{(2)} \\ q_{21}^{(2)} & q_{22}^{(2)} \end{bmatrix} = \begin{bmatrix} 0.4 & 0.6 \\ 0.25 & 0.75 \end{bmatrix}.$$

We used random starting points for $Q^{(g)}$ and $\delta^{(g)}$, $g = 1, 2,$; starting values for μ_j have been drawn from a standard Gaussian distribution $N(0, 1)$. Starting values for the covariance matrices are

$$\Sigma_1 = \Sigma_2 = \begin{bmatrix} 1 & 0 \\ 0 & 1 \end{bmatrix}.$$

Parameter estimates are shown in Table 2, together with corresponding variances (within the brackets).

As can be noted, results show a clear and consistent path with respect to both Markov and state-specific density parameters: parameter bias decreases and corresponding estimates show a lower variability as the sample size increases.

Table 1 Parameter estimates

n = 100	
$\mu_1 = \begin{bmatrix} 0.275\ (0.041) \\ 0.846\ (0.084) \end{bmatrix}$	$\mu_2 = \begin{bmatrix} 0.643\ (0.049) \\ 0.590\ (0.119) \end{bmatrix}$
$\Sigma_1 = \begin{bmatrix} 0.897\ (0.041) & 0.399\ (0.036) \\ 0.399\ (0.036) & 0.855\ (0.053) \end{bmatrix}$	$\Sigma_2 = \begin{bmatrix} 0.583\ (0.044) & 0.235\ (0.026) \\ 0.235\ (0.026) & 0.642\ (0.062) \end{bmatrix}$
$\delta^{(1)} = \begin{bmatrix} 0.636\ (0.099) \\ 0.364\ (0.099) \end{bmatrix}$	$\delta^{(2)} = \begin{bmatrix} 0.424\ (0.116) \\ 0.576\ (0.116) \end{bmatrix}$
$Q^{(1)} = \begin{bmatrix} 0.747(0.038) & 0.253(0.038) \\ 0.252(0.043) & 0.748(0.043) \end{bmatrix}$	$Q^{(2)} = \begin{bmatrix} 0.456(0.070) & 0.544(0.070) \\ 0.431(0.057) & 0.569(0.057) \end{bmatrix}$
$\pi = \begin{bmatrix} 0.538\ (0.048) \\ 0.462\ (0.048) \end{bmatrix}$	

(continued)

Table 1 (continued)

n = 500

$$\mu_1 = \begin{bmatrix} 0.221\ (0.012) \\ 0.836\ (0.022) \end{bmatrix} \qquad \mu_2 = \begin{bmatrix} 0.571\ (0.012) \\ 0.437\ (0.023) \end{bmatrix}$$

$$\Sigma_1 = \begin{bmatrix} 0.977\ (0.012) & 0.482\ (0.009) \\ 0.482\ (0.009) & 0.968\ (0.015) \end{bmatrix} \qquad \Sigma_2 = \begin{bmatrix} 0.526\ (0.012) & 0.172\ (0.007) \\ 0.172\ (0.007) & 0.531\ (0.015) \end{bmatrix}$$

$$\delta^{(1)} = \begin{bmatrix} 0.722\ (0.054) \\ 0.278\ (0.054) \end{bmatrix} \qquad \delta^{(2)} = \begin{bmatrix} 0.425\ (0.053) \\ 0.575\ (0.053) \end{bmatrix}$$

$$Q^{(1)} = \begin{bmatrix} 0.770(0.015) & 0.230(0.015) \\ 0.200(0.018) & 0.800(0.018) \end{bmatrix} \qquad Q^{(2)} = \begin{bmatrix} 0.413(0.041) & 0.587(0.041) \\ 0.314(0.025) & 0.686(0.025) \end{bmatrix}$$

$$\pi = \begin{bmatrix} 0.529\ (0.050) \\ 0.471\ (0.050) \end{bmatrix}$$

n = 1000

$$\mu_1 = \begin{bmatrix} 0.207\ (0.005) \\ 0.804\ (0.010) \end{bmatrix} \qquad \mu_2 = \begin{bmatrix} 0.590\ (0.005) \\ 0.414\ (0.010) \end{bmatrix}$$

$$\Sigma_1 = \begin{bmatrix} 0.989\ (0.005) & 0.491\ (0.004) \\ 0.491\ (0.004) & 0.985\ (0.007) \end{bmatrix} \qquad \Sigma_2 = \begin{bmatrix} 0.509\ (0.005) & 0.158\ (0.003) \\ 0.158\ (0.003) & 0.511\ (0.006) \end{bmatrix}$$

$$\delta^{(1)} = \begin{bmatrix} 0.748\ (0.035) \\ 0.252\ (0.035) \end{bmatrix} \qquad \delta^{(2)} = \begin{bmatrix} 0.436\ (0.038) \\ 0.564\ (0.038) \end{bmatrix}$$

$$Q^{(1)} = \begin{bmatrix} 0.785(0.012) & 0.215(0.012) \\ 0.208(0.018) & 0.792(0.018) \end{bmatrix} \qquad Q^{(2)} = \begin{bmatrix} 0.387(0.031) & 0.613(0.031) \\ 0.283(0.014) & 0.717(0.014) \end{bmatrix}$$

$$\pi = \begin{bmatrix} 0.511\ (0.052) \\ 0.489\ (0.052) \end{bmatrix}$$

6 Conclusion

A novel mixture clustering model is presented for the analysis of three-way data, where the third way represents time occasion. The proposed method presents a possible solution to the problem of time dependence in hierarchical mixture models;

particularly, it overcomes some limits of previous solutions proposed for time-dependent data, i.e. longitudinal data. Its structure allows class membership change over time through a hidden Markov chain. Here a bivariate case with two groups and two states has been discussed, but we are working on extension to multivariate, multigroups and multistates models.

References

Basford, K. E., & McLachlan, G. J. (1985). The mixture method for clustering applies to three-way data. *Journal of Classification, 2*, 109–125.

Cappé, O., Moulines, E., & Rydén, T. (2005). *Inference in hidden Markov models. Springer series in statistics*. Berlin: Springer.

Carroll, J. D., & Arabie, P. (1980). Multidimensional scaling. *Annaual Review of Psychology, 31*, 607–649.

Everitt, B. S. (1993). *Cluster analysis*. London: Edward Arnold.

Giudici, P., Rydén, T., & Vandekerkhove, P. (2000). Likelihood-ratio tests for hidden Markov models. *Biometrics, 56*, 742–747.

Hunt, L. A., & Basford, K. E. (2001). Fittinga a mixture model to three-mode three-way data with missing information. *Journal of Classification, 18*, 209–226.

McLachlan, G. J., & Basford, K. E. (1988). *Mixture models: Inference and applications to clustering*. New York: Marcel Dekker.

McLachlan, G. J., & Peel, D. (2000a). *Finite mixture models*. New York: Wiley.

McLachlan, G. J., & Peel, D. (2000b). Mixture of factor analyzers. In P. Langley (Ed.), *Proceedings of the seventeenth international conference on Machine Learnings*. San Francisco: Morgan Kauffmann.

Meulders, M., De Boeck, P., Kuppens, P., & Van Mechelen, I. (2002). Constrained latent class analysis of three-way three-mode data. *Journal of Classification, 19*, 277–302.

Pearl, J. (1988). *Probabilistic reasoning in intelligent systems: Networks of plausible inference*. San Matteo, CA: Morgan Kauffmann.

Rijmen, F., Vansteelandt, K., & De Boeck, P. (2008). Latent class models for diary method data: Parameter estimation by local computations. *Psychometrika, 73*(2), 167–182.

Vermunt, J. K. (2007). A hierarchical mixture model for clustering three-way data sets. *Computational Statistics and Data Analysis, 51*, 5368–5376.

Vermunt, J. K., Tran, B., & Magidson, J. (2008). Latent class models in longitudinal research. In S. Menard (Ed.), *Handbook of longitudinal research: Design, measurement, and analysis* (pp. 373–385). Burlington, MA: Elsevier.

Covariate Error Bias Effects in Dynamic Regression Model Estimation and Improvement in the Prediction by Covariate Local Clusters

Pietro Mantovan and Andrea Pastore

Abstract We consider a dynamic linear regression model with errors-in-covariate. Neglecting such errors has some undesirable effects on the estimates obtained with the Kalman Filter. We propose a modification of the Kalman Filter where the perturbed covariate is replaced with a suitable function of a local cluster of covariates.

Some results of both a simulation experiment and an application are reported.

1 Introduction

In regression analysis, covariate measurement errors occur in many applications. This happens both in economics and in social sciences, where covariates are often represented by difficultly measurable constructs. Also in other areas of application, like hydrology (see, for instance Chowdhury and Sharma 2007 and the references therein), covariates are often affected by instrumental or calibration errors. When dealing with parametric models, neglecting such errors has some undesirable effects on the estimates. In this context, many techniques have been proposed in order to face this problem (see, for instance, Fuller 1986; Cook and Stefanski 1994; Carroll et al. 2006), following different approaches, like regression calibration or likelihood-based corrections. In all these approaches, the covariate error variance needs to be known or estimated from covariate replicated measurements (if they are available).

The paper is organized as follows. In Sect. 2 we consider a dynamic regression model West and Harrison (1997), when covariates are affected by an additive error and report some results regarding the bias effects of those errors on the Kalman Filter (KF) equations. In Sect. 3, in order to attenuate these bias effects, we suggest a modification of the Kalman Filter based on local clusters of units by covariate values (in the follows simply *Local Cluster Kalman Filter* [LC-KF]). The results of a simulation experiment, reported in Sect. 4, allow to show and to evaluate the bias effects due to covariate errors and the performances of the proposed method. Moreover, Sect. 5 contains the results of an application to a problem of

A. Pastore (✉)
Department of Statistics, University Ca Foscari, S Giobbe, Cannaregio, 873 -I-30121 Venezia, Italy
e-mail: andrea.pastore@unive.it

F. Palumbo et al. (eds.), *Data Analysis and Classification*, Studies in Classification, Data Analysis, and Knowledge Organization, DOI 10.1007/978-3-642-03739-9_32,

Nitrogen Dioxide concentration forecasting from hourly data, where LC-KF gives appreciable improvements with respect to the standard KF. Finally, in Sect. 6 some conclusions are drawn.

2 Bias Effects in Dynamic Regression Models with Errors in the Covariate

For the aim of this work, we consider a dynamic multiple linear regression model West and Harrison (1997) written in the following state-space form:

$$y_t = \mathbf{x}_t'\boldsymbol{\theta}_t + v_t, \tag{1}$$

$$\boldsymbol{\theta}_t = \mathbf{M}\boldsymbol{\theta}_{t-1} + \mathbf{u}_t, \tag{2}$$

for $t = 1, 2, \ldots,$ where y_t is the (univariate) response variable, $\mathbf{x}_t \in \mathbb{R}^k$ is a vector or covariate (possibly including a constant term), $\boldsymbol{\theta}_t$ is the regression parameter vector and $\mathbf{M}$ is the system matrix, here assumed as known. For the random errors v_t and $\mathbf{u}_t$, we assume: $v_t \sim \mathrm{N}[0, q]$, $\mathbf{u}_t \sim \mathrm{N}_k[\mathbf{0}, \mathbf{Q}]$, with known variances q and $\mathbf{Q}$. Moreover, we assume that these errors are mutually and serially uncorrelated, and also uncorrelated with $\boldsymbol{\theta}_0$. For this model, and under these assumptions, the updating equations for the estimate $\mathbf{m}_{t|t}$ of $\boldsymbol{\theta}_t$ and its variance $\mathbf{P}_{t|t}$ are those known as the Kalman Filter (Kalman 1960):

$$\mathbf{P}_{t|t} = \left(\mathbf{P}_{t|t-1}^{-1} + \mathbf{A}_t\right)^{-1}, \tag{3}$$

$$\mathbf{m}_{t|t} = \mathbf{m}_{t|t-1} + q^{-1}\mathbf{P}_{t|t}\mathbf{x}_t\left(y_t - \mathbf{x}_t'\mathbf{m}_{t|t-1}\right), \tag{4}$$

with: $\mathbf{A}_t = q^{-1}\mathbf{x}_t\mathbf{x}_t'$, $\mathbf{m}_{t|t-1} = \mathbf{M}\ \mathbf{m}_{t-1|t-1}$ and $\mathbf{P}_{t|t-1} = \mathbf{M}\ \mathbf{P}_{t-1|t-1}\ \mathbf{M}' + \mathbf{Q}$, where $\mathbf{m}_{0|0}$ and $\mathbf{P}_{0|0}$ are known starting values.

We suppose now that a vector $\mathbf{z}_t = \mathbf{x}_t + \mathbf{e}_t$ is observed instead of $\mathbf{x}_t$, with: $\mathrm{E}[\mathbf{e}_t] = \mathbf{0}$, $\mathrm{VAR}[\mathbf{e}_t] = \boldsymbol{\Sigma}$ and $\mathrm{cov}[\mathbf{x}_i, \mathbf{e}_t] = \mathbf{0}$, for $i \le t$, where $\boldsymbol{\Sigma}$ is a positive definite matrix. We assume this covariate measurement error to be uncorrelated with any other random term in the state-space model (1)–(2).

If $\boldsymbol{\Sigma}$ is unknown, the use of the observed values of $\mathbf{z}_t$ in the Kalman Filter equations can produce some undesired effects, in terms of both bias and efficiency, on $\mathbf{m}_{t|t}$ and $\mathbf{P}_{t|t}$. Some characterizations for these effects can be studied considering the Kalman Filter equations where the perturbed covariate $\mathbf{z}_t$ is used instead of $\mathbf{x}_t$. Let us denote with $\mathbf{z}_t^0$ the observed value of $\mathbf{z}_t$ and define the set $\mathbf{Z}_t^* = (\mathbf{z}_1^0, \ldots, \mathbf{z}_{t-1}^0, \mathbf{z}_t)$ containing the observed values $\mathbf{z}_t^0$ up to time $t-1$ and the not yet observed value $\mathbf{z}_t = \mathbf{x}_t + \mathbf{e}_t$. Moreover, let us define $\mathbf{P}_{t|t}(\mathbf{Z}_t^*)$ and $\mathbf{m}_{t|t}(\mathbf{Z}_t^*)$, as the analogous of $\mathbf{P}_{t|t}$ and $\mathbf{m}_{t|t}$ in (3) and (4), and consider their expectations with respect to $\mathbf{e}_t$. It is possible to prove that, for any distribution of $\mathbf{e}_t$, these expectations depend from the distribution of $\mathbf{e}_t$ only by means of the variance matrix $\boldsymbol{\Sigma}$. We call $\mathrm{E}_{\boldsymbol{\Sigma}}\left[\mathbf{P}_{t|t}(\mathbf{Z}_t^*)|\mathbf{P}_{t|t-1}\right]$ *expected current variance matrix* (the subscript in $\mathrm{E}_{\boldsymbol{\Sigma}}[\cdot]$ denotes that the expectation depends from the matrix $\boldsymbol{\Sigma}$). Let us consider two variance matrix $\mathbf{\Sigma}_1$ and $\mathbf{\Sigma}_2$, such that $\mathbf{\Sigma}_1 \leq_{\mathrm{L}} \mathbf{\Sigma}_2$, where $\leq_{\mathrm{L}}$ denotes the Löwner partial ordering (Marshall and Olkin 1979). Under some reasonable conditions, we can prove that:

$$\mathbf{P}_{t|t} \leq_{\mathrm{L}} \mathrm{E}_{\mathbf{\Sigma}_1}\left[\mathbf{P}_{t|t}(\mathbf{Z}_t^*)|\mathbf{P}_{t|t-1}\right] \leq_{\mathrm{L}} \mathrm{E}_{\mathbf{\Sigma}_2}\left[\mathbf{P}_{t|t}(\mathbf{Z}_t^*)|\mathbf{P}_{t|t-1}\right]. \tag{5}$$

Moreover, let us define the *expected current bias*:

$$\delta_{\mathbf{\Sigma}} = ||\mathrm{E}_{\mathbf{\Sigma}}\left[\mathbf{m}_{t|t}(\mathbf{Z}_t^*)|\mathbf{m}_{t|t-1}\right] - \mathbf{m}_{t|t}||.$$

We can prove that, under general conditions:

$$\mathbf{\Sigma}_1 \leq_{\mathrm{L}} \mathbf{\Sigma}_2 \rightarrow \delta_{\mathbf{\Sigma}_1} \leq \delta_{\mathbf{\Sigma}_2}. \tag{6}$$

The proof of these results can be found in Mantovan and Pastore (unpublished).

3 A Local Cluster Kalman Filter

In this section, we propose a modification of the Kalman Filter in order to mitigate the effect of the covariate error. Essentially, the idea is to replace the observable values $\mathbf{z}_t$ with suitable values: $\overline{\mathbf{z}}_t = \overline{\mathbf{x}}_t + \overline{\mathbf{e}}_t$ such that: $\mathrm{E}\left[\overline{\mathbf{e}}_t\right] = \mathrm{E}\left[\mathbf{e}_t\right] = 0$ and $\mathrm{VAR}\left[\overline{\mathbf{e}}_t\right] \leq_{\mathrm{L}} \mathrm{VAR}\left[\mathbf{e}_t\right]$. The values $\overline{\mathbf{z}}_t$ are defined by a function of a cluster $\mathcal{I}_{\mathbf{z}_t}$ of the previous covariate vectors:

$$\overline{\mathbf{z}}_t = \sum_{i=1}^{t} w_{i,t}\mathbf{z}_i,$$

where $\mathcal{I}_{\mathbf{z}_t} \subseteq (\mathbf{z}_1, \ldots, \mathbf{z}_t)$ and $w_{i,t}$ are non negative weights such that $w_{i,t} = 0$ if $\mathbf{z}_i \notin \mathcal{I}_{\mathbf{z}_t}$ and $\sum_{i=1}^{t} w_{i,t} = 1$. For any choice of both the cluster $\mathcal{I}_{\mathbf{z}_t}$ and the weights $w_{i,t}$, we can write: $\overline{\mathbf{z}}_t = \overline{\mathbf{x}}_t + \overline{\mathbf{e}}_t$, where:

$$\overline{\mathbf{x}}_t = \sum_{i=1}^{t} w_{i,t}\mathbf{x}_i \qquad \text{and} \qquad \overline{\mathbf{e}}_t = \sum_{i=1}^{t} w_{i,t}\mathbf{e}_i.$$

In general, $\overline{\mathbf{x}}_t$ will be different from $\mathbf{x}_t$, nevertheless: $\mathrm{E}\left[\overline{\mathbf{e}}_t\right] = \mathrm{E}\left[\mathbf{e}_t\right] = 0$, and

$$\overline{\mathbf{\Sigma}}_t = \mathrm{VAR}\left[\overline{\mathbf{e}}_t\right] = \sum_{i=1}^{t} w_{i,t}^2\, \mathbf{\Sigma} \leq_{\mathrm{L}} \mathbf{\Sigma} = \mathrm{VAR}\left[\mathbf{e}_t\right].$$

The effect of replacing the observed value $\mathbf{z}_t$ with $\overline{\mathbf{z}}_t$ gives an improvement in the *expected current precision matrix*, for any choice of the weights vector $\mathbf{w}_t$. Moreover, a reduction in the *expected current bias* is possible under specific conditions on $\mathbf{w}_t$.

For the computation of $\bar{\mathbf{z}}_t$, we suggest a two-stage procedure, involving a clustering stage, for the definition of $\mathcal{I}_{\mathbf{z}_t}$, and a weighting stage, for the definition of the weights $w_{i,t}$. Let us assume to be at time t. Given the observed value of the perturbed covariate vector $\mathbf{z}_t$, in the clustering stage we define the cluster $\mathcal{I}_{\mathbf{z}_t}$ as

$$\mathcal{I}_{\mathbf{z}_t} = \{\mathbf{z}_i : ||\mathbf{z}_i - \mathbf{z}_t||_{\mathbf{L}} \leq \tau, \quad t - s \leq i \leq t\}, \tag{7}$$

where $\mathbf{L}$ is a positive definite matrix, $\tau > 0$ is a *threshold* and $s \in \{1, 2, \ldots, t\}$ is a *window length.* The cluster $\mathcal{I}_{\mathbf{z}_t}$ includes the s previous observations of the covariate such that the distance, defined by the metric $\mathbf{L}$, from $\mathbf{z}_t$ is below a suitable threshold τ. The weights $\mathbf{w}_t$ can be defined in a sufficiently general way, setting:

$$w^{\star}_{i,t} = \begin{cases} \exp\left\{-\frac{1}{2}|i - t|^{\alpha}||\mathbf{z}_i - \mathbf{z}_t||_{\mathbf{L}}\right\} & \text{if } z_i \in \mathcal{I}_{\mathbf{z}_t}, \\ 0 & \text{otherwise,} \end{cases} \tag{8}$$

with $i = 1, \ldots, t$, $\alpha \geq 0$, and: $w_{i,t} = w^{\star}_{i,t} \cdot \left(\sum_{i=1}^{t} w^{\star}_{i,t}\right)^{-1}$.

The two stages of clustering and weighting require to specify the following parameters: s, α, τ, $\mathbf{L}$. Reasonable possible values for α are within the set $\{0, 1, 2\}$, The parameter τ has to be set according to the measurement scale of the covariates. The choice of the matrix $\mathbf{L}$ allows us to define an appropriate distance between covariate vectors.

4 Simulation Experiments

In this section, we denote with $\mathbf{Z}_t^0 = (\mathbf{z}_1^0, \ldots, \mathbf{z}_t^0)$ the set of observed values of $\mathbf{z}_t$ up to time t. A first simulation experiment was set up in order to illustrate the effect described by (5) and (6) in Sect. 2. A simulated series of the model specified by (1)–(2) was considered with: $q = 0.1$, $\mathbf{Q} = 0.2\mathbf{I}_4$, $\mathbf{M} = 0.98\mathbf{I}_4$, $\mathbf{m}_{0|0} = \mathbf{0}$, $\mathbf{P}_{0|0} = 1000\mathbf{I}_4$, and covariate specified by:

$$x_{j,t} = \cos(2\pi t j/100) + \varepsilon_{j,t}, \qquad \text{with} \quad \varepsilon_{j,t} \sim \mathrm{N}[0, 0.1],$$

for $t = 1, \ldots, 100$ and $j = 1, \ldots, 4$. For this model (model A), the values of $\mathbf{P}_{t|t}$ and $\mathbf{m}_{t|t}$ were computed according to the Kalman Filter equations (3) and (4). Then, two other models (B1 and B2) have been obtained replacing the values of $\mathbf{x}_t$ generated for model A with $\mathbf{z}_t = \mathbf{x}_t + \mathbf{e}_t$ where: $\mathbf{e}_t \sim \mathrm{N}_4[\mathbf{e}_t \mid 0, \boldsymbol{\Sigma}]$ with: $\boldsymbol{\Sigma} = \boldsymbol{\Sigma}_1 = 0.1\mathbf{I}_4$ (model B1) and $\boldsymbol{\Sigma} = \boldsymbol{\Sigma}_2 = 0.3\mathbf{I}_4$ (model B2). Five hundred replications for each of the models B1 and B2 were considered. For each replication, the values $\mathbf{P}_{t|t}(\mathbf{Z}_t^0)$ and $\mathbf{m}_{t|t}(\mathbf{Z}_t^0)$ were computed according to the Kalman Filter equations (3) and (4), replacing $\mathbf{x}_t$ with $\mathbf{z}_t$.

In Fig. 1, the 5th, 50th and 95th percentiles of the sample distribution of the variance matrix bias $||\mathbf{P}_{t|t}(\mathbf{Z}_t^0) - \mathbf{P}_{t|t}||$ are reported for both the models B1 (solid lines) and B2 (dotted lines). In Fig. 2, the 5th, 50th and 95th percentiles of the sample

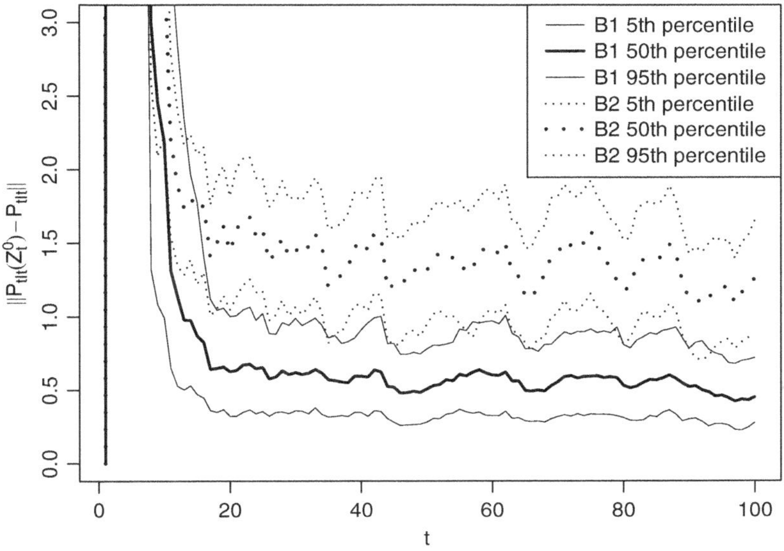

Fig. 1 Distribution of $||\mathbf{P}_{t|t}(\mathbf{Z}_t^0) - \mathbf{P}_{t|t}||$, for models B1 and B2

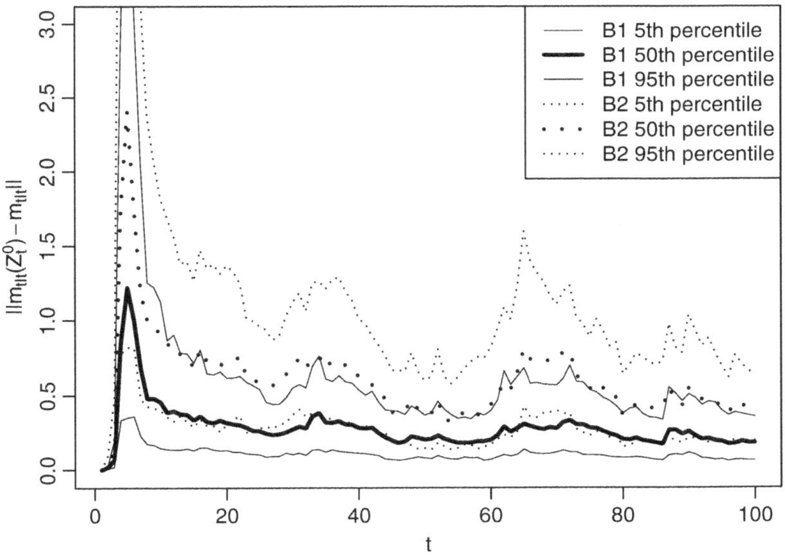

Fig. 2 Distribution of $||\mathbf{m}_{t|t}(\mathbf{Z}_t^0) - \mathbf{m}_{t|t}||$, for models B1 and B2

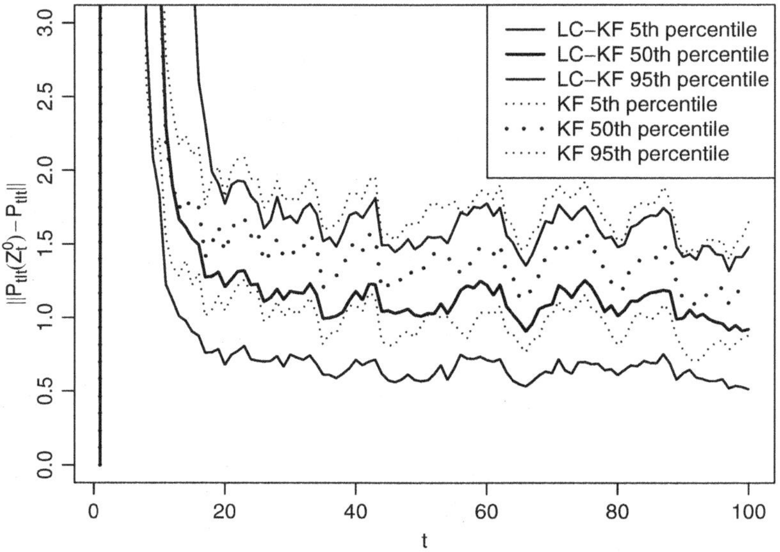

Fig. 3 Distribution of $||\mathbf{P}_{t|t}(\mathbf{Z}_t^0) - \mathbf{P}_{t|t}||$, with LC-KF and KF

distribution of the mean vector bias $||\mathbf{m}_{t|t}(\mathbf{Z}_t^0) - \mathbf{m}_{t|t}||$ are reported for both the models B1 (solid lines) and B2 (dotted lines). Both these figures highlight that the theoretical results (6) and (5) are confirmed by the results of the simulation.

We perform a second simulation experiment, in order to evaluate the performances of the filtering procedure proposed in Sect. 3. Here, for each replication of the model B2, the estimate of the state vector and its variance matrix have been obtained with the standard Kalman Filter (3)–(4), and with the filter procedure based on the covariate local cluster, setting: $\tau = 1$, $s = t - 1$, $\mathbf{L} = \mathbf{I}$ and $\alpha = 0$. The results are summarized in Figs. 3 and 4, where the same three percentiles of the sample distribution of the variance matrix bias and of the mean vector bias are reported respectively for LC-KF and KF. The effect of replacing the perturbed covariate with those computed with the local cluster procedure are quite evident in terms of the variance matrix bias, and less relevant for the mean vector bias.

5 An Application

We consider, as an illustrative application, a dataset containing 1,400 hourly-averaged measurements of Nitrogen Dioxide concentration and other related variables, from a sampling station of Venezia-Mestre in the period between january and march 1997. For more details on these data, see Mantovan and Pastore (2004). We consider a dynamic linear regression model for the Nitrogen Dioxide concentration,

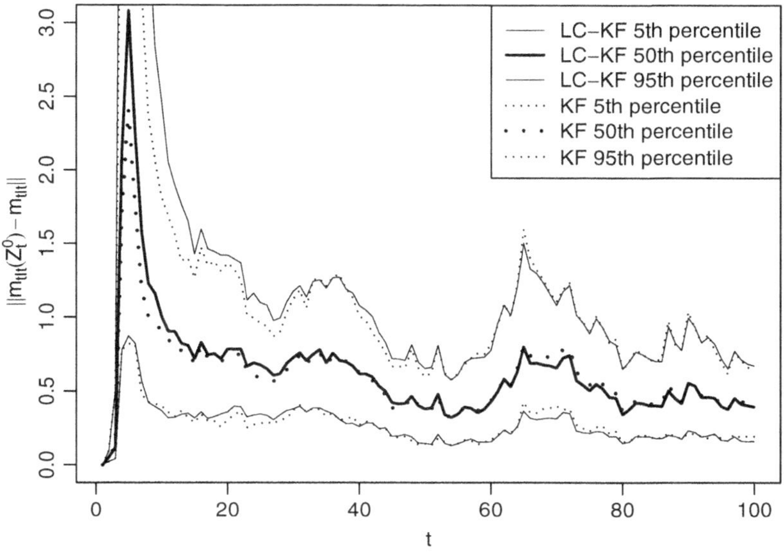

Fig. 4 Distribution of $||\mathbf{m}_{t|t}(\mathbf{Z}_t^0) - \mathbf{m}_{t|t}||$, with LC-KF and KF

taking as regressors: a weekly periodic component estimated by traffic data, the temperature (at the same time), the wind speed (1 h before) and the thermal gradient (8 h before). Generally, these regressors are measured with an instrumental error. We will assume that thee errors can be modelled by an additive model, but we don't need to know or estimate the variance of these errors. We adopt the algorithm proposed in Mantovan et al. (1999), which allows to estimate also the system matrix, the measurement error variance and the system error variance. We consider the d-hours ahead (with $d = 1, 2, 3, 4$) predictions of Nitrogen Dioxide concentration provided by this model, with the LC algorithm (setting: $\alpha = 1$, $s = 2$, $\tau = +\infty$ and $\mathbf{L} = \mathbf{I}$) and without the LC algorithm. We also consider the prediction from a standard (non-dynamic) Bayesian linear regression model (*std*, in short), as a benchmark. Denote with $\hat{y}_{t|t-d}$ the d-step ahead prediction of Nitrogen Dioxide concentration, and let t_1 and t_2 be two values of t such that $t_1 < t_2$. For evaluating the goodness of the d-step ahead prediction in the period from $t = t_1$ to $t = t_2$, we consider the following relative mean squared prediction error:

$$\text{rmspe}(d) = \frac{\sum_{t=t_1}^{t_2} \left(\hat{y}_{t|t-d} - y_t\right)^2}{\sum_{t=t_1}^{t_2} \left(\bar{y}_{[t1:t2]} - y_t\right)^2},$$

where $\bar{y}_{[t1:t2]}$ is the sample mean of y_t in the period from $t = t_1$ to $t = t_2$. The values of this statistic have been reported in Table 1, for the considered models and estimation methods, in the period from $t_1 = 150$ and $t_2 = 1,400$. Some considerations arise from these results. First, the performances of the non-dynamic model tends are similar for different values of d, while the goodness of prediction

Table 1 Nitrogen dioxide data: value of rmspe(d) for the considered algorithms

d	LC-KF	KF	std
1	0.0920	0.1027	0.6508
2	0.1969	0.2102	0.6595
3	0.2950	0.3070	0.6659
4	0.3860	0.4052	0.6708

in the dynamic model strongly depends on the value of d. Second, the LC algorithm seems to perform better than the standard algorithm, for all the values of d we have considered.

6 Conclusions

In this work we show how bias effects of the covariate errors in the estimation process by Kalman Filter of a dynamic linear regression model can be understood in terms of the Löwner partial ordering. The application of the proposed LC-KF algorithm to a problem of Nitrogen Dioxide air concentration forecasting shows how the proposed method may mitigate the negative effects of the covariate errors by improving the predictive performances of a given model.

References

Carroll, R., Ruppert, D., Stefanski, L. A., & Crainiceanu, L. M. (2006). *Measurement error in nonlinear models* (2nd edition). Boca Raton: Chapman and Hall/CRC.

Chowdhury, S., & Sharma, A. (2007). Mitigating parmeter bias in hydrological modelling due to uncertainty in covariates. *Journal of Hydrology, 340*, 197–204.

Cook, J. R., & Stefanski, L. A. (1994). Simulation-extrapolation estimation in parametric measurement error models. *Journal of the American Statistical Association, 89*(428), 1314–1328.

Fuller, W. A. (1986). *Measurement error models*. New York: Wiley.

Kalman, R. E. (1960). A new approach to linear filtering and prediction problems. *Journal of Basic Engineering, 82*, 35–45.

Mantovan, P., & Pastore, A. (2004). Flexible dynamic regression models for real-time forecasting of air pollutant concentration. In: H. H. Bock, M. Chiodi, & A. Mineo (Eds.), *Advances in multivariate data analysis* (pp. 265–276). Berlin: Springer.

Mantovan, P., Pastore, A., & Tonellato, S. (1999). A comparison between parallel algorithms for system parameter estimation in dynamic linear models. *Applied Stochastic Models in Business and Industry, 15*, 369–378.

Marshall, A. W., & Olkin, I. (1979). *Inequalities – Theory of majorization and its applications*. New York: Academic.

West, M., & Harrison, J. (1997). *Bayesian forecasting and dynamic models* (2nd edition). New York: Springer.

Local Multilevel Modeling for Comparisons of Institutional Performance

Simona C. Minotti and Giorgio Vittadini

Abstract We propose a general methodology for evaluating the quality of public sector activities such as education, health and social services. The traditional instrument used in comparisons of institutional performance is Multilevel Modeling (Goldstein, H., Multilevel statistical models, Arnold, London, 1995). However, rankings based on confidence intervals of the organization-level random effects often prevent to discriminate between institutions, because uncertainty intervals may be large and overlapped. This means that, in some situations, a single global model is not sufficient to explain all the variability, and methods able to capture local behaviour are necessary. The proposal, which is entitled Local Multilevel Modeling, consists of a two-step approach which combines Cluster-Weighted Modeling (Gershenfeld, N., The nature of mathematical modeling, Cambridge University Press, Cambridge, 1999) with traditional Multilevel Modeling. An example regarding the evaluation of the "relative effectiveness" of healthcare institutions in Lombardy region is discussed.

1 Introduction

In the 1990s, numerous authors proposed the use of Multilevel Models (Goldstein 1995) in institutional comparisons (see Bryk and Raudenbush 2002, Chap. 5), with rankings based on confidence intervals of the random effects associated with organizations. However, "an overinterpretation of a set of rankings where there are large uncertainty intervals can lead both to unfairness and inefficiency and unwarranted conclusions about changes in ranks" (Goldstein and Spiegelhalter 1996). This is the case of regional or national studies, where confidence intervals may be large and overlapped due to the heterogeneity of individuals within organizations and, more specifically, to non-homogeneity and non-linearity of individual relationships.

S.C. Minotti (✉)
Dipartimento di Statistica, Università degli Studi di Milano-Bicocca, Via Bicocca degli Arcimboldi 8, 20126 Milano, Italy
e-mail: simona.minotti@unimib.it

F. Palumbo et al. (eds.), *Data Analysis and Classification*, Studies in Classification, Data Analysis, and Knowledge Organization, DOI 10.1007/978-3-642-03739-9_33,

We argue that a single global model is not sufficient to interpret such a complexity, and methods able to capture local behaviour are necessary. We will show how the combination of Cluster-Weighted Modeling and traditional Multilevel Modeling represents a very flexible methodology for comparisons of institutional performance. The proposal, which is entitled Local Multilevel Modeling, provides local rankings that allow the policy makers to identify institutions which are above or below the average for specific groups of users.

Typical methods able to detect and model unknown patterns at individual-level are described in Sect. 2. The proposal of Local Multilevel Modeling will be introduced in Sect. 3. Then, in Sect. 4, an example regarding the evaluation of the 'relative effectiveness' of healthcare institutions in Lombardy region is discussed. Finally, in Sect. 5 we provide conclusions and discuss further research.

2 Capturing Local Behaviour

2.1 Mixture Modeling

A good approach to capture local behaviour at individual-level is given by Mixture Modeling (McLachlan and Basford 1988), which is closely related to splitting up a data set by clustering and is an example of unsupervised learning.

Given a set of multivariate data points $\{\mathbf{x}_n\}_{n=1}^N$, where $\mathbf{x}$ is a vector of real-valued predictors, Mixture Modeling factors the density $p(\mathbf{x})$ over multivariate class-conditional densities:

$$p(\mathbf{x}) = \sum_{k=1}^{K} p(\mathbf{x}, c_k) = \sum_{k=1}^{K} p(\mathbf{x}|c_k)p(c_k), \tag{1}$$

where $p(\mathbf{x}|c_k)$, $(k = 1, \ldots, K)$, is the k-th component density and is assumed to be multivariate Gaussian, and the prior probability $p(c_k)$, $(k = 1, \ldots, K)$, is the k-th mixing parameter (i.e. the fraction of the data explained by cluster c_k).

The posterior probability $p(c_k|\mathbf{x})$, $(k = 1, \ldots, K)$, which is given by

$$p(c_k|\mathbf{x}) = \frac{p(\mathbf{x}, c_k)}{p(\mathbf{x})} = \frac{p(\mathbf{x}|c_k)p(c_k)}{\sum_{k=1}^{K} p(\mathbf{x}|c_k)p(c_k)}, \tag{2}$$

indicates the fraction of a point to be associated with the k-th Gaussian; "as a Gaussian begins to have a high probability at a point, that point effectively disappears from the other Gaussians" (Gershenfeld 1999).

However, Mixture Modeling is not completely adequate to the problem under study, because it does not take into account the functional dependence between individual-level characteristics $\mathbf{x}$ and the outcome variable y; the n-th individual, in fact, contributes to each cluster k, $(k = 1, \ldots, K)$, with a posterior probability $p(c_k|\mathbf{x}_n)$ which depends on individual characteristics $\mathbf{x}_n$ only.

2.2 Cluster-Weighted Modeling

A better alternative is given by Cluster-Weighted Modeling (Gershenfeld 1999), which is a framework for supervised learning that combines a density estimation of the input data with a functional relationship to the output data. This leads to a number of local clusters, each containing its own model for describing the observed data.

Cluster-Weighted Modeling essentially tries to estimate the joint density $p(y, \mathbf{x})$ by means of a Gaussian Mixture Model. "But where conventional Gaussian Mixture Modeling only estimates the quantity $p(\mathbf{x})$, Cluster-Weighted Modeling includes an additional output term $p(y|\mathbf{x}, c_k)$ to capture the functional dependence of the output values y_n on the input vectors $\mathbf{x}_n$ as part of the density estimation" (Engster and Parlitz 2006).

Given a set of multivariate data points $\{y_n, \mathbf{x}_n\}_{n=1}^N$, where y is a scalar (the generalization to vector $\mathbf{y}$ is straightforward) and $\mathbf{x}$ is a vector of real-valued predictors, Cluster-Weighted Modeling factors the joint density $p(y, \mathbf{x})$ over multivariate class-conditional densities:

$$p(y, \mathbf{x}) = \sum_{k=1}^{K} p(y, \mathbf{x}, c_k) = \sum_{k=1}^{K} p(y, \mathbf{x}|c_k)p(c_k) = \sum_{k=1}^{K} p(y|\mathbf{x}, c_k)p(\mathbf{x}|c_k)p(c_k), \tag{3}$$

where $p(y|\mathbf{x}, c_k)$, $(k = 1, \ldots, K)$, indicates a dependence in the output space and is assumed to be multivariate Gaussian (with mean given by a function which reflects prior belief on the local relationship between $\mathbf{x}$ and y, for example locally linear (Gershenfeld 1999)); $p(\mathbf{x}|c_k)$, $(k = 1, \ldots, K)$, indicates a domain of influence in the input space and is also assumed to be multivariate Gaussian; $p(c_k)$, $(k = 1, \ldots, K)$, is the k-th prior probability.

If there are unordered, discrete-valued variables $\mathbf{x}_d$ in addition to real-valued predictors $\mathbf{x}_r$, the prior probability $p(c_k)$ in (3) is conditioned on $\mathbf{x}_d$ and hence is replaced by $p(c_k|\mathbf{x}_d)$. Moreover, if y indicates a target variable (in classification problems), the output term in (3) is simply a histogram of the probability to see each state for each cluster and we refer to Cluster-Weighted Classification (Gershenfeld 1999).

The posterior probability $p(c_k|\mathbf{x})$, $(k = 1, \ldots, K)$, is given by

$$p(c_k|y, \mathbf{x}) = \frac{p(y, \mathbf{x}, c_k)}{p(y, \mathbf{x})} = \frac{p(y|\mathbf{x}, c_k)p(\mathbf{x}|c_k)p(c_k)}{\sum_{k=1}^{K} p(y|\mathbf{x}, c_k)p(\mathbf{x}|c_k)p(c_k)}. \tag{4}$$

Analogously to Mixture Modeling, the parameters are estimated using an expectation maximization (EM) algorithm, leading to a local optimum in parameter space.

Cluster-Weighted Modeling is adequate to the problem under study, because it does take into account the functional dependence between user-level characteristics $\mathbf{x}$ and the outcome y; the n-th individual, in fact, contributes to each cluster k, $(k = 1, \ldots, K)$, with a posterior probability $p(c_k | y_n, \mathbf{x}_n)$ which depends on both y_n and $\mathbf{x}_n$.

3 The Proposal: Local Multilevel Modeling

We propose a general methodology for local comparisons of institutional performance. The proposal consists of a two-step approach which combines Cluster-Weighted Modeling with traditional Multilevel Modeling.

During the first step, we model the heterogeneity of individuals by means of Cluster-Weighted Modeling, by taking as response variable a continuous outcome y and as predictors the individual-level characteristics $\mathbf{x}$:

$$p(y, \mathbf{x}) = \sum_{k=1}^{K} p(y, \mathbf{x}, c_k) = \sum_{k=1}^{K} p(y, \mathbf{x} | c_k) p(c_k) = \sum_{k=1}^{K} p(y | \mathbf{x}, c_k) p(\mathbf{x} | c_k) p(c_k), \tag{5}$$

where $p(y | \mathbf{x}, c_k)$, $p(\mathbf{x} | c_k)$ and $p(c_k)$ are defined in (3).

If we assign each individual to the cluster with the highest posterior probability, groups of users characterized by the same relationship between individual-level characteristics and the outcome are identified.

In the second step, we apply a Multilevel Model for each cluster k, $(k = 1, \ldots, K)$, by taking as response variable the outcome y and as predictors the individual-level and organization-level characteristics, $\mathbf{x}$ and $\mathbf{z}$ respectively:

$$y_{ij} = \alpha_j + \sum_{g=1}^{G} \beta_{gj} x_{gij} + e_{ij}, \qquad \alpha_j = \sum_{h=1}^{H} \delta_{hj} z_{hj} + u_j, \tag{6}$$

where y_{ij} is an outcome regarding the i-th individual $(i = 1, \ldots, n_j; N = n_1 + \cdots + n_j + \cdots + n_Q)$ belonging to the j-th organization $(j = 1, \ldots, Q)$, α_j is a random coefficient associated with the j-th organization, β_{gj} is a fixed coefficient associated with individual-specific covariate x_{gij}, $(g = 1, \ldots, G)$, e_{ij} is a random disturbance associated with the i-th individual belonging to the j-th organization; δ_{hj} is a fixed coefficient associated with organization-specific covariate z_{hj}, $(h = 1, \ldots, H)$, u_j is a random effect associated with the j-th organization, adjusted for individual-level and organization-level characteristics.

The plot of confidence intervals for ordered second-level random effects provides, for each cluster k, a ranking of organizations. The advantage is that Local

Multilevel Modeling gives to the policy makers an instrument which is able to locally highlight differences among institutions, i.e. for specific groups of users.

4 An Example

We briefly discuss an example regarding the evaluation of the "relative effectiveness" of healthcare institutions in Lombardy region. The data involved in the study refer to a sample of 22,877 patients hospitalised in 168 hospitals in 2006 and are provided by the Regional Agency for Health Care. The variables utilised in the study are described in Table 1, where y is the outcome, x_1–x_7 denote the patient-level characteristics (1), while z_1–z_3 the hospital-level characteristics (2).

In Table 2 the odds ratios for the significative variables in the reference Logistic Multilevel Model are reported.

The results indicate that the odds of death considerably increases for hospitals with large levels of high medical case-mix and low medical case-mix, and increases also for patients with emergency and tumour diagnosis. Note that the null effect of the age variable is due to the particular nature of the sample, which is merely composed by old patients (Mode = 71 years, Median = 60 years, Mean = 55.39 years). For sake of brevity we do not include the figure regarding hospital ranking, which shows that confidence intervals of the second-level random effects are large and overlapped.

Table 1 Variables in the effectiveness study

Label	Variable	Type
y	Total mortality indicator	Dichotomous
x_1	Age (1)	Discrete; number of years
x_2	Comorbidity (1)	Discrete; six levels
x_3	Length of stay (1)	Discrete; number of days
x_4	Relative level of case severity (1)	Continuous
x_5	Cardiovascular diagnosis (1)	Dichotomous
x_6	Emergency diagnosis (1)	Dichotomous
x_7	Tumour diagnosis (1)	Dichotomous
z_1	Low surgical case-mix (2)	Continuous
z_2	High medical case-mix (2)	Continuous
z_3	Low medical case-mix (2)	Continuous

Table 2 Reference model. Odds ratios for the significative variables (p-value < 0.05)

Variable	Beta	OR	Std. error
Age (1)	0.060	1.063	0.012
Emergency diagnosis (1)	1.340	3.822	0.416
Tumour diagnosis (1)	1.182	3.262	0.384
High medical case-mix (2)	2.199	9.019	0.370
Low medical case-mix (2)	1.728	5.630	0.858

Table 3 Cluster 1. Emergency and tumour diagnosis frequency tables

	y = 0, 1		y = 0		y = 1	
	Frequency	%	Frequency	%	Frequency	%
No emergency	20,379	98.62	20,322	99.45	57	24.89
Emergency	285	1.38	113	0.55	172	75.11
No tumour	19,191	92.87	18,997	92.96	194	84.72
Tumour	1,473	7.13	1,438	7.04	35	15.28
Total	20,664	100.00	20,435	100.00	229	100.00

Table 4 Cluster 2. Emergency and tumour diagnosis frequency tables

	y = 0, 1		y = 0		y = 1	
	Frequency	%	Frequency	%	Frequency	%
No emergency	645	87.52	301	90.39	344	85.15
Emergency	92	12.48	32	9.61	60	14.85
No tumour	373	50.61	58	17.42	315	77.97
Tumour	364	49.39	275	82.58	89	22.03
Total	737	100.00	333	100.00	404	100.00

Table 5 Cluster 3. Emergency and tumour diagnosis frequency tables

	y = 0, 1		y = 0		y = 1	
	Frequency	%	Frequency	%	Frequency	%
No emergency	493	33.40	73	6.97	420	98.13
Emergency	983	66.60	975	93.03	8	1.87
No tumour	1,278	86.59	983	93.80	295	68.93
Tumour	198	13.41	65	6.20	133	31.07
Total	1,476	100.00	1,048	100.00	428	100.00

The main purpose of the example is to show how the results of Cluster-Weighted Modeling allow us to identify subpopulations of patients. In particular, three is the optimal number of clusters derived from AIC and BIC criteria. If each patient is assigned to the cluster with the highest posterior probability, we have 20,664 patients in cluster 1, 737 in cluster 2 and 1,476 in cluster 3. Some of the main characteristics of the three clusters are briefly described in the following (see Tables 3, 4 and 5).

Cluster 1 is characterized by a mortality rate equal to 1.11% (against the 4.64% in the entire sample); the 75.11% of the dead patients had an emergency diagnosis (against the 22.62% in the entire sample). Instead Cluster 2 is characterized by an high percentage of patients with tumour diagnosis (49.39% vs. 8.90% in the entire sample) and in particular which are not dead (the 82.58% of the living patients, against the 8.15% in the entire sample); moreover, the living patients had high levels of comorbidity. Here the mortality rate is 54.82%, where the 77.97% of the dead patients had no tumour diagnosis (near to the percentage of 75.78% in the entire sample). By the end, cluster 3 is characterized by patients with emergency diagnosis (66.60% vs. 5.94% in the entire sample) and in particular which are not dead (93.03% of the living patients, against 5.13% in the entire sample). Here the

Table 6 Local models. Odds ratios for the significative variables (p-value < 0.05)

Variable	Beta	OR	Std. error
Age (1)	0.065	1.067	0.038
Cardiovascular diagnosis (1)	−0.240	0.786	0.116
High medical case-mix (2)	2.239	9.389	0.171
Age (1)	0.066	1.068	0.040
Comorbidity (1)	0.177	1.195	0.074
Emergency diagnosis (1)	1.528	4.609	0.696
High medical case-mix (2)	2.607	13.558	0.861
Age (1)	0.055	1.057	0.033
Comorbidity (1)	0.164	0.089	1.178
Emergency diagnosis (1)	1.540	4.668	0.582
Tumour diagnosis (1)	1.205	3.335	0.313
High medical case-mix (2)	2.074	7.960	0.428
Low medical case-mix (2)	1.755	5.784	0.416

mortality rate is 29%, where the 98.13% of the dead patients had no emergency diagnosis (against the 77.38% in the entire sample).

In Table 6 the odds ratios for the significative variables in the three Local Logistic Multilevel Models are reported.

For sake of brevity we do not include the ranking of hospitals for each group of patients, where confidence intervals of the hospital-level random effects are less large and overlapped than in the entire sample.

The example shows that there are, in this case, three different populations of patients which require three different models. The choice of a single global model would have caused a large loss of information. For example, hospital-level variable z_2, high medical case-mix, is an important variable for both the reference and the local models. However, the much higher effect in the second model seems to indicate that patients assigned to cluster 2 are characterized by a higher level of complexity, as confirmed by previous analysis of frequency tables. Analogous considerations can be made regarding the other variables.

5 Conclusions and Further Research

We have proposed a general methodology for the evaluation of public sector activities, which is entitled Local Multilevel Modeling. The proposal applies to regional or national studies, where a single global model is not sufficient to describe situations typically characterized by heterogeneity. The idea underlying our proposal is that non-homogeneity and non-linearity may appear in the individual-level relationships. The two-step approach proposed is able, firstly, to capture and model individual-level relationships whatever they are and, secondly, to locally highlight differences among organizations, i.e. for specific groups of users.

Of course, precautions which are valid for the use of traditional Multilevel Modeling in institutional comparisons hold true also for Local Multilevel Modeling. First, "we should exert caution when applying statistical models to make comparisons between institutions, treating results as suggestive rather than definitive" (Goldstein and Spiegelhalter 1996). Secondly, "measurement of outcomes for research purposes is useful to help organisations to detect trends and spot extreme outliers, but league tables of outcomes are not a valid instrument for day-to-day performance management by external agencies" (Lilford et al. 2004).

Further research will regard the extension of Cluster-Weighted Modeling to multilevel data structures, as proposed in Galimberti and Soffritti (2007) for Mixture Models, in order to allow some of the parameters of the conditional densities to differ across second-level units (schools, hospitals, etc.). This would reinforce the first step of the proposal, by taking into account also eventual non-homogenous and non-linear second-level relationships.

Acknowledgements The authors would like to express their thanks to Maurizio Sanarico for his valuable advice.

References

Bryk, A. S., & Raudenbush, S. W. (2002). *Hierarchical linear models. Applications and data analysis methods*. Newbury Park, CA: Sage.

Engster, D., & Parlitz, U. (2006). Local and cluster weighted modeling for time series prediction. In B. Schelter, M. Winterhalder, & J. Timmer (Eds.), *Handbook of time series analysis. Recent theoretical developments and applications* (pp. 39–65). Weinheim: Wiley.

Galimberti, G., & Soffritti, G. (2007). Multiple cluster structures and mixture models: Recent developments for multilevel data. In *Book of short papers CLADAG 2007 "Meeting of the Classification and Data Analysis Group of the Italian Statistical Society"* (pp. 203–206), September 12–14, Università degli Studi di Macerata. EUM, Macerata.

Gershenfeld, N. (1999). *The nature of mathematical modeling*. Cambridge: Cambridge University Press.

Goldstein, H. (1995). *Multilevel statistical models*. London: Arnold.

Goldstein, H., & Spiegelhalter, D. J. (1996). League tables and their limitations: Statistical issues in comparisons of institutional performance. *JRSS A, 159*(3), 385–443.

Lilford, R., Mohammed, M. A., Spiegelhalter, D. J., & Thomson, R. (2004). Use and misuse of process and outcome data in managing performance of acute medical care: Avoiding institutional stigma. *The Lancet 363*, 1147–1154.

McLachlan, G. J., & Basford, K. E. (1988). *Mixture models: Inference and applications to clustering*. New York: Marcel Dekker.

Modelling Network Data: An Introduction to Exponential Random Graph Models

Susanna Zaccarin and Giulia Rivellini

Abstract A brief introduction to statistical models for complete network data is presented. An example is provided by the collaboration network of Italian scholars on Population Studies.

1 Introduction

Social network data can be viewed as a social relational system characterized by a set of actors – maybe with their attribute variables – and their social ties (Wasserman and Faust 1994). Two main types of social network data are distinguished: ego-centered or personal network, and complete or one-mode network. Ego-centered network data are usually collected from a sample of actors (egos) reporting on the ties with and between other people (alters). Complete network data, on the other hand, concern a well-defined group of actors who report on their ties with all other actors in the group.

This contribution provides an introductory outline for statistical modelling of relational data, summarizing in particular the current methodological developments of the exponential random graph models (p^*) for complete social networks (Sect. 2).

The statistical models applied in social network analysis are typically non-standard because the common assumption of independent observations does not hold: the multiple ties to and from the same actor are related (Rivellini and Zaccarin 2007). Moreover, the popular assumption of continuous normally distributed variables does not hold when tie variables are binary, nominal, ordinal, or count variables.

As an example of fitting such a model, data from the collaboration network of Italian scholars in Population Studies for the year 2001 will be used (Sects. 3–5). The description of network and node properties by the means of the usual network measures will also be given.

S. Zaccarin (✉)
Università di Trieste, Piazzale Europa 1, 34127 Trieste, Italy
e-mail: susanna.zaccarin@econ.units.it

F. Palumbo et al. (eds.), *Data Analysis and Classification*,
Studies in Classification, Data Analysis, and Knowledge Organization,
DOI 10.1007/978-3-642-03739-9_34,

2 Exponential Random Graph Models for Social Networks

Although many techniques (usually based on measures of the network properties -density, in-degree, out-degree, ... - and on methods for partitioning networks into subgroup components -clique detection, structural equivalence, and blockmodelling, ... - Wasserman and Faust 1994) for measuring network properties serve valuable purposes in describing and understanding network features that might bear on particular research questions, there are some good reasons that motivate the statistical modelling of an observed social network (Robins et al. 2007a, p. 173). Social behaviour is complex and it seems reasonable to suppose that social processes giving rise to network ties are stochastic, so statistical models are appropriate to understand if, and how, certain network characteristics – often represented in the model by the introduction of a small number of parameters – are more or less observed in the network than expected by chance. Further, sometimes there are competing explanations for a structural effect. For instance, the appearance of clustering in networks could occur through different situations: (a) a balance-type effect or (b) an homophily effect (people with similar attributes attract); only a model including both effects can help to assess whether one or both are important. In general, models, and especially statistical models that are estimable from data and explicitly recognize uncertainty may help to understand the range of possible outcomes for processes on networks.

The leading assumption in modelling is that the observed network is generated by some (unknown) stochastic process. The goal in formulating a model is to propose a plausible set of hypotheses for this process from the observed structural characteristics of the network. As outlined in Robins et al. (2007a, pp. 177–178), five steps are of main concern in the formulation of a statistical model for a network:

- *Regard each tie as a random variable.* For each i and j which are distinct members of a set of n actors, the random variable Y_{ij} is defined with $Y_{ij} = 1$ if there is a tie from actor i to actor j, and $Y_{ij} = 0$ if there is no tie. Y is the matrix of all variables with y the matrix of observed ties, the observed network. Y may be directed (in which case Y_{ij} is distinguished from Y_{ji}), non-directed (where $Y_{ij} = Y_{ji}$ and the two variables are not distinguished) or valued. Here we mainly refer to models for non-directed networks.
- *Specify a dependence hypothesis, leading to a dependence graph.* This is the crucial point concerning the hypothesis about the local social processes that are assumed to generate the network ties. For instance, ties may be assumed to be independent of each other or, more realistically, a reciprocity process can take place, implying some form of dyadic dependence. Ties may also depend on node-level attributes, with for instance possible homophily effects. Each of these processes can be represented as a small-scale graph configuration: for instance, a reciprocated tie, or a tie between two actors with the same attribute (sex, education, affiliation, ...).
- *Generate a specific model from the specified dependence graph.* Specified dependence assumptions imply a particular class of models and each parameter in the model is related to the presence of specific configurations (the structural characteristics of interest, e.g. single ties, transitive triads and a two-stars).

- *Introduce homogeneity or other constraints on the parameter space.* Usually, in order to make models identifiable some parameters should to be equal for all ties or should be constrained to other requirements.
- *Estimate and interpret model parameters.* This step is complicated if the dependence structure is complex and recently fruitful results have been obtained by the development of MCMC estimation algorithms (Snijders 2002; Snijders et al. 2006).

A general form which describes a probability distribution of graphs on n nodes can be written as

$$Pr(Y = y) = (1/\kappa)\exp\left(\sum_A \eta_A g_A(y)\right), \tag{1}$$

where η_A is the parameter corresponding to the configuration A; $g_A(y) = \prod_{y_{ij}\in A} y_{ij}$ is the network statistic corresponding to configuration A; κ is a normalizing constant to ensure that (1) is a proper probability distribution, and the summation is over all configurations A. The probability of observing the graph is dependent on the presence of various configurations (structural characteristics) introduced in the model. Because of the exponential term in (1), such distributions are referred as Exponential Random Graph Models (ERGM) or p^* models. Several dependence assumptions, along with the correspondent model formulation, can be derived from (1). Frank and Strauss (1986) introduced Markov dependence in which pairs of ties that share a node are conditionally dependent, given the rest of the graph. A Markov random graph model for a non-directed network that incorporates well-known structural regularities such as edge (or density), two-star, three-star, and triangle effects is given in (2):

$$Pr(Y = y) = (1/\kappa)\exp(\theta L(y) + \sigma_2 S_2(y) + \sigma_3 S_3(y) + \tau T(y)), \tag{2}$$

where θ is the $density$ or $edge$ parameter and $L(y)$ refers to the number of edges in the network y; σ_k and $S_k(y)$ refer to the parameters associated with k-star effects and the numbers of k-stars in y; τ and $T(y)$ are the parameter for triangles and the number of triangles, respectively. In a non-direct graph, k-stars are the k edges expressed by the one actor and triangles are cliques of three actors. Star parameters represent the propensities for individual actors to have connections with multiple network partners and triangle parameter represents the tendency for clustering, i.e. for transitivity relationships. For a given observed network y, parameter estimates indicate the strength of effects in the data. Frank and Strauss (1986) proved that a random graph has the Markov property and is invariant under relabeling of nodes if and only if formula (2) holds, when generalized to include k-star counts.

Markov models have been the most widely explored formulation so far because they represent a fairly natural and realistic way to generalize previous versions simply stating dyad independence based on Bernoulli distribution. Despite this, the Markov random graph showed problems in fitting empirical network data and recently new specifications including more complex dependence assumptions have been developed (Snijders et al. 2006; Robins et al. 2007b). New models are *higher order* models that consider configurations (along with the corresponding statistics and parameters associated) involving more than three nodes. The proposed three new parameters allow a more realistic representation of the degree distribution and the transitivity structures of the observed network. In particular, the *alternating k-star* parameter considers the inclusion of all star configurations (Markov model set up to 0 star parameters higher than 3) but with a linear constrain among parameter values such that $\sigma_{k+1} = -\sigma_k/\lambda$ for $k \geq 2$ and for some $\lambda > 1$. In (1) there is one parameter σ_2 (the *alternating k-star*) that takes into account all star effects simultaneously. A positive *alternating k-star* parameter suggests that most likely network contains some higher degrees nodes (hubs). The introduction of *alternating k-triangle* and *alternating k-2-path* parameters follows an analogous reasoning (Robins et al. 2007b). The inclusion of these last two parameters overcomes the Markov assumption accounting not only for triangulation among actors but also for the possibility of denser areas of "clumping" triangles in the network. In this case, the underlying dependence assumption refers to the notion of partial conditional dependence (Pattison and Robins 2002) between any two disjoint pairs of actor if ties are observed between the two actors within each pair. Positive values for the two parameters provide evidence for transitivity effects in the network (see Robins et al. 2007b for more details on parameter interpretation). In addition, dependence structures involving actor attributes can be formulated (Robins et al. 2007a, p. 185). Programs for estimation and goodness of fit evaluation are implemented in the SIENA procedure in the StOCNET suite of programs (Boer et al. 2006), in the R package Statnet (Handcock et al. 2008) and in the Pnet software (http://www.sna.unimelb.edu.au/pnet/pnet.html).

As an example of a higher order model application, run on observed network characteristics, the relational data on Italian scholars on Population Studies in Italy are considered.

3 The Collaboration Network of Italian Scholars on Population Studies

Since 1993, every two years a national workshop on Population Studies is organized by the Italian Association of Population Studies (AISP). From the abstract collection of the last four meetings (1999, 2001, 2003 and 2005) the collaboration networks of people presenting a contribution have been derived, along with information on authors and papers.

In Fig. 1, the 2001 network along with four attributes is shown. Ties (280) linking actors (200) are non-directed and dichotomous. The structure shows a pattern of low cohesion: 37 small components can be recognised with majority (60%) represented by isolated dyads. The density is quite low (0.014) and small research groups are clustered in different sub-graphs, except for the component one-node

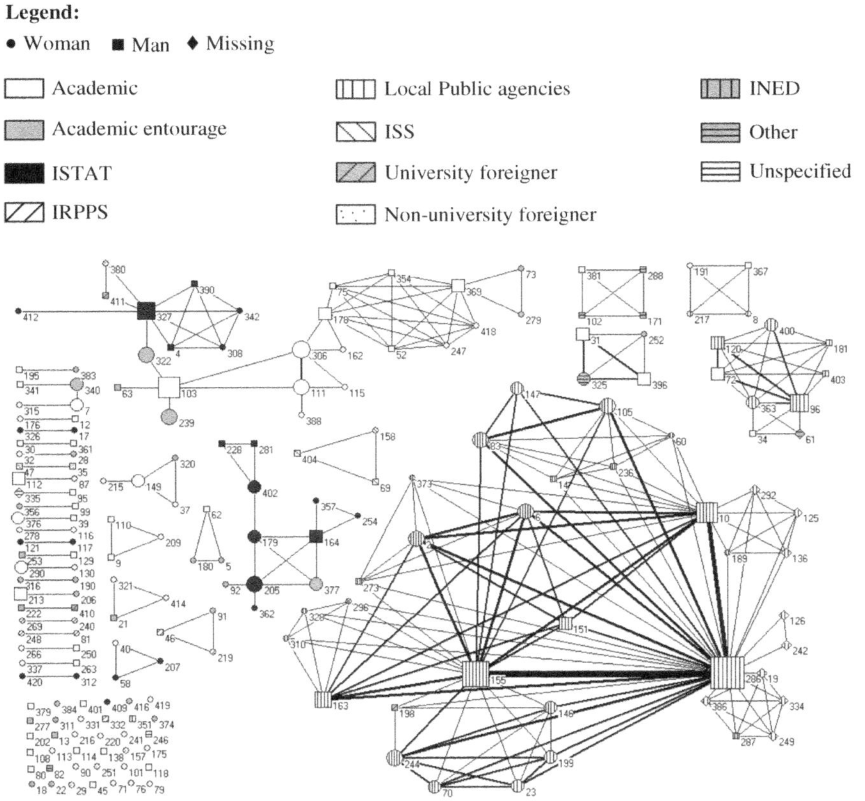

Fig. 1 Collaboration network by gender, institution, number of contributions (size node), number of collaborations inside the same research group (size line). Year 2001. Source: (Terzera and Rivellini 2006)

and two-lines connected that involve authors ($n = 35$) in local public agencies. The average degree (= 2.8) combines a general situation of low author-level degree across the whole network with very few authors with higher degrees in the larger component. From Fig. 1, a suggestion can also be derived that author productivity could be associated with network centrality.

When author gender and institution are taken into account (Table 1), a few potential patterns emerge. Despite no clear gender majority (54% of the network components are mixed), the smaller the component the greater the female homogeneity. Dyads exhibit the most female homogeneity (41%) while components of greater size share a more balanced mix of author gender.

A similar pattern follows for author institution: greater homogeneity in smaller components (especially for academic affiliation with 55% of dyads) but also prevalence of academic collaboration in the overall network.

Table 1 Gender and institution components

	Gender			
Component type	Homogeneity of males	Homogeneity of females	Mixed	Total
Dyad	4(18%)	9(41%)	8(36%)	22 (1 missing)
Triad	0	1(17%)	5(83%)	6
Quadrad	1(20%)	1(20%)	3(60%)	5
Other subgroups	0	0	4(100%)	4
Total	5(14%)	11(30%)	20(54%)	37

	Institution			
Component type	Homogeneity of academics	Homogeneity of not academics	Mixed	Total
Dyad	12(55%)	7(32%)	3(14%)	22
Triad	2(33%)	3(50%)	1(17%)	6
Quadrad	1(20%)	0	4(80%)	5
Other subgroups	0	0	4(100%)	4
Total	15(40%)	10(27%)	12(33%)	37

Taking into account the features observed in the 2001 network, a model of the form (1) including the new higher order specification and the two author attributes (gender and institution – aggregated in "academic" vs. "non-academic" – considered respectively as main effects as well as similarity effects) has been estimated to investigate how collaboration arises among scholars. For example, academic authors play a major role, although not exclusive, in stimulating cooperation but interaction can be made easier by "institutional proximity" (Rivellini et al. 2006), and the need for interdisciplinarity can be another factor to enhance collaboration.

4 Model Estimation Results

As first step, we tried to estimate a Markov model of the form (2) but this model did not fit well our observed data (Snijders et al. 2006). On the contrary, fit was reasonable good for the model incorporating the first two parameters of the new specification (alternating k-star and alternating k-triangle), with significant effects for all included parameters. To help convergence, the estimation kept the total number of ties fixed at the observed value (the analysis was carried out in Siena, according the procedure suggested in Lubbers and Snijders 2007; checks performed by Statnet provide comparable results). Parameter estimates and their standard errors are presented in Table 2.

The signs of alternating k-stars and alternating k-triangles are opposite, the former being negative and the latter being positive. This is not an uncommon pattern (Robins et al. 2007b, p. 205) suggesting the presence of two conflicting tendencies:

Table 2 Model for the 2001 collaboration network: parameter estimates and standard errors (*estimation conditional of the number of ties*)

Parameter	Estimate	Standard error
Alternating k-stars	−1.072	0.078
Alternating k-triangles	3.071	0.125
Gender (female = ref. cat.)	0.114	0.033
Same gender	0.446	0.138
Institution (academic = ref.cat.)	0.194	0.067
Same institution	0.624	0.105

Goodness of fit statistics

Graph counts	
2-Stars	2.099
3-Stars	4.393
Triangles	3.224
Alt-k-stars	0.092
Alt-k-triangles	−0.126
Sum of degree on gender	0.273
Similarity on gender	−4.061
Sum of degree on institution	0.108
Similarity on institution	0.045

one toward a triangulated core–periphery structure, and one against a degree based core–periphery structure. The overall outcomes would be several (often connected) small regions of overlapping triangles as it appears in network visualization of Fig. 1.

Contamination of research and "on the field" experience occurs rarely: co-authorship is more probable among authors of the same institution and same gender with academic authors more collaborative than scholars in national or local statistical offices. Males are also more collaborative than females. Further fitting of a complete model that combines Markov and higher order parameters has not been successful, confirming the specification reported in Table 2.

Once parameters have been estimated from an observed network, the goodness of model fit has to be checked by simulation of the resulting distribution of graphs (as suggested in Robins et al. 2007b) not only with the aim to reproduce the network observed statistics, but also to investigate the model's capability in replicating network characteristics not directly modelled.

Goodness of fit statistics for graph counts reproduced by the estimated model are reported in the bottom panel of Table 2. The statistics are calculated as the differences between the measure for the observed graph and the mean from a sample of graphs simulated from the parameter estimates; the resulting differences are scaled by the standard deviation. Small values for "goodness of fit statistics" indicate that the model "captures" that particular feature of the network well. The estimated higher order model does not seem to fit well on all of the features. In particular, a very high value of the goodness of fit statistics can be noted for two-stars, three-stars and triangle counts – the classical effects in the Markov model, although if included in model formulation, do not allow us to achieve convergence – as well for counts on gender similarity. A tentative explanation can be found in the extremely dense region that characterizes the big component involving authors in local public agencies. Moreover, in this component most of the authors are women but men have the highest degrees.

5 Concluding Remarks

The empirical example provided some evidence that new specification is more effective than the Markov random graph model, overcoming the convergence problem of classical Markov formulation. Parameters estimates seem reasonable with respect to network characteristics although results are not definitive requiring a deeper insight to adequately capture the process generating co-authorship in Population Studies in Italy.

Although our co-authorship network, actually being the one-mode projection of a two mode-network – with abstracts classified by authors –, has structural characteristics difficult to represent by an ERGM model for general graphs, model improvements can be made taking into account other authors attributes such as number of contributions and number of collaborations among authors. Further, stability of results over time and analysis of co-authorship evolution from different meetings have to be investigated.

References

Boer, P., Huisman, M., Snijders, T. A. B., Wichers, L. H. Y., & Zeggelink, E. (2006). *StOCNET: An open software system for the advanced anlysis of social networks. Version 1.7.* Groningen: ICS/Science Plus.

Frank, O., & Strauss, D. (1986). Markov graphs. *Journal of American Statistical Association, 81*, 832–842.

Handcock, M., Hunter, D., Butts, C., Goodreu, S., & Morris, M. (2008). statnet: Software tools for the representation, visualizations, analysis and simulation of network data. *Journal of Statistical Software, 24*. Retrieved from http://www.jstatsoft.org/v24/i01.

Lubbers, M. J., & Snijders, T. A. B. (2007). A comparison of various approaches to the exponential random graph model: A reanalysis of 102 student networks in school classes. *Social Networks, 29*, 489–507.

Pattison, P., & Robins, G. (2002). Neighbourhood-based models for social network. *Sociological Methodology, 32*, 301–337.

Rivellini, G., & Zaccarin, S. (2007). I modelli statistici nell'analisi di rete: evoluzione e utilizzo. In A. Salvini (Ed.), *Analisi delle reti sociali. Teorie, metodi, applicazioni* (pp. 141–173). Milan: Franco Angeli.

Rivellini, G., Rizzi, E., & Zaccarin, S. (2006). Science network in Italian population research: An analysis according the social network perspective. *Scientometrics, 67*, 407–418.

Robins, G., Pattison, P., Kalish, Y., & Lusher, D. (2007). An introduction to exponential random graph ($p*$) models for social networks. *Social Networks, 29*, 173–191.

Robins, G., Snijders, T. A. B., Wang, P., Handock, M., & Pattison, P. (2007). Recent developments in exponential random graph ($p*$) models for social networks. *Social Networks, 29*, 192–215.

Snijders, T. A. B. (2002). Markov chain Monte Carlo estimation of exponential random graph models. *Journal of Social Structure, 3*(2), 1–40.

Snijders, T. A. B., Pattison, P., Robins, G., & Handcock, M. (2006). New specifications for exponential random graph models. *Sociological Methodology, 36*, 99–153.

Terzera, L., & Rivellini, G. (2006). The analysis of the most recent national workshops on Italian population studies according a network perspective. In *EPC, poster session.*

Wasserman, S., & Faust, K. (1994). *Social network analysis. Methods and application*. Cambridge: Cambridge University Press.

Part VII
Latent Variables

An Analysis of Earthquakes Clustering Based on a Second-Order Diagnostic Approach

Giada Adelfio

Abstract A diagnostic method for space–time point process is here introduced and applied to seismic data of a fixed area of Japan. Nonparametric methods are used to estimate the intensity function of a particular space–time point process and on the basis of the proposed diagnostic method, second-order features of data are analyzed: this approach seems to be useful to interpret space–time variations of the observed seismic activity and to focus on its clustering features.

1 Introduction

Seismologists often require the definition of effective stochastic models to adequately describe the seismic activity of a fixed area; indeed a reliable description of earthquakes occurrence might suggest useful ideas on the mechanism of a such complex phenomena. For this reason in this paper a nonparametric estimation approach together with a diagnostic method for space–time point processes are introduced and used to interpret dependence features of seismic activity of a fixed region of Japan. The diagnostic tool here introduced allows to detect properties of clustering and inhibition of observed data and therefore may suggest directions for an improved fitting, even if the model is estimated by nonparametric methods. The diagnostic method is introduced in Sect. 2; the base of the model used to describe the seismic activity of the analyzed area is introduced in Sect. 3. In Sect. 4 the description of nonparametric estimation procedures for point processes and the use of the proposed diagnostic approach are showed. Conclusive remarks are provided in Sect. 5.

G. Adelfio (✉)
Department of Statistical and Mathematical Sciences, University of Palermo,
viale delle Scienze, ed 13, 90128, Palermo, Italy
e-mail: adelfio@unipa.it

F. Palumbo et al. (eds.), *Data Analysis and Classification*,
Studies in Classification, Data Analysis, and Knowledge Organization,
DOI 10.1007/978-3-642-03739-9_35,

2 Second-Order Residual Analysis

The definition of residuals for point processes is not unique. For example, some diagnostic methods are based on tests to verify whether the second-order properties of an observed point pattern are consistent with the stationary Poisson process (Diggle 1983; Baddeley and Silverman 1984). Other approaches, usually requires the transformation of data into a residual homogeneous Poisson process (result of a thinning or rescaling procedure, Meyer 1971) and the use of tests to assess the consistence of second-order statistics properties of the residual Poisson process with the theoretical ones (Schoenberg 2003). The weighted measures here introduced, is an extension of the method proposed in Veen and Schoenberg (2005), where the author focussed on the use of a weighted K-function and showed that the resulting weighted statistic is more powerful than residual methods. The proposed method deals with the definition and interpretation of a weighted version of second-order statistics (such as autocorrelation, K-function, spectrum, fractal dimension and R/S statistic), such that the contribution of each observed point is weighted by the inverse of the conditional intensity function that identifies the process. Weighted second-order statistics directly apply to data, since their definition does not require homogeneity assumption or previous transformation of data into residuals. The proposed diagnostic measures can be applied to processes of any dimension, provided the statistics discussed here can be computed, and gives to such second-order statistics a primary role in the diagnostics procedure, so that features such as clustering and inhibition can be easily interpreted.

Some original theoretical results (moments, asymptotical Normality, etc.) relatively to the weighted version of some second-order statistics (R/S statistic and correlation integral) are provided in Adelfio and Schoenberg (2008), by an easily generalizable approach based on the theory of martingales. In that work, it is shown that some weighted second-order statistics behave as the corresponding ones (not weighted) of a homogeneous Poisson process. Therefore, if weighted statistics are not consistent with the ones of a homogeneous Poisson process directions for a better fitting are provided.

In seismic data analysis, the proposed diagnostic method allows to interpret the clustering dependence of earthquakes, often observed near fault structures and close to large magnitude events (e.g. Schoenberg 2003), not accounted for by fitted models.

2.1 The Weighted Process and Its Second-Order Properties

Let N be a point process defined on $S \in \mathbb{R}^d$, $d \geq 1$. For any point s in S, let $\lambda(s|\mathcal{F})$ be the conditional intensity function of the process with respect to some filtration $\mathcal{F}$ on S, for simplicity denoted by $\lambda(s)$, assumed positive and bounded away from zero. For any set S, N_w is defined as a real-valued random measure such that:

$$N_w(S) = \int_S \frac{1}{\lambda_*(s)} dN$$

with $\frac{1}{\lambda_*(s)} = \frac{\lambda_{\min}}{\lambda(s)}$, and assuming the existence of the positive constant $\lambda_{\min} \leq \inf\{\lambda(s); s \in S\}$.

On the basis of the definition of the weighted point process N_w, its second-order statistics can be easily obtained, generalizing the definition of ordinary second-order statistics (Adelfio and Schoenberg 2008). The theoretical properties relative to these statistics are based on the result that a martingale can be obtained by the weighted process N_w, extending the known result that $N(S \times [0, t]) - \iint_S \int_0^t \lambda(s, t|\mathcal{H}_t) ds dt$ is a martingale (Daley and Vere-Jones 2003).

2.1.1 The Weighted Spectrum

Since $\lim_{|dt| \to 0} \frac{E[N_w(dt)]}{|dt|} = \lambda_{\min}$, the spectral density of the weighted process is here defined by

$$f_{N_w}(\omega) = \frac{1}{2\pi} \int_{-\infty}^{\infty} c_w^{(c)}(h) \exp(-i\omega h) dh, \tag{1}$$

where $c_w^{(c)}(h) = \lambda_{\min}\delta(h) + c_w(h)$ is the complete covariance density of the process N_w, with $\delta(\cdot)$ the Dirac delta function and $c_w(h)$ continuous at the origin. Since $c_w(h) = 0$ (see Adelfio and Schoenberg 2008) the spectral density of the weighted process (that is the spectral density calculated on the weighted points by using the inverse of the conditional intensity function multiplied by its minimum value) reduces to

$$f_{N_w}(\omega) = \frac{\lambda_{\min}}{2\pi}$$

that is the power spectrum of Poisson process with constant rate $\lambda_{\min}$.

2.1.2 The Weighted Correlation Integral

Let N be a time point process defined on $[0, T] \in \mathbb{R}$, with Lebesgue measure T and let $I_{ij}(\delta)$ be the indicator variable $I(|t_i - t_j| \leq \delta)$, with t_k $\forall k = 1, \ldots, n$ points of the state space.

The weighted correlation integral, for a time point process N with realizations $t_1, t_2, \ldots, t_n$ on $[0, T]$, can be written as

$$\hat{C}_W(\delta) = \frac{1}{(\lambda_{\min} T)^2} \sum_i^n \omega_i \sum_{j \neq i}^n \omega_j I(|t_i - t_j| \leq \delta) \tag{2}$$

with $\omega_k = \frac{\lambda_{\min}}{\lambda(t_k)}, \forall k$ and $\lambda(t)$ the conditional intensity function of the process with respect to some filtration $\mathcal{H}_t$ on $[0, T]$.

The asymptotic normality of the usual correlation integral for the i.i.d. case and under mixing conditions was proved in Denker and Keller (1986). In Adelfio and Schoenberg (2008) a martingale approach has been used to prove the asymptotical normality of the weighted version of the correlation integral and therefore that it is consistent with the one of a homogeneous Poisson process; this result has been extended to spatial point processes too.

The weighted correlation dimension, denoted with D_w, is defined as the slope of the plot of $\log \hat{C}_W(\delta)$ vs. $\log \delta$ for sufficiently small δ.

3 Space–Time ETAS Model

ETAS model (Ogata 1988) is a self-exciting point process, describing earthquakes activity, in a given region during a period of time, through a branching structure. ETAS model is completely characterized by its conditional intensity function, which provides a quantitative evaluation of future seismic activity and is proportional to the probability that an event with magnitude m (a logarithmic measure of earthquake strength) will take place at time t, in the epicenter of coordinates (x, y), that is a position on the surface of the earth expressed in latitude and longitude. It is defined as the sum of a term describing spontaneous activity (background) and one relative to the induced seismicity. The main hypothesis of the model states that all events, both mainshocks or aftershocks, may induce offsprings. The ETAS conditional intensity function is defined by

$$\lambda_{\boldsymbol{\theta}}(x, y, t, m) = J(m)\left[\mu(x, y) + \sum_{t_j < t} \nu_{\boldsymbol{\theta}}(t - t_j, x - x_j, y - y_j | m)\right] \tag{3}$$

sum of spontaneous activity and triggered one, with $\nu_{\boldsymbol{\theta}}(t - t_j, x - x_j, y - y_j | m) = g(t - t_j; \boldsymbol{\theta}) f(x - x_j, y - y_j | m, \boldsymbol{\theta})$ and $\boldsymbol{\theta}$ vector of parameters. ETAS model is a special case of a marked Hawkes process, with marks given by the magnitude values, such that $m \in \mathcal{M} = \mathbb{R}$; the marks distribution $J(m)$ corresponds to the Gutenberg-Richter frequency magnitude law (Gutenberg and Richter 1944).

In ETAS model, background seismicity $\mu(x, y)$ is assumed stationary in time, while time triggering activity is represented by a non stationary Poisson process according to the modified Omori's formula (Utsu 1961). In this model, the occurrence rate of aftershocks at time t following the earthquake of time τ, is described by

$$g(t - \tau) = \frac{K}{(t - \tau + c)^p}, \quad \text{with } t > \tau \tag{4}$$

with K a normalizing constant, c and p characteristic parameters of the seismic activity of the given region; p is useful for characterizing the pattern of seismicity, indicating the decay rate of aftershocks in time.

Ogata, in Ogata (1998), proposed three different models for the clustering spatial distributions conditioned to magnitude of the generating event, suggesting as the best one the following:

$$f(x - x_j, y - y_j | m_j) = \left\{ \frac{(x - x_j)^2 + (y - y_j)^2}{e^{\alpha(m_j - m_0)}} + d \right\}^{-q}. \tag{5}$$

In this model the occurrence rate of aftershocks is related to the mainshock magnitude; α is a measure of the influence on the relative weight of each sequence, m_0 is the completeness threshold of magnitude, i.e. the lower bound for which earthquakes with higher values of magnitude are surely recorded in the catalog, d and q are two parameters related to the spatial influence of the mainshock.

4 Nonparametric Estimation and Diagnostics

To find an adequate model describing the seismic activity of the east coast of the Tohoku District (Japan) from 1926 to 1995, with magnitude equal or larger than 4.5 (Fig. 1), second-order residuals are used. A flexible model, mostly useful in presence of several data identified by a complex generating process, is proposed; its intensity function is

$$\lambda(t, x, y; \mathbf{h}) = \sum_{i=1}^{n} g(t - t_i; h_t) f(x - x_i, y - y_i; h_x, h_y) \tag{6}$$

with f and g spatial and temporal densities, with parameters $\mathbf{h} = (h_t, h_x, h_y)$ respectively, estimated by kernel estimators. Therefore, the complex estimation issue of a self-exciting process now reduces to the estimation of the intensity function of an inhomogeneous Poisson process identified by a space–time Gaussian kernel intensity function as in (6), assuming independence between spatial and temporal components [as in (3)].

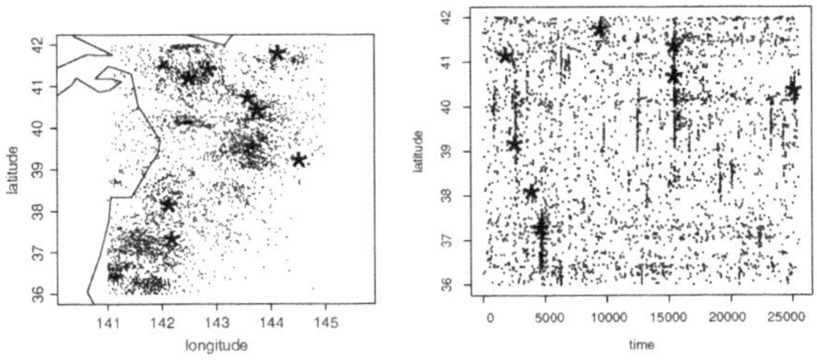

Fig. 1 Epicenters of earthquakes occurred in the east coast of the Tohoku District – Japan (fractal dimension 1.7), in the region defined by $36^\circ \sim 42^\circ$N and $141^\circ \sim 145^\circ$E for all depth and for the time span 1926–1995, with magnitude equal or larger than 4.5. (Big-stars: $m \geq 7.5$)

The kernel estimator of an unknown density f in $\mathbb{R}^d$ is defined as

$$\hat{f}(z_1, \ldots, z_d) = \frac{1}{nh_{z_1} \cdots h_{z_d}} \sum_{i=1}^{n} K\left(\frac{z - z_{i1}}{h_{z_1}}, \ldots, \frac{z - z_{id}}{h_{z_d}}\right),$$

where $K(z_1, \ldots, z_d)$ denotes a multivariate kernel function operating on d arguments centred at $(z_{i1}, \ldots, z_{id})$ and $(h_{z_1}, \ldots, h_{z_d})$ are the smoothing parameters of kernel functions. In this paper kernel estimators assuming constant bandwidth values for each point but different for space and time components (i.e. $h_x \neq h_y \neq h_t$) are first used to estimate the intensity in (6). The h values are estimated by the Silverman's rule, that for the one-dimensional case (Silverman 1986) is

$$h_{opt} = 1.06An^{-1/5} \tag{7}$$

with $A = \min\{\text{standard deviation, range-interquartile}/1.34\}$. It optimizes the estimator asymptotic behavior in terms of mean integrated square error and provides valid results on a wide range of distributions. This model is referred as model a. Afterwards, a nonparametric model with variable bandwidth values (referred as model b) is estimated. According to this procedure (see Zhuang et al. 2002), the bandwidth for the jth event, $j = 1, \ldots, n$ is $h_j = (h_x^j, h_y^j, h_t^j)$ that is the radius of the smallest circle centered at the location of the jth event (x_j, y_j, t_j) that includes at least a fixed number n_p of other events.

To study the goodness of fit of the two alternative models the weighted spectral density (Fig. 2) is analyzed. Looking at the plot on the right in Fig. 2 we could say that by fitting model b, the estimated weighted spectral density behaves as the one of a temporal homogeneous Poisson process. On the other hand, the estimated weighted spectral density of model a is almost always outside the 95% bounds of a homogeneous Poisson process. Indeed, because of the strong clustering nature of data (easily evident from Fig. 1), model a tends to smooth them out more than expected. On the other hand the nonparametric model with variable bandwidth values (model b) seems to account for the residual correlation (due to the presence of

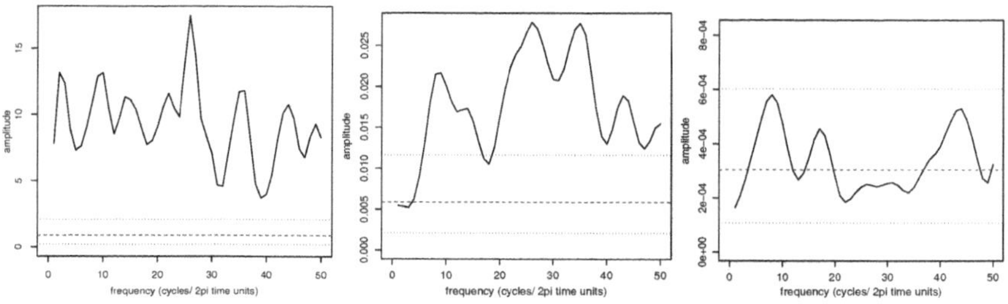

Fig. 2 Original periodogram (*on the left*), weighted periodogram for constant bandwidth model (model a; *in the middle*) and for varying bandwidth model (model b; *on the right*), with 95%-bounds of a homogeneous Poisson process (*dotted lines*)

aftershocks) not taken into account by model a. Moreover the estimated weighted correlation dimension for model b is $D_W = 2.0$, not significatively different from 2, that is the expected value of the correlation dimension for a homogeneous Poisson process in space. Such features were not observed for model a, since $D_W = 1.92$ with 95% c.i. (1.89, 1.95). From this simple analysis we could say that a constant smoothing value for all the observed region could provide a valid description of the activity of that area in case of low correlation; when events are strongly correlated, the introduction of a variable smoothing that accounts for the high production of offsprings seems to be preferable.

Moreover, the studied area presents an highly clustered seismicity identified by a complex intensity function. The nonparametric approach makes possible a reasonable characterization of it, since it does not constrain the process to have predetermined properties. The images in Figs. 3 and 4 show the three dimensional estimated densities considering just events occurred in a neighbor of the mainshock with magnitude 7.9 and epicenter 143.59°E – 40.73°N of May 16, 1968. As shown in those figures the variable kernel model provides a more realistic description of the seismic activity of the studied area than the constant bandwidth model and the ETAS one. Indeed model b seems to follow more adequately the seismic activity of the observed area, characterized by highly variable changes both in space and in time and because of its flexibility, it provides a better fitting to local space–time changes as just suggested by data. On the other hand, the parametric approach of

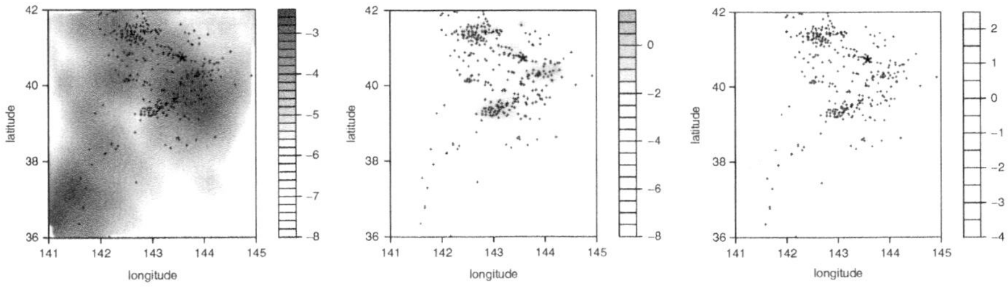

Fig. 3 Log-scale of space–time Silverman-based kernel density, ETAS intensity and their ratio, for points following the mainshock of 1968

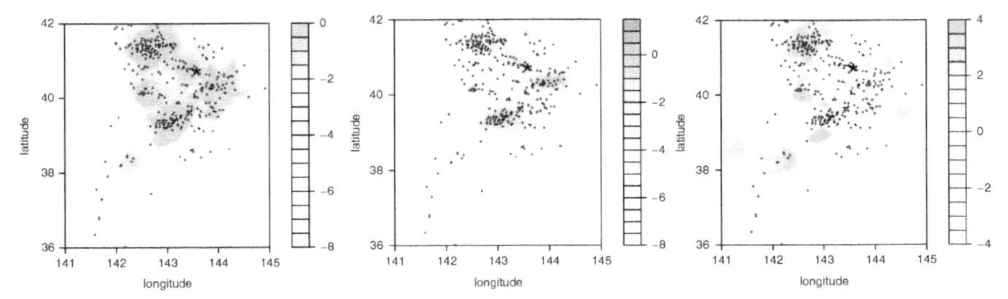

Fig. 4 Log-scale of space–time variable kernel density, ETAS intensity and their ratio, for points following the mainshock of 1968

ETAS model constrains the seismic process to predetermined properties, such as p-values of (4) constant in space, that induce a too quick decay of the intensity function in correspondence of events that are strongly clustered in time, or constant K-values of (4), that may not be reasonable when data are strongly clustered and higher value of K accounting for the high production of offsprings may be worthwhile.

5 Conclusion

In this paper a nonparametric estimation approach is proposed to estimate the three dimensional intensity function of the seismic process for an area of Japan, looking for a better model on the basis of the second-order diagnostic method here introduced. For this context, both the estimation procedure and the diagnostic tool here provided seem advantageous. Indeed, the proposed nonparametric approach provides a realistic description of seismicity fitting by a flexible procedure and in conjunction with the second-order residuals allows a reliable interpretation of dependence and inhibitive features of observed point patterns.

References

Adelfio, G., & Schoenberg, F. P. (2008). Point process diagnostics based on weighted second-order statistics and their asymptotic properties. *Annals of the Institute of Statistical Mathematics*. doi:10.1007/s10463-008-0177-1.

Baddeley, A. J., & Silverman, B. W. (1984). A cautionary example on the use of secondorder methods for analyzing point patterns. *Biometrics, 40*, 1089–1093.

Daley, D. J., & Vere-Jones, D. (2003). *An introduction to the theory of point processes* (2nd edition). New York: Springer.

Denker, M., & Keller, G. (1986). Rigorous statistical procedures for data from dynamical systems. *Journal of Statistical Physics, 44*, 67–94.

Diggle, P. J. (1983). *Statistical analysis of spatial point patterns*. London: Academic.

Gutenberg, B., & Richter, C. F. (1944). Frequency of earthquakes in California. *Bulletin of the Seismological Society of America, 34*, 185–188.

Meyer, P. (1971). Demonstration simplifee dun theoreme de knight. In *Seminaire de Probabilites V Universite de Strasbourg*. *Lecture Notes in Mathematics* (Vol. 191, pp. 191–195). Berlin: Springer.

Ogata, Y. (1988). Statistical models for earthquake occurrences and residual analysis for point processes. *Journal of the American Statistical Association, 83*(401), 9–27.

Ogata, Y. (1998). Space–time point-process models for earthquake occurrences. *Annals of the Institute of Statistical Mathematics, 50*(2), 379–402.

Schoenberg, F. P. (2003). Multi-dimensional residual analysis of point process models for earthquake occurrences. *Journal American Statistical Association, 98*(464), 789–795.

Silverman, B. W. (1986). *Density estimation for statistics and data analysis*. London: Chapman and Hall.

Utsu, T. (1961). A statistical study on the occurrence of aftershocks. *Geophysical Magazine, 30*, 521–605.

Veen, A., & Schoenberg, F. P. (2005). Assessing spatial point process models using weighted K-functions: Analysis of California earthquakes. In A. Baddeley, P. Gregori, J. Mateu, R. Stoica, & D. Stoyan (Eds.), *Case studies in spatial point process models* (pp. 293–306). New York: Springer.

Zhuang, J., Ogata, Y., & Vere-Jones, D. (2002). Stochastic declustering of space–time earthquake occurrences. *Journal of the American Statistical Association, 97*(458), 369–379.

Latent Regression in Rasch Framework

Silvia Bacci

Abstract Rasch-type measurement models are an important and widespread instrument in estimating latent variables. In this contribution the attention is set on a special type of Rasch models: the latent regression Rasch models. The interest is focused on two main problems: the latent regression modelling of longitudinal data and latent regression modelling with missing not at random responses. The empirical study concerns the measurement of Health related Quality of Life in cancer patients.

1 Introduction

In recent years Health related Quality of Life (HrQoL) assessment has become routine measure used to evaluate clinical research and policy outcomes and to support clinicians' decisional processes. The World Health Organization (WHO) in 1948 defined health as a state of physical, mental and social well-being, and not merely the absence of disease or illness. Furthermore, in 1995 WHO defined quality of life as the individuals' perceptions of their position in life in the context of their culture and value systems in which they live, also in relation to their goals, expectations, standards and concerns.

The main problem related with HrQoL concerns with its measurement. Since it is a latent variable, suitable measurement methods are needed to synthesize and to translate the qualitative information (on ordinal scale) coming from *ad hoc* questionnaires into quantitative information (on interval scale). This problem can be dealt by means of different theoretical frameworks. The present contribution is focused on the Rasch models (Fisher and Molenaar 1995) owing to their properties, such as the specific objectivity (Rasch 1960).

In the most common approach, a latent variable is usually analyzed in two different steps: firstly, it is quantified by means of a measurement model (e.g. a Rasch

S. Bacci (✉)
Department of Statistics "G. Parent", Viale Morgagni 59, 50134 Firenze, Italy
e-mail: s.bacci@ds.unifi.it

F. Palumbo et al. (eds.), *Data Analysis and Classification*,
Studies in Classification, Data Analysis, and Knowledge Organization,
DOI 10.1007/978-3-642-03739-9_36,

model) and *then* the estimates are used in a regression model as observed values of the response variable. However, some factors suggest caution toward an such approach: e.g. the bias and the inconsistency of the latent variable estimates, the underestimation of the true association between latent variable and covariates and, more generally, the lack of flexibility and integration between applied psychometricians and statisticians. A possible solution is constituted by a global approach that estimates the regression coefficients *directly* from the observed item responses. This is the case of latent regression (Andersen and Madsen 1977).

In this paper two extensions of latent regression Rasch models are proposed. In Sect. 2 the latent regression model is extended to longitudinal data, while in Sect. 3 the problem of missing not at random responses is analyzed. Assuming a distribution for the latent variable, these models are formally equivalent with non-linear mixed models (Rijmen et al. 2003).

The empirical analysis is performed on two data sets, both of them concerning with the measurement of HrQoL in cancer patients. The longitudinal latent regression analysis treats the item responses of 285 terminal patients under palliative care, interviewed both at baseline and once a week during the therapy. Instead, the study of missingness treatment in latent regression models is based on the item responses of 624 cancer patients under physical or psychological rehabilitation. In the former case, the analysis is based on the responses to eight binary items related to the psychological dimension of HrQoL measured by the Therapy Impact Questionnaire (Tamburini et al. 1992). In the latter case, the analysis is based on the responses to the 12 polytomous items of the Short Form 12 version 2 (Ware et al. 2002).

2 Longitudinal Latent Regression Model

The longitudinal latent regression model proposed in this section is an extension to longitudinal data of the linear latent regression Rasch model, which is known in literature since several years (for more details see (Bacci 2008)).

Let us denote by: $i = 1, 2, \ldots, N$ the individuals; t the measurement occasion; θ_{it} the "ability" parameter (i.e. the HrQoL level) for ith person at t time; $f(t)$ any function of t (e.g. a linear or a quadratic function); z_i the value of covariate z assumed by the ith person; α the regression coefficient of z_i; δ_{0i} and δ_{1i} the random intercept and random coefficient, respectively; γ_{00} and γ_{11} the fixed components of δ_{0i} and δ_{1i}, respectively; u_{0i} and u_{1i} the random components of δ_{0i} and δ_{1i}, respectively. In more detail, the random intercept u_{0i} explains how the initial level (i.e. for $t = 0$) of θ_{it} of each patient differs from the average population value. Additionally, the random coefficient u_{1i} represents how the individual time effect differs from the average population time effect. A linear multilevel model is described by the following equations, where the residuals u_{0i} and u_{1i} are normally distributed, with mean equal to 0, variances and covariances matrix denoted by $\mathbf{\Sigma}$:

$$\begin{cases} \theta_{it} &= \delta_{0i} + \delta_{1i} \cdot f(t) + \alpha \cdot z_i \\ \delta_{0i} &= \gamma_{00} + u_{0i} \\ \delta_{1i} &= \gamma_{11} + u_{1i} \end{cases} \tag{1}$$

With respect to other multilevel linear models, this is different because the response variable is not observed: hence, it has to be estimated by means of a Rasch model. For dichotomous items, the Rasch model is given by the following equation (Fisher and Molenaar 1995), where the subscript t is introduced here for coherence with (1):

$$logit[P(X_{jti} = 1|\theta_{it};\beta_j)] = \log\left[\frac{P(X_{jti}=1|\theta_{it},\beta_j)}{1-P(X_{jti}=1|\theta_{it},\beta_j)}\right] = \theta_{it} - \beta_j, \tag{2}$$

where: i is the ith person ($i = 1, 2, \ldots, N$); j is the jth item ($j = 1, 2, \ldots, J$); x_{jti} is the response category of the jth item, i.e. 0 or 1, chosen by the ith person at tth time; β_j is the difficulty parameter of jth item.

If θ_{it} is substituted in the Rasch model with the structural model of (1), the following longitudinal latent regression model is obtained:

$$\begin{aligned} &logit[P(X_{jti} = 1|z_i, u_{0i}, u_{1i}; \gamma_{11}, \alpha, \mathbf{\Sigma}, \beta_j)] = \\ &= (\gamma_{00} + \gamma_{11} \cdot f(t) + \alpha \cdot z_i - \beta_j) + (u_{0i} + u_{1i} \cdot f(t)), \end{aligned} \tag{3}$$

The longitudinal latent regression model has been implemented by the Nlmixed procedure in SAS. As outlined in the introduction, the model has been estimated on a data set referred to terminal cancer patients under palliative care. The main objective of the clinicians is the estimation of the effect of the time and the HrQoL-at-baseline on the HrQoL during the care. The HrQoL has been measured from eight binary items referred to psychological dimension: difficulty in concentrating (difconce), confusion feeling (confusion), difficulty in performing free time activities (diffree), insecurity feeling (insecurity), illness (illness), nervousness (nervousness), fatigue (fatigue) and sadness or depressed feelings (sad).

In Table 1 are reported the estimates of the significant parameters of the model. For each item is shown the difficulty parameter: smaller the estimated value is, more probable is to observe a patient with that symptom. As regards the covariates, the HrQoL-at-baseline is significant (with a positive effect) to predict the trend of HrQoL during the therapy; instead, none of the other individual characteristics observed at baseline give significant information on the future trend of HrQoL. As regards the time effect, it is negatively correlated (with a quadratic effect) with the *average* level of HrQoL. On the other hand, the second-level residuals (u_{1i}) show how much the time effect for the ith patient differs from the average value of the population. In particular, in Fig. 1 the residual value and the corresponding confidence interval are shown for every patient. As a small score on the items indicates a high level of HrQoL, patients at the bottom and at the left in the figure show a statistically significant improvement of HrQoL over the time, whereas patients at the top and at the right have a statistically significant worsening of HrQoL. To conclude, the individual characteristics that have a different distribution between the

Table 1 Longitudinal latent regression model: estimates, standard errors, p-values, confidence intervals ($\alpha = 0.05$)

	Estimate	Stand.error	p-value	Lower limit	Upper limit
Difconce	−0.546	0.088	< 0.0001	−0.719	−0.373
Confusion	0.308	0.873	0.0005	0.137	0.480
Diffree	−1.997	0.096	< 0.0001	−2.187	−1.808
Insecurity	−0.302	0.088	< 0.0001	−0.474	−0.129
Illness	−1.077	0.090	0.0007	−1.253	−0.901
Nervousness	−0.632	0.088	< 0.0001	−0.805	−0.458
Fatigue	−2.487	0.102	< 0.0001	−2.687	−2.287
Sad	−1.275	0.091	< 0.0001	−1.454	−1.097
$\sqrt{t}$	0.271	0.044	< 0.0001	0.186	0.357
$HrQoL_{base}$	0.825	0.062	< 0.0001	0.703	0.947
σ^2_{u0}	0.971	0.138	< 0.0001	0.699	1.242
σ^2_{u1}	0.272	0.041	< 0.0001	0.190	0.351
σ_{u01}	−0.139	0.060	0.0222	−0.257	−0.020

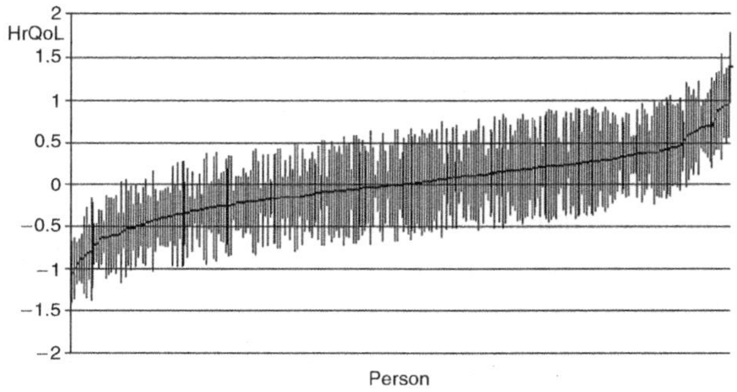

Fig. 1 Second-level residuals u_{1i}: estimates and confidence intervals

two groups of patients have been analyzed. The differences are statistically significant only in relation to the absence of metastasis (1% significativity level) and to being bed ridden (5% significativity level).

3 Latent Regression Rasch Model with Missing Data

One of the main problems of observational studies is the data incompleteness. The default method to treat the incompleteness does not take into account the non-response and analyzes only the complete cases. The consequences of this simple approach on the inferential analysis can be particularly negative, mainly because of the reduction of sample size. To overcome these drawbacks, several alternative techniques have been proposed. When the assumption of non-ignorable missingness (Little and Rubin 2002) is plausible, such as in surveys on the HrQoL, modelling the non-response process is considered the optimal solution.

Given any data process, the missingness can be described by a lot of several models. In a very recent approach (Holman and Glas 2005) the Rasch model represents a simple but meaningful solution. The original idea is based on the hypothesis that the presence or absence of X_{ij} values depend on a latent variable. One can think that every interviewee person has a specific level of propensity to (non-)response: higher the propensity to non-response, more probable observing a missing value. Propensity to non-response can be considered as a latent variable. Therefore, the answers to a set of items can be used as indicators of the latent variable ($X^*_{ij} = 1$ if the ith person answered to the jth item; else, $X^*_{ij} = 0$). When both the analysis model and the missingness model are Rasch models, the resulting joint model is a bi-dimensional Rasch model (for more details on the structure of multidimensional Rasch models see (Adams et al. 1997)), where the observed data depend directly on one latent trait θ_1 (e.g. HrQoL) and indirectly – through the correlation coefficient – on the other latent trait θ_2 (propensity to non-response), while non-responses depend only on θ_2:

$$P(X_{ij} = x|\theta_{di}, \beta_{dj}) = \frac{exp[x(\sum_{d=1}^{2} \gamma_{dj}\theta_{di} - \beta_{dj})]}{1 + exp(\sum_{d=1}^{2}(\gamma_{dj}\theta_{di}) - \beta_{dj})}, \tag{4}$$

where i is the ith person ($i = 1, 2, \ldots, N$), j is the jth item ($j = 1, 2, \ldots, J$), x is the response mode for the jth item, i.e. 0 or 1, θ_{di} is the person parameter for the dth latent trait and β_{dj} is the difficulty parameter of the jth item in the dth dimension. This model belongs to Rasch family only if matrix $\Gamma = \{\gamma_{dj}\}$ is composed by 0s and 1s, so that every response loads on one latent trait (see also Fig. 2):

$$\gamma_{dj} = \begin{cases} 0 \text{ if } [j \in (1, \ldots, J) \cap d = 2] \cup [j \in (J+1, \ldots, 2J) \cap d = 1] \\ 1 \text{ if } [j \in (1, \ldots, J) \cap d = 1] \cup [j \in (J+1, \ldots, 2J) \cap d = 2] \end{cases}$$

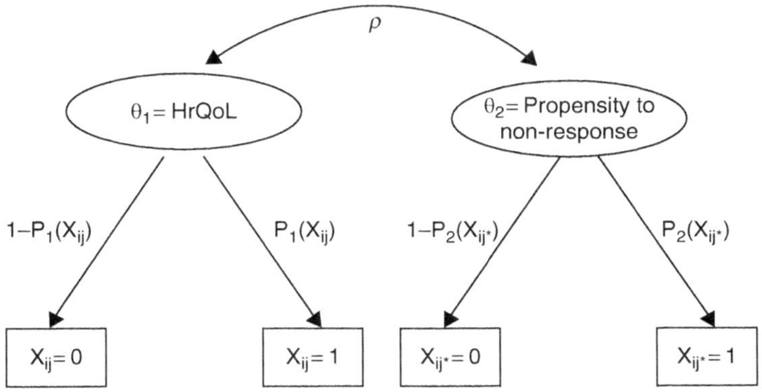

Fig. 2 Model A: Non-ignorable missing Rasch model

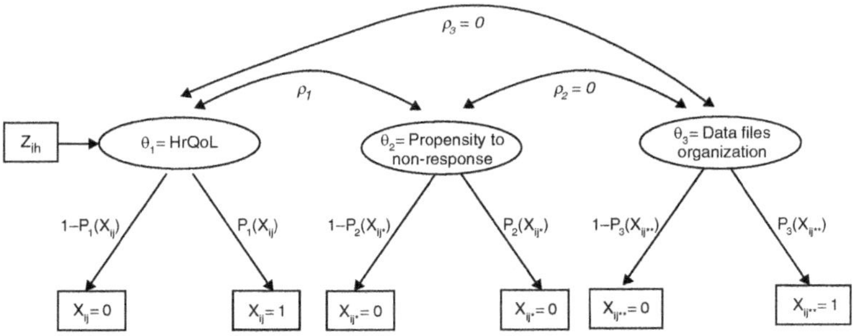

Fig. 3 Model B: Latent regression missing Rasch model

The non-ignorable missing Rasch model can be extended to take into account one or more explicative variables of θ_1. Here, it is proposed a non-ignorable missing Rasch model where the latent variable θ_1 is substituted, similarly to Sect. 2, by the related regression expression. As shown in Fig. 3, the introduction of a third latent variable θ_3 is useful to take into account also the missingness of the explicative variables. In this specific case we assumed that this specific kind of missing data are missing at random, so θ_3 is uncorrelated with both θ_1 and θ_2.

The latent regression missing Rasch model can be implemented by Nlmixed procedure by SAS. A partial credit version (to take into account the polytomous nature of items) has been estimated on the data set related with 624 cancer patients under physical or psychological rehabilitation and it has been compared with a complete cases analysis (Model 0). The 12 items that compose the physical and mental HrQoL and the two covariates used to explain the HrQoL are shown in Table 2, with the indication of the number and the percentage of missing values. In the last column of the table one can see the difficulty parameter estimates for the 12 items, that measure the propensity to non-response: smaller the estimated value is, greater the propensity to non-response is.

Several differences turned out between the two kinds of analysis. Table 3 highlights that in the latent regression missing Rasch model (Model B) the efficiency is better, the residual variance is reduced, and, most of all, the seriousness of tumor is statistically significant for the mental dimension of HrQoL. Moreover, some relevant results about the individual physical HrQoL estimates stand out from Fig. 4 (results about mental HrQoL are similar): the individual estimates are shrunk toward the average, the range of variability is smaller and the estimated HrQoL level of some patients (e.g. see patient number 92) is completely different than the estimate from complete case analysis (Model 0). This last element is especially significant from a practical point of view, because on the base of the level of HrQoL the clinician decides what kind of therapy is better for the patient.

Table 2 Data set SF12: number and percentage of missing values, propensity to non-response

	Missing values	% of missing values	Propensity to non-resp.
Physical dimension			
General health	21	3.4	3.093
Ability to do moderate activities	19	3.0	2.923
Ability to climb several flights of stairs	45	7.1	2.521
Limits in the regular daily activities as a result of physical health	33	5.2	3.384
Limits in the work as a result of physical health	41	6.4	2.215
Physical pain	26	4.1	3.292
Mental dimension			
Limits in the regular daily activities as a result of emotional problems	30	4.7	2.774
Limits in the work as a result of emotional problems	36	5.7	2.774
Felt calm and peaceful	40	6.4	3.093
Have a lot of energy	45	7.2	2.521
Felt depressed	27	4.3	3.093
Limits in social activities	14	2.2	4.259
Covariates			
Seriousness of tumor	399	63.9	–
Post-operation therapy	416	66.7	–

Table 3 Model 0 and Model B: estimate and standard errors of regression coefficients and variances

	Model 0		Model B	
	Estimate	SE	Estimate	SE
Physical dim.				
Constant	−0.053	0.469	−0.055	0.106
Seriousness	0.071	0.227	0.038	0.048
Post-operation therapy	−0.294	0.231	−0.138	0.051
Var(θ_P)	3.658	0.469	0.870	0.106
Mental dim.				
Constant	−0.368	0.452	−0.101	0.100
Seriousness	0.291	0.219	0.114	0.046
Post-operation therapy	−0.344	0.223	−0.168	0.048
Var(θ_M)	3.399	0.452	0.772	0.100

4 Conclusion

This contribution deals with some methodological aspects of latent regression Rasch modelling, with specific reference to HrQoL applications. Two latent regression models has been presented: one for longitudinal data sets and another for the treatment of missing not at random responses. With this contribution we want to highlight the advantages of a joint approach to the measurement and the analysis of latent variables. The consequences of ignoring latent regression can be particularly

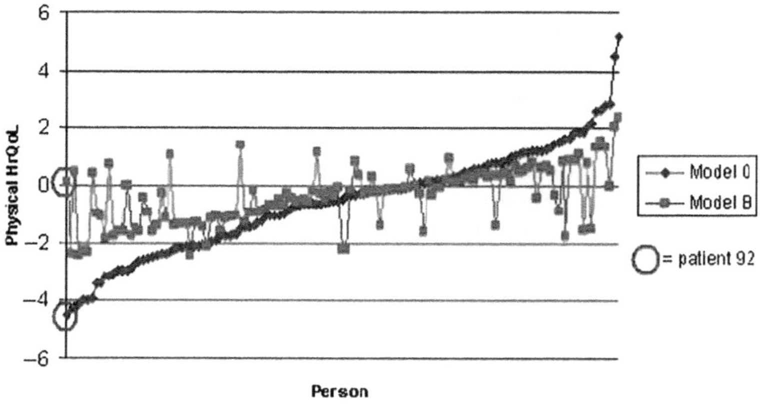

Fig. 4 Model 0 and Model B: individual Physical HrQoL estimates

negative for clinicians, both in relation to the assessment of the individual level of HrQoL and to the detection of HrQoL determinants.

Further improvements are still possible. As regards longitudinal latent regression model, some problems concern with the treatment of one or more latent covariates, the extension to multi-dimensional data structures and to informative drop out. As regards the non-ignorable missingness Rasch model, the extension to longitudinal framework and the treatment of missing questionnaires represent interesting aspects, which wait for a deepening.

References

Adams, R., Wilson, M., & Wang, W. (1997). The multidimensional random coeffcients multinomial logit model. *Applied Psycholigical Measurement, 21*, 1–23.

Andersen, E. B., & Madsen, M. (1977). Estimating the parameters of the latent population distribution. *Psychometrika, 42*, 357–374.

Bacci, S. (2008). Analysis of longitudinal Health related Quality of Life using latent regression in the context of Rasch modelling. In C. Huber, N. Limnios, M. Mesbah, & M. Nikuline, (Eds.), *Mathematical Methods for Survival Analysis, Reliability and Quality of Life* (pp. 277–292). Hermes.

Fisher G. H., & Molenaar, I. W. (1995). *Rasch models. foundations, recent developments and applications*. New York: Springer-Verlag.

Holman, R., & Glas, C. A. W. (2005). Modelling non-ignorable missing-data mechanisms with item response theory models. *British Journal of Mathematical and Statistical Psychology, 58*, 1–17.

Little, R. J. A., & Rubin, D. B. (2002). *Statistical analysis with missing data*. Hoboken: Wiley Series in Probability and Statistics.

Rasch, G. (1960). *Probabilistic models for some intelligence and attainment tests*. Copenhagen: Danish Institute for Educational Research.

Rijmen, F., Tuerlinckx, F., De Boeck, P., & Kuppens, P. (2003). A nonlinear mixed framework for item response theory. *Psychological Methods, 8*, 185–205.
Tamburini, M., Rosso S. et al. (1992). A therapy impact questionnaire for quality-of-life assessment in advanced cancer research. *Annals of Oncology, 5*, 565–570.
Ware, J. E., Kosinski, M., Turner-Bowker, D. M., & Gandek, B. (2002). *SF-12v2. How to score version 2 of the SF-12 health survey*. Lincoln: QualityMetric Incorporated.

A Multilevel Latent Variable Model for Multidimensional Longitudinal Data

Silvia Bianconcini and Silvia Cagnone

Abstract We propose a latent variable approach for modeling repeated multiple continuous responses. First the item correlation over time is explained by using latent growth curves, then the variability among items is accounted for by an overall latent variable. An EM algorithm is implemented to obtain maximum likelihood estimation of the model parameters.

1 Introduction

Multidimensional longitudinal data occur in social and educational researches when a number of different response variables are observed over time. Very often, these multiple responses are measures of a latent outcome not directly observable (e.g. costumer satisfaction, student ability). Methods for multidimensional longitudinal data must account for the autocorrelation that exists across time within each response variable and the cross-correlation that exists between different responses both across time and at the same time. Several approaches have been developed for the study of this kind of data.

Roy and Lin (2000) proposed a 2-step linear mixed model applied to multiple continuous outcomes. They use time-dependent factors to account for correlations of items within time and a random intercept to account for correlations of items across time. They also assume a linear model that allows for time-dependent covariates and a random effect to affect the time-dependent latent variables. That random effect accounts for correlations of latent variables across time.

Dunson (2003) introduced a dynamic latent trait model for multidimensional longitudinal data in the context of the Generalized Linear Latent Variable Model (GLLVM) so that different kinds of observed multidimensional data can be considered. An autoregressive structure that allows for covariates is used to model the structural part. The model is estimated by using the MCMC procedure.

S. Bianconcini (✉)
Department of Statistics, University of Bologna, Via Belle Arti, 41 - 40126 Bologna, Italy
e-mail: silvia.bianconcini@unibo.it

F. Palumbo et al. (eds.), *Data Analysis and Classification*, Studies in Classification, Data Analysis, and Knowledge Organization, DOI 10.1007/978-3-642-03739-9_37,

A very general framework is represented by multilevel models (Goldstein 2003; Skrondal and Rabe-Heskett 2004) since they allow to deal with longitudinal and/or multidimensional data. For the former measurement occasions are viewed as first level units whereas respondents are the second level units. For the latter first level units are represented by items nested within individuals that are second level units.

Multidimensional and longitudinal data are also treated within the traditional structural equation modeling approach (SEM). They are modeled in two different ways.

According to the first one (Jöreskog and Sörbom 2001), a standard confirmatory factor model is considered where the main feature is that the corresponding error terms are correlated over time. Moreover the latent variable are identified by fixing the same loading to one over time.

The second approach (Bollen and Curran 2006) is the so called latent variable growth models. The aim of this approach is to model the behavior of individuals over time. At this regard, two random effects are included into the model. The random intercept accounts for the individual differences in the initial status, the random slope captures the individual differences in the rate of growth. The main feature of this approach is that the random effects can be viewed as latent variables and treated in the traditional SEM approach. However, so far just unidimensional data have been considered.

The objective of this paper is to propose a latent variable approach for repeated multiple outcomes over time. The main idea of this proposal is firstly to model the item trajectories by means of growth curve models and then to specify an overall latent factor to account for the variability among items and individuals. An EM algorithm is implemented to obtain maximum likelihood estimates of model parameters. Finally, an application to real data set is presented.

2 Model for Continuous Responses

Suppose that J items are measured over time on N subjects. Let Y_{tji} denote the jth response variable for the ith individual at the tth time point, where $j = 1, 2, \ldots, J$, $i = 1, 2, \ldots, N$, $t = 1, 2, \ldots, T_{ji}$. The observation time points for item j and individual i are denoted by T_{ji} to stress that balanced data are not a requirement for efficient estimates.

2.1 Model Specification

The proposed methodology combines two different approaches, latent curves and factor analysis, with the aim of characterizing and estimating a global latent variable $\theta_i, i = 1, 2, \ldots, N$, by taking into account both the cross-sectional and longitudinal correlations of the J items.

One way to view this problem is that, for each individual, the observed outcomes $y_{tji}, t = 1, 2, \ldots, T_{ji}$ are measurements repeated on the same items. Several models can be used to study repeated measures, and we focus here on latent growth curves. These models are based on the basic idea that items differ in their growth over time. The items are likely to show differences in their temporal behavior as a function of differences in particular characteristics, defined in terms of item-specific, time-invariant covariates and/or in terms of unit-specific outcomes.

In order to model the within-item longitudinal correlations, we assume that the item series are generated by a nonstationary process, where the nonstationarity property results from a stochastic growth curve

$$\alpha_{ji} + \beta_{1ji}t + \cdots + \beta_{rji}t^r + \cdots + \beta_{pji}t^p, \quad t = 1, 2, \ldots, T_{ji} \tag{1}$$

where generally $p \leq 3$, α_{ji} and $\beta_{rji}, r = 1, 2, \ldots, p$, are random parameters that vary over items. Besides polynomials in time, other suitable mathematical functions could be used to represent nonlinear stochastic curves, such as the modified exponential, the Gompertz, and the logistic functions. However, given the small number of waves generally available for longitudinal studies in social and educational survey, a linear polynomial can be considered suitable for our analysis. Hence, the model becomes

$$y_{tji} = \alpha_{ji} + \beta_{ji}t + \varepsilon_{tji}, \quad t = 1, 2, \ldots, T_{ji} \tag{2}$$

where the random parameters α_{ji} and β_{ji} are used to model the within-item correlation of the repeated measures y_{tji}, and the ε_{tji}'s are assumed to be normally distributed with zero mean and variance σ_j^2, that is each observed variable is homoscedastic over time. This model implies the conditional independence of the outcomes given the latent variables α_{ji} and β_{ji}, that is for each item j, the T_{ji} outcomes y_{tji} are uncorrelated. In addition, it assumes that the relationships between the random coefficients and observed responses are constant over items, or in other words the error covariance matrix $\sigma_j^2 I_{T_{ji}}$ is the same for every j.

The variability among items' initial level and slopes is accounted for by a general latent factor θ_i, loaded on the random parameters, as follows

$$\begin{aligned} \alpha_{ji} &= \mu_{\alpha_j} + \lambda_j \theta_i + \zeta_{\alpha_{ji}}, \\ \beta_{ji} &= \mu_{\beta_j} + \lambda_j \theta_i + \zeta_{\beta_{ji}} \end{aligned} \tag{3}$$

where μ_{α_j} and μ_{β_j} are the mean intercept and mean slope of each item, $\lambda_j, (j = 1, 2, \ldots, J)$ are the factor loadings, θ_i is the individual latent outcome, and $\zeta_{\alpha_{ij}}$ and $\zeta_{\beta_{ij}}$ are residuals assumed to be normally distributed with zero means, variances equal to $\psi^2_{\alpha_j}$, $\psi^2_{\beta_j}$ respectively, and covariance given by ψ_{α_j,β_j}. For the sake of identifiability, we suppose that $\theta_i \sim N(0, 1)$.

Equations (2) and (3) account for both the variability of each item over time and the variability between items at the same and different time points. In more detail, these different sources of variability can be expressed as function of the model parameters as follows

$$
\begin{aligned}
V(y_{tji}) &= (1+t)^2\lambda_j^2 + \psi_{\alpha_j}^2 + t^2\psi_{\beta_j}^2 + \sigma_j^2 + 2t\psi_{\alpha_j\beta_j} \\
Cov(y_{tji}, y_{tj'i}) &= \lambda_j\lambda_{j'}(1+t)^2 \quad j \neq j' \\
Cov(y_{tji}, y_{t'ji}) &= \lambda_j^2(1+t)(1+t') + \psi_{\alpha_j}^2 + tt'\psi_{\beta_j}^2 + (t+t')\psi_{\alpha_j\beta_j} \quad t < t' \\
Cov(y_{tji}, y_{t'j'i}) &= \lambda_j\lambda_{j'}(1+t)(1+t') \quad t < t' \quad j \neq j'
\end{aligned}
$$

Once the serial and sectional correlations of the items have been modeled, we gain information on a global latent outcome of interest. The model can be viewed as a multilevel model where (2) and (3) can be thought as the first and second stage. First, for each individual, we considered separately each item series modeling their time path, then we accounted for the fact that these responses are measuring the same underlying quantity in view of describing the variability between items. Hence, time is the unit of the first level, item is the second level unit, individual is the unit of the last level.

Let $\mathbf{y}_i = (\mathbf{y}_{1i}, \ldots, \mathbf{y}_{Ji})^T$ represent the whole response pattern for a randomly selected individual. The marginal distribution of $\mathbf{y}_i$ is given by

$$
f(\mathbf{y}_i) = \int_{-\infty}^{+\infty} \cdots \int_{-\infty}^{+\infty} g(\mathbf{y}_i|\boldsymbol{\alpha}_i, \boldsymbol{\beta}_i, \theta_i) h(\boldsymbol{\alpha}_i, \boldsymbol{\beta}_i, \theta_i) d\boldsymbol{\alpha}_i d\boldsymbol{\beta}_i d\theta_i \qquad (4)
$$

where $\boldsymbol{\alpha}_i = (\alpha_{1i}, \ldots, \alpha_{Ji})^T$, $\boldsymbol{\beta}_i = (\beta_{1i}, \ldots, \beta_{Ji})^T$, $g(\mathbf{y}_i|\boldsymbol{\alpha}_i, \boldsymbol{\beta}_i, \theta_i)$ is the conditional density function of $\mathbf{y}_i$ given $\boldsymbol{\alpha}_i, \boldsymbol{\beta}_i, \theta_i$, and $h(\boldsymbol{\alpha}_i, \boldsymbol{\beta}_i, \theta_i)$ is the joint density function of the latent variables.

Under the assumption of conditional independence of $\mathbf{y}_i$ with respect to $\boldsymbol{\alpha}_i, \boldsymbol{\beta}_i$ and θ_i we obtain:

$$
g(\mathbf{y}_i|\boldsymbol{\alpha}_i, \boldsymbol{\beta}_i, \theta_i) = \prod_{j=1}^{J}\prod_{t=1}^{T_{ji}} g(y_{tji}|\alpha_{ji}, \beta_{ji}, \theta_i) \qquad (5)
$$

Parameter estimation are obtained using the maximum likelihood via EM algorithm.

2.2 Estimation

A full maximum likelihood method is implemented with the aim of estimating all the parameters simultaneously. The parameters to be estimated are $\sigma_j^2, \mu_{\alpha_j}, \mu_{\beta_j}, \psi_{\alpha_j}^2, \psi_{\beta_j}^2, \psi_{\alpha_j\beta_j}$, and λ_j, $j = 1, 2, \ldots, J$. The joint density function of the random variables is given by

$$
f(\mathbf{y}_i, \boldsymbol{\alpha}_i, \boldsymbol{\beta}_i, \theta_i) = g(\mathbf{y}_i|\boldsymbol{\alpha}_i, \boldsymbol{\beta}_i, \theta_i) h_2(\boldsymbol{\alpha}_i, \boldsymbol{\beta}_i|\theta_i) h_1(\theta_i). \qquad (6)
$$

Using (5) and (6), for the sample of N individuals, the complete log-likelihood is written as

$$L = \sum_{i=1}^{N} \log f(\mathbf{y}_i, \boldsymbol{\alpha}_i, \boldsymbol{\beta}_i, \theta_i) \tag{7}$$
$$= \sum_{i=1}^{N} \left\{ \sum_{j=1}^{J} \left[\sum_{t=1}^{T_{ji}} \log g(y_{tji}|\alpha_{ji}, \beta_{ji}, \theta_i) + \log h_2(\alpha_{ji}, \beta_{ji}|\theta_i) \right] + \log h_1(\theta_i) \right\}.$$

Since the latent variables $\boldsymbol{\alpha}_i$, $\boldsymbol{\beta}_i$ and θ_i are unobserved we use the E-M algorithm to maximize the log-likelihood. The E-M starts with initial values of the parameters. The algorithm consists of an expectation and a maximization step. In the expectation step the expected score function from the complete likelihood $(\boldsymbol{\alpha}_i, \boldsymbol{\beta}_i, \theta_i)$ given the observed variables $(\mathbf{y})$ is set equal to zero. In the maximization step updated parameter estimates are obtained from the equations obtained in the E-step. The whole procedure is repeated until convergence. From simplicity, from now on we assume $T_{ij} = T$, that is we assume balanced data even if the results do not change for the case of unbalanced data. The estimates of the 1-level error variances σ_j^2 are obtained by computing the expected score functions of $g(y_{tji}|\alpha_{ji}, \beta_{ji}, \theta_i)$ as follows

$$E\left[\frac{\partial \log g(y_{tji}|\alpha_{ji}, \beta_{ji}, \theta_i)}{\partial \sigma_j^2}\right] = \int \cdots \int \frac{\partial \log g(y_{tji}|\alpha_{ji}, \beta_{ji}, \theta_i)}{\partial \sigma_j^2} \times p(\boldsymbol{\alpha}_i, \boldsymbol{\beta}_i, \theta_{ji}|\mathbf{y}_{ji}) d\boldsymbol{\alpha}_i d\boldsymbol{\beta}_i d\theta_i$$

Solving $\sum_{i=1}^{N} \sum_{t=1}^{T} E\left[\frac{\partial \log g(y_{tji}|\alpha_{ji}, \beta_{ji}, \theta_i)}{\partial \sigma_j{}^2}\right] = 0$ and approximating the integrals with Gauss–Hermite quadrature points (Stroud and Secrest 1966), we get an explicit solution for the maximum likelihood estimation of $\sigma_j{}^2$. It is given by

$$\hat{\sigma_j}^2 = \frac{\begin{array}{c}\sum_{i=1}^{N} \sum_{t=1}^{T} \sum_{q_1,\ldots,q_{2J+1}} (y_{tji} - \alpha_{q_j i} - \beta_{q_j i} t)^2 \\ p(\alpha_{q_1 i}, \ldots, \alpha_{q_J i}, \beta_{q_{J+1} i}, \ldots, \beta_{q_{2J} i}, \theta_{q_{2J+1} i}|\mathbf{y}_i)\end{array}}{T \cdot N}. \tag{8}$$

On the other hand, the estimation of the 2-level parameters $\mathbf{a}_j = (\mu_{\alpha_j}, \mu_{\beta_j}, \lambda_j, \psi_{\alpha_j}^2, \psi_{\beta_j}^2, \psi_{\alpha_j,\beta_j})$ depends only on $h_2(\alpha_{ji}, \beta_{ji}|\theta_i)$. The expected score function with respect to the parameters $\mathbf{a}_j$, $j = 1, 2, \ldots, J$, takes the form

$$E\left[\frac{\partial \log h_2(\alpha_{ji}, \beta_{ji}|\theta_i)}{\partial \mathbf{a}_j}\right] = \int \cdots \int \frac{\partial \log h_2(\alpha_{ji}, \beta_{ji}|\theta_i)}{\partial \mathbf{a}_j} \times p(\boldsymbol{\alpha}_i, \boldsymbol{\beta}_i, \theta_{ji}|\mathbf{y}_{ji}) d\boldsymbol{\alpha}_i d\boldsymbol{\beta}_i d\theta_i. \tag{9}$$

As before, solving $\sum_{i=1}^{N} E\left[\frac{\partial \log h_2(\alpha_{ji}, \beta_{ji}|\theta_i)}{\partial \mathbf{a}_j}\right] = 0$ and approximating the integrals with Gauss–Hermite quadrature points, we get an explicit solution for the maximum likelihood estimators for each of the elements of $\mathbf{a}_j$. In particular, for μ_{α_j} and μ_{β_j}, we get

$$\hat{\mu}_{\alpha_j} = \frac{\sum_{i=1}^{N} \sum_{q_1,\ldots,q_{2J+1}} (\alpha_{q_j i} - \lambda_j \theta_i) \, p(\alpha_{q_1 i},\ldots,\alpha_{q_J i},\beta_{q_{J+1} i},\ldots,\beta_{q_{2J} i},\theta_{q_{2J+1} i}|\mathbf{y}_i)}{N}. \quad (10)$$

$$\hat{\mu}_{\beta_j} = \frac{\sum_{i=1}^{N} \sum_{q_1,\ldots,q_{2J+1}} (\beta_{q_j i} - \lambda_j \theta_i) \, p(\alpha_{q_1 i},\ldots,\alpha_{q_J i},\beta_{q_{J+1} i},\ldots,\beta_{q_{2J} i},\theta_{q_{2J+1} i}|\mathbf{y}_i)}{N}. \quad (11)$$

The expression of λ_j is given by

$$\hat{\lambda}_j = \frac{1}{2} \frac{\sum_{i=1}^{N} \sum_{q_1,\ldots,q_{2J+1}} [(\alpha_{q_j i} - \mu_{\alpha_j}) + (\beta_{q_j i} - \mu_{\beta_j})] \theta_{q_{2J+1}} \, p(\alpha_{q_1 i},\ldots,\alpha_{q_J i},\beta_{q_{J+1} i},\ldots,\beta_{q_{2J} i},\theta_{q_{2J+1} i}|\mathbf{y}_i)}{N}. \quad (12)$$

Deriving with respect to $\psi^2_{\alpha_j}$, and $\psi^2_{\beta_j}$, we obtained the following expressions for the corresponding maximum likelihood estimates

$$\hat{\psi}^2_{\alpha_j} = \frac{\sum_{i=1}^{N} \sum_{q_1,\ldots,q_{2J+1}} (\alpha_{q_j i} - \lambda_j \theta_{q_{2J+1}})^2 \, p(\alpha_{q_1 i},\ldots,\alpha_{q_J i},\beta_{q_{J+1} i},\ldots,\beta_{q_{2J} i},\theta_{q_{2J+1} i}|\mathbf{y}_i)}{N} \quad (13)$$

$$\hat{\psi}^2_{\beta_j} = \frac{\sum_{i=1}^{N} \sum_{q_1,\ldots,q_{2J+1}} (\beta_{q_j i} - \lambda_j \theta_{q_{2J+1}})^2 \, p(\alpha_{q_1 i},\ldots,\alpha_{q_J i},\beta_{q_{J+1} i},\ldots,\beta_{q_{2J} i},\theta_{q_{2J+1} i}|\mathbf{y}_i)}{N} \quad (14)$$

Finally, the expression of the maximum likelihood estimation of ψ_{α_j,β_j} is given by

$$\hat{\psi}_{\alpha_j,\beta_j} = \frac{\sum_{i=1}^{N} \sum_{q_1,\ldots,q_{2J+1}} (\alpha_{q_j i} - \lambda_j \theta_{q_{2J+1}})(\beta_{q_j i} - \lambda_j \theta_{q_{2J+1}}) \, p(\alpha_{q_1 i},\ldots,\alpha_{q_J i},\beta_{q_{J+1} i},\ldots,\beta_{q_{2J} i},\theta_{q_{2J+1} i}|\mathbf{y}_i)}{N} \quad (15)$$

2.3 *Application to a Real Data Set*

The data are from a longitudinal study by Wheaton et al. (1977) concerning the stability over time of attitudes such as alienation due to the effects of industrial development in a rural region in Illinois. The data set consists of six attitude scales collected from 932 persons in two rural regions at three points in time: 1966, 1967, and 1971. This example uses data from 1967 and 1971 only.

The variables used as indicators of Alienation are:

- *Anomia subscale*. A person who has difficulty remembering names of people and objects in everyday speech has a condition called anomia.

Table 1 Descriptive statistics

Item	*Mean*	*St dev*	Covariance matrix			
Anomia67	3.56	3.59	12.89			
Power67	3.09	3.04	7.06	9.24		
Anomia71	3.56	3.55	7.42	5.21	12.59	
Power71	3.08	3.17	5.08	5.09	7.38	10.04

Table 2 Parameter estimates and their standard errors (first level)

Item	$\hat{\sigma}_j^2$
Anomia	4.86 (0.22)
Power	2.05 (0.12)

Table 3 Parameter estimates and their standard errors (second level)

Item	$\hat{\mu}_{\alpha_j}$	$\hat{\mu}_{\beta_j}$	$\hat{\lambda}_j$	$\hat{\psi}_{\alpha_j}^2$	$\hat{\psi}_{\beta_j}^2$	$\hat{\psi}_{\alpha_j\beta_j}$
Anomia	3.21(0.02)	−0.01(0.03)	1.27(0.03)	5.07(0.21)	1.81(0.04)	−2.68(0.03)
Power	3.25(0.02)	−0.05(0.03)	1.20(0.03)	5.37(0.23)	2.02(0.05)	−2.93(0.04)

- *Powerlessness subscale*. Indicates the powerlessness of people due to the industrial development.

In Table 1 the means, the standard deviations and the covariance matrix of the two variables in the two time points are reported.

We can notice that both the means and variances are different between items but they do not change sensibly over time. In more detail, the moments of the item Anomia are slightly higher than those corresponding to the item Power. Thus the items are characterized by different magnitude.

In Tables 2 and 3 the results of the estimates of the first level and second level parameters are shown (standard errors in brackets). The model has been estimated with Fortran 95.

As for the first level parameters, the variance of Anomia is more than double the variance of Power but they are both significant.

The results concerning the second level parameters are very interesting. The mean intercepts and mean slopes indicate the initial level are very similar for both the items and they are also quite close to the observed means of the items. On the contrary, the mean slopes are almost 0 and not significant. Thus there is no significant growth of the two items in the period considered, that is the two indicators remain stable over time. The loadings are both greater than one and significant for both the items, so that we can conclude that the item contributes significantly to measure the general factor Alienation. The estimates of the variances of the random effects are all significant but, as expected, they are higher for the α_j's than for the β_j's even if they are all significant. Finally the covariances between α_j and β_j are negative, indicating that there is a negative relation between the initial level of the item and its rate of growth.

3 Conclusion

In this paper we discuss the specification and the estimation of a latent growth model for multidimensional longitudinal data. We show how this model can be viewed as a multilevel model. Indeed it consists of three level equations. The first level equation allows to model the variability of each observed variable within each time point by means of a latent growth model. The main components of this model are two random effects, the random intercept that accounts for the item differences in the initial status and the random slope that captures the item differences in its rate of growth.

The second level equation accounts for the remaining variability, that is the variability that exists between items due just to the effect of a general latent factor. Thus, the second level model can be viewed as a one-factor model.

The third level equation involves only the general factor. In this particular case we assume that it is standardized normally distributed.

Different generalizations can be considered as further developments of the model.

First of all, the possibility of including more than one latent factors as well as including covariates in the model.

Moreover, it can be important to evaluate the behavior of the parameter estimates in different conditions like, for example, in the case in which the time is modeled by means of a non linear curves or a polynomial with degree $p > 3$.

Finally the goodness of fit problem should also be discussed in the analysis to evaluate the performance of the model.

References

Bollen, K. A., & Curran, P. J. (2006). *Latent Curve Models: a structural equation perspective*. New Jersey: Wiley Series in Probability and Statistics.

Goldstein, H. (2003). *Multilevel statistical models*. Kendalls Library of Statics. London: Arnold.

Jöreskog, K., & Sörbom, D. (2001). *LISREL 8: Users' reference guide*. Chicago: SSI.

Stroud, A., & Secrest, D. (1966). *Gaussian quadrature formulas*. Englewood Cliffs, NJ: Prentice-Hall.

Skrondal, A., & Rabe-Heskett, S. (2004). *Generalized latent variable modeling: multilevel, longitudinal, and structural equation models*. Boca Raton, FL: Chapman and Hall/CRC.

Dunson, D. B. (2003). Dynamic latent trait models for multidimensional longitudinal data. *Journal of American Statistical Association, 98*(4), 555–563.

Roy, J., & Lin, X. (2000). Latent variable models for longitudinal data with multiple continuous outcomes. *Biometrics, 56*(4), 1047–1054.

Wheaton, B., Muthén, M., Alwin, D. F., & Summers, G. F. (1977). Assessing reliability and stability in panel models. *Sociological Methodology, 8*, 84–136.

Turning Point Detection Using Markov Switching Models with Latent Information

Edoardo Otranto

Abstract A crucial task in the study of business cycle is the detection of the turning points, indicating the beginning (end) of a phase of growth (recession) of the economy. The dating proposed by experts are generally evaluated with the support of some empirical procedure. Many efforts have been devoted to propose statistical models able to detect the turning points and to forecast them. A class of models largely used for these purposes is the Markov Switching one. In this work we use a new version of the Markov Switching model, named *Markov Switching model with latent information*, which seems particularly able to detect the turning points in real time and to forecast them. In particular this model uses a latent variable, representing the cycle of the economy, as information to drive the transition probabilities from a state to another. Two examples are provided: in the first one we use Japanese GDP data, where our model and the classical Markov Switching model have a similar performance, but the first provides more information to forecast the turning points; in the second one we analyze USA GDP data, where the classical Markov Switching model fails in the turning points detection, whereas our model fits adequately the data.

1 Introduction

The business cycle analysis is largely supported by statistical procedures to extract latent signals, representing the cycle indicators, and to detect turning points. For example, in Italy, ISAE (*Istituto di Studi ed Analisi Economica*) has developed statistical procedures to obtain a coincident indicator of the economy (see Altissimo et al. (2000)), whereas, in USA, NBER (*National Bureau of Economic Research*) utilizes the Bry and Boschan procedure (Bry and Boschan 1971) to support the decision to fix a turning point in official dating. In the economic and statistical literature many models were developed to extract the latent cycle and to detect automatically

E. Otranto
Dipartimento di Economia, Impresa e Regolamentazione, Via Torre Tonda 34, 07100 Sassari, Italy
e-mail: eotranto@uniss.it

F. Palumbo et al. (eds.), *Data Analysis and Classification*,
Studies in Classification, Data Analysis, and Knowledge Organization,
DOI 10.1007/978-3-642-03739-9_38,

the turning points (see, for example, Bruno and Otranto 2008). In particular the class of Markov Switching (MS hereafter) models (Hamilton 1989) has been largely utilized for its flexibility and for the capability to provide simultaneously a statistical representation of the analyzed time series and a dating, based on the estimated probability of a certain state of a Markov Chain (the state represents the regime of recession or growth). Many extensions of MS models have been proposed in literature; in particular the use of time varying transition probabilities (Diebold et al. 1994; Filardo 1994) increases the flexibility of this model. In this case we need some independent observable variables which drive the dynamics of the transition probabilities. The identification of these variables is not easy and they are often artificial; furthermore, they can cause estimation problems and can not be available in real time. We propose a new approach in which the variable driving the changes in the state is a latent variable depending on the same state variable. In particular, the latent variable used in our applications is a lagged indicator of the business cycle. The reason to use a lagged latent variable is double: from an economic point of view, the most recent value of the cyclical indicator can help to forecast the state of the economy at the next time; from a technical point of view, Filardo (1998) shows that, under mild conditions, it is possible to use the Hamilton filtering to estimate time varying MS models if the transition probabilities are driven by lagged information variables. In this paper we extend the Generalized Hamilton model proposed in (Lam 1990) and apply it to the Gross Domestic Product (GDP) of USA and Japan, comparing the results in terms of real time detection of turning points and one-step ahead forecasts with respect to the original model. The model proposed has a nonlinear state-space form and can be estimated using an extended Kalman filter; for details about this model we refer to (Otranto 2008). As the classical MS models, our model has the advantage to provide a prediction of turning points, but provides other useful information, described at the end of Sect. 2; moreover, to drive the transition probabilities it uses an information which is always available, being a sub-product of the filtering. In the next section we present the new model, whereas in Sect. 3 the estimation results are shown.

2 The Generalized Hamilton Model with Latent Information

Lam (1990) proposes a generalization of the two-states MS Hamilton model (Hamilton 1989), allowing for the possibility of permanent and transitory shocks; in the Hamilton model the business cycle asymmetry shows up in the growth rate of the output, whereas in the Lam model it shows up in the permanent component of the output. The Hamilton model is given by:

$$\begin{aligned} y_t &= n_t + z_t \qquad t = 1, \ldots, T \\ n_t &= n_{t-1} + \mu_{s_t} \\ \varphi(L) z_t &= w_t \\ Pr(s_t &= 0|s_{t-1} = 0) = q; \quad Pr(s_t = 1|s_{t-1} = 0) = 1 - q \\ Pr(s_t &= 0|s_{t-1} = 1) = 1 - p; \quad Pr(s_t = 1|s_{t-1} = 1) = p \end{aligned} \tag{1}$$

where y_t is the variable of interest at time t, which is decomposable into a stochastic trend n_t, subject to switches in the level μ_{s_t}, and an autoregressive component z_t; $\varphi(L)$ is an AR polynomial, w_t are independent Normal disturbances with zero mean and variance equal to σ^2, T is the length of the time series. The variable s_t is an unobservable discrete variable, assuming two possible values (labeled with 0 and 1), representing the state of recession and growth respectively. The distribution of s_t is unknown, but we suppose that its dynamics is driven by a Markov chain with transition probabilities expressed by the last two equations in (1).

Hamilton (1989) hypothesizes that $\varphi(L)$ has a unit root and rewrites the model in terms of first differences of y_t, obtaining an autoregressive model with a Markov-switching mean. Lam (1990) considers the possibility that the roots of $\varphi(L)$ lie outside the unit circle; in this case the first equation in (1), in terms of first differences, is given by:

$$\begin{aligned} \Delta y_t &= y_t - y_{t-1} \\ \Delta y_t &= \mu_{s_t} + (z_t - z_{t-1}) \end{aligned} \tag{2}$$

In this specification the variable z_t can be interpreted as a transitory deviation (cycle) of y_t from its trend n_t.

The model (1) with the specification (2) and using for $\varphi(L)$ a polynomial of order 2, can be written in a state-space form as:

$$\begin{aligned} y_t &= \mathbf{a}\boldsymbol{\xi}_t + \mu_{s_t} \\ \boldsymbol{\xi}_t &= \boldsymbol{\gamma} + \mathbf{H}\boldsymbol{\xi}_{t-1} + \mathbf{v_t} \end{aligned} \tag{3}$$

where:

$$\mathbf{a} = [1 \quad -1] \qquad \xi_t = [z_t \quad z_{t-1}]'$$

$$\gamma = [0 \quad 0]' \qquad \mathbf{H} = \begin{bmatrix} \varphi_1 & \varphi_2 \\ 1 & 0 \end{bmatrix} \qquad \mathbf{v}_t = [w_t \quad 0]'$$

and φ_1 and φ_2 are the autoregressive coefficients.

It is very plausible that the latent transitory component z_t can drive the transition from a state to another, because it represents the oscillation of y_t around the trend, so that largest is the deviation from trend, more probable is the persistence in the same state at the next time, whereas when the level of the transitory component decreases this probability would decrease.

In this case the transition probabilities are given by:

$$\begin{aligned} p_{ii,t} &= \frac{\exp(\phi_i + \xi'_{t-1}\vartheta_i)}{1 + \exp(\phi_i + \xi'_{t-1}\vartheta_i)} \\ p_{ij,t} &= 1 - \frac{\exp(\phi_i + \xi'_{t-1}\vartheta_i)}{1 + \exp(\phi_i + \xi'_{t-1}\vartheta_i)} \end{aligned} \tag{4}$$

where $i, j = 0, 1$ $(i \neq j)$ and $\vartheta_i = [\theta_i \quad 0]'$. Interpreting the two states as recession (state 0) and growth (state 1), we expect that the sign of θ_0 is negative (the probability to stay in the recession period increases when z_{t-1} is negative) while for θ_1 is positive (the probability to stay in a growth period increases when z_{t-1} is positive). We name this model the Lam MS model with latent information (MSLI); a general formulation of this model is given in Otranto (2008). In the same paper the procedure to obtain the estimation of such a model is described; it is an extension of the well known Kim filter for state-space MS models (Kim 1994), which consists in applying the Kalman filter and the Hamilton filter (Hamilton 1990) to the state-space model. In practice we apply the Kim algorithm to the state-space model obtained linearizing the model (3)–(4); in fact the latent variable ξ_t enters the transition probabilities of the MS process in a nonlinear form, making not feasible the Kalman filter. To avoid this problem we use the extended Kalman filter (see, for example, Harvey and Shephard (1993)), including the transition probabilities in the latent vector and linearizing the second equation (transition equation) in (3) using a fist order Taylor expansion. Following Otranto (2008) it is given by:

$$\begin{bmatrix} \xi_t \\ p_{00,t} \\ p_{11,t} \end{bmatrix} = \begin{bmatrix} \gamma \\ c_{0,t} \\ c_{1,t} \end{bmatrix} + \begin{bmatrix} \mathbf{H} & 0 & 0 \\ d_{0,t} & 0 & 0 \\ d_{1,t} & 0 & 0 \end{bmatrix} \begin{bmatrix} \xi_{t-1} \\ p_{00,t-1} \\ p_{11,t-1} \end{bmatrix} + \begin{bmatrix} \mathbf{v}_t \\ 0 \\ 0 \end{bmatrix} \tag{5}$$

where $d_{i,t} = \frac{\delta \pi_i(\xi_{t-1})}{\delta \xi_{t-1}}|_{\xi_{t-1|t-1}}$, $c_{i,t} = \pi_i(\xi_{t-1|t-1}) - d_{i,t}\xi_{t-1|t-1}$ and $\pi_i(\boldsymbol{\xi}_{t-1})$ represents the logit function in (4). The Kim algorithm can be applied directly to the extended state-space model, obtaining the likelihood function of the MSLI model (see Otranto (2008) for details). A nice characteristic of the MSLI model is the possibility to establish a threshold value for $\boldsymbol{\xi}_t$ which denotes an almost sure change in regime at the next time; of course this property is particularly interesting in forecasting turning points. For example, considering a scalar latent variable ξ_t, from (4), fixing $p_{ii,t}$, by simple algebra, we can find the value of ξ_{t-1} which provides the probability $p_{ii,t}$:

$$\xi^*_{t-1} = \left[log\left(\frac{p_{ii,t}}{1 - p_{ii,t}}\right) - \phi_i \right] / \vartheta_i \tag{6}$$

For example, if $p_{00,t}$ is fixed equal to 0.01, ξ^*_{t-1} represents the value of the latent variable at time $t-1$ which will indicate a likely switch from state 0 to state 1.

3 Identifying and Forecasting USA and Japanese Turning Points

We apply the model described in the previous section to the USA and Japanese GDP to identify and forecast the turning points. For the evaluation of the model we use the official NBER dating for USA and the dating proposed by ECRI (Economic Cycle Research Institute) for Japan. Moreover we compare the results of the

Table 1 Estimation results (standard errors in parentheses), Quadratic Probability Scores and threshold values for the Lam model: MS and MSLI cases

	MS Japan	MSLI Japan	MS USA	MSLI USA
μ_0	0.322 (0.046)	0.276 (0.028)	0.778 (0.025)	−0.811 (0.253)
μ_1	0.867 (0.129)	0.784 (0.038)	2.792 (0.527)	1.858 (0.263)
σ	0.741 (0.055)	0.755 (0.048)	0.834 (0.043)	0.804 (0.039)
φ_1	0.909 (0.120)	0.941 (0.083)	1.346 (0.065)	1.251 (0.055)
φ_2	−0.007 (0.122)	−0.045 (0.097)	−0.386 (0.067)	−0.391 (0.034)
p_{00}	0.983 (0.018)	0.985 (0.013)	0.985 (0.011)	0.679 (0.102)
p_{11}	0.955 (0.040)		0.365 (0.242)	
ϕ_1		5.370 (2.861)		5.578 (1.670)
θ_1		15.093 (5.368)		2.028 (0.729)
QPS_0	0.362	0.343	0.785	0.092
QPS_1	0.365	0.350	0.776	0.132
z^*_{t-1} for $p_{11,t}$=0.99[a]		−0.051		−0.485
z^*_{t-1} for $p_{11,t}$=0.01[b]		−0.660		−5.017

[a]It represents the value of z_{t-1} which indicates an almost sure permanence in state 1 at the next time, following (6).
[b]It represents the value of z_{t-1} which indicates an almost sure change in regime (from 1 to 0) at the next time, following (6).

MSLI model with those obtained with the original MS Lam model (1)–(2). The data available cover the period from 1947-II to 2005-II for the USA GDP and the period from 1980-II to 2005-II for Japanese GDP (quarterly data; source Eurostat). In Table 1 we show the estimation of the Lam and MSLI models for the two time series.

In both MSLI models the coefficient θ_0 relative to the transition probability $p_{00,t}$ is not significant, so we have estimated a model with only the transition probability $p_{11,t}$ as time-varying. The Japanese case shows similar estimates for the common parameters; as a result the inference on the regime is very similar. This fact arises comparing the graphs of the filtered probabilities of a certain state obtained from the two alternative models. The filtered probabilities of the state s_t ($Pr[s_t = i|\Psi_t]$, $i = 0, 1$) are obtained from the Hamilton filter; Ψ_t is the information available at time t. In other words we can calculate the probability that at time t the business cycle is in a phase of growth or recession. Notice that this information is given in real time, in the sense that we use the set of observations available at time t. Alternatively, we could use the smoothed probabilities, calculating $Pr[s_t = i|\Psi_T]$; in practice we evaluate the probability of the state i ex post, conditional on the full data set, but, of course, this is a less useful result in practical cases, when we are interested to evaluate in real time the state of economy. Moreover, calculating the probabilities $Pr[s_t = i|\Psi_{t-1}]$, we can express the one-step ahead probability of a certain state at time t. In Fig. 1 the filtered real-time probabilities of the MS and MSLI models are shown. We can note that the recession periods, detected by ECRI, are correctly identified in the second half of the series; on the other hand, the MS and MSLI models identify some periods of recession also in the first half of the series,

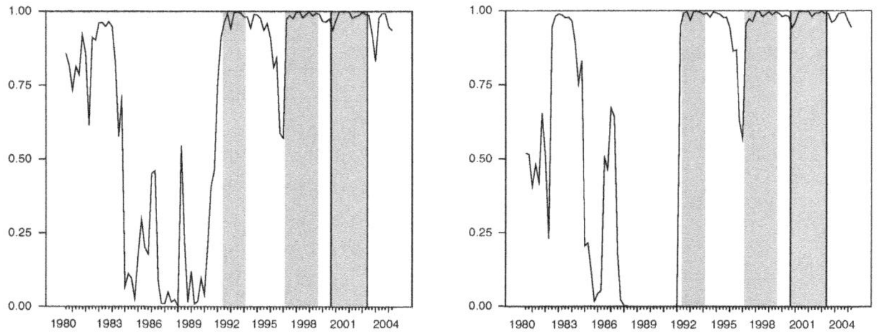

Fig. 1 Japanese GDP: filtered probabilities of state 0 ($Pr[s_t = 0|\Psi_t]$) obtained by MS model (*left figure*) and MSLI model (*right figure*). The shadow bars indicate the recession periods identified by ECRI

whereas the ECRI considers the full period as growth (anyway we recall that the ECRI dating is not official for the Japanese economy). The behavior of the one-step ahead filtered probabilities is similar, so we do not show them. The similarities in terms of inference on the regime between the two models can be confirmed using a simple index, proposed in (Diebold and Rudebusch 1989) and diffusely used in business cycle literature (see, for example, Otranto (2001); Layton and Katsuura (2005)). It is named *Quadratic Probability Score* (QPS) and is given by:

$$QPS_d = \frac{\sum_{t=1}^{T}(Pr[s_t = 1|\Psi_{t-d}] - D_t)^2}{T} \tag{7}$$

where D_t is the state at time t detected by ECRI. The index falls in the interval [0, 1] and it is equal to 0 in the case of correct assignment of the state without uncertainty for each t, and one in the opposite case. At the bottom of Table 1 the QPS's in real time ($d = 0$) and one-step ahead ($d = 1$) are shown. The value of the index is very similar for the MS and MSLI cases (slightly better in the MSLI case). Our conclusion is that there are not clear differences between the two models, so that the MSLI model does not add advantages in terms of inference on the regime. Anyway, it can provide an information about possible change in regime, using expression (6). The last columns of Table 1 show these values: if the state at time $t-1$ is the growth one and when the cyclical signal is around zero, we are almost sure to stay in the same state at time t; when the cyclical signal is below -0.66 we are almost sure to change state at the next time.

The US case is more interesting for the different behavior of the two models. The classical MS model shows very different estimates from the MSLI model; in particular, the switching intercepts are very different and the transition probability p_{00} is equal to 0.985 for the MS case and 0.679 for the MSLI case. This implies a strong persistence on average in state 0 using the MS model, equal to 65.8 quarters (it is calculated as $1/(1-p_{00})$; see Hamilton (1994), Chap. 22), whereas it is only three quarters using the MSLI model. The MS model fails to fit adequately the data

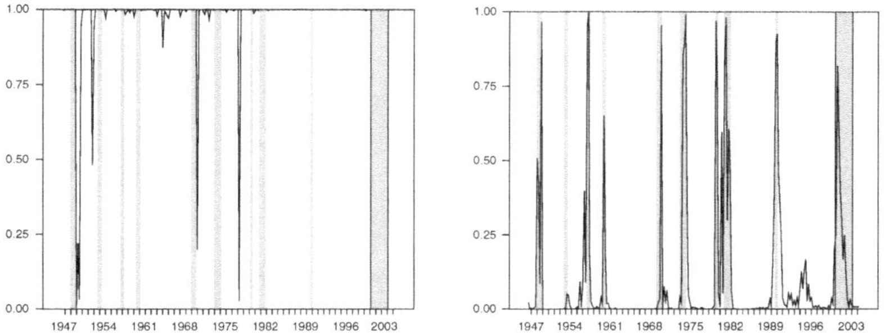

Fig. 2 USA GDP: filtered probabilities of state 0 ($Pr[s_t = 0|\Psi_t]$) obtained by MS model (*left figure*) and MSLI model (*right figure*). The shadow bars indicate the recession periods identified by NBER

Fig. 3 USA GDP: filtered probabilities of state 0 ($Pr[s_t = 0|\Psi_{t-1}]$) obtained by MSLI model. The shadow bars indicate the recession periods identified by NBER

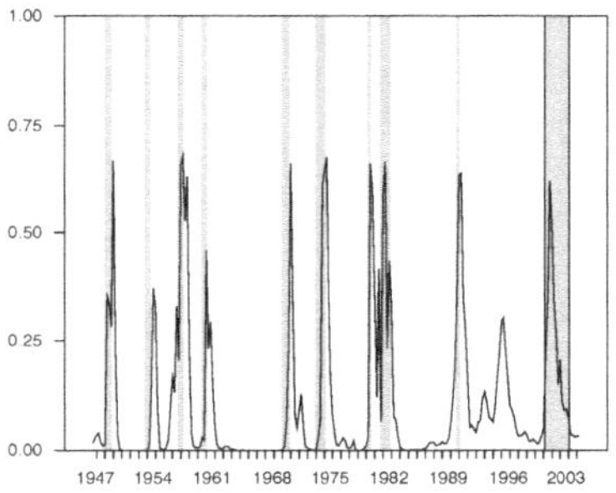

because the period studied does not include so long periods of recession. This is confirmed in Fig. 2, where the real time filtered probabilities are shown, with the gray bars representing the official recession periods detected by NBER. Moreover the MSLI model captures all the recessions periods indicated by NBER, except for the brief recession in 1954. These different behaviors are well synthesized by the QPS's (Table 1); it is equal to 0.785 in the MS case and 0.092 in the MSLI case. Recalling that the QPS is optimal when it is equal to 0 and very poor when it is 1, we deduce that the MSLI represents in a very good way the cyclical behavior of the USA economy. Similar considerations hold for the one-step ahead forecasts (the performance is expressed by QPS_1). In Fig. 3 we show the one-step ahead filtered probabilities derived from the MSLI model. We can note that they signal the possible change of state at the next time in correspondence of the NBER recession periods.

We conclude with some remarks. The MSLI model can be applied in the cases where a state-space MS model makes sense. In particular, it is preferable to adopt it when the latent variable has some interpretation. In the example of Sect. 3 it represents the cyclical fluctuations, but it can be used in many fields (climatological models, multicamera tracking experiments, etc.; see Otranto (2008)). Moreover it can be adopted both for univariate and multivariate models, using a general state-space form as:

$$\mathbf{y}_t = \mathbf{A}_{s_t}\xi_t + \mathbf{B}_{s_t}\mathbf{x}_t + \mathbf{e}_t$$
$$\xi_t = \gamma_{s_t} + \mathbf{H}_{s_t}\xi_{t-1} + \mathbf{v}_t$$

where $\mathbf{y}_t$ is a vector of data and $\mathbf{A}_{s_t}$, $\mathbf{B}_{s_t}$, $\boldsymbol{\gamma}_{s_t}$, $\mathbf{H}_{s_t}$ are matrices of appropriate dimension containing unknown coefficients; they can switch according to the state variable s_t, which can follow a generical $k-$ states Markov chain.

Acknowledgements Financial support from Italian MIUR under grant 2006137221_001 is gratefully acknowledged.

References

Altissimo, F., Marchetti, D. J., & Oneto, G. P. (2000). The Italian business cycle: coincident and leading indicators ans some stylized facts. *Temi di Discussione del Servizio Studi–Banca d'Italia, 377*.

Bry, G., & Boschan. C. (1971). Cyclical analysis of time series: selected procedures and computer programs. NBER Technical Paper, 20.

Bruno, G., & Otranto, E. (2008). Models to date the business cycle: the Italian case. *Economic Modelling, 25*, 899–911.

Diebold, F. X., Lee, J. H., & Weinbach, G. C. (1994). Regime switching with time-varying transition probabilities. In P. Hargreaves, (ed.), *Nonstationary Time Series Analysis and Cointegration* (pp. 283–302). Oxford: Oxford University Press.

Diebold, F. X., & Rudebusch, G. D. (1989). Scoring the leading indicators. *Journal of Business, 60*, 369–391.

Filardo, A. J. (1994). Business-cycle phases and their transitional dynamics. *Journal of Business and Economic Statistics, 12*, 299–308.

Filardo A. J. (1998). Choosing information variables for transition probabilities in a time-varying transition probability Markov switching model. Federal Reserve Bank of Kansas City, RWP 98–109.

Hamilton, J. D. (1989). A new approach to the economic analysis of nonstationary time series and the business cycle. *Econometrica, 57*, 357–384.

Hamilton, J. D. (1990). Analysis of time series subject to change in regime. *Journal of Econometrics, 45*, 39–70.

Hamilton, J. D. (1994). *Time series analysis*. Princeton: Princeton University Press.

Harvey, A. C., & Shephard, N. (1993). Structural time series models. In: G. S. Maddala, C. R. Rao, & H. D. Vinod (Eds.), *Handbook of statistics Vol. 11* (pp. 261–302). Amsterdam: Elsevier Science Publishers B.V.

Layton, A. P., & Katsuura, M. (2005). Comparison of regime switching, probit and logit models in dating and forecasting US business cycles. *International Journal of Forecasting, 17*, 403–417.

Kim, C.-J. (1994). Dynamic linear models with Markov-switching. *Journal of Econometrics, 60*, 1–22.

Lam, P.-S. (1990). The Hamilton model with a general autoregressive component: estimation and comparison with other models of economic time series. *Journal of Monetary Economics, 26*, 409–32.

Otranto, E. (2001). The Stock and Watson model with Markov switching dynamics: an application to the Italian business cycle. *Statistica Applicata, 13*, 413–429.

Otranto, E. (2008). A time varying hidden Markov model with latent information. *Statistical Modelling, 8*, 347–366.

Part VIII
Knowledge Extraction from Temporal Data

Statistical and Numerical Algorithms for Time Series Classification

Roberto Baragona and Salvatore Vitrano

Abstract Cluster analysis of time series is usually performed by extracting and comparing relevant interesting features from the data. Quite a few numerical algorithms are available that search for highly separated data sets with strong internal cohesion so that some suitable objective function is minimized or maximized. Algorithms developed for classifying independent sample data are often adapted to the time series framework. On the other hand time series are dependent data and their statistical properties may serve to drive allocation of time series to groups by performing formal statistical tests. Which is the class of methods that has better chance of fulfilling its task is an open question. Some comparisons are presented concerned with the application of different algorithms to simulated time series. The data recorded for monitoring the visitors flow to archaeological areas, museums, and other sites of interest for displaying Italy's cultural heritage are examined in some details.

1 Introduction

Time series classification has become an important topic in many research fields. Several circumstances motivate this interest. We may mention availability of large data bases, widespread use of automatic data collection devices, fast computing machines resources and inexpensive small size recording devices with practically unlimited capacity. Quite a few data are collected over time regularly. Examples may be found in meteorology, seismology, ecology, finance, economics and demography. Understanding the data, extracting knowledge and useful information, producing accurate and reliable forecasts call for development of high complexity methods, models and algorithms. The development of methods for effective time series data classification may help these procedures. As a matter of fact, recognizing

R. Baragona (✉)
Department of Sociology and Communication,
Sapienza University of Rome, Via Salaria 113, 00198 Rome, Italy
e-mail: roberto.baragona@uniroma1.it

F. Palumbo et al. (eds.), *Data Analysis and Classification*,
Studies in Classification, Data Analysis, and Knowledge Organization,
DOI 10.1007/978-3-642-03739-9_39,

either similar or related time series may highlight useful information, contribute to explain the behavior of complex systems and simplify data analysis and model building.

In this paper we limit ourselves to unsupervised classification of time series. Cluster analysis is among the favorite playgrounds for heuristic methods and algorithms. Usually mixture distributions are the theoretical framework for studying the properties of partitions of objects into groups. This approach is believed somewhat unpractical in most cases as a priori information is not available and computations are rather heavy. On the other hand, heuristic methods do not need detailed assumptions while are usually able to assign objects to groups to achieve greatest cohesion within groups and highest separation between groups.

Comparison of time series in order to assess similarity/dissimilarity may be based either on geometrical properties of trajectories from time series graphical display or on features related to the properties of the underlying stochastic process that is assumed to generate the data. Features may consist of measurements that characterize the individual time series structures and relations between pairs of time series. Measurements may be taken either in the time or in the frequency domain. Comprehensive surveys on methods for time series classification are Liao (2005) and Keogh and Kasetty (2003). Sets of different features have been taken into account as well, for instance by Wang et al. (2005).

We confine our attention on two characteristic features that may be computed in the time domain, namely the coefficients of the (truncated) infinite autoregressive time series representation (Piccolo 1990) and the cross correlations between pairs of time series (Zani 1983). The procedure for unsupervised classification that we use in the present paper may be summarized as follows. The features extracted from each and every time series in a given set allow us to build a distance matrix that accounts for similarity/dissimilarity between pairs of times series. Then a quick partition algorithm (Hartigan 1975) is designed constrained on statistical testing of pairwise relations. An objective function is defined on the set of all admissible partitions that assigns to each and every partition a positive real value according to the internal cohesion and the external separation criteria. The set of potential solutions, that is the set of all admissible partitions, is searched for the optimal partition.

The set of potential solutions is a discrete space which is largest even for time series sets of moderate size. We consider genetic algorithms as the basis to develop a stochastic algorithm which may search this space efficiently. Exhaustive search and most deterministic algorithms do not seem appropriate because of the size of the problem and the discreteness of the solution space. On the other hand, genetic algorithms require little assumptions and proved to be effective to optimize objective functions in large discrete space. Murthy and Chowdhury (1996) addresses applications of genetic algorithms for solving partitioning problems. Genetic algorithms examine iteratively subsets of the set of admissible solutions. In each iteration the current subset is evaluated by computing the objective function, and a new subset is built by replicating the existing solutions as more times as higher the objective function value. Then recombination and random mutation take place to improve the solutions in the current subset. The partition with largest objective function after the last iteration is assumed as the optimal solution.

For any pair of time series $\{x_{it}, x_{jt}\}$ in a given data set $(t = 1, \ldots, n)$ we considered the squared autoregressive distance

$$\hat{d}_{ij}^2 = \sum_{k=1}^{m} \left(\hat{\pi}_{ik} - \hat{\pi}_{jk}\right)^2,$$

where $\{\hat{\pi}_{ik}, \hat{\pi}_{jk}\}$ denote the least squares estimates of the parameters of the autoregressive models of order m (AR(m)). The choice $m = n^{1/3}$ was recommended. The distributional properties of this distance as summarized in Sarno (2001) have been used for similarity statistical hypothesis testing. This way, given the appropriate significance level α, we could assess the thresholds required by the quick partition algorithm to decide whether a time series may aggregate to an existing cluster or has to start a new one. This procedure was proposed by Maharaj (2000) as well, though parameters estimation was suggested in a seemingly unrelated regression framework. A similar distribution, that is a mixture of chi-square random variables, was found by Piccolo (1989) and Corduas (1996) if autoregressive moving average (ARMA) models were fitted to the time series and the parameters were estimated by maximum likelihood. In this case, the coefficients $\{\hat{\pi}_{ik}, \hat{\pi}_{jk}\}$ may be computed from the ARMA parameter estimates by using recursive equations. The computational burden is considerable, so either chi-square or log-normal approximations may be tried. In our extensive simulation experiment we found all these procedure, to our implementation accuracy, fairly equivalent. Similar results were obtained as well by assessing empirically the thresholds by simulation. Least squares estimation and chi-square approximation yielded slightly better results by involving limited computing effort so that this was our choice as regards the simulation results to be described in the next Section.

The alternative similarity measure was considered the residuals cross correlations

$$\hat{\varrho}_{ij}(h) = \frac{1}{n} \sum_{t=1}^{n-h} \hat{a}_{it} \hat{a}_{j,t+h},$$

where h is any integer usually less than m and $\{\hat{a}_{it}\}$ and $\{\hat{a}_{jt}\}$ denote the residuals computed from suitable models fitted to time series $\{x_{it}\}$ and $\{x_{jt}\}$ respectively. Under the null hypothesis $\hat{\varrho}_{ij}(h)$ is asymptotically normally distributed with zero mean and variance n^{-1}. The thresholds to be used in the quick partition algorithm were computed accordingly given the significance level α.

The paper is organized as follows. In the next Section we give a short account of a simulation experiment where some artificial sets of time series data are generated and different procedures based on statistical testing and numerical algorithms are compared. The application of some cluster analysis procedures to a set of real data is described in Sect. 3. The data set is concerned with the visitors flow recorded in

Italy, in state owned museums with paid admission, on a regular monthly frequency between years 1996 and 2005. In the last Section some conclusions are drawn.

2 A Simulation Experiment

An extensive simulation experiment was carried out to evaluate the algorithms performance. We shall present only the results concerned with 3 sets of time series A, B and C. The three sets were generated by some univariate autoregressive moving average (ARMA) models (Table 1), with either correlated or uncorrelated residuals. Each cluster was generated by generating instances of the ARMA model

$$x_t - \phi_1 x_{t-1} - \phi_2 x_{t-2} = a_t - \theta_1 a_{t-1} - \theta_2 a_{t-2},$$

where $\{a_t\}$ was a sequence of zero mean and unit variance uncorrelated Gaussian random variables. Possibly the a_t's are allowed to be simultaneously correlated across time series that belong to the same artificial cluster. Let g be the number of clusters and k the cluster size. We assumed $g = 6$ and $k = 11$ for set A and $g = 12$ and $k = 5$ for set B, and correlated residuals (unity variance and cross covariances all equal to 0.75). For set C $g = 6$ and $k = 11$ was assumed, with unit variance uncorrelated residuals. For every set 1000 replications were performed in each of which the best partition found by the algorithms was recorded and the corrected Rand index (R) (Hubert and Arabie 1985) and the estimated number of

Table 1 Univariate models for artificial time series generation

Data	Model	Orders		Coefficients			
set	label	p	q	ϕ_1	ϕ_2	θ_1	θ_2
A,C	1	2	2	0.7	−0.5	0.9	−0.3
	2	2	1	1.2	−0.5	−0.7	
	3	2	0	1.0	−0.3		
	4	1	2	−0.8		1.2	−0.5
	5	1	1	−0.9		−0.7	
	6	0	2			1.4	−0.7
B	1	2	2	0.7	−0.5	0.9	−0.3
	2	2	2	−0.7	0.2	−0.4	0.3
	3	2	1	1.2	−0.5	−0.7	
	4	0	1			−0.7	
	5	2	0	1.0	−0.2		
	6	2	0	−0.6	−0.2		
	7	1	2	−0.7		1.1	−0.6
	8	1	2	−0.7		−1.2	−0.5
	9	0	2			0.4	0.4
	10	0	2			−0.4	−0.4
	11	1	0	0.8			
	12	1	0	−0.8			

Table 2 Classification of time series generated from univariate ARMA models[a]

Data	Eval.	Cross correlations		Autoregressive coef.		K-means algorithm	
set	index	$n = 100$	$n = 200$	$n = 100$	$n = 200$	$n = 100$	$n = 200$
A	R	0.9990	0.9940	0.9313	0.9608	0.4964	0.5495
		(0.0136)	(0.0103)	(0.0731)	(0.0342)	(0.0879)	(0.1257)
	$\hat{g}$	6.003	6.000	7.246	7.557	19.089	17.864
		(0.0547)	(0.0000)	(1.0047)	(1.0202)	(2.3889)	(3.4971)
B	R	0.9891	0.9908	0.7665	0.9096	0.6790	0.7987
		(0.0303)	(0.0271)	(0.0774)	(0.0651)	(0.0707)	(0.0624)
	$\hat{g}$	12.008	11.996	12.407	12.898	19.499	18.775
		(0.1262)	(0.1183)	(1.1964)	(0.9938)	(1.5518)	(1.9946)
C	R	0.0025	0.0023	0.7581	0.7679	0.5215	0.6283
		(0.0151)	(0.0152)	(0.0649)	(0.0516)	(0.1329)	(0.1521)
	$\hat{g}$	27.700	26.770	12.261	13.485	17.375	15.515
		(1.0954)	(1.0517)	(1.4625)	(1.5349)	(3.8620)	(4.0985)

[a] Significance level $\alpha = 0.01$. Standard errors of the estimates are enclosed in parentheses.

clusters ($\hat{g}$) were computed as indexes for evaluating the agreement between the estimated and true partition. The number of observations was either $n = 100$ or $n = 200$.

In Table 2 computations were performed by using the k-means algorithm also for comparison. Best results are obtained for $n = 200$. This may suggest the validity of the consistency property for the procedures. The ARMA models used to generate the artificial time series do not seem to cause apparent differences on the quality of the results. The procedures seem more sensible to the cluster size, as results were better for $k = 11$ than for $k = 6$. The procedures based on cross correlations and AR coefficients yielded fairly equivalent results if residuals were correlated, while, as expected, the structure dissimilarity could not be recognized by using the cross correlations-based measure. The threshold constraints were important, as the k-means algorithm, that was unconstrained, produced results apparently worse in almost all cases.

As far as computing time is concerned, only a gross estimate may be done because overall implementation may be decisive to determine this aspect. We used for computation a desktop equipped with a 1.4 GHz Intel Pentium 4 processor and 512Mb random access memory, and Windows XP operating system. The computer programs were written in Fortran and compiled with Microsoft Developer Studio. The fastest procedure was the cross correlations-based procedure which required 1.8 s to find a partition of a data set with 66 time series of length 200. The autoregressive metric-based procedure had to perform a test more complicated than the test on cross correlations and to estimate a time series model. The computing time was slightly larger, either 2.88 or 4.92 s whether least squares or Marquardt (1963) algorithm were used. The seemingly unrelated regression-based procedure was by far the slowest one as it required about 6 min and 17 s to find a partition of the same set of 66 time series with 200 observations each.

3 Application to Real Data

This study on real data is concerned with the monthly time series collected on a regular basis from the records of the tickets sold in Italy for visiting the state owned museums with paid admission. We considered 111 sites, that is 76 museums, 16 monuments and 19 archaeological areas scattered all over Italy. For each site a detailed description is available, but we confine our attention to their territorial location and their main field of interest (modern art, archaeology, etc.). The time series data set covers January 1996–December 2005. So we had 111 time series for each of which 120 observations were available. There were no missing observations.

The package TRAMO-SEATS (Gómez and Maravall 1996) was used for preliminary analysis and model building. Full automatic identification procedure accounted for outlier correction, pre-testing for trading day and Easter effects. Then identification was refined if needed and ARIMA models were estimated by maximum likelihood. ARIMA coefficients were used to compute the autoregressive weights by recursive equations while residuals provided by the estimation procedure were used to compute the matrix of the pairwise cross correlations. The cluster analysis was performed by using both the distance matrices that included the autoregressive distances and the cross correlations.

The autoregressive distance-based procedure yielded a partition of the time series data set into 14 clusters. Clusters $6 - 14$ included only 19 time series. These latter were concerned with sites that are not usually included in the best known tourist routes. Most of them recorded a large number of visitors, though. The special time structure of the visitors flow in these sites may be explained by the fact that they are of interest on their own, and are considered to be worth of a visit though they are not located in areas crowded with tourist facilities and attractions. Clusters $1-5$ included 92 out of the 111 time series and had opposite characteristics as they are all located in areas of interest other than for cultural relevance. For instance, most of the sites included in the first clusters were located in Lazio, Tuscany and Emilia and Romagna, and often adjacent areas cluster together. Some clusters included sites where combined tickets too are sold, that is tickets that allow the tourist to visit other sites as well. Detailed data are not available for a close examination of that part of visitors who gain access to a site by a combined ticket. A hierarchical classification algorithm based on average linkage was used too for comparison. The autoregressive coefficients were taken as the set of variables for each time series. We chose the cutting level of the dendrogram along the usual guidelines and obtained a partition of 13 clusters. The partition was similar to the previous one and the results obtained were essentially confirmed.

The cross correlations-based method produced 14 cluster but the assignment of time series to clusters looks different than in previous classification. In this case the clusters resulted all approximately of equal size. Moreover, the regions that are of particular cultural interest were equally represented across the clusters. All 14 clusters included museums, while monuments and archaeological areas did not cluster together. This latter circumstance did not come out the autoregressive coefficients-based procedure and was a further interesting fact to supplement our analysis.

4 Concluding Remarks

The classification based on cross correlations works well unless the dependence structure is too weak, and no special care is needed as regards the estimation procedure. Even the gross approximation of thresholds given by $z_{\alpha/2}/\sqrt{n}$ seems to perform well both in practice and in simulation experiments. We tried several implementations of the classification methods based on the autoregressive metric. In all cases extreme care was needed as far as parameter estimation was concerned. If we want the procedure to perform well, then extremely accurate algorithm design is needed, specially as far as the identification stage is concerned (namely, how many autoregressive and moving average parameters have to be assumed). The simulation experiments seem to point out that significant improvement of the clustering procedures may be achieved if (1) statistical testing is made more reliable by accurately assessing the statistical distribution of the test statistics, and, on the other hand, if (2) more efficient fast and reliable numerical algorithms are designed.

Both methods require a dissimilarity/similarity matrix to be built and this is likely to limit somewhat the number of time series allowed. We may note that this matrix is needed anyway for the cross correlations-based method, while in case of the autoregressive metric we may renounce to statistical testing so avoiding the preliminary computation of the distance matrix. Then clustering may proceed as usual starting from the familiar units × variables data set. Nevertheless our results seem to point out that this method is not able to ensure good solutions to be achieved and at this stage we do not have definite suggestions to overcome this problem, to be deferred to further research.

Acknowledgements This work was financially supported by grants from MIUR, Italy. R. Baragona gratefully acknowledges financial support from the EU Commission through MRTN-CT-2006-034270 COMISEF.

References

Corduas, M. (1996). Uno studio sulla distribuzione asintotica della metrica autoregressiva, *Statistica, 56*, 321–332.

Gómez, V., & Maravall, A. (1996). Programs TRAMO and SEATS: instructions for users, Technical Report 9628, The Banco de España, Servicios de Estudios.

Hartigan, J. A. (1975). *Clustering algorithms*. New York: Wiley.

Hubert, L., & Arabie, P. (1985). Comparing partitions. *Journal of Classification, 2*, 193–218.

Keogh, E., & Kasetty, S. (2003). On the need for time series data mining benchmarks: A survey and empirical demonstration. *Data Mining and Knowledge Discovery, 7*, 349–371.

Liao, T. W. (2005). Clustering of time series data – a survey. *Pattern Recognition, 38*, 1857–1874.

Maharaj, E. A. (2000). Clusters of time series. *Journal of Classification, 17*, 297–314.

Marquardt, D. W. (1963). An algorithm for least squares estimation of nonlinear parameters. *Journal of the Society of Industrial and Applied Mathematics, 11*, 431–441.

Murthy, C. A., & Chowdhury, N. (1996). In search of optimal clusters using genetic algorithms. *Pattern Recognition Letters, 17*, 825–832.

Piccolo, D. (1989). On the measure of dissimilarity between ARIMA models. In *Proceedings of the A. S. A. Meetings, Business and Economic Statistics Sect.*, Washington D. C., 231–236.

Piccolo, D. (1990). A distance measure for classifying ARIMA models. *Journal of Time Series Analysis, 11*, 153–164.

Sarno, E. (2001). Further results on the asymptotic distribution of the Euclidean distance between MA models. *Quaderni di Statistica, 3*, 165–175.

Wang, X., Smith, K. A., & Hyndman, R. J. (2005). Characteristic-based clustering for time series data. *Data Mining and Knowledge Discovery, 13*, 335–364.

Zani, S. (1983). Osservazioni sulle serie storiche multiple e l'analisi dei gruppi. In D. Piccolo (Ed.), *Analisi Moderna delle Serie Storiche* (pp. 263–274). Milano: Franco Angeli.

Mining Time Series Data: A Selective Survey

Marcella Corduas

Abstract Time series prediction and control may involve the study of massive data archive and require some kind of data mining techniques. In order to make the comparison of time series meaningful, one important question is to decide what similarity means and what features have to be extracted from a time series. This question leads to the fundamental dichotomy: (a) similarity can be based solely on time series *shape*; (b) similarity can be measured by looking at time series *structure*. This article discusses the main dissimilarity indices proposed in literature for time series data mining.

1 Introduction

Prediction and control are typical objectives of time series analysis and many applications in real-life involve the study of massive data archive and require some kind of data mining techniques. The real challenge is the large amount of data available which makes any traditional "ad hoc" procedure useless.

For this reason, data mining implies a strong role of data processing and the related research field has been significantly occupied by researchers working on database management and machine learning who have often rediscovered known statistical techniques and rarely considered the inferential problem (see Keogh and Kasetty 2003, for a review).

In this article the attention will be focused on the *indexing* problem which, given a time series (a query sequence), finds the nearest matching time series in a database. The solution is achieved in two steps: firstly, a subset of series is selected by means of a crude or approximate dissimilarity criterion; secondly, a refined search is performed. In this respect, the leading idea is that data mining techniques have to discover objects that move similarly or closely follow certain given pattern. This

M. Corduas
Dipartimento di Scienze Statistiche, Università di Napoli Federico II, Via L.Rodino, 80138, Napoli(I), Italy
e-mail: corduas@unina.it

F. Palumbo et al. (eds.), *Data Analysis and Classification*,
Studies in Classification, Data Analysis, and Knowledge Organization,
DOI 10.1007/978-3-642-03739-9_40,

concept is typical of shape based dissimilarity measures. However, as this article will discuss, the final objective of a statistical analysis may lead to different approaches where time series modelling assumes a definite role. The article is organized as follows: in Sect. 2, the dissimilarity measures based on shape comparison are introduced; then, in Sect. 3 the problem of features extraction will be examined both in time and frequency domain. Finally, some distance criteria which compare time series dynamics by looking at the underlying generating processes are considered.

2 Comparing Time Series Shape

The most common device used in practice for data mining purposes is the Euclidean distance between the observations:

$$D_E(x_t, y_t) = \{\sum_{t=1}^{n} [x_t - y_t]^2\}^{1/2}, \tag{1}$$

where x_t and y_t, $t = 1, 2, \ldots n$, are zero mean time series. The distance may be referred to standardized data $\widetilde{x}_t$ and $\widetilde{y}_t$ leading to a more meaningful criterion which is invariant for linear transformation of data. In such a case, D_E^2 is just a linear transformation of the correlation coefficient of the two series $r_{xy}(0)$, being $D_E(\widetilde{x}_t, \widetilde{y}_t) = \{2n(1 - r_{xy}(0)\}^{1/2}$.

From a computational point of view, the distance (1) is very simple to implement since a possible matching candidate to a given time series is dismissed as soon as the distance between the first k observations is larger than a fixed threshold. However, the use of data-base archives may be inefficient since only time series with the same length can be considered. Moreover, the criterion is very sensitive to outliers and to distortion in time axis. The latter implies that the similarity of sequences which are locally out of phase is not detectable. For this reason, in spite of the computational complexity, the Dynamic Time Warping (DTW), originally introduced for speech processing (Sakoe and Ciba 1978; Berndt and Clifford 1994) was reconsidered for data mining purposes. In a certain sense, DTW generalizes the concept of dissimilarity between time series trajectories since it allows non-linear alignments of data. Specifically, given two data sequences $x = \{x_i, i = 1, 2, \ldots, m\}$, and $y = \{y_j, j = 1, 2, \ldots, n\}$, the procedure starts by constructing the $m \times n$ matrix Δ where the (i, j) element is the distance (or dissimilarity) $\delta(x_i, y_j)$ between two points x_i and y_j. The best matching is found by searching a path through this matrix such that the total cumulative distance between the aligned elements of the two time series is minimized.

We denote by $w = \{(i(k), j(k)), k = 1, \ldots, K, i(1) = j(1) = 1, i(K) = m, j(K) = n\}$ with $\max(m, n) \leq K \leq m + n - 1$, a warping path connecting $(1, 1)$ and (m, n). The alignment between the time series is obtained by searching for the path through the matrix Δ which minimizes a cost function such as:

$$\mathcal{C}(x, y, w) = \sum_{k=1}^{K} \delta(x_{i(k)}, y_{j(k)}) r(k), \tag{2}$$

where $r(k)$ is an appropriate non negative weighting function (this is often set to $1/k$). Of course, the choice of the cost function determines the warping result.

Some constraints are imposed in order to reduce the number of paths considered:

- *Boundary*: $i(1) = j(1) = 1, i(K) = m, j(K) = n$
- *Monotonicity*: $i(k) \leq i(k+1)$ and $j(k) \leq j(k+1)$
- *Continuity*: $i(k+1) - i(k) \leq 1$ and $j(k+1) - j(k) \leq 1$
- *Window*: the path is allowed to move within a definite region around the matrix diagonal (see Sakoe and Chiba 1978, for the rectangular window; Itakura 1975, for the parallelogram window)
- *Slope*: the path should be neither too steep nor too shallow.

At the end of the optimizing process, the optimal path provides a measure of the *dynamic warping distance* between the two time series:

$$DTW(x, y) = \inf_{w} \mathcal{C}(x, y, w). \tag{3}$$

The main disadvantages of this technique are the computing burden, which limits its usage in practice, and the sensitivity to extreme data. As a matter of facts, DTW can be severely affected by outliers since it tends to adjust extreme data in one of the time series by relating them to extreme values of the other. Various developments of this technique have been proposed such as, among the others, the derivation of a lower bound for DTW using different types of windows (Keogh 2002; Ratanamahatana and Keogh 2004), the extension to multidimensional data (Vlachos et al. 2006b), the use of smoothing for noisy data (Morlini 2005), the joint use of DTW and *S*elf *O*rganizing *M*ap (SOM) algorithm for improving time series clustering (Romano and Scepi 2006), the study of new techniques for approximating DTW (Chu et al. 2002).

3 Criteria Based on Fourier and Wavelet Analysis

Fourier and wavelet analysis provide a good framework in order to extract time series features which become object of the subsequent comparison.

Firstly, Agrawal et al. (1994) considered the *D*iscrete *F*ourier *T*ransform (DFT) of the data:

$$x(\omega_j) = n^{-1/2} \sum_{t=0}^{n-1} x_t \exp(-\imath \omega_j t), \tag{4}$$

where $\omega_j = 2\pi j/n$, $j = 0, 1, \ldots, (n-1)$ and introduced the criterion:

$$D^2_{A,n} = \sum_{j=0}^{n-1} |x(\omega_j) - y(\omega_j)|^2. \tag{5}$$

Of course, $D^2_{A,n} = D^2_E(x_t, y_t)$, but for indexing purposes, only the first k coefficients of each time series DFT are stored so that a selection of potential candidates for the final matching is simply given by: $D_{A,k} < \epsilon$. Standardizing the data first will allow for differences in level and scale. In this respect, the mentioned indexing strategy has two critical issues: the selection of the threshold (which is data dependent) and, above all, the assumption that low frequencies will be in general more informative about the temporal dynamics.

Other criteria in frequency domain have been investigated such as the Euclidean distance between periodograms (Caiado et al. 2006), periodograms of the standardized series at dominant frequencies (Vlachos et al. 2006a) or between smoothed periodograms (Wang and Wang 2000).

The assumption of stationarity of the data generating process is, in our opinion, a critical issue for all methods which rely on time series features such as periodogram, spectrum or autocorrelation functions. In case of non stationary series, these criteria are still useful whenever the non stationarity is removed from each time series by means of the same differencing operator or detrending technique.

In order to find a valid alternative to the traditional Fourier analysis several contributions have explored the use of wavelet analysis (see, for instance, Struzik and Sieber 1999; Li et al. 2002 for an extensive review). Wavelet transforms (or coefficients) are, in fact, characteristic of the local behaviour of a function whereas Fourier transforms relate to the global behaviour.

Any function $f(t) \in L^2(\Re)$ can be written as a wavelet series expansion:

$$f(t) = \sum_{k=-\infty}^{\infty} \sum_{j=-\infty}^{\infty} w_{j,k} \psi_{j,k}(t), \tag{6}$$

where the set of basis functions:

$$\{\psi_{j,k}(t) = 2^{(j/2)} \psi(2^j t - k), \ j,k \in Z\} \tag{7}$$

are obtained by dilations and translations of a mother wavelet $\psi(t)$ and the coefficients are given by: $w_{j,k} = \int_{-\infty}^{\infty} f(t)\psi_{j,k}(t)dt$.

Data mining mainly relies on the use of the Haar *D*iscrete *W*avelet *T*ransform (DWT) which simply extracts the underlying pattern of a time series by a recursive pairwise averaging and differencing of data. Chan and Fu (1999) showed the preservation of Euclidean distance in both time and Haar domain and, by analogy to other mining techniques mentioned above, they proposed to retain only the first few coefficients of the transformed sequences in order to perform a similarity search. Later, the approach was improved by taking the "best" (that is, largest) Haar coefficients into account.

4 Structural Dissimilarity

The interest for the dynamic structure inevitably conveys the investigation to the stochastic generating process that has originated the observed trajectory.

In this respect, the class of Gaussian $ARIMA$ processes provides a useful parsimonious representation (Box and Jenkins 1976) for linear time series. Specifically, $Z_t \sim ARIMA(p,d,q)$ is defined by:

$$\phi(B)\nabla^d Z_t = \theta(B)a_t, \tag{8}$$

where a_t is a Gaussian White Noise (WN) process with constant variance $\sigma^2_{a_z}$, B is the backshift operator such that $B^k Z_t = Z_{t-k}$, $\forall k = 0, \pm 1, \ldots$, the polynomials $\phi(B) = 1 - \phi_1 B - \ldots - \phi_p B^p$ and $\theta(B) = 1 - \theta_1 B - \ldots - \theta_q B^q$, have no common factors, and all the roots of $\phi(B)\theta(B) = 0$ lie outside the unit circle. Moreover, we assume that the time series has been preliminary transformed in order to improve Gaussianity, to deal with non-linearities, to reduce asymmetry, and to remove any outlier or deterministic components (such as deterministic seasonality, trading days, calendar effects, mean level, etc.).

First of all, we will introduce a distance criterion based on the *cepstral coefficients*, $c_{x,j}$, of zero mean stationary series determined by the following expansion:

$$\ln f_x(\omega) = \sum_{j=-\infty}^{\infty} c_{x,j} \exp(-\imath\omega j), \tag{9}$$

where $f_X(\omega)$, $\omega \in (-\pi, \pi]$ is the spectrum of the process X_t (Bogert et al. 1962).

For a pure stationary $AR(p)$ model, a simple expression of the cepstral coefficients in terms of the AR parameters can be derived (Gray and Markel 1976). For this reason, for several decades, the cepstral distance:

$$D_{C,k} = \sqrt{\sum_{j=1}^{k} \left[c_{x,j} - c_{y,j}\right]^2} \tag{10}$$

had been widely applied to signal processing (see for instance Markel and Gray 1976; Kang et al. 1995), and, more recently, it was used for data mining purposes (Kalpakis et al. 2001).

Note that, in the expression (12), the term $(c_{x,0} - c_{y,0})^2 = \ln(\sigma^2_{a_x}/\sigma^2_{a_y})$ is omitted since it is the log of the White Noise variance ratio and hence it simply represents a scale factor. Moreover, the cepstral coefficients quickly decay to zeros, and then, by analogy to previous methods, just a few number of cepstral coefficients, M, have to be stored for indexing purposes so that the Euclidean distance will be computed on the truncated series of cepstral coefficients.

Several improvements have been proposed for speech recognition purposes such as the use of Mahalanobis distance or the introduction of a weighted Euclidean distance in which each coefficient is simply weighted by the inverse of its variance in order to enhance the contribution of weights with lower variability (Tohkura 1987).

Furthermore, Piccolo (1984, 1990) proposed a distance criterion which compares the forecasting functions of two $ARIMA$ models given a set of initial values. In particular, assuming that Z_t is a zero mean invertible process which admits the $AR(\infty)$ representations: $\pi(B)Z_t = a_t$, the π-weights sequence and the WN variance completely characterize Z_t (given the initial values). Hence, a measure of structural diversity between two $ARIMA$ processes with given orders, X_t and Y_t, can be defined as:

$$D_{AR} = \sqrt{\sum_{j=1}^{\infty} \left(\pi_{xj} - \pi_{yj}\right)^2}. \tag{11}$$

As before, the WN variances are not included in the distance formulation since they depend on the units of measurement. The criterion has been widely experimented (see Piccolo 2007 for a review) and the asymptotic properties has been derived under general assumptions (Corduas 2000; Corduas and Piccolo 2008). Moreover, Baragona and Vitrano (2007) compared the performance of the AR metric with a criterion based on cross-correlations for data mining purposes.

Recently, Bagnall and Janaceck (2005) suggested to translate a time series into binary sequences in order to reduce the amount of storage and computational resources for time series comparison, and then, to apply the AR metric for subsequent clustering. The technique achieved a clustering accuracy equivalent to that obtained by cepstral distance and proved to be of help in presence of outliers.

In the same vein, the Mahalanobis distance between AR processes was proposed and, as we will discuss in the next section, the related distributional properties were investigated (see Thomson and De Souza 1985, and references therein reported). Xiong and Yeung (2004), instead, introduced a model-based clustering approach based on mixtures of $ARMA$ models.

5 Final Remarks

Concluding this brief review, we illustrate some results which helps the set up of time series comparison within an inferential framework. Assuming that X_t and Y_t are independent Gaussian and stationary zero mean processes, the following results hold.

- *Mahalanobis distance between $AR(p)$ processes*
  The null hypothesis is $H_0 : \boldsymbol{\phi}_x = \boldsymbol{\phi}_y = \boldsymbol{\phi}, \sigma^2_{a_x} = \sigma^2_{a_y} = \sigma^2$. Under H_0, $(\hat{\boldsymbol{\phi}}_x - \hat{\boldsymbol{\phi}}_y) \sim N_p(0, 2n^{-1}\sigma^2\boldsymbol{\Gamma}^{-1})$, being $\hat{\boldsymbol{\phi}}_x$ and $\hat{\boldsymbol{\phi}}_x$ the ML estimator vector of AR parameters, and Γ the p-order Toeplitz matrix of the common generating process. Hence,

$$\widehat{M}^2(X_t, Y_t) = \frac{n}{2}(\hat{\boldsymbol{\phi}}_x - \hat{\boldsymbol{\phi}}_y)'\sigma^{-2}\boldsymbol{\Gamma}(\hat{\boldsymbol{\phi}}_x - \hat{\boldsymbol{\phi}}_y)$$

is asymptotically distributed a $\chi_{(p)}$ random variable. When the matrix $\sigma^{-2}\mathbf{\Gamma}$ is unknown, it will be replaced by the corresponding pooled estimator. Also, the criterion can be generalized to time series with different length (Thomson e De Souza 1985).

- *AR metric*
  The null hypothesis $H_0 : D^2_{AR} = 0$, is equivalent to: $H_0 : \boldsymbol{\pi}_x - \boldsymbol{\pi}_y = \mathbf{0}$. For $ARMA$ processes, the m-lag truncated ML estimator $\widehat{D}^2_m$ is asymptotically distributed as a linear combination of independent χ^2_1 random variables; the weights are the non zero eigenvalues of $\mathbf{C}_0 = \left(\frac{1}{n_x} + \frac{1}{n_y}\right)\mathbf{BVB}'$, being $\mathbf{V}$ the covariance matrix of the ML estimators of $ARMA$ parameters and $\mathbf{B} = \{b_{ij}\}$ with $b_{ij} = \left\{\partial\widehat{\pi}_i / \partial\widehat{\beta}_j\right\}_{\widehat{\beta}=\beta}$, where $\beta_j = \phi_j$, $j = 1, \ldots, p$, and $\beta_{j+p} = \theta_j$, $j = 1, \ldots, q$. The result can be easily generalized to $ARIMA$ processes (Corduas 2000; Corduas and Piccolo 2008).

Other dissimilarity criteria have been proposed in the statistical literature which have not found a clear role so far, although the fields of application, which they were originally designed for, typically require the use of large data archives. For instance, we refer to the Kullback-divergence (Shumway 1982), the Bhattacharrya distance (Kazakos and Papantoni-Kazakos 1980) which were largely applied for signal recognition. Moreover, any distance criterion has to be related to an adequate clustering technique in order to produce sensible and useful results. Then, the study of efficient methods for clustering is a further and crucial research topic for data mining (Scepi and Milone 2007).

Acknowledgements This research was supported by Dipartimento di Scienze Statistiche, University of Naples Federico II, and CFEPSR (Portici).

References

Agrawal, R., Faloutsos, C., & Swami, A. (1994). Efficient similarity search in sequence databases. 4th F.O.D.O. *Lecture notes in Computer Science* (Vol. 730, pp. 69–84). New York: Springer

Bagnall, A. J., & Janaceck, G. J. (2005). Clustering time series from ARMA models with clipped data. *Machine Learning*, *58*, 151–178

Baragona, R., & Vitrano, S. (2005). Statistical and numerical algorithms for time series classification. *Proceedings of CLADAG 2007* (pp. 65–68). EUM, Macerata

Berndt, D., & Clifford, J. (1994). Using dynamic time warping to find patterns in time series. Proceedings of the AAAI-94 workshop of SIGKDD, pp. 229–248

Box, G. E. P., & Jenkins, G. M. (1994). *Time series analysis: Forecasting and control* (rev. ed.). San Francisco: Holden-Day

Caiado, J., Crato, N., & Peña, D. (2006). A periodogram-based metric for time series classification. *Computational Statistics & Data Analysis*, *50*, 2668–2684

Chan, K., & Fu, A. W. (1999). *Efficient time series matching by wavelets*, ICDE (pp. 126–133)

Chu, S., Keogh, E., Hart, D., & Pazzani, M. (2002). Iterative deepining dynamic time warping for time series. Proceedings of SIAM KDD, electronic edition

Corduas, M. (2000). La metrica Autoregressiva tra modelli ARIMA: una procedura in linguaggio GAUSS. *Quaderni di Statistica*, *2*, 1–37

Corduas, M., & Piccolo, D. (2008). Time series clustering and classification by the Autoregressive Metric. *Computational Statistics & Data Analysis*, *52*, 1860–1872

Gray, A. H., & Markel, J. D. (1976). Distance measures for speech recognition. *IEEE Transactions on Acoustics, Speech, & Signal Processing, ASSP-24*, 380–391

Itakura, F. (1975). Minimum prediction residual principle applied to speech recognition. *IEEE Transactions on Acoustics, Speech, & Signal Processing, ASSP-23*, 67–72

Kalpakis, K., Gada, D., & Puttagunta, V. (2001). Distance measures for effective clustering of ARIMA time series. *IEEE International Conference on Data Mining*, 273–280

Kang, W., Shiu, J., Cheng, C., Lai, J., Tsao, H., & Kuo, T. (1995). The application of cepstral coefficients and maximum likelihood method in EGM pattern recognition. *IEEE Transactions on Biomedical Engineering, 42*, 777–785

Kazakos, D., & Papantoni-Kazakos, P. (1980). Spectral distances between Gaussian processes. *IEEE Transactions on Automatic Control, AC-25*, 950–959

Keogh, E. (2003). Exact indexing of dynamic time warping. *28th International Conference on VLDB* (pp. 406–417). Hong Kong

Keogh, E., & Kasetty, S. (2003). On the need for time series data mining benchmarks: A survey and empirical demonstration. *Data Mining and Knowledge Discovery, 7*, 349–371

Li, T., Li, Q., Zhu, S., & Ogihara, M. (1997). A survey on wavelet applications in data mining. *SIGKDD Explorations, 4*, 49–68

Markel, J. D., & Gray, A. H. (1976). *Linear prediction of speech*. New York: Springer

Morlini, I. (2005). On the dynamic time warping for computing the dissimilarity between curves. In M. Vichi, P. Monari, S. Mignani, & A. Montanari (Eds.), *New developments in classification and data analysis* (pp. 63–70). Berlin: Springer

Piccolo, D. (1984). Una topologia per la classe dei processi ARIMA. *Statistica, XLIV*, 47–59

Piccolo, D. (1990). A distance measure for classifying ARIMA models. *Journal of Time Series Analysis, 11*, 153–164

Piccolo, D. (2007). Statistical issues on the AR metric in time series analysis. *Proceedings of the SIS Intermediate Conference* (pp. 221–232). Cleup, Padova

Ratanamahatana, C. A., & Keogh, E. (2004). Making time-series classification more accurate using learned constraints. *4-th SIAM International Conference on Data Mining* (pp. 1–20)

Romano, E., & Scepi, G. (2004). Integrating time alignment and Self-Organizing Maps for classifying curves. Proceedings of KNEMO COMPSTAT 2006 Satellite Workshop, Capri

Sakoe, H., & Chiba, S. (1978). Dynamic programming algorithm optimization for spoken word recognition. *IEEE Transactions on Acoustics, Speech, & Signal Processing, 26*, 143–165 (1978).

Scepi, G., & Milone, G. (2007). Temporal data mining: clustering methods and algorithms. *Proceedings of CLADAG 2007* (pp.73–76). EUM, Macerata

Shumway, R. H. (1982). Discriminant analysis for time series. In Krishnaiah,P. R. & Kanals, L.N. (Eds.), *Handbook of Statistics* (Vol. 2, pp. 1–46). New York: North Holland

Struzik, Z. R., & Siebes, A. (1999). The Haar wavelet in the time series similarity paradigm. *3rd European Conference on Principles of Data Mining and Knowledge Discovery* (pp. 12–22). Prague: Springer

Thomson, P. J., & De Souza, P. (1985). Speech recognition using LPC distance measures. In E. J. Hannan, P. R. Krishnaiah, & M. M. Rao (eds.), *Handbook of Statistics* (Vol. 5, pp. 389–412). Amsterdam: Elsevier

Tohkura, Y. (1987). A weighted cepstral distance measure for speech recognition. *IEEE Transactions on Acoustics, Speech, & Signal Processing, ASSP-35*, 1414–1422

Vlachos, M., Yu, P., Castelli, V., & Meek, C. (2006a). Structural periodic measures for time series data. *Data Mining and Knowledge Discovery, 12*, 1–28

Vlachos, M., Hadejieleftheriou, M., Gunopulos, D., & Keogh, E. (2006b). Indexing multidimensional time series. *The VBDL Journal, 15*, 1–20

Wang, C., & Wang, X. S. (2000). Supporting content-based searches on time series via approximations. 12th International Conference on Scientific and Statistical Database Management (pp. 69–81)

Xiong Y., & Yeung D. (2004). Time series clustering with ARMA mixtures. *Pattern Recognition, 37*, 1675–1689

Predictive Dynamic Models for SMEs

Silvia Figini

Abstract Considering the fundamental role played by small and medium sized enterprises (namely SMEs) in the economy of many countries and the considerable attention placed on them in the new Basel Capital Accord, we analyze a set of classical and Bayesian longitudinal models to predict SME default probability. In this contribution we present a real application based on a panel data set of German SMEs provided in our Musing European project by Creditreform, which is one of the major rating agencies for SMEs in Germany. Creditreform deals with balance sheet services, credit risk and portfolio analyses as well as consultation and support for the development of internal rating systems.

1 Introduction

Many SMEs in Germany have to rethink their financing structure, for instance against the background of Basel II. The credit-standing index is, as it were, a rating verdict in miniature – but how can default probabilities be forecast even more accurately than before? For a rating agency this means: how high is the probability that a company in Germany will slip in the course of a year into the risk class of firms with massive payment delays or even insolvency?

Classical Credit Scoring procedures use a representative sample to estimate the credit risk and the probability that a customers will not repay the credit (probability of default). Several statistical methods are used to develop credit scoring systems like, logit models, probit models, and discriminate analysis models.

Seminal contributions to default prediction are Altman (1968) and Beaver (1967), and recent ones include Altman et al. (2006).

Statistical methods for evaluating default probability estimates are discussed in Sobehart and Keenan (2001). Giudici (2003) measures the power of scoring

S. Figini (✉)
Department of Statistics and Applied Economics L. Lenti, University of Pavia, Italy
e-mail: silvia.figini@unipv.it

F. Palumbo et al. (eds.), *Data Analysis and Classification*, Studies in Classification, Data Analysis, and Knowledge Organization,
DOI 10.1007/978-3-642-03739-9_41,

models using accuracy ratios, and attach a monetary (i.e. dollar) value to a bank's application of a model that is more powerful.

Therefore the uncertainty of the estimate is reduced if the number of sample elements grows. In the general case complex models are required to capture the relation between the features of the borrowers and the credit risk. Most current procedures are not capable to estimate the uncertainty of the predicted credit risk. We propose a new methodology based on Bayesian Credit Scoring Models. The Bayesian approach is a mixture between the experimentation, that come from the data (likelihood) and the Old Knowledge, that is a measure of a priori Knowledge (prior). Bayesian estimators effectively combine the available information.

The objective of a Bayesian Model for SMEs is to have characteristics which discriminate between Good and Bad SMEs with a very high probability and a set of procedures which allow to derive the predictive uncertainty for a new case. Not only a single best risk is estimated, but in addition also the plausible range of the risk is determined.

2 Default Estimation: A Methodological Proposal

Most rating agencies, included our data provider, usually analyze each company on site and evaluates the default risk on the basis of different financial criteria considered in a single year. However, as the IMF pointed out with regard to sovereign default modelling, "temporal stability and country homogeneity that are assumed under probit estimation using panel data might be problematic" (see Oka 2003). The same consideration can be directly extended to SMEs risk modelling, too.

Therefore, based on our unbalanced dataset, we present the following specification for longitudinal models: for observation i, $(i = 1, \ldots, n)$, time t, $(t = 1, \ldots, T)$ and sector j, $j = 1, \ldots, J$, let Y_{itj} denote the response solvency variable, let X_{itj} denote a $p \times 1$ vector of candidate predictors (fixed effects), and let Z_{itj} denote a $q \times 1$ vector of candidate predictors (random effects), where $\varsigma_i \sim N(0, \Sigma)$ are random effects for SME_i and Σ is the covariance matrix. The elements of $Y_{itj} = (y_{1tj}, \ldots, y_{ntj})'$ are modelled as conditionally independent random variables from a simple exponential family:

$$\pi(Y_{itj}|X_{itj}, Z_{itj}, \varsigma_i) \propto exp\left\{\frac{Y_{itj}\theta_{itj} - b(\theta_{itj})}{a_{itj}(\phi)} + c(Y_{itj}, \phi)\right\},$$

where θ_{itj} is the canonical parameter related to the linear predictor $\eta_{itj} = X'_{itj}\beta + Z'_{itj}\varsigma_i$ with a $p \times 1$ vector of fixed effects regression coefficient β and a $q \times 1$ vector of subject specific random effects $\varsigma_i \sim N_q(0, \Sigma)$, ϕ is a scalar dispersion parameter and a_{itj}, b, c are known functions with $a_{itj}(\phi) = \frac{\phi}{\omega_{itj}}$, where ω_{itj} is a known weight.

We are interested in predicting the expectation of the response as a function of the covariates. The expectation of a simple binary response is just the probability that the response is 1:

$$E(Y_{itj}|X_{itj}, Z_{itj}, \varsigma_i) = \pi(Y_{itj} = 1|X_{itj}).$$

In linear regression, this expectation is modelled as a linear function $\beta' X_{itj}$ of the covariates. For binary responses, as in our case, this approach may be problematic because the probability must lie between 0 and 1, whereas regression lines increase (or decrease) indefinitely as the covariate increases (or decreases). Instead, a nonlinear function is specified in one of two ways:

$$\pi(Y_{itj} = 1|X_{itj}) = h(\beta' X_{itj}),$$

or

$$g\{\pi(Y_{itj} = 1|X_{itj})\} = \beta' X_{itj} = \nu_i,$$

where ν_i is referred to as the linear predictor. These two formulations are equivalent if the function $h(.)$ is the inverse of the link function $g(\cdot)$.

We have introduced two components of a generalized linear model: the linear predictor and the link function. The third component is the distribution of the response given the covariates. For binary response, this is always specified as Bernoulli (π_i). Typical choices of link function g are the logit or probit links.

The logit link is appealing because it produces a linear model for the log of the odds, $ln\left\{\frac{\pi(Y_{itj}=1|x_i)}{1-\pi(Y_{itj}=1|x_i)}\right\}$, implying a multiplicative model for the odds themselves (for more details, see e.g. Dobson 2002).

To relax the assumption of conditional independence among the firms given the covariates, we can include a subject-specific random intercept $\varsigma_i \sim N(0, \psi)$ in the linear predictor:

$$g\{\pi(Y_{itj} = 1|X_{itj}, \varsigma_i)\} = \beta' X_{itj} + \varsigma_i.$$

This is a simple example of a generalized linear mixed model because it is a generalized linear model with both fixed effects X_{itj} and a random effects ς_i. In particular we propose the following models (logit link):

$$g\{\pi(Y_{itj} = 1|X_{itj})\} = \beta' X_{itj} + \varsigma_i$$
$$\varsigma_i \sim N(0, \sigma^2_{\varsigma,1}) \quad (1)$$

$$g\{\pi(Y_{itj} = 1|X_{itj})\} = \beta' X_{itj} + \varsigma_j$$
$$\varsigma_j \sim N(0, \sigma^2_{\varsigma,2}) \quad (2)$$

$$g\{\pi(Y_{itj} = 1|X_{itj})\} = \beta' X_{itj} + \varsigma_j + \varsigma_i$$
$$\varsigma_i \sim N(0, \sigma^2_{\varsigma,1}) \quad (3)$$
$$\varsigma_j \sim N(0, \sigma^2_{\varsigma,2}) \quad (4)$$

for firm $i = 1, \ldots, n$, and for business sector $j = 1, \ldots, J$.

3 Data Sources

The empirical analysis is based on annual 1996–2004 data from Creditreform, which is one of the major rating agencies for SMEs in Germany, for 1,003 firms belonging to 352 different business sectors. For a classical rating system only quantitative data is needed. Our rating system need to focus on all types of information and data. To explain the required data in detail there are three parts: The quantitative data of the system: The data will be from annual accounts which are provided by companies themselves. Addicted to the company's size the complexity of the information will be different. Data on which will be focused anyway are the profit, the equity, the loans and the different types of assets. The qualitative data of the system: The data will be provided by the company themselves and will be purified be analysts. There will be a questionnaire which helps the information split into useful and useless data. The quanti-qualitative data of the system: In combination with the questionnaire every information will be rated in form of a scale from 1 to 10. This leads to the number-system of qualitative data.

In particular, our data set consists of a binary response variable Y_{itj} and a set of explanatory variables given by financial ratios and time dummies. The financial ratios used to build our rating models are based on total assets, total liabilities and equity. We cannot to describe these financial ratios, since they are exclusive property of Creditreform. For more details, see Figini and Fantazzini (2007).

4 Application

Considering the longitudinal models (logit) implemented, only three financial ratios are statistically significant: the *Equity ratio*, the *Liabilities ratio* and the *Result ratio*. This evidence confirms business practice and empirical literature using similar ratios (see e.g. Altman and Sabato 2006 and references therein). Concerning the signs of the three ratios, we observe that while the ones for the Equity ratio and Results ratio are reasonably negative (i.e. the higher the Equity the less probable the default), the negative sign for the Liabilities ratio seems counter intuitive. Nevertheless, our business data provider has explained us that the majority of debts in our datasets were covered by external funds provided by the owners of the firms. This is usually done for tax savings purposes. Therefore, a high Liabilities ratio can signal a very wealthy firm. As for the random effects, we see that the most important one is the business sector, while the firm-specific one is not significant, once the business sector is taken into account. Instead, all the three random coefficient models show significant random variances, thus highlighting a strong degree of heterogeneity in financial ratios as well. To better assess the predictive performance for each model, we also implemented a cross-validation procedure. We used as training set observations ranging between 1996 and 2003, while we use the year 2004 as validation set (Table 1). Given the context of our study, the entries in the confusion matrix have the following meaning: a is the number of correct predictions that a SME is insolvent,

Table 1 Theoretical confusion matrix

$Observed/Predicted$	$EVENT$	$NON-EVENT$
$EVENT$	a	b
$NON-EVENT$	c	d

b is the number of incorrect predictions that a SME is insolvent, c is the number of incorrect of predictionsthat a SME is solvent, and d is the number of correct predictions that a SME is solvent. Therefore, the more complex formulations such as random effects and random coefficient models that allow for unobserved heterogeneity across firms and business sector tend not to predict well out-of-sample despite the fact that they describe the data quite well in-sample. In contrast, the parsimonious pooled logit and probit regression forecasts relatively well. Similarly, the random effects logit and probit also work reasonably well as early warning devices for firm default. Perhaps the reduction in forecast uncertainty and accuracy gains from simple models outweigh the possible misspecification problems associated with heterogeneity. It may be that simple models like the logit yield more robust forecasts because the available data for relevant financial ratios across firms are rather noisy. It is a well known fact that balance sheet data for SMEs are much less transparent and precise than for quoted stocks. Furthermore, it has been documented in a variety of scenarios that heterogeneous estimators forecast worse than simple pooling approaches (for a summary of this issue, see Baltagi 2005). Our findings, although novel in the credit risk literature, are consistent with this accepted wisdom.

5 Conclusion

We developed a set of classical and Bayesian panel data models to predict the default probabilities for Small and Medium Enterprizes. Our research can be improved following two directions: first, larger datasets should be considered to confirm the evidence emerging out of this work. Second, new financial ratios could be considered and derived from the balance sheet, Finally, another avenue of future research is to include different qualitative information, such as questionnaire as well as external analysts recommendations. For SMEs, it is well known that subjective information is an important supplementary tool for credit scoring.

Acknowledgements The author acknowledge financial support from the MIUR-FIRB 2006–2009 and Musing.

I am grateful to Prof. Paolo Giudici for the useful suggestions.

References

Altman, E. (1968). Financial ratios, discriminant analysis and the prediction of corporate Bankruptcy. *Journal of Finance*, *23*(4), 589–609

Altman, E., & Sabato, G. (1968). Modeling credit risk for SMEs: Evidence from the US market. *Abacus*, *19*(6), 716–723

Baltagi, B. (2005). *Econometric analysis of panel data*. London: Wiley

Beaver, W. (1967). Financial ratios predictors of failure. *Journal of Accounting Research*, *4*, 77–111

Dobson, A. J. (2002). *Introduction to Generalised Linear Model*. London: Chapman Hall

Figini, S., & Fantazzini, D. (2007). Bayesian panels models to predict credit default for SMEs. http://www.unipv.it/dipstea/workingpapers.htm

Giudici, P. (2003). *Applied data mining*. London: Wiley

Oka, W. (2003). *Anticipating arrears to the IMF: Early earnings systems*, (IMF 18)

Sobehart, J. R., & Keenan, S. C. (2001). A practical review and test of default prediction models. *RMA Journal*, *84*(3), 54–59

Clustering Algorithms for Large Temporal Data Sets

Germana Scepi

Abstract Temporal Data Mining is a rapidly evolving and new area of research that is at the intersection of several disciplines, including statistics, temporal pattern recognition, optimisation, visualisation, high-performance computing, and parallel computing. This paper is intended to serve a discussion on a specific Temporal Data Mining task: Temporal Cluster Analysis. Most clustering algorithms of the traditional type are severely limited in dealing with large temporal data sets. Therefore we discuss the applicability of clustering algorithms to these data sets. This paper is enriched with an application of a new algorithm on a real sequential database.

1 The Framework: Temporal Data Mining

The importance of *Temporal Data Mining* (TDM) problems can be seen within a process of *Knowledge Discovery in Temporal Databases*(KDT). Indeed, temporal databases are historical databases and TDM is concerned with data mining of large sequential data sets. In particular, the ultimate goal of temporal data mining is to discover hidden relations between sequences and sub-sequences of events.

The discovery of relations between sequences (and sub-sequences) of events can be divided into three phases: the representation and modelling of the data sequence, the definition of similarity measures between sequences and the application of models and representations to the actual mining problems (Antunes and Oliveira 2001; Roddick and Spiliopoulou 2001).

By sequential data, we mean data that is ordered with respect to some index. Time series constitute a popular class of sequential data, where records are indexed by time. Other examples of sequential data could be text, gene sequences, protein sequences, lists of moves in a chess game etc. Here, although there is no notion of time as such, the ordering among the records is central to the data description (modelling).

G. Scepi
University of Naples, Via Cinthia, Monte Sant'Angelo (NA), Italy
e-mail: scepi@unina.it

F. Palumbo et al. (eds.), *Data Analysis and Classification*,
Studies in Classification, Data Analysis, and Knowledge Organization,
DOI 10.1007/978-3-642-03739-9_42,

TDM methods are different from the classical ones (classical time series analysis, for example) for the size, the nature of data sets and the manner in which the data is collected. Another important difference lies in the scope of TDM. The estimate of the exact model parameters may be of little interest in data mining context while may be useful to discover trends (often unexpected) or patterns in the data. In briefly, Temporal Data Mining methods must be capable of analysing data sets that are prohibitively large for conventional methods as well as dealing with sequences that may be nominal-valued or symbolic and they must be capable of analysing data, often collected for some entirely different purpose, with little or no control over the data gathering process.

According to Lin et al. (2002) Temporal Data Mining is: *"a single step in the process of Knowledge Discovery in Temporal Databases that enumerates structures (temporal patterns or models) over the temporal data, and any algorithm that enumerates temporal patterns from, or fits models to, temporal data is a Temporal Data Mining Algorithm"*.

A relevant and important question is how to apply data mining techniques on a temporal database. According to techniques of data mining and theory of statistical time series analysis, the possible aims of TDM, which are often called tasks of temporal data mining, may be classified into some broad groups: *(a)* Temporal data characterization and comparison, *(b)* Classification, *(c)* Clustering, *(d)*Search and retrieval, *(e)*Pattern discovery.

In this paper, we focus our attention on Clustering (in Sect. 2) and in particular we provide a more detailed account of the advantages and the problems of applying the classical clustering algorithms on temporal large data sets (in Sect. 3). Finally, we show an application of a new clustering algorithm on a sequential radar satellite database (in Sect. 4).

2 Temporal Cluster Analysis

Clustering is perhaps the most frequently used data mining algorithm, being useful in its own right as an exploratory technique, and also as a subroutine in more complex data mining algorithms such as rule discovery, indexing, summarization, anomaly detection, and classification. TDM deals with large databases that impose on clustering analysis additional severe computational requirements.

Clustering of sequences or time series is concerned with grouping a collection of time series (or sequences) based on their similarity. It is well known that data clustering is inherently a more difficult task if compared to supervised classification, in which classes are already identified, so that a system can be adequately trained. This intrinsic difficulty worsens if sequential data are considered: the structure of the underlying process is often difficult to infer, and typically different length sequences have to be dealt with.

Sequential data clustering methods could be generally classified into three categories: *proximity-based methods*, *feature-based methods* and *model-based methods*.

In the *proximity-based* approaches, the main effort of the clustering process is in devising similarity or distance measures between sequences. With such measures, any standard distance-based method (as agglomerative clustering) can be applied.

Feature-based methods extract a set of features from each individual data sequence that captures temporal information. The problem of sequence clustering is thus reduced to a more addressable point (vector of features) clustering.

Finally, *model-based approaches* assume an analytical model for each cluster, and the aim of clustering is to find a set of such models that best fit the data.

In the clustering approaches based on similarity an important problem appears: the existence of a meaningful similarity measure between sequences. If the sequence elements are feature vectors (with real vectors), standard metrics such as Euclidean distance may be used. Sometimes, Euclidean norm could be inappropriate. This is in the case, for example, of speech or audio signals, where similar sounding patterns may give feature vectors that have large Euclidean distance and vice versa. Log spectral distances and other transformed measures have been proposed for dealing with this type of data (Juang and Rabiner 1993). Similarity measures invariant under various transformations (like shifting, amplitude scaling) have been explored as well (Perng et al. 2000). When the sequences consist of symbolic data, the different proposes have been determined by application aims ((Ewens and Grant 2001) for applications in biology).

In most application the sequences would be of different lengths. *Time warping methods* have been used for sequence classification for many years. In recent times, Dynamic Time Warping (DTW) and its variants are being used. DTW can, in general, be used for sequence alignment even when the sequences consist of symbolic data. Many problems in bioinformatics relate to the comparison of DNA or protein sequences, and time-warping-based alignment methods are well suited for such problems.

Another approach is to regard two sequences as similar if they have enough non-overlapping time-ordered pairs of subsequence that are similar (Agrawal and Srikant 1995). In some applications it is possible to locally estimate some symbolic features in real-valued time series and match the corresponding symbolic sequences. Approaches like this are particularly relevant for data mining since there is considerable efficiency to be gained by reducing the data from real-valued time series to symbolic sequences, and by performing the sequence matching in this new higher level of abstraction.

There are a variety of proposes for clustering sequences based on the above-mentioned measures of similarity (or their variants). Kalpakis and Puttagunta (2001) compare results obtained by clustering ARIMA time series with various similarity measures. They use a *k*-medoid method, the PAM (Partitioning Around Medoids) algorithm. PAM is iterative optimization that combines relocation of points between perspective clusters with re-nominating the points as potential medoids. The guiding principle for the process is the effect on an objective function, which, obviously, is a costly strategy. Another Partitioning Relocation Method, the *k*-means algorithm, is frequently used for clustering sequences. In some papers, this algorithm is used jointly with the DTW distance. Nevertheless the DTW algorithm's time complexity causes a problem in a way that "... *performance on very large databases may be a limitation*". Morlini (2005) proposes a modification of this algorithm that considers a smoothed version of the data.

Another fundamental problem in clustering temporal sequences consists on the discovery of a number of clusters, says *K*, able to represent the different sequences. Considering that *K* is known, if a sequence is viewed as being generated according to some probabilistic model, clustering may be viewed as modeling the data sequences as a finite group of *K* sequences in the form of a finite mixture model. *Model-based* clustering *approaches* (also called generative approaches) have been successfully applied in the context of TDM. These methods assume some form of the underlying generating process, estimate the model from each data then cluster based on similarity between model parameters. The most commonly assumed model forms are: polynomial mixture models (e.g. Ewens and Grant (2001)); ARMA (e.g. Piccolo (1990), Xiong and Yeung (2002)); Markov Chain and Hidden Markov models (e.g.Cadez et al. (2000); Alon et al. (2003)). Some techniques use both model-based as well as alignment based methods (Oates et al. 2001). In some papers the mixtures are estimated via an EM formulation. Learning the value of *K*, if it is unknown, may be accomplished by a Monte-Carlo cross validation approach (Smyth 1997). Different approaches propose to use hierarchical clustering algorithm like COBWEB (Ketterlin 1997). COBWEB has an important quality, it utilizes incremental learning. Instead of following divisive or agglomerative approaches, it dynamically builds a dendrogram by processing one data point at a time. Furthermore, *Model-based* probabilistic and conceptual algorithms, as COBWEB, have better scores in regard to cluster interpretability.

3 Clustering Algorithms: Applicability to Large Temporal Datasets

In spite of the great wealth of clustering algorithms, the rapid accumulation of large databases of increasing complexity poses a number of new problems that traditional algorithms are not equipped to address. One important feature of modern data collection is the ever increasing size of a typical database: it is not so unusual to work with databases containing from a few thousands to a few millions of individuals and hundreds or thousands of variables. Now, most clustering algorithms of the traditional type are severely limited as to the number of individuals they can comfortably handle (from a few hundred to a few thousands). The scalability of the clustering algorithms to large data sets is a very important property but not "the only one". Handling outliers, time and space complexity, interpretability of results, ability to find clusters of irregular shape, are some of the desirable proprieties concerned with clustering algorithms in temporal data mining.

Hierarchical algorithms are widely used for their flexibility (handling of any forms of similarity or distance) and applicability to any attribute types. One advantage of hierarchical algorithms is that the number of clusters is not required to be provided as a parameter. Furthermore, they have legible results in terms of dendograms. However, the quadratic computational complexity restricts their application to small data sets. Algorithms like BIRCH (Zhang et al. 1996) and CURE (Guha et al. 1998) were proposed to improve the scalability of hierarchical agglomerative clustering and the quality of the discovered partitions. K-means clustering is the most commonly used algorithm, with the number of clustering K specified by the user. K-means is a faster method compared to hierarchical clustering but the preassignment of the number of clusters does not help to obtain natural cluster results. Furthermore this algorithm is not robust in presence of outliers and noise in the data and it does not find irregular shapes.

This algorithm is used jointly with different measures of distance, such as Euclidean and Dynamic Time Warping (DTW) distance measure, in temporal data mining applications. A peculiar measure, the Longest Common Subsequence (LCSS) distance, is used in one interesting application of this algorithms (Gaudin et al. 2006). This application is about data of a survey carried out by the French National Institute for Statistics and Economic Studies on the quality of life for each year of 8403 disable people. This application deals with very heterogeneous data: time series with very different size, characterized by temporal gaps, with both numerical and categorical values.

A K-medoid method, such as the PAM algorithm, is often used for solving the difficulty of clustering data in presence of outliers and noise. Unfortunately this algorithms does not scale well for very large databases because it is characterized by time and space complexity. The algorithm CLARA (Clustering LARge Applications) (Kaufman and Rousseeuw 1990) tries to solve this difficulty by applying PAM on several samples. Further progress is associated with Ng and Han (1994) who introduced the algorithm CLARANS (Clustering Large Applications based upon RANdomized Search) in the context of clustering in temporal spatial databases and with Ester et al. (1995) who extended CLARANS to Very Large Data Bases.

The Self Organizing Map (SOM) algorithm can be used for clustering. The versatile properties of SOM make it a valuable tool in data mining and knowledge discovery. In particular, the SOM is a class of neural network algorithms in the unsupervised learning category, originally proposed by Kohonen in 1981–1982. The central property of SOM is that it forms a nonlinear projection of a high-dimensional data manifold on a regular, low-dimensional (usually 2D) grid. The clustered results can show the data clustering and metric-topological relations of the data items. It has a very powerful visualization output and is useful to understand the mutual dependencies between the variables and data set structure. Like in K-means algorithm you must choose the number of clusters to fit the data into, in SOM you choose the shape and size of a network of clusters. Unlike K-means, SOM will not force there to be exactly as many clusters as there are nodes and it is possible for a node to end up without any associated cluster items when the map is complete. Furthermore, SOM automatically provides some information on the similarity between nodes. SOM has some interesting properties: the advantage of being quite robust to parameter selection, producing a natural clustering results, and providing superior visualization compared to other clustering methods, such as hierarchical and

K-means. Several applications of SOM for clustering large data sets, in particular for spatio-temporal data, have been proposed (for example, on image data sequences (Honda et al. 2002)).

Peculiar algorithms have been applied for real-world applications deal with huge amounts of temporal data such as alarms/events and performance measurements generated by distributed computer systems and by telecommunication networks, the web server logs, online transaction logs, financial data, workflow process logs, and so on. An example is the clustering method proposed by Verde and Lechevallier (2003) on symbolic objects. Very interesting its application to complex data from Web in e-commerce domain (Chelcea et al. 2005).

4 An Example of Application on a Radar Satellite Data Base

The radar satellite databases used for our application comes from: "Progetto TELLUS Unit di Supporto Locale n.6 Campania-Progetto Operativo Difesa Suolo (PODiS)". In this project an innovative technique for the remote assessment of ground displacements, based on satellite radar interferometry, has been used for monitoring mass movements and land subsidence. The technique involves interferometer phase comparison of 72 radar images of the same scene taken at different time along the same orbit by the radar sensors of ERS-1 and ERS-2 satellites, during the period June 1992 – December 2000. The technique analyses points on the ground that strongly reflect radar signal (e.g. buildings, streets, rock outcrops…). These points are called "permanent scatters".

The area studied includes the areas near of the city of Benevento in Campania. The database is formed by over 18.000 records. Each point is defined by a code, by two coordinates (latitude = north, longitude = east), average velocity of ground deformation expressed in mm/year (vel; negative rates indicate subsidence, positive rates indicate uplift), coherence (reliability index of series), and time-series of displacement data where each number at the top of the column corresponds to a given satellite image. The distance is given in millimetres (negative values indicate subsidence, positive values indicate uplift and is centred for each point with respect to the distance that the same point has with the satellite in the image termed as master, which in our case is the number 32).

In order to estimate ground deformation velocity trends an innovative clustering approach, based on the implementation of the DTW distance in a Self Organizing Map algorithm, has been used (methodological details in (Romano 2006)). The clustering analysis allows to recognize six different types (in Fig. 1) of ground deformation trends referred to the central sector of Campania Region. Three classes describe a subsidence trend characterised by different rates: slow for class A, medium for class E and fast for class C. Class B shows an uplift trend with variable rates, while classes D and F show a composite variable trend. In particular class D shows a trend characterized by stability or light subsidence followed by fast uplift, while class F shows fast uplift followed by subsidence. All these ground

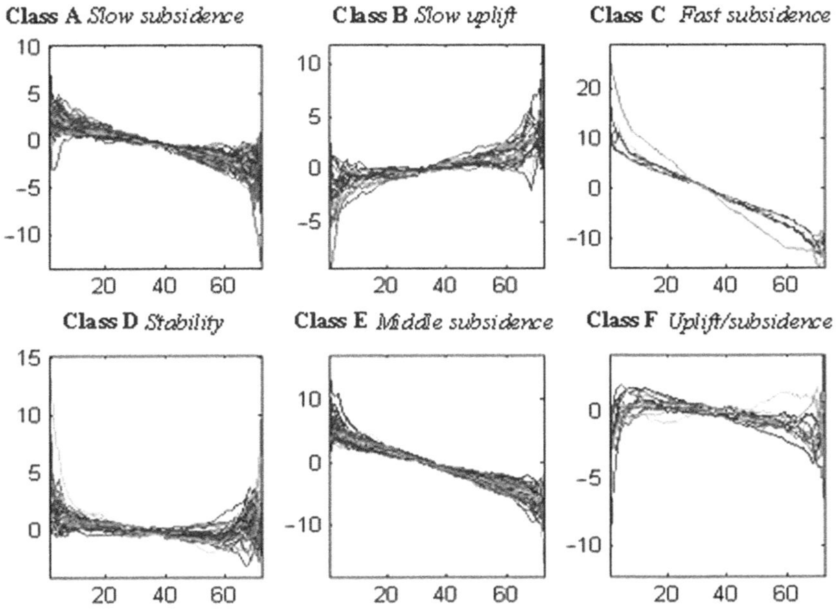

Fig. 1 Clustering results

deformation trends are the results of combined morphological and tectonic processes acting on the studied territory, and the method allows the evaluation of their rates with a precision never gained in the past.

5 Concluding Remarks

As emerging in the KDD2006 Workshop Report, new aspects of temporal data deserve theories and algorithms of their own. Some of these new aspects are: (a) *Irregularity*: many types of numerical temporal data are not equally spaced; (b) *Asynchronism*: in distributed computing environments like sensor networks, data from different sources tend to be not aligned and hence can not apply synchronous methods; (c) *Streaming Data*: some temporal data is stored only temporally and requires near real-time analysis; (d) *Heterogeneous data types*: it is very common that temporal data is partly categorical events and partly numerical time series; (e) *Huge Volume*: the stream of data can be huge for a long, continuous observation period.

The field of temporal data mining is relatively young and one expects to see many new developments in the future. Improving time and space complexities of algorithms is a problem that must continue to attract attention.

References

Agrawal, R., & Srikant, R. (1995). *Mining sequential patterns*. ICDE (pp. 3–14)

Alon, J., Sclaroff, S., Kollios, G., & Pavlovic V. (2003). Discovering clusters in motion time series data. In *Proceedings of IEEE Computer Society Conference on Computer Vision and Pattern Recognition* (pp. 375–381)

Antunes, C. M., & Oliveira, A. L. (2001). Temporal Data Mining: An overview. In *Proceedings of KDD'01 Workshop on Temporal Data Mining* (pp. 1–13). San Francisco

Cadez, I., Heckerman, D., Meek, C., Smyth, P., & White, S. (2000). Visualization of navigation patterns on a web site using model-based clustering. In *Knowledge discovery and data mining* (pp. 280–284).

Chelcea, S., Da Silva, A., Lechevallier, Y., Tanasa, D., & Trouss, B. (2005). Pre-Processing and clustering complex data in E-commerce domain. In *Proceedings of the First International Workshop on Mining Complex Data'05*. Houston

Ester, M., Kriegel, H. P., & Xu, X. (1995). A database interface for clustering in large spatial databases, In *Proceedings of the 1st International Conference on KDD*. Montreal: AAAI

Ewens, W. J., & Grant, G. R. (2001). *Statistical methods in bioinformatics: An introduction*. Berlin: Springer

Gaffney, S., & Smyth, P. (2003). Curve clustering with random effects regression mixtures. In Bishop et al.(eds.), *Proceedings of the Ninth International Workshop on Artificial Intelligence and Statistics*

Gaudin, R., Barbier S., Nicoloyannis, N., & Banens, M. (2006). Clustering of bi-dimensional and heterogeneous time series: Application to social sciences data. In S. F. Crone, S. Lessmann, & R. Stahlbock (eds.), *DMIN* (pp. 10–16). USA, NV: CSREA

Guha, S., Rastogi, R., & Shim, K. (1998). CURE: An efficient clustering algorithm for large databases. In:*Proceedings of ACM SIGMOD*.

Honda, R., Wang, S., Kikuchi, T., & Konishi, O. (2002). Mining of moving objects from time-series images and its application to satellite weather imagery. *Journal of Intelligent Information System*, *19*(1), 79–93

Juang, B. H., & Rabiner, L. (1993). *Fundamentals of speech recognition*. New Jersey: Prentice Hall

Kalpakis, K., & Puttagunta, D. G. V. (2001). Distance measures for effective clustering of ARIMA time series. In *Proceedings of IEEE'01 International Confernce on Data Mining*. San Jose, CA.

Kaufman, L., & Rousseeuw, P. J. (1990). *Finding groups in data: An introduction to cluster analysis*. New York: Wiley

Ketterlin, A. (1997). Clustering sequences of complex objects. In *Proceedings of KDD'97* (pp.215–218)

Lin, W., Orgun, M. A., & Graham, J. W. (2002). An overview of temporal data mining. In *Proceedings of the Australasian Data Mining Workshop* (pp. 83–90). Sydney

Ng, R., & Han, J. (1994). Efficient and effective clustering methods for spatial data mining. In *Proceedings of the 20th Conference Very Large Databases* (pp. 144-155)

Morlini, I. (2005). On the dynamic time warping for computing the dissimilarity between curves. In Vichi et al. (eds.), *New developments in classification and data analysis*, Proceedings of CLADAG, Bologna.

Oates, T., Firoiu, L., & Cohen, P. R. (2001). Using dynamic time warping to bootstrap HMM-based clustering of time series, *Lecture Notes in Computer Science* (Vol. 1828, pp. 35–52)

Perng, C. S., Wang, H., Zhang, S. R., & Parker, D.S. (2000). A new model for similarity-based pattern querying in time series databases. In *Proceedings of the 16th International Conference on Data Engineering*, 33

Piccolo, D. (1990). A distance measure for classifying ARIMA models. *Journal of Time Series Analysis*, *11*(2), 153–164

Roddick, J., & Spiliopoulou, M. A. (2001). Survey of temporal knowledge discovery paradigms and methods. In *IEEE Transactions of Knowledge and Data Engineering*, 13

Romano, E., & Scepi, G. (2006). Integrating time alignment and self organizing maps for classifying curves. In *Electronic Proceedings of Knowledge Extraction and Modeling*. Anacapri

Smyth, P. (1997). Clustering sequences with hidden markov models. In Tesauro et al. (eds.), *Advances in neural information processing systems* (Vol. 9, pp. 648–654). Cambridge: MIT

Verde, R., & Lechevallier, Y. (2003). Crossed clustering method on symbolic data tables. In M. Vichi, P. Monari, S. Migneni, & A. Montanari (eds.), *New developments in classification and data analysis* (pp. 87–96). Heidelberg: Springer

Xiong, Y., & Yeung, D. Y. (2002). Mixtures of ARMA models for model-based time series clustering. In *Proceedings of IEEE'02 International Conference on Data Mining* (pp. 717–720)

Zhang, T., Ramakrishnan, R., & Livny, M. (1996). BIRCH: An efficient data clustering method for large databases. In *Proceedings of the 1996 ACM SIGMOD* (pp. 103–114). Montreal, Canada.

Part IX
Outlier Detection and Robust Methods

Robust Clustering for Performance Evaluation

Anthony C. Atkinson, Marco Riani, and Andrea Cerioli

Abstract The evaluation of the effectiveness of organisations can be aided by the use of cluster analysis, suggesting and clarifying differences in structure between successful and failing organisations. Unfortunately, traditional methods of cluster analysis are highly sensitive to the presence of atypical observations and departures from normality. We describe a form of robust clustering using the forward search that allows the data to determine the number of clusters and so allows for outliers. An example is given of the successful clustering of customers of a bank into groups that are decidedly non-normal.

1 Introduction

The evaluation of the effectiveness of organisations has become an important strategic element in both the public and private sectors. Successful organisational structures need to be studied and emulated, whilst those that are failing need to be identified as early as possible so that preventive measures can be put in place and the waste of resources minimized. If organisations can be appropriately classified into homogeneous groups their differences in structure become more certainly identifiable and the number of special cases that has to be studied is dramatically reduced. The clustering of data is being increasingly used as a method of evaluation in public administration, see Peck (2005), and as a strategic element of political and administrative action, partly because it falls within the range of methods which has been deemed appropriate by the EU and the OECD (see for example the working papers contained in the web site http://www.oecd.org).

There are many statistical methods for the classification of multivariate observations such as those that describe the properties of an organisation. But, as is well known, at least to statisticians, the traditional methods of cluster analysis are highly sensitive to the presence of atypical observations and to incorrectly specified

A. Cerioli (✉)
Dipartimento di Economia, University of Parma, Via Kennedy 6, Italy
e-mail: andrea.cerioli@unipr.it

F. Palumbo et al. (eds.), *Data Analysis and Classification*,
Studies in Classification, Data Analysis, and Knowledge Organization,
DOI 10.1007/978-3-642-03739-9_43,

structures. Despite this sensitivity, robust statistical methods that are unaffected by outliers and model-misspecification are little used. It is the purpose of the present paper to extend and apply robust cluster analysis using the forward search as introduced in Chapter. 7 of Atkinson et al. (2004). This graphics-rich robust approach to clustering uses the data to identify the number of clusters, to confirm cluster membership and to detect outlying observations that do not belong to any cluster. More specifically, our analyses rely on forward plots of robust Mahalanobis distances. In order to provide sensitive inferences about the existence of clusters it is necessary to augment such graphs with envelopes of the distributions of the statistics being plotted. Examples of such envelopes and their use in the forward search for clustering moderate sized data sets are presented by Atkinson et al. (2006) and Atkinson and Riani (2007), in which the largest example has 1,000 observations. The theoretical results of Riani et al. (2009) provide the tools for extending our methodology to larger data sets, where indeed inspection of the trajectory of a single minimum Mahalanobis distance, defined in (3), greatly simplifies the cluster identification process. In Bini et al. (2004) we applied earlier versions of these methods to the analysis of a complicated set of data on the performance of Italian universities. Here we exemplify our method with a simpler example from banking. Other successful applications of the forward search to classification problems with several clusters and outliers are described by Cerioli et al. (2006) and Riani et al. (2008).

2 Mahalanobis Distances and the Forward Search

The main tools that we use are plots of Mahalanobis distances. The squared distances for the sample of n v-dimensional observations are defined as

$$d_i^2 = \{y_i - \hat{\mu}\}^T \hat{\Sigma}^{-1} \{y_i - \hat{\mu}\}, \tag{1}$$

where $\hat{\mu}$ and $\hat{\Sigma}$ are the unbiased moment estimators of the mean and covariance matrix of the n observations and y_i is $v \times 1$.

In the forward search the parameters μ and Σ are estimated from a subset $S(m)$ of m of the n observations $Y^{n \times v}$, with element y_{ij}. The parameter estimates are $\hat{\mu}(m)$ with

$$\hat{\mu}(m)_j = \sum_{i \in S(m)} y_{ij}/m, \qquad j = 1, \ldots, v$$

and $\hat{\Sigma}(m)$ where

$$\hat{\Sigma}(m)_{jk} = \sum_{i \in S(m)} \{y_{ij} - \hat{\mu}(m)_j\}\{y_{ik} - \hat{\mu}(m)_k\}/(m-1), \qquad j,k = 1, \ldots, v.$$

From this subset we obtain n squared Mahalanobis distances

$$d_i^2(m) = \{y_i - \hat{\mu}(m)\}^T \hat{\Sigma}^{-1}(m) \{y_i - \hat{\mu}(m)\}, \qquad i = 1, \ldots, n. \tag{2}$$

To start the search for cluster identification we take a random sample of $m_0 = v + 1$ observations, the minimum size for which Σ can be estimated. We require this subset to be as small as possible to maximize the probability that all members of $S(m_0)$ come from the same cluster. This subset of m_0 observations grows in size during the search in such a way that non-cluster members will be excluded with high probability. When a subset $S(m)$ of m observations is used in fitting we order the squared distances and take the observations corresponding to the $m + 1$ smallest as the new subset $S(m + 1)$. Usually this process augments the subset by one observation, but sometimes two or more observations enter as one or more leave.

To detect outliers we examine the minimum Mahalanobis distance amongst observations not in the subset

$$d_{\min}(m) = \min d_i(m) \quad i \notin S(m). \tag{3}$$

If this observation is an outlier relative to the other m observations, this distance will be "large" compared to the maximum Mahalanobis distance of observations in the subset. All other observations not in the subset will, by definition, have distances greater than $d_{\min}(m)$ and will therefore also be outliers.

For small datasets we can use envelopes from bootstrap simulations to determine the threshold of our statistic during the forward search. For moderate sized datasets we can instead use the polynomial approximations of Atkinson and Riani (2007). For larger samples, Atkinson et al. (2007) rescale a paradigmatic curve obtained by simulation to have the correct sample size and number of variables. Riani et al. (2009) use arguments from order statistics and estimation in truncated samples to obtain envelopes without requiring simulation.

For cluster definition, as opposed to outlier identification, several searches are needed, the most informative being those that start in individual clusters and continue to add observations from the cluster until all observations in that cluster have been used in estimation. There is then a clear change in the Mahalanobis distances as units from other clusters enter the subset used for estimation. This strategy seemingly requires that we know the clusters, at least approximately, before running the searches. But we, as do Atkinson and Riani (2007), instead use many searches with random starting points to provide information on cluster existence and definition.

3 Example

To illustrate our methodology we look at an example with a dataset of customers from a bank operating in Italy. The variables that we consider are:

y_1: Direct debts to the bank;
y_2: Assigned debts from third parties;
y_3: Amount of funds deposited;
y_4: Total amount invested in government securities.

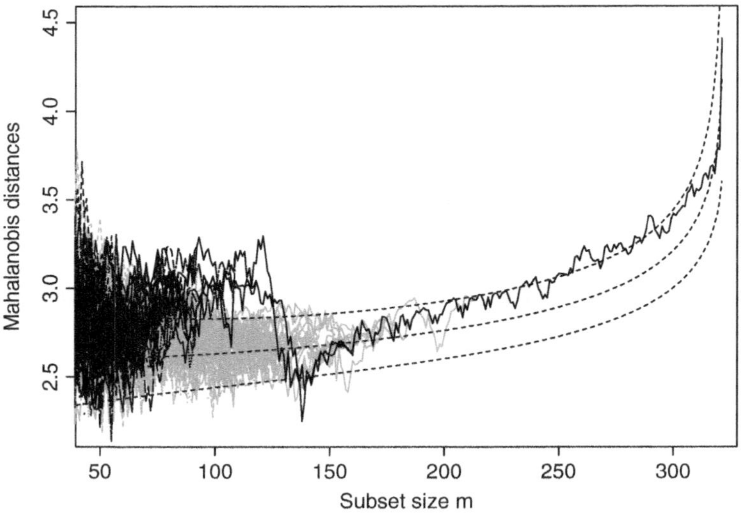

Fig. 1 Logged Banking data. Forward plot of minimum Mahalanobis distances, indicating two clusters; the trajectories in grey always include units from both of our final groups

The bank under study had just undertaken a thorough restructuring of all its activities. The purpose of the data analysis was to classify into homogeneous groups only those customers who had positive values for these four variables, of whom there were 322. Because the data were highly asymmetric, logs were taken to achieve approximate symmetry. In order to avoid singularity problems the logged data were also slightly jittered by adding a small normal noise.

Figure 1 shows a forward plot of minimum Mahalanobis distances from 200 random starts with 1 and 99% bounds. The structure of this plot is similar to that seen in Fig. 5 of Atkinson and Riani (2007), in which the simulated data consisted of two overlapping clusters.

As m increases the number of different subsets found by the forward search decreases, as is shown in the panels of Fig. 2. For m greater than 215 all searches follow the same trajectory. Earlier, around $m = 110$–130, there are two sets of trajectories lying clearly outside the envelopes (the black lines in the figure) and a large number of trajectories, represented in grey, within or close to the envelopes. The two sets of black trajectories in this range correspond to searches in which all the units in the subset are likely to come from a single cluster. If we identify the units in the subsets at $m = 118$ we obtain two initial clusters of observations. The largest value of $d_{\text{min}}(m)$ gives a cluster with 118 observations and the second largest value a cluster of 115 observations, once three observations that might be in either cluster are removed. At this point we have preliminary clusters with a total of 233 observations and 89 observations to be clustered.

The scatterplot of the values of y_3 and y_4 for these two initial clusters are shown in the left-hand panel of Fig. 3. The two groups are clearly separated as they are

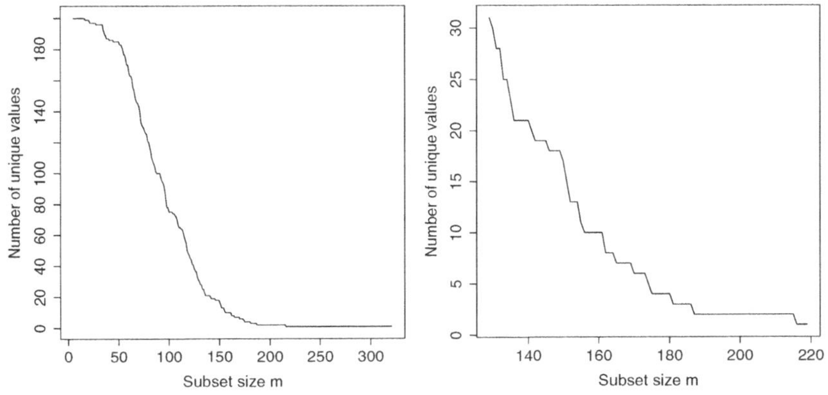

Fig. 2 Logged Banking data. Forward plots of number of unique Mahalanobis distances from 200 random starts. *Left-hand panel*, from 200 to 1; *right-hand panel*, zoom of plot where clusters become apparent

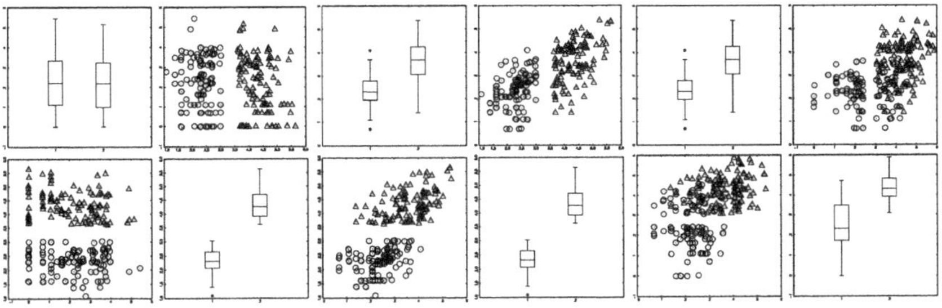

Fig. 3 Logged banking data: Scatterplot matrices of the two initial clusters of 118 and 115 observations found from Fig. 1. Reading across: y_3 and y_4, y_2 and y_4 and, *right-hand panel*, y_1 and y_2

in the centre panel, which is the scatterplot of y_2 and y_4. However they overlap in the right-hand panel, the scatter plot for y_1 and y_2. We have thus found two clear clusters, which plausibly have a multivariate normal structure, together with 89 observations which may perhaps belong to one of the groups, or to other groups, or that may be unstructured outliers.

To explore these possibilities we now run a forward search with two clusters starting with the cluster centres we have already found. In an extension of (2) we now assess two Mahalanobis distances for each unit

$$d_i^2(l, m) = \{y_i - \hat{\mu}_l(m)\}^T \hat{\Sigma}_l^{-1}(m)\{y_i - \hat{\mu}_l(m)\}, \qquad (l = 1, 2), \qquad (4)$$

where $\hat{\mu}_l(m)$ and $\hat{\Sigma}_l(m)$ are the estimates of the mean and covariance matrix based on the observations in group l, $l = 1$ or 2, and $m = m_1 + m_2$ is the total number of observations in the subsets for both groups. As before we start with a subset of

$m_0 = m_{01} + m_{02}$ observations. But now we want to preserve the cluster structure we have already established. So, for each m, we only consider the properties of the $2(n - m_0)$ squared Mahalanobis distances for the units that are not in the initial subset. We repeat the process several times for increasing values of m_0 that we take as 75% of the numbers of units which are indicated as correctly classified.

For each value of m we can use the values of $d_i^2(l, m)$ to allocate each unit not in m_0 to the cluster to which it is closest. We monitor how this allocation changes as the search proceeds. Those units that are firmly clustered stay in the same cluster throughout. Usually only those units about which there is some doubt have an allocation that changes as the search progresses. We ran one such search with the initial subset formed from the central 75% of units yielding our initial clusters of 118 and 115 units, that is the first 75% of this new set of units to enter these clusters in the individual searches shown in Fig. 1. We then obtained a set of units the allocations of which remained constant throughout the search. 75% of this new set of units resulted in an increased value of 204 for m_0. Figure 4 shows a forward plot of the allocation of the seven units that changed allocation during this two-cluster search. The bottom two lines serve as a key. The next band of two lines is for units 118 and 124. The classification of these units in the first cluster was not in doubt in our previous analyses, but they briefly become closer to the second group as the parameter estimates change with the inclusion of new units in the subsets used in fitting. The remaining seven lines, working upward, show the allocation, from $m = 240$, of units 110, 134, 135, 145, 178, 179 and 211. All other units, excluded from the plot, would have a single symbol throughout. As we shall see, these seven units lie between our two groups, so we refer to them as a "bridge". If we repeat the two-group search with the larger value of 268 for m_0 indicated by the results of Fig. 4 we find that the units in the bridge are, indeed, the only ones whose classification changes during the search.

The three panels of Fig. 5 show our proposed classification into two groups, of 145 and 177 units, with seven bridge units. The left-hand panel of the figure shows

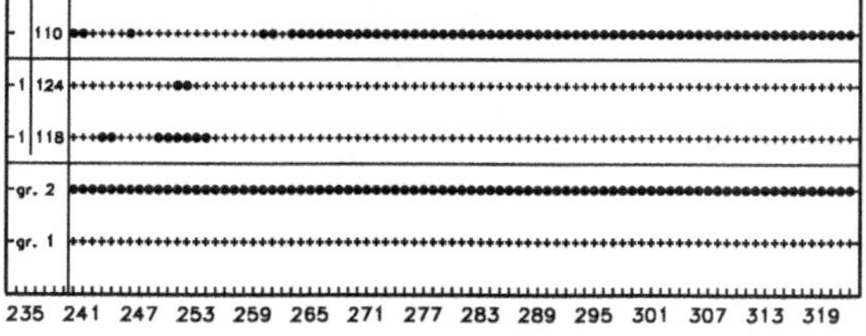

Fig. 4 Logged banking data. Cluster membership during a confirmatory search with two clusters starting with $m_0 = 204$. The bottom two lines serves as a key; the next two lines are units whose classification has never previously been in doubt, whereas the top seven lines give membership for the units that change classification during the search

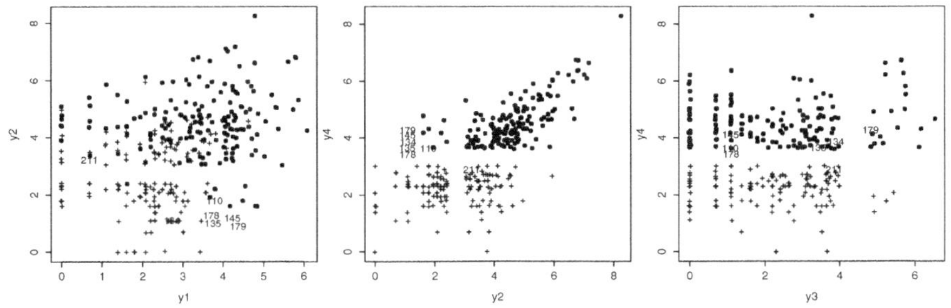

Fig. 5 Logged banking data: scatterplot matrices of the two final clusters with numbering for the seven units from Fig. 4. Reading across: y_2 and y_1, y_4 and y_2 and, *right-hand panel*, y_4 and y_3

the plot of y_2 against y_1, with the two clusters plotted with different symbols and the seven bridge units numbered. The separation of the two groups is not complete in these dimensions, with some interpenetration. Here the bridge units, apart from 211, seem to lie in Group 1, the crosses. The second panel is the plot of y_4 against y_2. There is a clear division into two groups on the values of y_4 and the bridge units seem to cluster in Group 2, again apart from unit 211. The final plot of y_4 against y_3 again shows the clear separation on values of y_4, but now the bridge units are dispersed.

These plots seem to indicate that we have satisfactorily clustered nearly all the data. But this has been achieved without any reference to the statistical properties of our procedure.

The classification of units shown in Fig. 5 is obtained by comparing Mahalanobis distances calculated using parameter estimates from the two groups. A potential difficulty, discussed by Atkinson et al. (2004), [p. 370], arises if the variances of the two clusters are very different. Then Euclidean distances and Mahalanobis distances are very different. As measured by Mahalanobis distance, an observation on the edge of a tight cluster may have a large distance for that cluster, but a smaller distance from a cluster with a larger variance. It will then be assigned to the cluster with a large variance. Due to the inclusion of this unit, the estimate of variance of the cluster with larger variance will increase and other units in the cluster with small variance become increasingly less remote from the cluster with larger variance as the search progresses. As a result the cluster with the looser structure absorbs units from the tighter cluster.

A solution to this problem, suggested by Atkinson et al. (2004), is to use instead distances standardised by the determinant of the estimated covariance matrix. These distances behave more like Euclidean distances and avoid the particular problem of loose clusters absorbing observations from tight clusters. However, these problems arise when the variances of the groups are very different. As a result of taking logarithms of the data, we have broken the relationship between the means and variances of our observations and, as Fig. 3 indicates, have obtained two groups with roughly equal variances. In fact, here a search with standardised distances yields the same classification as that found using unstandardized distances.

In Fig. 1 we used envelopes derived from the multivariate normal distribution to establish preliminary clusters. We now repeat this procedure to confirm the two clusters that we have found. If we look at the scatterplots of the final clusters in Fig. 5 and compare them with the preliminary clusters in Fig. 3, we see that our final clusters have become appreciably less elliptical in outline and so can be expected to be relatively poorly described by a multivariate normal distribution. This feature is revealed in the confirmatory forward plots of minimum Mahalanobis distance for the two separate groups.

Figure 6 shows the forward plot from the 145 units we finally classified in Group 1, together with 0.1 and 99.9% envelopes. We have taken these broader envelopes as a way of allowing for the very approximate normality of our groups. As the figure shows, the 200 random searches settle down as the search progresses to give a trajectory that lies towards the upper part of the distribution but without any systematic peak and trough of the sort that indicated the presence of clusters in Fig. 1.

The similar Fig. 7 shows the plot for the 170 units of Group 2, together with the 7 units in the "bridge". Here again there is no clear indication of any presence of clusters. The general shape of this plot, lying rather high in the envelope and then gradually decreasing is an indication of slight non-normality; Fig. 11 of Riani and Atkinson (2007) shows a more dramatic example of a plot with a related structure for regression with beta distributed error. The jump in the plot around $m = 120$ corresponds, as we saw in Fig. 1, to the end of the normally distributed central part

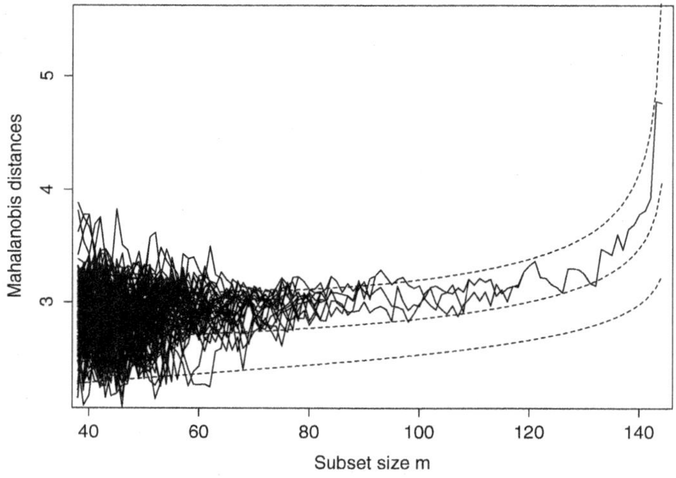

Fig. 6 Logged Banking data. Validation of Group 1. Forward plot of minimum Mahalanobis distances for the 145 units included in Group 1

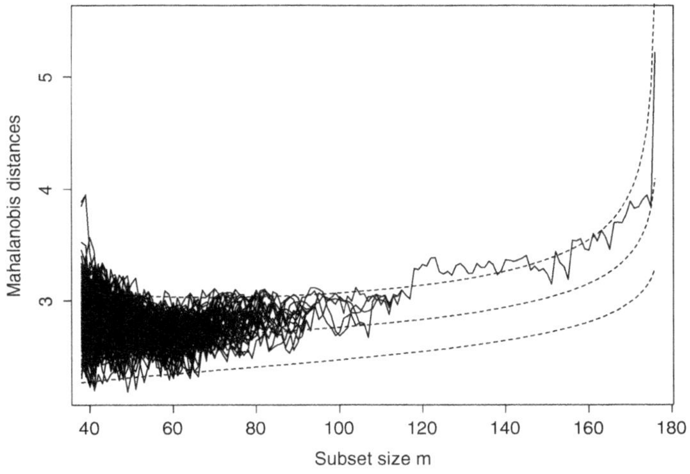

Fig. 7 Logged Banking data. Validation of Group 2. Forward plot of minimum Mahalanobis distances for the 170 units included in Group 2 and the seven"bridge" units

of the cluster in the scatterplots of Fig. 3. At the end of this search there is one extreme observation, 211, that has already been identified as the least well grouped.

An alternative method of clustering is the `mclust` procedure of Fraley and Raftery (2006) in which a mixture of normal distributions is fitted to the data. Atkinson and Riani (2007) provide examples in which `mclust` incorrectly finds more clusters than our robust method. The "incorrectness" is a feature of the analysis of simulated data in which, of course, we know the true number of clusters. In the example of the current paper, the BIC plot from `mclust` indicates five clusters. The forward plots of Figs. 6 and 7 however give no indication of such a structure. These forward plots can also be produced for the five tentative clusters. The searches do not at all lie within the envelopes, indicating that these five clusters are far from satisfactorily homogeneous. There are two conclusions from these analyses. One is that the data consist mostly of two rather non-normal clusters. The other is that we have found another example in which `mclust` indicates an excessive number of clusters.

Acknowledgements This work was supported by the grants "Metodi statistici multivariati per la valutazione integrata della qualità dei servizi di pubblica utilità: efficacia-efficienza, rischio del fornitore, soddisfazione degli utenti" and "Metodologie statistiche per lanalisi di impatto e la valutazione della regolamentazione" of Ministero dell'Università e della Ricerca PRIN 2006.

References

Atkinson, A. C., & Riani, M. (2007). Exploratory tools for clustering multivariate data. *Computational Statistics and Data Analysis*, *52*, 272–285 doi:10.1016/j.csda.2006.12.034

Atkinson, A. C., Riani, M., & Cerioli, A. (2004). *Exploring multivariate data with the forward search*. New York: Springer

Atkinson, A. C., Riani, M., & Cerioli, A. (2006). Random start forward searches with envelopes for detecting clusters in multivariate data. In S. Zani, A. Cerioli, M. Riani, & M. Vichi (eds.), *Data analysis, classification and the forward search* (pp. 163–171). Berlin: Springer

Atkinson, A. C., Riani, M., & Laurini, F. (2007). Approximate envelopes for finding an unknown number of multivariate outliers in large data sets. In S. Aivazian, P. Filzmoser, & Y. Kharin (eds.), *Proceedings of the Eighth International Conference on Computer Data Analysis and Modeling* (pp. 11–18). Russian Federation : Artia, Minsk

Bini, M., Riani, M., Atkinson, A., & Cerioli, A. (2004). Analisi di efficienza e di efficacia del sistema universitario italiano attraverso nuove metodologie statistiche multivariate robuste. Research report 03, Comitato Nazionale per la Valutazione del Sistema Universitario (CNVSU), MIUR, Ministero dell'Istruzione dell'Universit e della Ricerca. RDR document produced on behalf of CNVSU. http://www.cnvsu.it/_library/downloadfile.asp?id=11265

Cerioli, A., Riani, M., & Atkinson, A. C. (2006). Robust classification with categorical variables. In A. Rizzi & M. Vichi (eds.), *COMPSTAT 2006: Proceedings in Computational Statistics* (pp. 507–519). Heidelberg: Physica

Fraley, C., & Raftery, A. E. (2006). MCLUST version 3: an R package for normal mixture modeling and model-based clustering. Tech. Rep. 504, University of Washington, Department of Statistics, Seattle, WA

Peck, L. (2005). Using cluster analysis in program evaluation. *Evaluation Review*, *29*, 178–196

Riani, M., & Atkinson, A. C. (2007). Fast calibrations of the forward search for testing multiple outliers in regression. *Advances in data analysis and classification*, *1*, 123–141

Riani, M., Atkinson, A. C., & Cerioli, A. (2009). Finding an unknown number of multivariate outliers. *Journal of the Royal Statistical Society, Series B*, *71*, 447–466

Riani, M., Cerioli, A., Atkinson, A., Perrotta, D., & Torti, F. (2008). Fitting mixtures of regression lines with the forward search. In F. Fogelman-Soulié, D. Perrotta, J. Piskorski, & R. Steinberger (eds.), *Mining massive data sets for security* (pp. 271–286). Amsterdam: IOS

Outliers Detection Strategy for a Curve Clustering Algorithm

Balzanella Antonio, Elvira Romano, and Rosanna Verde

Abstract In recent years curve clustering problem has been handled in several applicative fields. However, most of the proposed approaches are sensitive to outliers. This paper aims to deal with this problem in order to make a partition, obtained by using a Dynamic Curve Clustering Algorithm with free knots spline estimation, more robust. The approach is based on a leave-some-out strategy, which defines a rule on the distances distribution of the curves from the barycenters, in order to identify outliers regions. The method is validated by an application on real data.

1 Introduction

In many applicative fields data can be real functions defined on a common interval in $\Re$ rather than on a finite dimensional space. These are commonly called *functional data* Ramsay and Silverman (2005). In many cases data come from sensors systems where low fidelity and frequent failures can lead to outliers presence. Only in recent years, strategies to detect outliers, in the framework of clustering algorithms for functional data, have received attention. The main proposed procedures are based on the concept of impartial trimming Gordaliza (1991) and on its extension to the k-means algorithm Cuesta-Albertos et al. (1997). One of these methods Cuesta-Albertos and Fraiman (2007) considers the case of a sample of elements extracted from a random process taking values in a Banach space. The goal is to estimate the center of a cluster only on a portion of its elements. According to this aim, the authors change the function to be minimized in the *k-means* strategy introducing an $\alpha \in (0, 1)$ parameter so that a proportion of individuals, less or equal than α, is trimmed. In the minimizing function, at each individual is given a weight in (0,1) in order to obtain the global α trimming level.

A method, also based on impartial trimming, is proposed by Garcia et al. (2005). This clustering strategy is an improvement of k-means where each observation

B. Antonio (✉)
Università degli Studi di Napoli Federico II, Via Cinthia I-80126 Napoli, Italy
e-mail: balzanella2@alice.it

F. Palumbo et al. (eds.), *Data Analysis and Classification*,
Studies in Classification, Data Analysis, and Knowledge Organization,
DOI 10.1007/978-3-642-03739-9_44,

is smoothed by means of B-spline basis functions with fixed knots. The minimized criterion is based on the coefficients resulting from the smoothing process. In such context, they introduce an impartial trimming strategy to detect the centers of clusters.

Febrero et al. (2007) suggested a further interesting strategy for detecting outliers in functional data externally from the clustering process. According to their approach a curve is an outlier if it is generated by a different stochastic process. They assume that the curves which are not outliers are identically distributed. The method consists in using the likelihood ratio test to detect differently distributed curves.

In the present work we introduce an approach to detect anomalous behaviors in curves data coming from sensors that are often loosely synchronized. The crucial point consists in avoiding to discover as outliers, curves that are only time-shifted.

The approach is based on a partitioning structure achieved through the Dynamic Curves Clustering Algorithm with Free knots Spline Estimation (DCC&FSE) Romano (2006). The basic idea is the specification of a re-phasing step in the DCC&FSE and of an external criterion for the outliers detection.

In the following section we provide a background of DCC&FSE. The main steps of our strategy are explained in Sects. 3 and 4. In Sect. 5, the applicative results are shown. Finally, in Sect. 6 perspectives and open problems are discussed.

2 Dynamical Curves Clustering with Free knots Spline Estimation

The DCC&FSE strategy is an extension of the Dynamical Clustering Method (DCM) Diday (1971), proposed on functional data structures. Let E be a set of n functions such that each function i is given by the $(t^i_j, y^i_j)_{1\leq j\leq J}$ list of J pairs, with $t^i_j \in T$, a compact subset of $\Re^J$, and $y^i_j \in \Re$. The core of the methodology consists in optimizing the following criterion:

$$\Delta(P, G) = \sum_{c=1}^{C} \sum_{i \in P_c} \delta^2(\mathbf{y^i}, \mathbf{g_c}(\boldsymbol{\xi}_c)) \quad P_c \in P, \; g_c \in G \tag{1}$$

where $\delta(\mathbf{y^i}, \mathbf{g_c}) = \|\mathbf{y^i} - \mathbf{g_c}\|$ is an L_2 distance and $\mathbf{y^i}$, $\mathbf{g_c}$ $(c = 1, \ldots, C)$ are respectively the vectors of the functions y^i and of the prototypes g_c. The obtained functions prototype are local models representative for each cluster, identified from a spline with a best set of knots ζ_c. The dimension of the knots vector which characterizes the function prototype is such to minimize the loss of variety of *true functions*, where the loss is defined to be the square root of mean of squared distance from the truth to the estimated functions prototype at the design points. According to this criterion, each cluster can be described by a function prototype expressed by:

$$g_c = \mathbf{B}(\boldsymbol{\zeta}^{\mathbf{c}})\boldsymbol{\alpha}_l \quad \forall i \in C \tag{2}$$

where the vector $\boldsymbol{\zeta}^{\mathbf{c}}$ is the Jupp transformation of $\boldsymbol{\xi}^{\mathbf{c}}$, defined as: $\zeta_m^c = log \frac{\xi_{m+1}^c - \xi_m^c}{\xi_{M+1}^c - \xi_0^c}$ for $m = \{1, \ldots, M\}$ $\xi_0^c = a$ and ${\xi^c}_{M+1} = b$. Let $\mathbf{B}(\zeta^c)$ be the matrix of $J \times (H + M)$ whose generic row is the vector $B^T \in R^{H+M}$ of the B-spline basis function. Using this transformation the adequacy criteria can be written as

$$\phi(g) = \sum_{i \in P_c} \delta^2(\mathbf{y^i}, \mathbf{g_c})_{\alpha_1^c}, \boldsymbol{\zeta}^{\mathbf{c}} = \sum_{i \in P_c} \left\| \mathbf{y}^i - \mathbf{B}(\boldsymbol{\zeta}^{\mathbf{c}})\boldsymbol{\alpha}_{\mathbf{l}}^{\mathbf{c}} \right\|^2 \tag{3}$$

and it's possible to find the best approximation of g_c by the least squares, such that:

$$(\hat{\boldsymbol{\alpha}}_{\mathbf{l}}^{\mathbf{c}}, \hat{\boldsymbol{\zeta}}^{\mathbf{c}}) = \min_{(\alpha_1^c, \zeta^c)} \sum_{i \in P_c} \left\| \mathbf{y^i} - \mathbf{B}(\boldsymbol{\zeta}^{\mathbf{c}})\boldsymbol{\alpha}_{\mathbf{l}}^{\mathbf{c}} \right\|^2 \tag{4}$$

and $\hat{\mathbf{g}}_c = \mathbf{B}(\hat{\boldsymbol{\zeta}}^{\mathbf{c}})\hat{\boldsymbol{\alpha}}_{\mathbf{l}}^{\mathbf{c}}$.

The Dynamical Curves Clustering Algorithm is performed iteratively by altering a representation and an allocation step. At each iteration of the algorithm, a new partition P with a new set of prototypes is found. The decreasing of the criterion Δ can be proved under the following condition:

- *Uniqueness* of the affectation cluster for each element of E;
- *Uniqueness* of the function prototype g_c which minimizes the criterion for all the cluster P_c (for $c = 1, \ldots, C$) of the partition P of E.

Among the existent curves clustering methods, DCC&FSE behaves like more an estimation of cluster shape, since it is an approach that allows to classify curves without a priori specification of the functional form.

The clusters elements, in this way, may be summarized by the functional structure of their relative prototype.

3 An Improvement of DCC&FSE Algorithm

A first challenge in clustering methods can arise from time shifted curves, since curves even having a similar shape, can be allocated to different clusters or can be detected as anomalous. In order to overcome this problem, we propose a criterion to re-phase the curves. The key idea is to introduce in the DCC&FSE algorithm a *re-phasing* step, between the representation and the allocation step. In such way, each curve, before being allocated, is time shifted according to the chosen criterion, which is optimized to determinate an optimal shifting for each curve in a cluster. Especially, for each y^i curve, the optimal time shifting τ_*^i is the shifting value which allows to minimize the Euclidean distance between the y^i curve and the g_c prototypes.

Formally the function to be minimized is:

$$\varphi(\tau) = \sum_{c=1}^{C} \sum_{i=1}^{n} \delta^2 \left(\mathbf{y^i} \left(\mathbf{t^i} + \boldsymbol{\o}^{\mathbf{i}} \right), \mathbf{g_c} \right) \tag{5}$$

By introducing the *re-phasing* step, the DCC&FSE algorithm becomes able to find, not only curves more similar in terms of best fitting, but also in in term of phase mismatching.

4 The Outliers Selection Process

In the clustering process, some of the analyzed curves can present anomalous behaviors which are not a consequence of time shifting problems. For instance, these anomalies could come from sensors failure or perturbations of the studied phenomenon.

With the aim to improve the partitioning quality and the prototypes robustness, the detection of anomalous curves is a challenging problem.

In the functional framework, we can refer to two types of outliers. The first type consists of curves presenting some isolated outlying points; the second type consists of curves which are entirely outliers.

In this paper we will focus on the second type of anomalies, since the smoothing process is able to attenuate the impact of isolated outlying points on the clustering and on the prototypes computation. In order to detect outlying behaviors, we propose a strategy based on the Mahalanobis distance. The chosen metric allows to compute the distance between the tested curve and a curve prototype. It keeps into account the covariance structure of the curves whose prototype is representative. Especially the covariance matrix accounts for the covariance among the time records of curves in the cluster.

The outlying curve will be a curve that satisfies the following definition:

Definition 1 (Outlier). Let $\hat{y}^i$ $(i = 1 \dots n)$ be the set of *true functional forms* coming from the smoothing process according to the best set of knots ζ_c (for the cluster $(c = 1 \dots C)$ which the curve belongs to) and $\mathbf{g}_{c^-}$ a functional prototype, $\hat{y}^i$ is an outlier for the cluster if:

$$d^2(\hat{\mathbf{y}}^i, \mathbf{g}_{c^-}) > \nu_c \tag{6}$$

where $\mathbf{g}_{c^-}$ and Σ_{c^-} are respectively the robustified prototype and covariance matrix for the cluster c and ν_c is a threshold value defined for each cluster;

d^2 is the squared Mahalanobis distance defined as:

$$d^2 = (\hat{\mathbf{y}}^{\mathbf{i}} - \mathbf{g}_{c^-})' \Sigma_{c^-}^{-1} (\hat{\mathbf{y}}^{\mathbf{i}} - \mathbf{g}_{c^-}) \tag{7}$$

Assuming that data in a cluster are Normally distributed, the d^2 is approximately χ^2-distributed thus the threshold value ν_c for each cluster can be chosen as $\chi^2_{j,1-\alpha_{nc}}$.

The value of α_{nc} can be fixed arbitrary for all clusters, as well as it can be chosen (as we assume) according to the proposal of Becker and Gather (2001). Then, for c cluster, this value delimits the outlier region defined as:

$$out(\alpha_{nc}, g_{c}-, \Sigma_{c}-) = \hat{\mathbf{y}}^{i} \in R^{J} : d^{2} > \chi^{2}_{j,1-\alpha_{nc}} \tag{8}$$

where $\alpha_{nc} = 1 - (1 - \alpha)^{\frac{1}{nc}}$ for some given value $\alpha \in (0, 1)$, such that

$$Pr(no\ observation\ in\ the\ cluster\ of\ size\ n_c\ lies\ in\ out(\alpha_{nc}, g_{c}-, \Sigma_{c}-)) = 1 - \alpha \tag{9}$$

The robust prototype and the robust covariance matrix are obtained through an iterative procedure.
It consists in:

- Detecting the curves, potential outliers, such that $d^2(\hat{\mathbf{y}}^i, \mathbf{g}_c-) > \nu_c$;
- Removing these from the computation of the prototype and of the covariance matrix;
- Computing the distances again;

The steps are repeated until the same set of potential outlying curves is found in two consecutive iterations.

The whole outliers detection strategy can be summarized as follows:

The $DCC\&FSE$ strategy is performed to get a set of prototypes $\mathbf{g_c}$ ($c = 1 \ldots C$), a partition of data into C clusters and a set of knots for each cluster;
For each cluster $c = 1 \ldots C$
    Repeat
        Compute the reference prototype $\mathbf{g_c}-$ and the covariance matrix Σ_c- of the reduced dataset
        For each curve in the c cluster ($c = 1 \ldots C$)
            The distance $d^2(\hat{\mathbf{y}}^i, \mathbf{g}_c-)$ is computed;
            If $d^2(\hat{\mathbf{y}}^\mathbf{i}, \mathbf{g}_c-) > \nu_c$ then $\hat{\mathbf{y}}^\mathbf{i}$ is removed from the current dataset.
        End for each
    Until no further outliers are found
End

5 Main Results

To demonstrate the effectiveness of the proposed approach, we have performed an analysis on real data. A first tested dataset comes from an experiment conducted by the Department of Hydraulic and Environmental Engineering "Girolamo Ippolito",

University of Naples Panisi et al. (2006), on the impact of the introduction of submerged breakwaters on the shape of sea waves. To perform the analysis, a detection system made by 90 sensors has been placed to detect the rupture process of each sea wave on the breakwaters. In the conducted experiments, the sensors are arranged on a wide area and the monitoring activity has been performed without human assistance. These two peculiarities rise both, time delay problems as consequence of the different time needed to signals to reach the storage unit, and outliers problems like transducers failures, noise or curves not belonging to the studied phenomenon. The object of the analysis consists in 90 curves, each of them recorded from the different sensors.

A second dataset originates from different cash machines at different, randomly selected locations within England. These are 111 time series representing activity performed starting on March 18, 1996 and running until March 22, 1998, providing two years of daily data NN5 (2008). The data may contain a number of time series patterns including multiple overlying seasonality, local trends, outliers and missing values. These are often driven by unknown and unobserved casual forces.

For measuring the performance of our strategy, we have compared the results obtained by the DCC&FSE strategy with and without alignment and outliers removal.

For each dataset we have at first, run the DCC&FSE algorithm, then, we have run again the procedure introducing the re-phasing step. This is to evaluate the impact of the curves alignment on the prototypes computation and on the achived partitioning.

Starting from the two clustering structures, we have tried the outliers detection procedure. The main objective is here to show the effectiveness in the outliers detection and how misaligned curves impact on this. The outliers detection has been tested at first using strictly our procedure and then we have computed the squared Mahalanobis distance without removing curves from the cluster which these have been allocated to. This is to evaluate the effectiveness of the procedure to make the prototypes and the covariance matrix robust.

The improvements in terms of clusters homogeneity are evaluated by means of a quality index δ_s defined on the basis of the partitioning criterion. Especially it compares the within variability of the partition obtained using DCC&FSE to the one obtained removing the outliers.

For DCC&FSE strategy we need, at first, to choose the order of B-spline functions, the minimum and maximum values of the knots, and then the number of the clusters. We have chosen B-spline of order 4, since these are the most flexible and numerical stable basis functions. To analyze the sea waves dataset, the number of knots have been fixed to be in the range [3, 12] while the number of clusters is 4 according to the AIC (Akaike Information Criterion).

In a situation where non informative phase displacements are present, although $DCC\&FSE$ strategy is able to discover the basic clustering structure, some of the data are not correctly allocated. This makes the oultiers detection problematic.

Furthermore, an index based on the optimized criterion is proposed to compare the two methodologies.

Figure 1(a) illustrates our outliers detection procedure on not re-phased data. As we can see, even though the procedure detects the most outlying curve, a curve that is not anomalous is still identified as outlier. By introducing the re-phasing

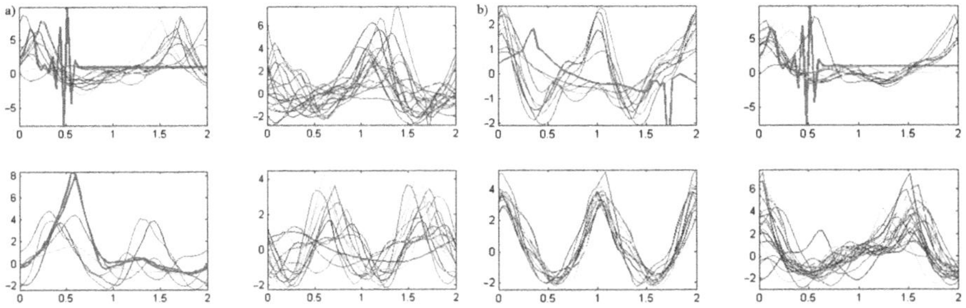

Fig. 1 (**a**) Outliers detection on sea waves data without re-phasing step; (**b**) Outliers detection on sea waves data with re-phasing step

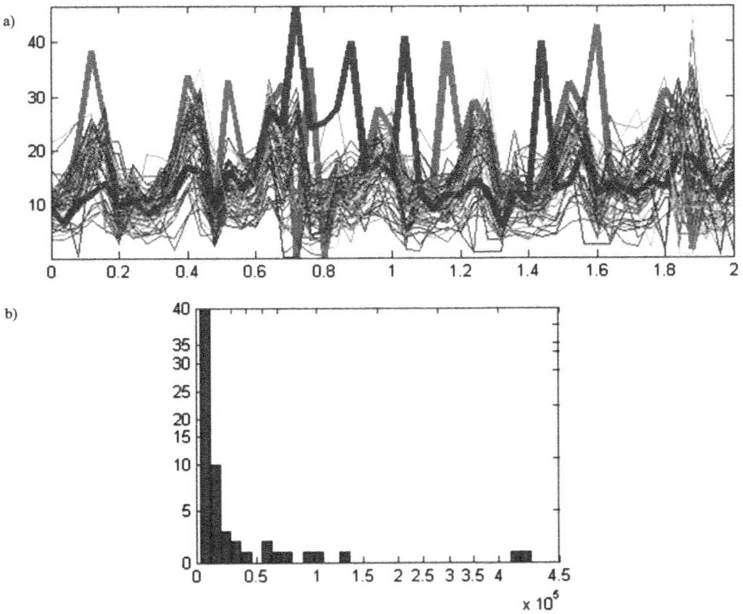

Fig. 2 (**a**) Outliers detection on cash machines dataset; (**b**) Mahalanobis distances distribution

step, the outliers detection is more effective Fig. 1(b), since it is able to find out the real anomalies. By repeating the experiment ten times, so to mitigate the impact of the initial random partitioning, we have observed an improving of the quality index quantifiable in about 15%. As shown in Fig. 2(a), the testing on the cash machines dataset, confirms the effectiveness in detecting outliers. The clustering process has been performed partitioning data into two groups and using knots in the range [5, 20]. We have introduced two outlying curves that have been detected using $1 - \alpha_{nc} = 0.975$ for the χ^2 probability distribution. The value of Δ_s has been 12%.

In Fig. 2(b) the empirical distribution of the squared Mahalanobis distances highlights the presence of a gap in the distances between anomalous behaviors and the real data.

6 Conclusion and Future Work

In this paper, we have proposed a new approach to make a clustering procedure robust. Our main concerns is not only to reduce the influence of outliers, but also to clearly identify them. Subject of future work will be to asses the impact of this on other functional datasets and to test other kind of measures for the proposed strategy.

References

Becker, C., & Gather, U. (2001). The largest nonidentifiable outlier: A comparison of multivariate simultaneous outlier identification rules. *Computational Statistics and Data Analysis, 36*, 119–127

Cuesta-Albertos J. A., & Fraiman R. (2007). Impartial trimmed k-means for functional data. *Computational Statistic and Data Analysis, 51*, 4864–4877

Cuesta-Albertos, J. A., Gordaliza, A., & Matrn, C. (1997). Trimmed k-Means: An attempt to robustify quantizers. *The Annals of Statistics, 25*, 553–576

Diday, E. (1971). La Mthode des nues dynamiques. *Revue de statistique applique, 19*(2), 19–34

Febrero, M., Galeano, P., & Gonzáleg Manteiga, W. (2007). A functional analysis of NOx levels: Location and scale estimation and outlier detection. *J. Computational Statistics*. doi 10.1007/s00180-007-0048-x

Garca-Escudero, L. A., & Gordaliza, A. (2005). A proposal for robust curves clustering. *Journal of classification, 22*, 185–201

Gordaliza, A. (1991). Best approximations to random variables based on trimming procedures. *Journal of Approximation Theory, 64*(2), 162–180

Panisi, F., Calabrese, M., & Buccino M. (2006). Breaker types and free waves generation at submerged breakwaters. In *Proceeding of XXIX of HYDRA06*, Roma

Ramsay, J. O., & Silverman, B. W. (2005). *Functional Data Analysis* (2nd ed.). New York: Springer

Romano, E. (2006). Dynamical curves clustering with free knots spline estimation. PHD Thesis, University of Federico II, Naples

NN5 (2008). Time series forecasting competition for computational intelligence. http://www.neural-forecasting-competition.com/index.htm

Robust Fuzzy Classification

Matilde Bini and Bruno Bertaccini

Abstract One of the most important problems among the methodological issues discussed in cluster analysis is the identification of the correct number of clusters and the correct allocation of units to their natural clusters. The most widely used index to determine the optimal number of groups is the Calinski Harabasz index. As shown in this paper, the presence of atypical observations has a strong effect on this index and may lead to the determination of a wrong number of groups. Furthermore, in order to study the degree of belonging of each unit to each group it is standard practice to apply a fuzzy k-means algorithm. In this paper we tackle this problem using a robust and efficient approach based on a forward search algorithm. The method is applied on a data set containing performance evaluation indicators of Italian universities.

1 Introduction

In a cluster analysis of a multivariate data set, it may happen that one or two observations have a disproportionately large effect on the analysis in the sense that their removal causes a dramatic change to the results. This effect is apparel for example when the Calinski and Harabasz index, one of the most widely used indicators to determine the right number of groups, is used in the cluster analysis Milligan and Cooper (1985). This index is defined as:

$$\frac{\text{trace} B/(k-1)}{\text{trace} W/(n-k)} \tag{1}$$

where n is the total number of items which have been considered and k is the number of clusters in the solution. The B and W terms are the between and within clusters sum of squares and product matrices. According to this index the optimum

M. Bini (✉)
Department of Statistics "G. Parenti", Viale Morgagni, 59, 50134 Firenze, Italy
e-mail: bini@ds.unifi.it

F. Palumbo et al. (eds.), *Data Analysis and Classification*, Studies in Classification, Data Analysis, and Knowledge Organization, DOI 10.1007/978-3-642-03739-9_45,

number of clusters is obtained when (1) reaches its maximum as a function of k. Unfortunately, as shown in the next section, this index is highly influenced by the presence of atypical observations and may lead to wrong conclusions about the real structure of the data.

Once the number of groups has been decided, one relevant issue in cluster analysis is the determination of the degree of belonging of each unit to each cluster. The traditional approach uses a k-means fuzzy approach which gives us as output the probability that each unit belongs to a cluster. However the output is "static" and non robust in the sense that we do not know whether the high (low) probability of a unit to belong to a particular group is due to the presence of particular observations. In this paper we propose a way to tackle this problem using a forward search through the data to monitor the cohesion inside groups and their degree of overlapping. The final purpose is to scrutinize in a robust and efficient way the output of k-means clustering algorithm after removing the atypical observations.

As an illustration of the suggested approach, we use a data set referred to performance university measurements. Data considered in this analysis include some performance indicators (see details in Biggeri and Bini (2001)), deriving from data of surveys conducted by the National University Evaluation Committee during these past four years (2000–2003), concerning 57 public universities during the academic years 1998–1999 until 2001–2002. Among the large number of proposed indicators (29), we have deleted those which were highly correlated. The final number of considered indicators is 12.

2 Robust Fuzzy Cluster Analysis

Before performing cluster analysis, given the highly skewed distributions of some variables, we proceeded to estimate the values of the Box-Cox transformation parameters, using the robust procedure described in Riani and Atkinson (2001). Given that the data appropriately transformed satisfy the multivariate normality assumption, cluster analysis based on k-means algorithm has been performed using the standardized transformed data. Figure 1 shows the behaviour of the Calinki-Harabasz index as a function of the number of groups in the 4 years.

This index suggests that the optimum number of groups varies considerably among the years, from 4 to 9. Our second step of the analysis consisted in the detection of multivariate outliers for each year. We used a forward search based on Mahalanobis (or Euclidean) distance to detect the presence of atypical observations. From the monitoring of forward plot of these distances Atkinson et al. (2004) we easily detected three atypical observations in year 2000 (units 8, 12, 57), three in year 2001 (units 8, 31, 57); two in year 2002 (units 29 and 57), and finally one (unit 57) in year 2003. Figure 2 shows the trajectories of the Calinki-Harabasz index after the removal of these outliers. The index, computed on clean data, shows that the optimum number of groups in each year is stable and equal to 4.

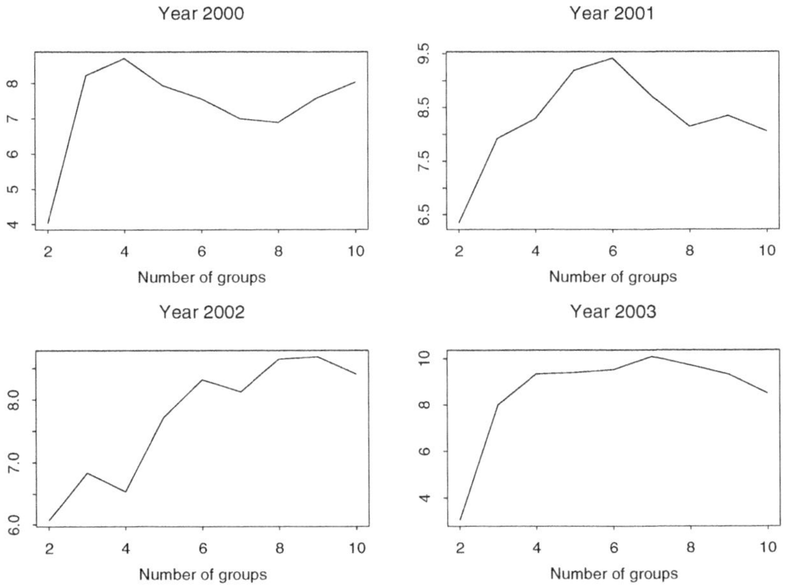

Fig. 1 Calinski Harabasz index versus number of groups (with the outliers)

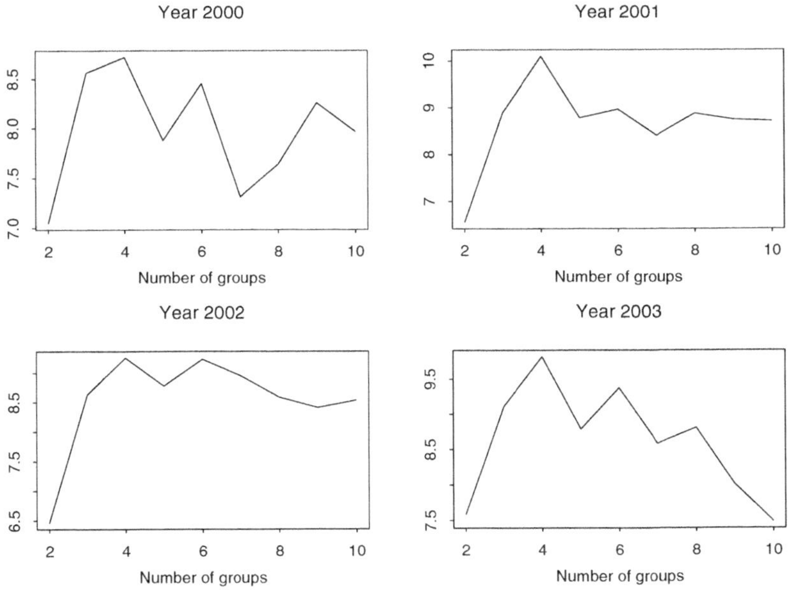

Fig. 2 Calinski Harabasz index versus number of groups (without the outliers)

The interpretation of the obtained groups is with intent omitted since the purpose of this paper is to show the "ability" of the forward search to solve the problem of determining the degree of overlapping of the different clusters.

3 Confirmatory Analysis

The Forward Search algorithm proposed by Atkinson and Riani (2000) is a general approach for detecting the presence of outliers and assessing their influence on the estimates of the model parameters. The method was firstly applied to regression analysis, but it could as well be applied to almost any multivariate method. The procedure starts out by fitting the model to an outlier free subset of the observations, say m observations, which is chosen in some robust way. The observations of the entire set are then ordered by their closeness to the estimated model. The model is then refitted using the subset of the $(m + 1)$ observations which are closest to the previously estimated model. The observations are ordered again, the model is refitted to a larger subset and the process is continued until all the data have entered. At every step the subset size is increased by one unit (usually one case is added to the previous subset, but sometimes two or more are added as one or more leave the subset), bringing about an ordering of all the observations. In the multivariate analysis, as in this case where a cluster analysis is performed, the closeness measure is represented by the Mahalanobis (or Euclidean) distance. Once the classification of units in groups is obtained from the exploratory analysis of the k-means method, we proceed with another analysis aimed to support the validity of the partitioning of all units into the detected groups. Atkinson et al. (2004) tackle the issue of the degree of overlapping among groups analyzing the evolution of the Mahalanobis distances. Given the high number of variables compared to the number of units we used a forward search based on Euclidean distances. More specifically, the analysis here presented refers to a particular step the forward search analysis applied in the cluster analysis which is called *confirmatory cluster analysis* (see details of this procedure in Atkinson et al. (2004), p. 371). In this confirmatory step, we detect the presence of some units which are uncertain to belong to the clusters previously assigned; then, we include those unassigned units in either obtained (called tentative) cluster; and finally, we establish to which clusters the units would be assigned if we insist on clustering all units, as well as to see how this assignment changes during the forward search. In this section we show the most significant plots yielded during forward search when we want to confirm the results obtained from grouping procedure.

Figure 3 presents Euclidean distances from centroid for the tentative group 1. In particular, the figure reports four panels, each one for each specific year, showing the monitoring of these distances for the first tentative cluster (cluster 1), as illustrative case. Similar plots are obtained when we repeat the search for the other tentative clusters. By observing the trajectories of the units into the panels, it is possible to see how much units agree into groups. In every year the trajectories of majority of units belonging to their groups are quite similar, so that we can conclude that the

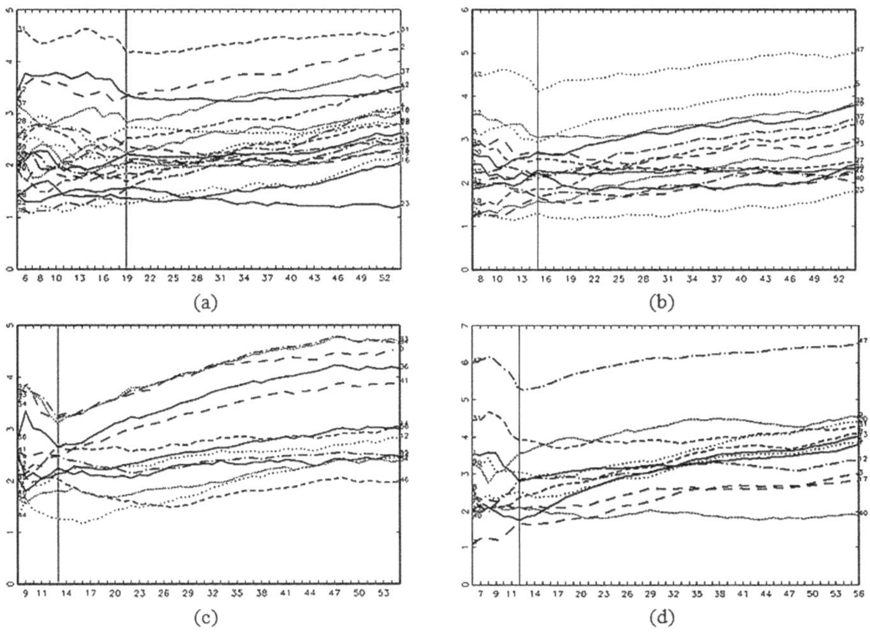

Fig. 3 Forward plot of Euclidean distances of all units, after removing the outliers, when the search starts in tentative cluster 1. Panels: (**a**) Year 2000. Uncertain units in tentative cluster 1: 23, 31, 42. (**b**) Year 2001. Uncertain units in tentative cluster 1: 27,40. (**c**) Year 2002. Uncertain units in tentative cluster 1: 46. (**d**) Year 2003. Uncertain units in tentative cluster 1: 47

groups are inside homogeneous. It is also possible to see the presence of units about which we are undecided for a search through all the data starting from cluster 1.

Throughout the search the largest distances generally come from the unassigned observations. Looking at the last year (fourth panel (d)), for example, the plot shows the presence of two separate subgroups and a unit, the number 47, having a trajectory (the highest shape in the panel) quite different from those of units of cluster 1. Again, looking at another example like year 2000 (first panel (a)), we could quite uncertain about the belonging of units 32, 31 and 42 to their group (cluster 1), since their trajectories have shapes quite different from those of units of this cluster. This monitoring of distances must be performed again considering as tentative cluster (each one in turn) the other clusters obtained from the previous k-means analysis.

The detection of outliers can also be highlighted by Fig. 4 that shows in the top left panel the trajectories of the Euclidean distances of the subset formed by 11 of the 12 units defined by the k-means algorithm as belonging to group 1.

This distribution of distances is used to judge the behaviour of successive units as they are included in the subset. Starting from the second panel, figures show the percentage points of the estimated distribution of these 11 distances. For example, in the second panel, the superimposed solid line is the trajectory of unit 47 which enters when $m = 12$. The shape of this unit is out the bands but it is in line

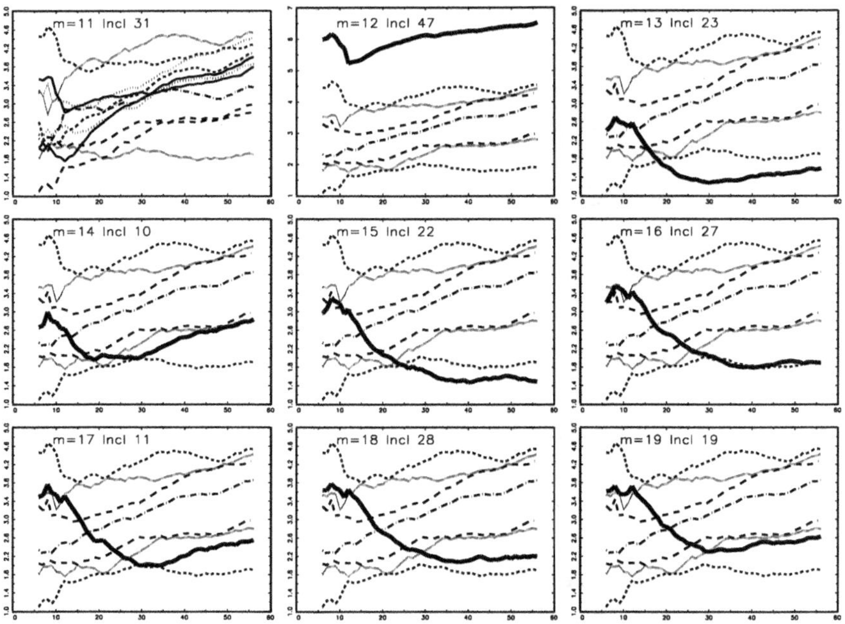

Fig. 4 Nine panel forward euclidean distances for specific units starting with a subset of size $m = 11$. Year 2003

with those of group 1. Unit 47 is indeed the most distant from the centroid among those which belong to group 1. The shapes of the other units which enter from steps $m = 13$ until $m = 19$ (with the exception of unit 10) are very different in the central part of the search from those belonging to group 1. Their distance decreases almost monotonically from steps 12 to 30. On the other hand, in these steps the distance of the units belonging to group 1 tends to increase. This figure clearly shows the amount of information on group membership which can be found by looking at individual profiles against a background profile from a known group, and the advantages of this approach with respect to the static output of fuzzy clustering algorithms.

Let focus now on Fig. 5 which represents the monitoring of Euclidean distances of all units from the centroid of clusters they belong to. For lack of space, we here refer only to year 2003. The units are divided in four groups separately into different panels, three for each group, at the step 11 of the search.

The situation about the coherence of the groups is well displayed in the figure. Groups 2 and 3 are quite homogeneous, while Group 4 shows trajectories similar to ones of Group 1, so then these two clusters are rather overlapped.

Figure 5 is obviously repeated also for other years involved in this study. The analysis of all these plots leads us to have a complete frame about the uncertain units for all the four years. More specifically, we found that there are several uncertain

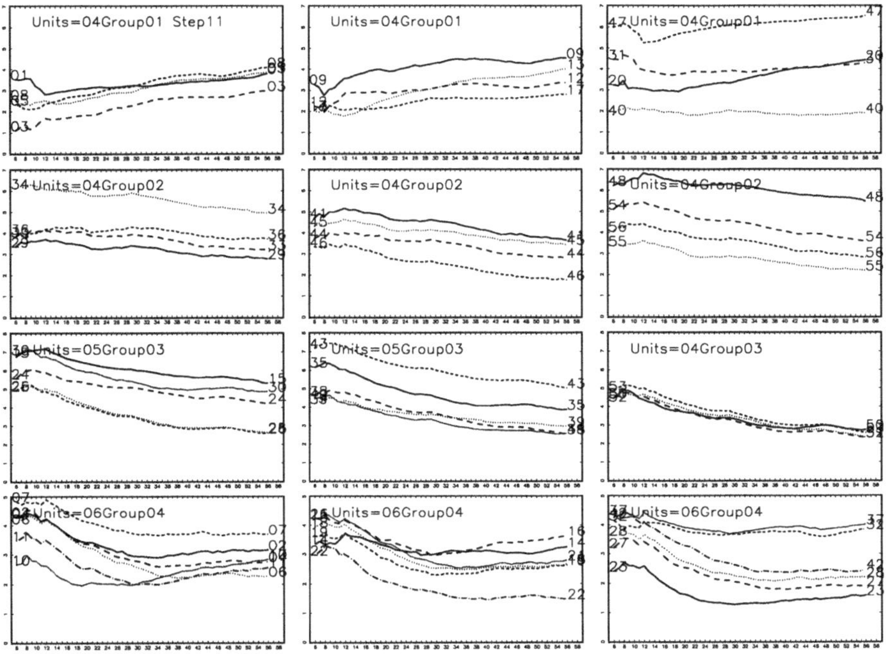

Fig. 5 Monitoring of distances of all units from centroid of 4 groups they belong, from a search starting in cluster 1. Step size $m = 11$ of the search, after removing outliers. Year 2003

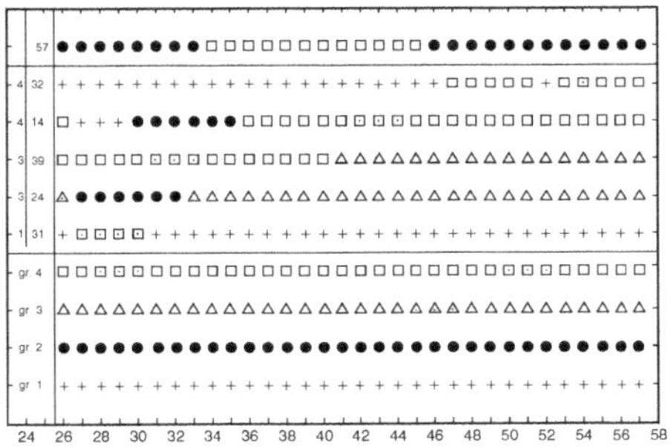

Fig. 6 Cluster membership of outliers and uncertain units during the last steps of the confirmatory search with four clusters using Euclidean distances

units for each year: units 24, 32, 13, 5, 42 for year 2000; units 52, 2, 1 for year 2001; units 32, 17, 39, 23, 19, 18, 11, 7, 2, 1 for year 2002; and units 32, 14, 39, 24 e 31 for the last year.

Figure 6 shows the allocation of units to populations for all units about which we were uncertain, together those for which the forward search detected as outliers in the previous preliminary analysis.

Since the closest cluster to each unit may change during the forward search, it is helpful to monitor the potential allocation of units as the search progresses. Units for which allocation remain stable can be certainly allocated. But sometime we can find that for few units it is impossible to cluster with certainty. We end the confirmatory analysis by monitoring the variation of the allocation of the units during the last steps of search.

Two different type of allocations can be identified. The certain allocation for which the certain lines include all units which have the same allocation, with only occasional changes, throughout the search. The uncertain allocation presents uncertain lines for which the allocation changed appreciably during the search. To make an example of our study, we show a plot only for year 2003. In this case we found that units 32 and 39 still have uncertain allocation, and probably they do not belong to any cluster identified; hence, these units can be classified as uncertain. While the variations we can observe for units 24, and 27 of year 2003 are not so relevant, therefore their allocations are definitely confirmed into the clusters the previous k-means did.

Finally, the analysis has been carried out also when outliers are included in the data set. As regards the membership of these units, surprisingly to what one could expect, they have certain allocation, also in year 2003 where they seem to show some occasional changes but not so relevant to be considered uncertain. This means that the allocation made by clustering approach is uncorrelated to the fact that these particular units affect (more or less) the number of clusters. These units present, for one or more indicators, some features so different to alter the the Calinski Harabasz index.

Acknowledgements We are grateful to National University Evaluation Committee of Ministry of Research for giving us the data set used in this work.

References

Atkinson, A. C., & Riani, M. (2000). *Robust diagnostic regression analysis*. New York: Springer

Atkinson, A. C., Riani, M., & Cerioli, A. (2004). *Exploring multivariate data with the forward search*. New York: Springer

Biggeri, L., & Bini, M. (2001). Evaluation at university and state level in Italy: Need for a system of evaluation and indicators. *Tertiary Education and Management*, *7*, 149–162

Milligan, G. W., & Cooper, M. C. (1985). An examination of procedures for determining the number of clusters in a data set. *Psycometrika*, *50*, 159–179

Riani, M., & Atkinson, A. C. (2001). A unified approach to outliers, influence, and transformations in discriminant analysis. *Journal of Computational and Graphical Statistics*, *60*, 87–100

Weighted Likelihood Inference for a Mixed Regressive Spatial Autoregressive Model

Carlo Gaetan and Luca Greco

Abstract As well as in the case of independence and by paralleling, in some sense, what happens in time series analysis, in spatial linear models the presence of anomalous observations can badly affect likelihood based inference, both on the significance of any large scale parameter and the strength of the spatial dependence. In this paper we look for a valuable robust procedure which, on the one hand, allows us to take into account possible departures of the data from the specified model, and on the other hand, can help in identifying spatial outliers. This procedure is based on weighted likelihood methodology. The effectiveness of the proposed procedure is illustrated through a small simulation study and a real data example.

1 Introduction

Let $Y(s_i)$ be a Gaussian random variable measured on different spatial locations $s_i \in S \subset \mathbb{R}^d$, $i = 1, \ldots, n$. As an illustrative example of our proposal, we focus on the mixed regressive spatial autoregressive model introduced by Cliff and Ord (1981):

$$Y(s_i) = X(s_i)^{\top}\beta + \rho \sum_{j \neq i} w_{ij} Y(s_j) + \sigma\varepsilon(s_i)\,, \qquad \sigma > 0\,, \tag{1}$$

$X(s_i)$ is a p-dimensional vector of covariates and $\varepsilon(s_i) \sim \mathrm{N}(0, 1)$, $\mathrm{Cov}(\varepsilon_i, \varepsilon_j) = 0$, $i \neq j$. The spatial relationships between sites are specified on the basis of a known connectivity matrix $n \times n$ matrix $W = [w_{ij}]$, where the diagonal elements are set to zeros. For instance, the entries $w_{i,j}$ may be taken to be a certain function of some distance, deemed to be relevant for the phenomenon under study, between the s_i and the s_j sites. The unknown vector parameter is $\theta = (\beta^{\top}, \sigma, \rho)^{\top}$ and the likelihood

L. Greco (✉)
Department PE.ME.IS. – Section of Statistics, University of Sannio, Benevento, Italy
e-mail: luca.greco@unisannio.it

F. Palumbo et al. (eds.), *Data Analysis and Classification*,
Studies in Classification, Data Analysis, and Knowledge Organization,
DOI 10.1007/978-3-642-03739-9_46,

function can be easily derived provided that the matrix $(I_n - \rho W)$ is invertible, where I_n is the $n \times n$ identity matrix.

This model plays an important role in spatial econometrics, see for instance Anselin (1988), and the proliferation of applications has been accompanied by contributions to a rigorous theory of inference for this model (see Lee (2004, 2007) for recent contributes). Here we consider an inferential aspect undertaken.

As well as in the case of independence or in time series, even in the analysis of spatial data the presence of anomalous observations can badly affect likelihood based inference, both on the significance of any trend parameter and the strength of the spatial dependence. In particular, the underlying dependence structure may be drastically altered. The presence of outliers in spatial data can give rise to problems hard to handle, basically because they can be of different types, that never come alone. Spatial outliers are observations which are not consistent with respect to the surrounding neighbors, even though they may not be significantly different from the rest of the data. They may be isolated or grouped and it may reveal difficult to detect them since they are prone to masking effects. They might be erroneous measurements or true values measured in different conditions.

Therefore, there is the need of a valuable robust procedure which, on the one hand, allows us to take into account possible departures of the data from the specified model and on the other hand can help in identifying anomalous situations. Methodologies are available to compute robust estimates of β and σ (Militino 1997; Militino and Ugarte 1997), whereas the problem of robustly estimating ρ and testing hypothesis on its value has not been considered so far. In this paper, we propose to apply the weighted likelihood based algorithms for linear regression models with independent errors (Agostinelli and Markatou 2001) to spatial linear models. In particular, it appears relevant the possibility of making robust inference on the dependence parameter ρ simultaneously with the coefficient vector β. An attempt in this direction has been made by a forward search methodology (Cerioli and Riani 2003).

Weighted likelihood is presented in Sect. 2; the finite sample performance of weighted likelihood estimators is investigated by a numerical study in Sect. 3 and through an application to real data in Sect. 4; final remarks are given in Sect. 5.

2 A Weighted Likelihood Approach

One single outlier can give rise to a dramatic bias in the estimation of ρ and badly affect testing procedures. Basically, this is due to the fact that the effect of a single anomalous value spreads across its neighbor system. Actually, in the four nearest neighbors structure, a single outlier at location s_j spoils five residuals: ε_j and those corresponding to its four neighbors. In view of this, we look for a robust procedure leading to resistant inference which allows us to put a bound, automatically, on the influence of a single outlying observation, both when it plays the role of response and when it appears as a neighbor. To this end we focus on weighted likelihood.

This robust procedure is based on a suitable reweighting of the likelihood equations. The method is designed to achieve optimal model efficiency and to provide robust estimators and tests under departures from the model assumptions as well as a diagnostic tool to automatically detect anomalous observations.

In order to find the weighted likelihood estimate (WLE) of the vector parameter $\theta = (\beta, \sigma, \rho)$, by paralleling standard maximum likelihood estimation, we follow a profile approach. For a fixed value of ρ, the data are transformed using a spatial filter,

$$Y_\rho^*(s_i) = Y(s_i) - \rho \sum_{j \neq i} w_{ij} Y(s_j).$$

Given any observed value at location s_i, $y(s_i)$, we construct a weight function $h_\rho(s_i) = h(r_\rho^*(s_i); F_\theta, \hat{F}_n)$, that depends on the constrained residual $r_\rho^*(s_i) = y_\rho^*(s_i) - X(s_i)^\top \beta$, the assumed theoretical model F_θ, with ρ held fixed, and the empirical distribution function $\hat{F}_n$. The weight function $h(u; F_\theta, \hat{F}_n)$ is defined as

$$h(u; F_\theta, \hat{F}_n) = \min(1, A[\delta(u; F_\theta, \hat{F}_n) + 1]^+ / \delta(u; F_\theta, \hat{F}_n) + 1) ,$$

with $[\,]^+$ denoting the positive part. The quantity $\delta(u; F_\theta, \hat{F}_n) = \hat{f}^*(u)/f^*(u) - 1$ is the so-called Pearson residual. This is based on the comparison between a non-parametric kernel density estimate $\hat{f}^*(u)$ and a smoothed model density $f^*(u)$, obtained by applying the same smoothing to the assumed model. The function $A(\cdot)$ is the residual adjustment function introduced by Basu and Lindsay (1994) in the context of minimum disparity estimation. For an extensive discussion on the adopted weighting scheme see Markatou et al. (1998) and references therein. The weight function takes values in the interval [0, 1]: weights near 0 identify points not conform with the model, while weights tend to unity according to the degree of accordance between the theoretical model and $\hat{F}_n$. The weights $h_\rho(s_i)$ are used to compute constrained WLEs $\tilde{\beta}_\rho, \tilde{\sigma}_\rho^2$, that are the solution to the system of estimating equations

$$\sum_{i=1}^{n} h_\rho(s_i) \ell_\beta(r^*(s_i); \theta) = 0; \qquad \sum_{i=1}^{n} h_\rho(s_i) \ell_{\sigma^2}(r^*(s_i); \theta) = 0 \tag{2}$$

where $\ell_\beta(\cdot)$ and $\ell_{\sigma^2}(\cdot)$ are the score functions under model (1) corresponding to β and σ^2, respectively. The solution of (2) can be found by an iterative reweighting scheme. Anyway, this solution may not be unique. This happens especially in those cases in which data deviate much from the model. In our subsequent numerical studies and applications, we stated to search two roots, according to the algorithm based on bootstrap resampling discussed in Markatou et al. (1998), and to select the root giving the minimum value of the scale.

Afterward, the estimates $\tilde{\beta}_\rho$ and $\tilde{\sigma}_\rho^2$ are used to find a robust estimate $\tilde{\rho}$ of ρ, by maximizing the following objective function:

$$\tilde{\ell}_P(\rho) = \ell(\tilde{\beta}_\rho, \tilde{\sigma}_\rho, \rho) = -\frac{n}{2} \log \tilde{\sigma}_\rho^2 + \log |I_n - \rho W| . \tag{3}$$

The function (3) is obtained by replacing the constrained MLEs of β and σ^2, with the corresponding constrained WLEs in the full log-likelihood associated to model (1). Therefore, it can be thought as a generalized profile likelihood, as defined in Severini (1998), since $\tilde{\beta}_\rho$ and $\tilde{\sigma}^2_\rho$ are consistent estimators, when ρ is known.

Finally, the WLEs for β and σ^2 are obtained by solving (2) for $\rho = \tilde{\rho}$ and the final fitted unconstrained weights $h_{\tilde{\rho}}(s_i)$ indicates which points are more down-weighted. As a rule of thumb, the less the weight attached to the observation the more that value might be anomalous and likely to badly influence standard inferential conclusions.

The generalized profile log-likelihood (3) shares the same first-order properties of the ordinary profile log-likelihood and can be used for inference about ρ in a standard fashion. In particular, one can set confidence intervals for ρ and test hypothesis on its value by using a statistic with the standard asymptotic behavior, namely

$$\tilde{W}_{\mathrm{P}}(\rho) = 2\left\{\tilde{\ell}_{\mathrm{P}}(\tilde{\rho}) - \tilde{\ell}_{\mathrm{P}}(\rho)\right\}. \tag{4}$$

It is clear that robust estimation of the spatial dependence parameter depends heavily on the availability of a robust estimate of scale. The robustness of $\tilde{\rho}$ descends from the structure of

$$\tilde{\sigma}^2_\rho = n^{-1}\tilde{r}^*_\rho(s)^{\mathsf{T}}\tilde{H}^{\mathsf{T}}_\rho \tilde{H}_\rho \tilde{r}^*_\rho(s),$$

with $\tilde{r}^*_\rho(s) = y^*(s_i) - X(s_i)^{\mathsf{T}}\tilde{\beta}_\rho$ and $\tilde{H}_\rho = \mathrm{diag}[\tilde{h}_\rho(s_i)]$, with $\tilde{h}_\rho(s_i) = h(\tilde{r}^*_\rho(s_i); F_\theta, \hat{F}_n)$. Actually, the weights $\tilde{h}_\rho(s_i)$ avoid large residuals to inflate the constrained estimate $\tilde{\sigma}^2_\rho$. Anyway, we can expect only a limited degree of robustness, for the reasons outlined at the beginning of this section. Furthermore, it is worth noting that the degree of robustness depends not only on the number of anomalous values in the sample but also on the structure of the neighborhood system: the highest the number of neighbor sites the largest the bias due to one anomalous value.

3 A Small Simulation Study

By paralleling the framework of time series (see Maronna et al. (2006), Chap 8 for a survey), we can consider different probability models for spatial outliers. In this paper, we focus on the additive outliers model. Under this model we observe a contaminated version $Y(s)$ of the underlying true value $Z(s)$, i.e.

$$Y(s) = Z(s) + \eta(s), \tag{5}$$

where the processes $Z(s)$ and $\eta(s)$ are independent. For instance, we can assume

$$\eta(s) = (1-\pi)\delta_0 + \pi \mathrm{N}(\mu_\eta, \sigma_\eta),$$

where δ_0 is a point mass distribution in zero and π is the probability of occurrence of one outlier in the sample. This model will generate isolated outliers when $\eta(s)$ is an independent and identically distributed process with location or scale or both much larger than that of $Z(s)$. According to the same model, grouped outliers can be obtained by assuming that the anomalous patch refers not only to one single point but to its neighborhood in a certain subregion of S.

A small simulation study, based on 1,000 Monte Carlo trials, was run to asses the finite sample properties of the weighted likelihood estimators of the parameters of model (1), both when the specified model and an additive effects model (5) holds. We generated data on a 10×10 square lattice according to model (1), setting $\beta = (1, 1, 1)^{\top}$, $\rho = 0.5$ and $\sigma = 1$. The vector $X(s)$ was generated from a multivariate normal distribution with independent unit variance components and mean vector $(4, -2, 0)^{\top}$. Spatial relationships between pairs of locations have been represented by a row standardized binary connectivity matrix in which each site has four neighbors, according to the classical nearest neighbors definition. Moreover, the matrix was embedded onto a torus to take into account edge effects. Two outliers' configurations were considered: the first in which 4 outliers are dispersed on the grid (c1) and the second in which 9 outliers form a 3×3 cluster in the top left corner (c2). In both cases, additive outliers were generated by a $N(\mu_\eta, 1)$ with $\mu_\eta = 10$ (a1) or $\mu_\eta = 50$ (a2): in the former scenario outliers take values different from the neighbors but not necessary from the rest of the data, while in the latter we obtain values larger then any other on the grid. Results are given in Table 1. Under the true model, the WLEs behaves well at the cost of very little efficiency loss. In the presence of outliers, the mean bias and the rmse of $\tilde{\beta}$ are restrained in all cases, whereas the bias of $\tilde{\rho}$ is small under case (a2) but non negligible in cases (a1), even if still better than that of the MLE. According to empirical evidence, when two different solutions of (2) were found, one moved in the direction of the MLE, whereas the other was the robust one agreeing only with the bulk of the data. On the contrary, the effect of additive outliers on MLEs $(\hat{\beta}, \hat{\sigma}, \hat{\rho})$ is quite evident;

Table 1 MLEs and WLEs (with rmse) for model (1) under different scenario

	TRUE		c1				c2			
			a1		a2		a1		a2	
	MLE	WLE	MLE	WLE	MLE	WLE	MLE	WLE	MLE	WLE
β_1	1.004	1.005	1.245	1.098	1.736	0.963	0.787	1.111	0.680	0.957
	(0.075)	(0.087)	(0.276)	(0.133)	(0.884)	(0.108)	(0.222)	(0.138)	(0.324)	(0.081)
β_2	0.989	0.988	0.830	0.920	0.516	1.020	1.086	0.905	0.980	1.023
	(0.097)	(0.102)	(0.258)	(0.136)	(1.052)	(0.124)	(0.129)	(0.139)	(0.100)	(0.107)
β_3	1.007	1.008	1.072	1.045	1.108	1.000	0.828	1.038	0.428	0.987
	(0.105)	(0.106)	(0.267)	(0.124)	(1.096)	(0.126)	(0.204)	(0.116)	(0.582)	(0.117)
σ	0.993	0.974	2.231	0.985	9.728	1.006	1.942	1.039	7.488	0.963
	(0.074)	(0.092)	(1.240)	(0.079)	(8.731)	(0.198)	(0.949)	(0.108)	(6.489)	(0.193)
ρ	0.491	0.490	0.255	0.340	0.018	0.542	0.797	0.329	0.920	0.552
	(0.073)	(0.090)	(0.267)	(0.184)	(0.499)	(0.110)	(0.299)	(0.186)	(0.420)	(0.074)
$\overline{O}$		0		4		18		9		21

in particular $\hat{\sigma}$ is seriously inflated when outliers are global, whereas $\hat{\rho}$ is clearly not stable among all considered scenarios. The MLE $\hat{\rho}$ is very biased downward in case (c1) (the bias is dramatic in the presence of massive outliers), while it tends to get larger when outliers are clustered (c2). In particular, even if not reported in Table 1, the presence of massive (a2) random (c1) outliers make the LRT not to reject the hypothesis of independence on average, so resulting in very misleading inference; this does not happen by using (4). The last entries $\overline{O}$ give the mean number of weights ≤ 0.1, which are supposed to identify anomalous observations. When additive effects are of type (a1), each outlier is detected, whereas in case (a2) the downweighting also involves neighboring sites, meaning that all the neighborhood is considered as anomalous.

4 A Real Example

We consider the data set on 49 neighborhoods in Columbus (Ohio) in 1980 reported in Anselin (1988). This data set includes observations of the residential burglaries and vehicle thefts per thousand households (CRIME), household income (INC) and housing value (HOVAL) in each neighborhood. Spatial relationships between sites are summarized by a row standardized connectivity matrix. The spatial distribution

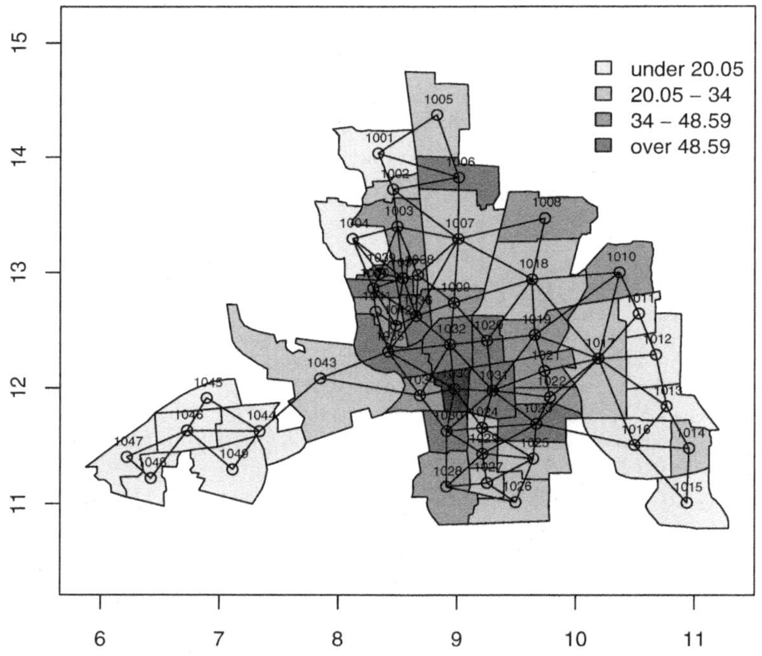

Fig. 1 49 Columbus neighborhoods with contiguities overimposed

Table 2 Columbus data: estimates of the parameters of model (6)

	MLE	WLE	Skew-t
β_0	45.079 (0.000)	44.943(0.000)	42.427 (0.000)
β_1	−1.032 (0.001)	−1.562 (0.000)	−1.456 (0.000)
β_2	−0.266 (0.002)	−0.055 (0.463)	−0.066 (0.380)
σ	9.974	7.259	7.041
ρ	0.431 (0.002)	0.441 (0.000)	0.574 (0.000)

of the variable `CRIME` is plotted in Fig. 1, where the spatial links are overimposed. In his analysis (Anselin 1988, p. 193) gives evidence that a mixed spatial autoregressive model is appropriate in modeling linear relationship between `CRIME` and the other variables i.e.

$$\mathtt{CRIME}(s_i) = \beta_0 + \beta_1 \mathtt{INC}(s_i) + \beta_2 \mathtt{HOVAL} + \rho \sum_{j \neq i} w_{ij} \mathtt{CRIME}(s_j) + \sigma \varepsilon(s_i)\,. \quad (6)$$

We compared likelihood and weighted likelihood estimation, aiming at discovering eventual spatial outliers and their effects on fitting. Moreover we performed a likelihood analysis by assuming that the error components $\varepsilon(s_i)$ of model (6) have a Skew-t distribution (Azzalini and Capitanio 2003). The Skew-t distribution has been suggested as a valid alternative to robust methods when dealing with possible anomalous features of the data (Azzalini and Genton 2007). In fact, the direct interpretation of its parameters, especially those controlling asymmetry and tail thickness, can give more information about the direction of the departures of the data from the central normal model. Entries in Table 2 give MLEs and WLEs under the assumption of normality and the MLEs under the Skew-t distribution. P-values are reported between parenthesis. The weighted likelihood based procedure identifies neighborhood `1004` as anomalous, giving a weight close to zero. Actually, an inspection of Fig. 1 shows that neighborhood `1004` is connected with sites in which the registered percentage of crimes is markably different. When comparing WLEs and MLEs, we note that the main effect of downweighting is that the estimate of the regression coefficient β_2 is no longer significant. Moreover, we obtain a lower WLE for σ than the MLE, whereas the estimate of ρ does not change indeed. The fitted model under the Skew-t distribution supports the robust analysis. Actually, the small estimate of the degrees of freedom $\hat{\nu} = 3.683(1.825)$ reflects heavy tails in the error distribution due to the single outlier. Furthermore, the estimate of the skewness parameter is $\hat{\alpha} = -0.596$ with a standard error 0.081, hence it is not significant.

5 Final Remarks

In this paper we stress the need of robust procedures for inference in spatial linear models. The weighted likelihood methodology provides a reliable first answer to the problem even if the propagation of additive effects over the lattice gives place to serious questions about the degree of robustness that can be achieved.

The properties of the WL estimator in the context of model (1) deserve further investigation. In particular, its consistency is conjectured under the assumptions outlined in Markatou et al. (1998) and Severini (1998) but it should be studied more in details.

The same algorithm described for model (1) can be extended to other spatial models, such as the simultaneous autoregressive model (SAR). In this case, spatial dependence is present in the error process and the data can be transformed by using a different spatial filter. It is worth noting that the pure SAR model is obtained when $\beta = 0$ in (1).

Other robust proposals, as those based on M-, GM-, MM-estimators, may be considered as an alternative to WL-estimators. As well as for WL-estimation, simple and highly efficient algorithms are automatically available. Furthermore, computational time is almost the same as that requested by maximum likelihood, for both WL and the other robust methods.

Acknowledgements The authors wish to thank Claudio Agostinelli for helpful discussion.

References

Agostinelli, C., & Markatou, M. (2001). Test of hypothesis based on the weighted likelihood methodology. *Statistica Sinica, 11*, 499–514

Anselin, L. (1988). *Spatial econometrics: Methods and models*. Boston: Kluwer

Azzalini, A., & Capitanio, A. (2003). Distributions generated by perturbation of symmetry with emphasis on a multivariate skew t distribution. *Journal of the Royal Statistical Society – Series B, 65*, 367–38

Azzalini, A., & Genton, M. G. (2007). Robust likelihood methods based on the skew-t and related distributions. *International Statistical Review, 65*, 367–389

Basu, A., & Lindsay, B. G. (1994). Minimum disparity estimation for continuos models: Efficiency, distribution and robustness. *Annals of the Institute of Statistical Mathematics, 46*, 683–705

Cerioli, A., & Riani, M. (2003). Robust methods for the analysis of spatially autocorrelated data. *Statistical Methods and Applications, 11*, 335–358

Cliff, A. D., & Ord, J. K. (1981). *Spatial process-models and applications*. London: Pion

Lee, L. (2004). Asymptotic distributions of quasi-maximum likelihood estimators for spatial autoregressive models. *Econometrica, 72*, 1899–1925

Lee, L. (2007). GMM and 2SLS estimation of mixed regressive, spatial autoregressive models. *Journal of Econometrics, 137*, 489–514

Markatou, M., Basu, A., & Lindsay, B. G. (1998). Weighted likelihood equations with bootstrap root search. *Journal of the American Statistical Association, 93*, 740–750

Maronna, A. R., Martin, R. D., & Yohai, V. J. (2006). *Robust statistics: Theory and methods*. Chichester: Wiley

Militino, A. F. (1997). M-estimator of the drift coefficients in a spatial linear model. *Mathematical Geology, 29*, 221–229

Militino, A. F., & Ugarte, M. (1997). A GM estimation of the location parameters in a spatial linear model. *Communications in Statistics: Theory and Methods, 26*, 1701–1725

Severini, T. A. (1998). Likelihood functions for inference in the presence of a nuisance parameter. *Biometrika, 85*, 507–522

Detecting Price Outliers in European Trade Data with the Forward Search

Domenico Perrotta and Francesca Torti

Abstract We describe empirical work in the domain of clustering and outlier detection, for the analysis of European trade data. It is our first attempt to evaluate benefits and limitations of the *forward search* approach for regression and multivariate analysis Atkinson and Riani (Robust diagnostic regression analysis, Springer, 2000), Atkinson et al. (Exploring multivariate data with the forward search, Springer, 2004), within a concrete application scenario and in relation to a comparable *backward* method developed in the JRC by Arsenis et al. (Price outliers in eu external trade data, Enlargement and Integration Workshop 2005, 2005). Our findings suggest that the automatic clustering based on Mahalanobis distances may be inappropriate in presence of a high-density area in the dataset. Follow up work is discussed extensively in Riani et al. (Fitting mixtures of regression lines with the forward search, Mining massive data sets for security, IOS, 2008).

1 Introduction

In this paper we describe clustering and outlier detection problems in the analysis of European trade data. We introduce with an example the application context, the available datasets and two well specified tasks or statistical patterns. We try an heuristic comparison between a solution based on the forward search (FS) Atkinson and Riani (2000), Atkinson et al. (2004) and a backward approach in use in the JRC Arsenis et al. (2005). So far the backward solution was used to treat bivariate datasets without major masking issues, affected by one or few outliers. In the conclusions we interpret the practical results obtained on the specific data, which are operationally relevant. The main contribution is in Sect. 5, where we show that the automatic clustering procedure based on Mahalanobis distances proposed by Atkinson et al. (2006) may be inappropriate when the populations of interest give

F. Torti (✉)
Università Milano Bicocca, Facoltà di Statistica, Milano, Italy
e-mail: f.torti1@campus.unimib.it,francesca.torti@jrc.it

F. Palumbo et al. (eds.), *Data Analysis and Classification*,
Studies in Classification, Data Analysis, and Knowledge Organization,

DOI 10.1007/978-3-642-03739-9_47,

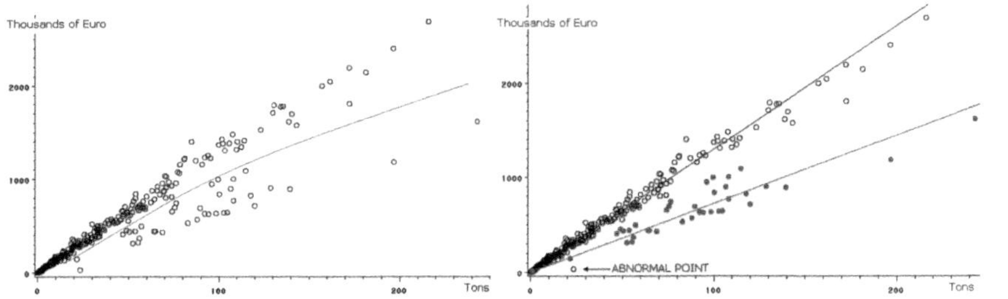

Fig. 1 Quantities and values of 677 monthly imports of a fishery product from a third country into the EU, in a period of 3 years. On the *left* the data are unclassified but two groups are visible. On the *right* the data are classified: flows to MS1 (*solid dots*) and flows to the other Member States (*black circles*) form distinct groups following different regression *lines*

rise to data with highly-dense areas. This opens new research issues that we are currently addressing in the regression context Riani et al. (2008).

2 Application Context, Data and Statistical Patterns

The data in the left plot of Fig. 1 are unclassified, i.e. there is no information on the division of the observations into categories. However we can recognise by eye that two groups of observations, separated by a curve in the plot, follow different linear distributions. In the plot on the right we use a variable present in the data to classify the observations in two categories. Then we fit two linear regression models using the observations in the two groups. There is an observation, the black circle on the bottom-left part of the plot, that does not follow the regression line fitted to the observations of the same category. Rather, it appears in the distribution of the solid dots. Is this just a chance occurrence in the group of the black circles? Is this observation classified in the wrong category? Is something unusual going on?

The data in this example refer to the quantity (x axis) and the value (y axis) of the monthly import flows of a fishery product into the EU from a third country. The solid dots are the flows to a given Member State, say MS1, and the black circles are the flows to the other Member States. The "abnormal" black circle is a single flow to a Member State that we identify with MS2. The *unit value* of this flow, obtained by dividing the value by the corresponding quantity, is so small ($\sim$1.27 €/Kg) compared to the market price of this fishery product (12.5 €/Kg in 2005) that we may suspect a data recording error. To investigate a trade flow of such volume ($\sim$20 Tons) may be scarsely relevant considering its economical value. On the contrary, the distribution of the solid dots indicates that the imports of the MS1 are systematically underpriced in comparison with the imports of the other Member States, which economically is very relevant considering that in this reference period the MS1 has imported about the 20% ($\sim$3.300 Tons) of the total EU imports of this product.

This example has introduced two patterns, *outliers* and *mixtures of linear models*, which can be used to reveal in trade data anomalies of various nature (e.g. recording errors), specific market price dynamics (e.g. discounts for big trading quantities) and cases of unfair competition or fraud. Outliers are among the statistical patterns that the JRC detects in trade data and presents for evaluation and feedback to subject matter experts of other services of the European Commission and of the Member States. Tools to identify and fit groups of observations with mixtures of linear models have been also explored Riani et al. (2008), with emphasis on *automatic* and efficient procedures. In fact Fig. 1 plots data taken from a dataset including over a million observations grouped in thousands of small to moderate size samples, that must be treated mechanically to come up with a reduced set of statistically relevant cases. The dataset was extracted from COMEXT, a EUROSTAT database which contains foreign trade data as reported by the Member States of the European Union.

3 Application of the Forward Search

We have analysed the above example with several standard clustering methods, with results which were dissimilar or difficult to interpret. Here, we concentrate on results obtained with the FS using functions in the R/Splus libraries `forward` and `Rfwdmv` implemented by Riani, Corbellini and Konis (`fwdmv` runs the FS).

The method starts from a subset of data free of outliers (selected using robust methods, e.g. least median of squares) and fits subsets of increasing size with a search that at each step tests the "outlyingness" of the remaining observations relative to the model fit on the current subset. The method orders the observations by closeness to the assumed model and possible outliers are included in the last steps of the search. The FS can be used for clustering data following the idea that the observations in a cluster are outliers of the model fitted to the observations of a different cluster. A natural diagnostic tool to test the outlyingness of one or more observations from a cluster (relying on multivariate normality) is the Mahalanobis distance (MD). If at a given step of the search the mean of the subset $S(m)$ of m observations is $\hat{\mu}(m)$ and the covariance matrix estimate is $\hat{\Sigma}(m)$, then the squared MD of the observations in the dataset are

$$d_i^2(m) = \{y_i - \hat{\mu}(m)\}^T \cdot \hat{\Sigma}^{-1}(m) \cdot \{y_i - \hat{\mu}(m)\} \qquad (i = 1, \ldots, n) \qquad (1)$$

and the $m + 1$ observations with smallest distances form the new subset $S(m + 1)$.

We have first treated all observations assuming no clusters, i.e. a single population. We run the FS starting from various initial subsets chosen within robustly centered ellipses, or within robust bi-variate boxplots, and sometimes by selecting manually either flows of the MS1 or flows from the other Member States. We monitored the search by plotting the progress of the MD of the n observations, scaled

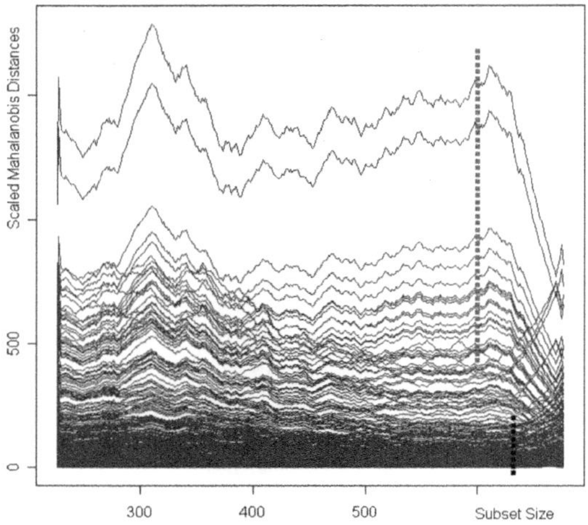

Fig. 2 The forward plot of the scaled Mahalanobis distances of the observations in the dataset. In evidence, with *vertical dashed lines*, two groups of MD curves: one of rather stable and small MD values that slightly increase at the end, the other of higher and fluctuating MD values that decrease considerably in the last part of the search

by the square root of the determinant of the covariance matrix at the end of the search:

$$d_i(m) \times \left(\left| \hat{\Sigma}(m) \right| / \left| \hat{\Sigma}(n) \right| \right)^{\frac{1}{4}} \qquad (i = 1, \ldots, n \quad m = 2, \ldots, n) \qquad (2)$$

The scaling is to give more emphasis on the last part of the search, when the MD of outlying observations would drastically decrease and the structure of the data would be more difficult to appreciate. Independently from the initial subset the *forward plots* of the scaled MD revealed two clear tentative groups, which we have roughly identified in Fig. 2 with two vertical dashed lines. The curves in the upper group, which correspond to the most outlying observations, show high fluctuating MD values which decrease considerably at the end of the search. The lower group is formed by a very dense set of low MD values, which slightly increase at the end when the outlying observations enter in the subset. The two tentative groups, identified by black circles and solid dots in the scatter plot of Fig. 3, were consisting of 641 and 21 observations respectively. In almost all runs the observations in the smaller group were identified in correspondence of the last steps of the search. The remaining 15 observations, identified by "×" symbols, were of more difficult allocation. Not surprisingly, they are located between the two tentative groups.

At this point, assuming the two populations, we run again the FS to fit the observations to the two tentative groups. In this phase the uncertain observations are assigned by the FS to the closest group in terms of Mahalanobis distance. can be easily interpret using the plot of the function `fwdmvConfirmPlot`:

- In the final part of the search (say, the last 200 steps) the 15 observations so far unassigned are systematically closer to the group of the solid dots.
- However of these 15 unassigned observations, four alternate at some point from one group to the other, but only slightly and between steps ~500 and 600. These

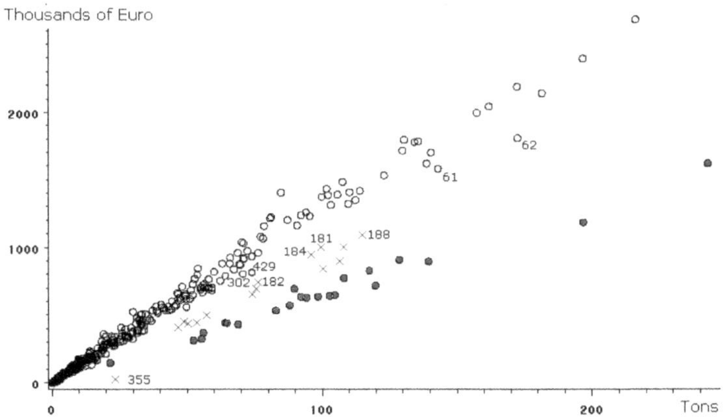

Fig. 3 Two tentative groups (*black circles and solid dots*) and a set of unassigned observations ("×" symbols), selected on the basis of the forward plots. Some relevant observations discussed in the text have been labelled by their position in the dataset

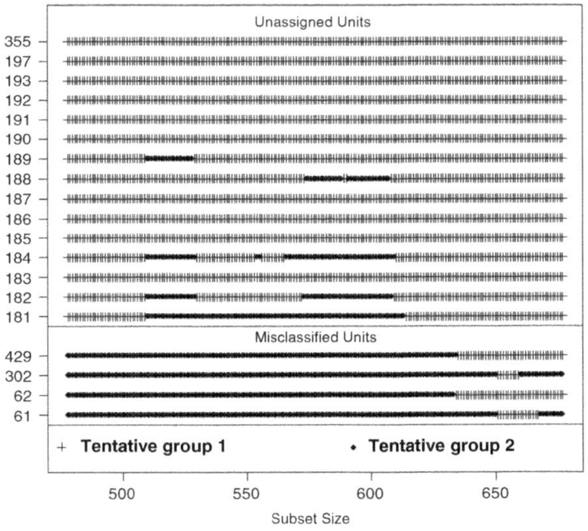

Fig. 4 The confirmation plot for the last 200 steps, based on the two tentative groups. The 15 unassigned observations are allocated to the tentative group 1 (the *solid dots* in Fig. 3). Some uncertainty remains for flows 181, 182, 184, 188, 189. The attribution of observations 62, 429, 61 and 302 to the tentative group 2 (the *black circles* in Fig. 3), deserves some attention

slightly ambiguous cases are the records number 181, 182, 184 and 188 of the dataset (see Fig. 4). According to the FS plot, the most uncertain case is 181. Figure 3 shows the position of these observations in the scatter plot.

- Four observations of the big tentative group show up as misclassified in the last 30 steps of the search. These are the records number 62, 429, 61 and 302, again in order of uncertainty following the FS plot.

4 Heuristic Comparison with the "Backward" Outliers

The FS suggests merging the small tentative group (solid dots) with the 15 unassigned observations ("×" symbols), with some attention to the slightly ambiguous case 181. The four observations that the FS hesitate to keep in the big tentative group (the "misclassified" 62, 429, 61 and 302) need more consideration. We have verified if the outliers detected with our backward method Arsenis et al. (2005) are consistent with these first conclusions. The method starts from a regression model fitted to the data and proceeds backward with deletion procedures associated with regression diagnostics. The statistic that we test to verify the agreement of an observation with the regression model fitted on the remaining observations is the *deletion residual*[1]. The diagnostic tests were made at 10% significance level, corrected with Bonferroni to account for multiple comparisons. We used the Cook's distance to assess the influence of an observation on the regression parameter estimates.

- All 15 unassigned observations were detected as low price outliers when added to the big tentative group (black circles). Consistently with the FS, the 4 misclassified observations were also detected as outliers.
- On the contrary, no outlier was detected by merging the 15 unassigned observations with the small tentative group (solid dots). Still no outliers if we also add the four misclassified observations, although they show up as the most critical in terms of P-values of the deletion residuals (i.e. they deviate more than the others from the regression model), followed by the slightly ambiguous case 181.
- The observation n. 355 was detected as an extreme low price outlier of the set of flows of the MS2. Note that this observation was assigned to MS1 by the confirmation phase of the FS (Fig. 4), but under the tentative hypothesis of two sub-populations of data.

5 Towards an Automatic Procedure

So far we have discussed a tentative clustering obtained by visual inspection of the trajectories of the MD in the forward plot of Fig. 2. Following Atkinson et al. (2006) we also tried to infer the clusters automatically on the basis of the distribution of the minimum MD computed at each step of the FS among the observations that are outside the current subset. The idea is that when the current subset corresponds to a cluster, the minimum MD of observations outside the subset in a different cluster will be large, and will decrease again as such observations join the subset. Thus,

[1] Unfortunately in the literature there are different terms for this form of standardised residual. Cook and Weisberg (1982) use "externally studentized residual" in contrast to "internally studentized residual" when the context refer to both forms of standardisations, with the current observation deleted or not. Belsley et al. (1980) use "studentized residual" or "RSTUDENT". The terms "deletion residual" or "jackknife residual" are preferred by Atkinson and Riani (2000).

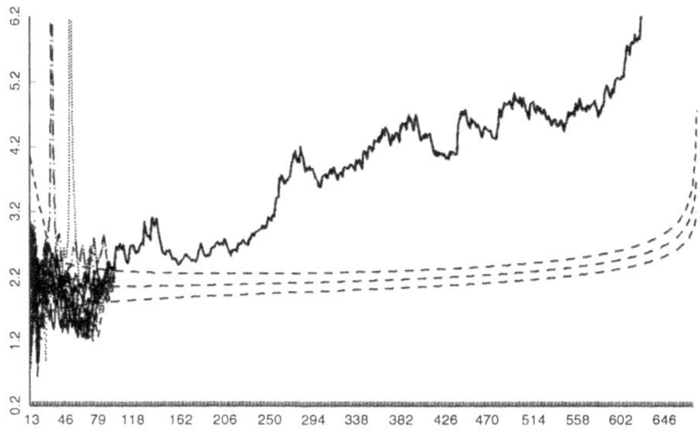

Fig. 5 The forward plot of the minimum Mahalanobis distance among observations not in the subset, for many forward searches starting from different initial subsets. Independently from the start, the distances degenerate to a unique search path which departs from the envelops already in the first steps of the search

significant minimum MD values may indicate the presence of clusters or correspond to isolated outliers.

Atkinson, Riani and Cerioli described how to assess the significance of extreme minimum MD values using envelopes for the minimum MD values distribution. The exact distribution and envelopes are hard to derive analytically, being the sequence of minimum MD values not independent. However the envelopes can be simulated by running N times the FS on data generated from a normal distribution and by plotting, for each subset size m, the desired quantiles (e.g. the 5 and 95%) of the set of the N minimum MD. Atkinson, Riani and Cerioli also proposed computationally convenient approximations for the envelopes. In the tests with our dataset we used simulated as well as approximated envelopes.

Unfortunately, in practice we obtained minimum MD curves of difficult interpretation and which do not reflect the tentative clustering that was determined rather naturally in Sect. 3. The forward plot of Fig. 5 superimposes the minimum MD curves obtained by running several hundreds forward searches from different initial subsets: the curves depart from the envelopes in the very first steps of the search and degenerate rather early (after about eighty steps) to the same search path. This behaviour can be explained by two concomitant factors. First, most likely the structure of the data is more complex than the two or three normal clusters that we argued in Sect. 3 on the basis of the MD plot and the confirmation plot. Second, independently from the choice of the initial subset the FS falls in the very dense area of observations that is visible near the origin of the scatter plot (Fig. 3) and remains confined until all observations in that area are included. Sub-populations generating observations in the dense area or spannig over the dense area even partially, cannot be detected using plots of minimum MD.

An approach to circumvent the problem may be to trim the dense area and repeat the analysis on the remaining observations and the dense area separately. We are experimenting several trimming possibilities, including ellipsoidal trimming and various types of convex hulls.

However so far we obtained the best results outside the multivariate approach based on MD, by exploiting the regression structure in the dataset that is well visible in the scatter plot of Fig. 3. Instead of (1) and (2), this approach uses the squared regression residuals for progressing in the search and the minimum deletion residual among the observations not in the subset to monitor the search and infer departures from linearity. In fact, using a method which applies iteratively this approach, we could detect in the dataset five linear mixture components of rather clear interpretation Riani et al. (2008).

6 Discussion and Main Conclusions

The forward search has identified accurately the group of import flows by MS1 with a cluster, which is however contaminated by the observation n. 355. This being a clear low price outlier of the flows into MS2, the case deserves further consideration. The cases of difficult attribution between the two tentative clusters have limited practical relevance and we have not treated them as a third distinct cluster.

We have verified in our dataset that the unassigned "×" observations (all except the extravagant 355) correspond to import flows that took place in the first 14 consecutive months of the period analysed. The estimated unit price for this group of flows is 9.17 €/Kg, while the import flows in the group of the solid dots took place in the following 21 months with an estimated unit price of 6.55 €/Kg. Note that the estimated unit price for the group of the imports in the other Member States (black circles) is of 13 €/Kg. In short, in the period analysed the MS1 gradually lowered the import price of this fishery product, up to half of the import price reported by the other Member States. Initially this type of pattern was not considered. The clustering suggested by the forward search is therefore useful for highlighting unexpected patterns of this kind.

We have shown that the clusters cannot be inferred on the basis of the distribution of the minimum Mahalanobis distance computed at each step of the search among the observations outside the current subset. We argue that this is a general limitation when the clusters in a dataset intersect in a high-density area. This would restrict the possibility to detect *automatically* clusters or outliers with the forward search based on Mahalanobis distances. This limitation should be therefore studied in the numerous application contexts where Atkinson, Riani, Cerioli and other authors have already shown the remarkable potential of the forward search, e.g. feature selection, discriminant analysis, spatial statistics, categorical classification, multivariate data transformations and time series analysis.

References

Arsenis, S., Perrotta, D., & Torti, F. (2005). Price outliers in eu external trade data. Internal note, presented at "Enlargement and Integration Workshop 2005", http://theseus.jrc.it/events.html.

Atkinson, A. C., & Riani, M. (2000). *Robust diagnostic regression analysis*. New York: Springer

Atkinson, A. C., Riani, M., & Cerioli, A. (2004). *Exploring multivariate data with the forward search*. New York: Springer

Atkinson, A. C., Riani, M., & Cerioli, A. (2006). Random start forward searches with envelopes for detecting clusters in multivariate data. In S. Zani, A. Cerioli, M. Riani, & M. Vichi (eds.), *Data analysis, classification and the forward search* (pp. 163–172). Berlin: Springer

Belsley, D. A., Kuh, E., & Welsch, R. E. (1980). *Regression diagnostics: Identifying influential data and sources of collinearity*. New York: Wiley

Cook, R., & Weisberg, S. (1982). *Residuals and Influence in Regression*. New York: Chapman & Hall. Out of print, available at http://www.stat.umn.edu/rir/.

Riani, M., Cerioli, A., Atkinson, A., Perrotta, D., & Torti, F. (2008). Fitting mixtures of regression lines with the forward search. In F. Fogelman-Soulie, D. Perrotta, J. Piskorski, & R. Steinberger (eds.), *Mining massive data sets for security* (pp. 271–286). Amsterdam: IOS

Part X
Statistical Methods for Financial and Economics Data

Comparing Continuous Treatment Matching Methods in Policy Evaluation

Valentina Adorno, Cristina Bernini, and Guido Pellegrini

Abstract The paper evaluates the statistical properties of two different matching estimators in the case of continuous treatment, using a Monte Carlo experiment. The traditional generalized propensity score matching estimator is compared with a new 2-steps matching estimator for the continuous treatment case, recently developed (Adorno et al., 2007). It compares treatment and control units similar in terms of their observable characteristics in both selection processes (the participation decision and the treatment level assignment), where the generalized propensity score matching estimator collapses the two processes into one single step matching. The results show that the 2-steps estimator has better finite sample properties if some institutional rules define the level of treatment with respect to the characteristics of treated units.

1 Introduction

The interest on the generalization of the programme evaluation framework from a binary treatment setting to a more general structure for the treatment has increased rapidly in the last years (Hirano and Imbens 2004; Imai and van Dyk 2004). The policy mechanism can be away from an experimental data framework because of the presence of multiple non random selection processes, related not only to the participation decision but also to the treatment level assignment. In these cases, the selection bias problem cannot be tackled using the estimation methods developed for the binary treatment case. The literature proposes few matching estimators for continuous treatment.[1] In all cases the analysis does not concern on the comparison

[1]Hirano and Imbens (2004) concentrate on treated individuals and estimate the average treatment effects on all treated individuals for different treatment levels, conditioning on the GPS. Behrman et al. (2004) develop a generalized matching estimator to control for nonrandom selectivity into the program and into exposure durations. Evaluation of the program effects is carried out by comparing

V. Adorno (✉)
Department of Economics, University of Bologna, Piazza Scaravilli, 2 Bologna
e-mail: valentina.adorno@unibo.it

F. Palumbo et al. (eds.), *Data Analysis and Classification*,
Studies in Classification, Data Analysis, and Knowledge Organization,
DOI 10.1007/978-3-642-03739-9_48,

between treated and untreated units and on the selection process related to the treatment level assignment. We have recently developed a novel 2-steps matching approach to estimate the causal treatment effects as a function of the doses (Adorno et al., 2007). It matches treatment and comparison units similar in terms of their observable characteristics both for the participation decision process and the treatment level assignment. This is the main difference with respect to the generalized propensity score matching procedure which collapses the two processes into one single step matching.

The main empirical advantage of our method is its aptitude to incorporate in the matching procedure some recognized restrictions on the relation between the two selection processes, as in many applications, where policy instruments have institutional restrictions. This is the case of public subsidies to private capital accumulation: in the European Union the (maximum) amount of subsidies is strictly linked to the firm dimension and its geographical localization. An important application of this institutional rules is the Law 488/1992 (L.488) in Italy, the most important policy intervention to subsidize private capital accumulation in the poorest regions in the last decades. Moreover, subsidies by L.488 are allocated by mimicking an auction mechanism: the firm can choose the amount of subsidy, and lower the amount, higher the probability to receive it. This procedure generates heterogeneity in the amount of subsidy allocated to similar firms. Therefore, the L.488 case is an interesting experimental framework in order to test the statistical properties of continuous treatment matching estimators.

The aim of the paper is to explore the finite sample properties of a 2-steps matching estimator in the presence of a system of external constraints in the continuous treatment case. The comparison is based on a Monte Carlo experiments, mimicking the allocation mechanism of L.488 and using different simulation settings.

2 Simulating a Subsidies Allocation Mechanism: The Case of L.488

The L.488 allocates subsidies through a "rationing" system based on an auction mechanism which guarantees compatibility of subsidies demand and supply. In each regional auction investment projects are ranked by a score assigned on the base of five indicators (among them the share of the subsidy requested and the highest subsidy applicable, given the rules determined by the UE Commission). Rankings are drawn up through the decreasing order of the score of each project and subsidies are allocated until financial resources granted to each region are exhausted. Then, the amount of allocated resources in every auction is different across regions, i.e.

different groups of untreated and treated with different level of exposure. Hornik et al. (2001) propose a propensity score with doses which entails a single scalar propensity score for all dose levels.

for every auction exists a specific regional allocation threshold. By pooling all firms together, an overlapping area of firms with the same propensity to be subsidised (treated and not) is available and matching estimators can be correctly applied.

The institutional framework of L.488 is important for the treatment level choice. There are institutional constrains related to the level of subsidies received: the maximum amount of incentive (relative to the level of investment) allowable to a project depends on both the region where the investment is localised and the size of the firm. This aspect can be fully exploited in the estimation of the treatment level decision. Furthermore, the amount of subsidy relative to the ceiling established by institutional rules is a choice variable for the firm: lower the aid applied for by the firm, higher the probability to receive the subsidy. This is the key indicator transforming the allocation procedure to an auction mechanism. On the other side, different amount of subsidies are allocated to similar firms, allowing for a matching procedure implementation on the subsidy level selection assignment.

The selection procedure of L.488 can help out simulation exercise. The procedure uses indicators as selection variables: they explain the most part of the difference between subsidized and not subsidized firms. Then, indicators can be very helpful in the construction of the counterfactual scenario. Moreover, different regional auctions (with different thresholds) can be easily replicated, generating a data set with treated and untreated firms having overlapping probability to be treated.

3 The Matching Methods in the Continuous Framework

In the continuous treatment framework $Y(T)$ represents the set of potential outcomes, for each unit i, given a random sample indexed by $i = 1 \dots N$, and T represents the continuous variable indicating the treatment level. The observed outcome Y can be written as $y_i = d_i y_i(t_i) + (1 - d_i) y_i(0)$. D is a dummy variable indicating the treatment status ($d_i = 1$ if the individual has been treated); $y_i(t_i)$ is the potential outcome at the observed level t_i. The participation decision assignment will determine the treatment status d_i, while the treatment level process will determine the dose t_i. Even if they can occur simultaneously, we suppose they can be logically separated in the analysis. Let assume assignment to treatment is made on the basis of:

$$t_i = \begin{cases} g(Z_i) + u_i & \text{if} \quad d_i = 1 \\ 0 & \text{otherwise} \end{cases}, \qquad d_i = \begin{cases} 1 \text{ if} & I_i > 0 \\ 0 \text{ otherwise} \end{cases}, \qquad I_i = h(w_i) + v_i \tag{1}$$

where W, Z and U, V represent a set of observable and unobservable variables available at a fixed time, when the selection processes occur. This structure represents the basis of our approach: differently from the previous literature, it specifies separately the two selection processes. Adopting different specifications for the two processes may be helpful for considering different influencing factors and for estimating the selection process in a more efficient way. The first step identifies the participation decision rule and units will be matched on the basis of similar value of the set of covariates, using the propensity score function $p(w) = P(d_i|W_i)$, instead of the full set of covariates W and Z. Among matched units in the first step, the next matching procedure will pair units with similar value in the covariates identifying the treatment level assignment process. Let $\theta(Z) = E(T|Z, p(W))$ the parameter that uniquely represents the propensity function (Imai and van Dyk 2004), then matching on it can be easily accomplished by matching on θ. Instead, the 1-step matching procedure is based on the propensity score function $p(w,z) = P(d_i|W_i, Z_i)$, exploiting together the full set of observable variables W and Z.

As regards parameters of interest, a natural development in the continuous case of the traditional treatment effect estimation (the average treatment effect on the treated ATT $= E[Y(1)-Y(0)|D = 1]$), is what we named the average treatment level effect on the treated ATLE $= \alpha(T) = E[Y(T) - Y(0)|T = t]$ for a person randomly drawn from the subpopulation of the participants at the level t. ATLE estimates for each observed treatment level allow to evaluate the relation between effects and levels, $\hat{\alpha} = f(t, \varepsilon)$, that is the entire function of average treatment effects over all possible values of the treatment doses.[2] It is important to note that this function is not a "true" dose-effect function (like the one in Hirano and Imbens (2004)) because our estimator compares treated versus untreated units (i.e. units at different levels of treatment might have dissimilar characteristics).

4 The Monte Carlo Experiment

There are very few papers which use Monte Carlo simulations to evaluate matching estimators (Frolich 2004; Zhao 2004, 2008). All the studies investigate different aspects of matching estimators (finite sample properties, sensitiveness to propensity score specifications, etc.) but they restrict their attention to the binary treatment case and do not consider the small-sample property in the continuous treatment case. This is the main contribution of the paper: our focus is on comparing one step and two steps matching estimators by Monte Carlo experiments in the case of continuous treatment. The Monte Carlo simulation mimics the L.448 two steps allocation mechanism:

1. Generating different datasets by unit, including for each unit the value of 3 indicators (I_1, I_2, I_3); these covariates affect the participation decision, the treatment level assignment, the outcome for treated $(Y(1))$ and untreated units $(Y(0))$. The indicators are generated as random numbers drawn from a standardized normal distribution $(N(0, 1))$.

[2]The relation between estimated effects and treatment levels is estimated in our simulations by a parametric approach, using a OLS regression. We compare a simple linear regression model and a regression model with a quadratic term in the treatment level variable, in order to better detect effect heterogeneity.

2. Generating different thresholds for each dataset and creating treated and control groups. We generate τ as a random number drawn from a uniform distribution (0.3; 0.8).[3]
3. Calculating the amount of treatment (T), as a function of two indicators (3). The treatment level indicator T enters in the participation selection function D, as in the L.488 selection mechanism (2).

The selection mechanism is defined by an index function, depending on indicators I_1, I_2 and the treatment level T (2). In each dataset ten thresholds τ are generated. Assignment to treatment of each unit is made on the basis of the following rule:

$$D = \begin{cases} 1 \text{ if } & \beta_1 I_1 + \beta_2 I_2 + \beta_3(1/T) > \tau \\ 0 \text{ otherwise} & \end{cases} \tag{2}$$

$$T = \beta_4 I_2 + \beta_5 I_3 + \varepsilon_0 \tag{3}$$

$$Y(0) = \beta_6 I_1 + \beta_7 I_3 + \varepsilon_1 \quad \text{if} \quad D = 0 \tag{4}$$

$$Y(1) = \begin{cases} \beta_6 I_1 + \beta_7 I_3 + \gamma_1 T + \varepsilon_1 & \text{in the linear case} \\ \beta_6 I_1 + \beta_7 I_3 + \gamma_2 T + \gamma_3 T^2 + \varepsilon_1 & \text{in the non linear case} \end{cases} \quad \text{if} \quad D = 1 \tag{5}$$

Treatment level depends also on I_2 and on the other index I_3, creating a positive correlation with the selection mechanism. A standardized normally distributed error term ε_0 is added (3). The outcome variable $Y(i)$ is observed after the treatment, for the control and treated group ($i = 0$ and $i = 1$, respectively). In the untreated state, $Y(0)$ depends on indicators I_1 and I_3, that are also in the two selection processes, and on a standardized normally distributed error term ε_1. In the treated state $Y(1)$ is generated by adding the effect of the treatment to the outcome of the untreated state. We are interested in capturing differences in the treatment effect by the treatment level. Therefore a linear and a quadratic treatment effect are experimented. In the last case we impose that the maximum of the curve lies inside the range of the generated treatment level. Then, the outcome of the treated units is defined as in (5). We allow perfect correlation between the error term in the outcome equation, but we do not enable correlation between the error term in the selection equation and the error term in the outcome equation, i.e. $cov(\varepsilon_0, \varepsilon_1) = 0$. Then, unconfoundedness is satisfied. β_i are fixed parameters[4] and γ_i are the policy impact coefficients, explaining the relation between treatment level and outcome.

We are interested in comparing the two continuous matching estimators with respect to the usual ATT and parameters γ_i, estimated by an OLS regression of the ATLE on treatment level. We investigate this issue with different designs, changing

[3]This regional specific threshold is introduced to mimic the L. 488 allocation setting. The presence of different thresholds widen the overlapping area for the matching experiment, although it is not essential. In fact, the randomness of the error term in (3) is sufficient to match similar units.

[4]The set of parameters we adopt in our simulation is: $\beta_1, \beta_2, \beta_3 = 0.33; \beta_4, \beta_5 = 0.4; \beta_6, \beta_7 = 0.1$.

both impact coefficients and the selection processes error variance. In the linear experiment we impose γ_1 equal to 0.2 and 1.2. In the quadratic case we impose $(\gamma_1; \gamma_2) = (6; -0.3), (9; -0.5)$. The variance of ε_0 assumes values 0.5, 1 and 2 in both experiments. For each combinations, 100 datasets are simulated, each of 10,000 observations, coming from the simulation of 1,000 observations replicated for ten different thresholds.

Among the matching algorithm proposed in literature we choose the stratification matching, properly adapted for the continuous case. In particular, for the 2-steps estimation, we first compute the stratification with respect to the propensity function that identifies the participation process ($p(w) = P(d_i|I_1, I_2, T)$) and then with respect to the treatment level assignment T. We compute the ATLE (for each stratum of the treatment level) as the weighted average among the mean differences between the outcome of treated and untreated units, for each stratum of the propensity function $p(w)$. Indeed, for the 1-step case, we compute a stratification matching on the basis of a unique propensity function $p(w) = P(d_i|I_1, I_2, T, I_3)$. The ATLE are computed at the same stratum of the treatment level used in the 2-steps case.

5 Results and Conclusions

Tables 1 and 2 report estimates of ATT and γ_i (Mean column), bias (difference between estimated and true effect/parameter) and mean square error (MSE). γ_i are estimated by an OLS regression of ATLE values on treatment levels. Figure 1 plots these regressions.

In the linear case both the 1-step and the 2-steps estimator show always a small upward biased ATT (Table 1). In both cases higher the error variance, higher the MSE. However, the MSE is always lower in the 2-steps case than in the 1-step one.

Table 1 Sensitivity to the treatment level effect: linear case

		2-STEPS CASE				1-STEP CASE			
		ATT		$\hat{\gamma}_1$		ATT		$\hat{\gamma}_1$	
γ_1	σ^2	Mean	S.D.	Mean	S.D.	Mean	S.D.	Mean	S.D.
0.2	0.5	2.029	0.052	0.198	0.033	2.038	0.071	0.007	0.012
	1	2.028	0.058	0.207	0.036	2.037	0.064	0.006	0.010
	2	2.024	0.062	0.194	0.038	2.028	0.063	0.006	0.010
1.2	0.5	12.195	0.305	1.203	0.043	12.213	0.323	0.032	0.029
	1	12.183	0.331	1.201	0.045	12.207	0.330	0.036	0.034
	2	12.159	0.354	1.200	0.036	12.195	0.355	0.035	0.030
γ_1	σ^2	Bias	MSE	Bias	MSE	Bias	MSE	Bias	MSE
0.2	0.5	0.029	0.004	−0.002	0.001	0.038	0.006	−0.193	0.037
	1	0.028	0.004	0.007	0.001	0.037	0.005	−0.194	0.038
	2	0.024	0.004	−0.006	0.002	0.028	0.005	−0.194	0.038
1.2	0.5	0.195	0.131	0.003	0.002	0.213	0.150	−1.168	1.366
	1	0.183	0.143	0.001	0.002	0.207	0.152	−1.164	1.356
	2	0.159	0.150	0.000	0.001	0.195	0.164	−1.165	1.358

Table 2 Sensitivity to the treatment level effect: non-linear case

			2-STEPS CASE						1-STEP CASE					
			ATT		$\hat{\gamma}_1$		$\hat{\gamma}_2$		ATT		$\hat{\gamma}_1$		$\hat{\gamma}_2$	
γ_2	γ_3	σ^2	Mean	S.D.	Mean	S.D.	Mean	S.D.	Mean	S.D.	Mean	S.D.	Mean	S.D.
6	−0.3	0.5	29.91	0.032	6.044	1.111	−0.302	0.055	29.90	0.045	0.032	0.203	−0.002	0.010
		1	29.90	0.035	6.174	1.057	−0.309	0.053	29.90	0.049	0.074	0.150	−0.004	0.007
		2	29.90	0.040	5.969	1.386	−0.299	0.067	29.89	0.054	0.029	0.285	−0.002	0.013
9	−0.5	0.5	39.70	0.297	9.423	1.172	−0.521	0.058	39.66	0.296	0.032	0.353	−0.003	0.017
		1	39.69	0.304	9.126	1.025	−0.506	0.051	39.65	0.309	0.016	0.392	−0.002	0.019
		2	39.68	0.304	9.066	0.973	−0.504	0.048	39.64	0.294	0.028	0.379	−0.003	0.018
γ_2	γ_3	σ^2	Bias	MSE	Bias	MSE	Bias	MSE	Bias	MSE	Bias	MSE	Bias	MSE
6	−0.3	0.5	−0.092	0.010	0.044	1.237	−0.002	0.003	−0.103	0.013	−5.968	35.662	0.298	0.089
		1	−0.096	0.010	0.174	1.148	−0.009	0.003	−0.100	0.012	−5.926	35.136	0.296	0.088
		2	−0.098	0.011	−0.031	1.922	0.001	0.004	−0.110	0.015	−5.971	35.736	0.298	0.089
9	−0.5	0.5	−0.300	0.178	0.423	1.553	−0.021	0.004	−0.336	0.200	−8.968	80.557	0.497	0.247
		1	−0.312	0.190	0.126	1.066	−0.006	0.003	−0.347	0.216	−8.984	80.860	0.498	0.248
		2	−0.325	0.198	0.066	0.952	−0.004	0.002	−0.356	0.213	−8.972	80.643	0.497	0.248

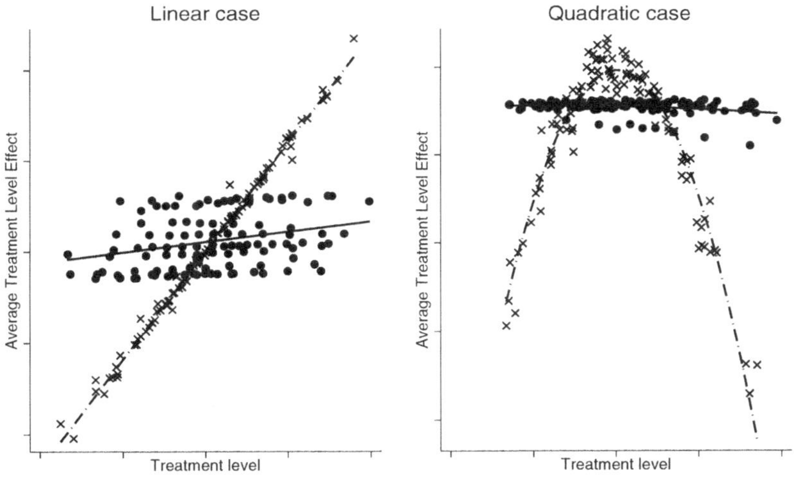

[redacted] [redacted]

Fig. 1 OLS regression of ATLE on treatment level.
Linear case ($\gamma_1 = 1.2, \sigma^2 = 0.5$) and quadratic case ($\gamma_2 = 6, \gamma_3 = -0.3, \sigma^2 = 0.5$)

The bias on γ_1 is substantially higher in the 1-step case, as well as the MSE. Figure 1 shows that the downward biased coefficient generates a flatter regression line for the 1-step estimator: even if the estimated ATT is close to the true one, the treatment effect is less affected by changes in the treatment level. The non linear case is very similar: there is a small downward biased average ATT in both cases, with a slight higher MSE for the 1-step estimator. However, the coefficients γ_2 and γ_3 are poorly estimated by the 1-step matching procedure in every simulation. The quadratic curve is much more flatter than the 2-steps case (Fig. 1), and it does not capture the strong heterogeneity of the treatment outcome with respect to different treatment levels. To understand the finding, we note that the 2-steps procedure enhances the quality of the matching: if the treatment effect depends on T, comparing units with the same potential amount of treatment improves the accuracy of the ATLE estimation. The result is more evident in presence of some institutional rules that relate T to the characteristics of treated units. This is the case of our experiment, where variables influencing the selection rule and the treatment level assignment are different. By adopting different specifications for the two processes we improve the selection process estimation, incorporating information on the institutional framework.

To conclude, the major finding from our simulations is that, even if the statistical performances of the two matching procedures are similar in the estimation of the ATT, results are deeply different on the effect-treatment level relationship estimation. In fact, the treatment impact coefficients are poorly estimated in the generalized (1-step) propensity score procedure, in particular in the non linear case. Monte Carlo results show an overall underestimation of the elasticity of the treatment effect to changes in the treatment level in the case of the 1-step estimator. The reason is that the 2-steps estimator sharps the matching procedure, comparing units with the same potential amount of treatment. The finding can be empirically relevant if there are strict rules relating the amount of treatment to the characteristics of treated units, as in several economic policies. Both methods may have a wide application field. Nevertheless, the 2-steps matching method allows to improve the empirical instrument evaluation: the comparison between treated and untreated units is more homogeneous with respect to the treatment level, and a less biased measure of the impact of all different treatment levels to treated units can be derived.

References

Adorno, V., Bernini, C., & Pellegrini, G. (2007). The impact of capital subsidies: New estimations under continuous treatment. *Giornale Degli Economisti e Annali di Economia*, *66*(1), 67–92

Behrman, J., Cheng, Y., & Todd, P. (2004). Evaluating preschool programs when length of exposureto the program varies: A non parametric approach. *The Review of Economics and Statistics*, *86*(1), 108–132

Frolich, M. (2004). Finite sample properties of propensity-score matching and weighting estimators. *The Review of Economics and Statistics*, *86*, 77–90

Hirano, K., & Imbens, G. (2004). The propensity score with continuous treatment. In *Missing data and bayesian methods in practise: Contributions by Donald Rubin's Statistical family*. Wiley

Hornik, R., Rosenbaum, P. R., Lu, B., & Zanutto, E. (2001). Matching with doses in an observational study of a media campaign against drug abuse. *Journal of The American Statistical Association (Applications and Case Studies)*, *96*(456), 1245–1253.

Imai, K., & van Dyk, D. (2004). Causal inference with general treatment regimes: Generalizing the propensity score. *Journal of The American Statistical Association*, *99*(467), 854–866

Zhao, Z. (2004). Using matching to estimate treatment effects: Data requirements, matching metrics and Montecarlo evidence. *The Review of Economics and Statistics*, *86*(1), 91–107

Zhao, Z. (2008). Sensitivity of propensity score methods to the specifications. *Economics Letters* *98*(3), 309–319.

Temporal Aggregation and Closure of VARMA Models: Some New Results

Alessandra Amendola, Marcella Niglio, and Cosimo Vitale

Abstract In this paper we examine the effects of temporal aggregation on Vector AutoRegressive Moving Average (VARMA) models. It has relevant implications both in theoretical and empirical domain. Among them we focus the attention on the main consequences of the aggregation (obtained from point in time sampling) on the model identification. Further, under well defined conditions on the model parameters, we explore the closure of the VARMA class (with respect to the temporal aggregation) through theoretical results discussed in proper examples.

1 Introduction

Time aggregation is commonly used in economic domain where, for example, the interest of researchers can be related to weakly, monthly, quarterly time series even if the time frequency of available data is higher (such daily, hourly and so on).

In that case a proper aggregation of the original data X_t can allow to obtain new time series with the desired time frequency. As expected, the aggregation of X_t can imply heavy consequences on the aggregated process whose stochastic structure can show relevant differences with respect to the generating process of X_t. These consequences have been differently investigated in literature, both in univariate and multivariate context (see Tiao (1972), Brewer (1973), Wei (1981), Weiss (1984), Stram and Wei (1986), Lütkephol (1987), Marcellino (1999), Jordà and Marcellino (2004), McCrorie and Chambers (2006) among the others).

In this wide theoretical context Jordà and Marcellino (2004) distinguish four types of aggregations. Our interest relies on the so called *Type I* aggregation where both the original time series and the aggregated one are considered regularly spaced in time with frequency of aggregation $k \in \mathcal{N}$. We discuss the effect of the temporal aggregation when $X_t \sim VARMA(p,q)$ and we propose an alternative way, with respect to that established in literature, to give evidence of the closure of this class

M. Niglio (✉)
Di.S.E.S. Università degli Studi di Salerno, Via Ponte Don Melillo 84084 Fisciano (SA), Italy
e-mail: mniglio@unisa.it

F. Palumbo et al. (eds.), *Data Analysis and Classification*,
Studies in Classification, Data Analysis, and Knowledge Organization,
DOI 10.1007/978-3-642-03739-9_49,

of models when time aggregation is performed. The proposed procedure is based on the so called "markovian" representation of the VARMA model and allows to obtain the autoregressive (AR) and moving average (MA) orders of the aggregated series using arguments that generalize what is done in the univariate context.

In particular, in Sect. 2 we present some features of the temporal aggregation and we give a brief presentation of the VARMA model to introduce the notation. In Sect. 3 we discuss some issues given in literature on time aggregation and we propose new results on this topic that are further explored in Sect. 4 where theoretical issues and proper examples are given.

2 Temporal Aggregation

Given a n-variate time series $\mathbf{X}_t = (\mathbf{X}_{1t}, \mathbf{X}_{2t}, \ldots, \mathbf{X}_{nt})$, with $n \geq 1$ and $t = 1, 2 \ldots,$ a *Type I* aggregated time series is obtained from a proper linear transformation of the original data

$$\mathbf{X}_t^{(k)} = \underset{(n \times nk)}{\mathbf{F}'_{(k)}} \underset{(nk \times 1)}{\mathbf{X}_{k(t-1)}}, \tag{1}$$

with $\mathbf{F}'_{(k)} = (\underset{(n \times n)}{\mathbf{F}'_1}, \mathbf{F}'_2, \ldots, \mathbf{F}'_k)$ and $\mathbf{X}_{k(t-1)} = (\underset{(n \times 1)}{\mathbf{X}_{k(t-1)+1}}, \ldots, \mathbf{X}_{k(t-1)+k})$.

Assigning different structures to the aggregation matrix $\mathbf{F}'_{(k)}$ in (1), we can distinguish three kinds of temporal aggregation:

1. *Point-in-time sampling*. Given $\mathbf{F}'_{(k)} = (0, 0, \ldots, \mathbf{I})$, with 0 a null matrix and $\mathbf{I}$ the identity matrix, the aggregated process, $\mathbf{X}_t^{(k)} = \mathbf{X}_{kt}$, is obtained systematically sampling $\mathbf{X}_t$ at frequency $k \in \mathcal{N}$.
2. *Average sampling*. Starting from $\mathbf{F}'_{(k)} = (\mathbf{I}, \mathbf{I}, \ldots, \mathbf{I})$, the aggregated time series, $\mathbf{X}_t^{(k)} = \sum_{j=1}^{k} \mathbf{X}_{k(t-1)+j}$, is the sum of k consecutive values of $\mathbf{X}_t$.
3. *Phase-averaged sampling*. In this case $\mathbf{F}'_{(k)} = \frac{1}{k}(\mathbf{I}, \mathbf{I}, \ldots, \mathbf{I})$ and the aggregated time series, $\mathbf{X}_t^{(k)} = \frac{1}{k} \sum_{j=1}^{k} \mathbf{X}_{k(t-1)+j}$, is the mean of k consecutive values of $\mathbf{X}_t$.

Note that the three enumerated procedures can be used in different contexts. A point-in-time sampling can be preferred in presence of stock variables whereas the second and the third procedures are most suitable with flow variables whose values, collected for example at daily frequency, can be aggregated to obtain monthly or quarterly data.

In all three cases the generating process of $\mathbf{X}_t^{(k)}$ can show substantial differences with respect to $\mathbf{X}_t$ and, in this regard, we focus the attention on the main consequences of the point-in-time sampling.

An interesting starting result in this context is related to the (weak) stationarity of the aggregated process. In fact, it is easy to show that in presence of a stationary process $\mathbf{X}_t$ with

$$E(\mathbf{X}_t) = \mu_{(n\times 1)} \quad \text{and} \quad Cov(\mathbf{X}_t, \mathbf{X}_{t-h}) = \Gamma(h)_{(n\times n)}, \quad \text{for} \quad h = 0, \pm 1, \pm 2, \dots,$$

even $\mathbf{X}_t^{(k)}$ is stationary with

$$E\left(\mathbf{X}_t^{(k)}\right) = \mathbf{F}'_{(k)}\boldsymbol{\mu}_k, \quad \text{and} \quad \underset{(1\times nk)}{\boldsymbol{\mu}_k} = (\mu, \mu \dots, \mu),$$

$$\Gamma_{(k)}(h) = Cov(\mathbf{X}_t^{(k)}, \mathbf{X}_{t-h}^{(k)}) = \mathbf{F}'_{(k)} \underset{(nk\times nk)}{\Gamma(k,h)} \mathbf{F}_{(k)}, \qquad h = 0, \pm 1, \pm 2, \dots, \tag{2}$$

where the element (i, j) in $\Gamma(k, h)$ is the matrix $\Gamma(kh - (j - i))_{(n\times n)}$.

Starting from the given definitions and preliminary results, our aim is to investigate the temporal point-in-time aggregation when $\mathbf{X}_t$ follows a stationary $VARMA(p, q)$ process. This n-variate model, widely used in many empirical domains, is given by

$$\Phi(B)\mathbf{X}_t = \Theta(B)\mathbf{a}_t, \qquad t = 1, 2, \dots, \tag{3}$$

where $\Phi(B) = \mathbf{I} - \Phi_1 B - \dots - \Phi_p B^p$ (with $|\Phi(B)| \neq 0$ for $|B| \leq 1$), $\Theta(B) = \mathbf{I} - \Theta_1 B - \dots - \Theta_q B^q$, $\mathbf{a}_t \sim WN(0, \Sigma_a)$, $\mathbf{I}$ is the identity matrix of order n, 0 is a null vector and B^r is the lag operator $B^r\mathbf{X}_t = \mathbf{X}_{t-r}$. In addition, following Reinsel (1993), the model identifiability is ensured assuming that $\Phi(B)$ and $\Theta(B)$ have no common roots and the rank$([\Phi_p, \Theta_q]) = n$, for small values of p and q.

As stated in Sect. 1, the temporal aggregation in this multivariate domain has been differently faced. We further discuss these results even proposing an alternative procedure given in the following sections.

3 Temporal Aggregation and VARMA Models

Given a time series that follows the VARMA model (3), Marcellino (1999) shows that the point-in-time aggregated process $\mathbf{X}_t^{(k)} \sim VARMA(s, r)$

$$\Phi^{(k)}(L)\mathbf{X}_t^{(k)} = \Theta^{(k)}(L)\mathbf{a}_t^{(k)}, \qquad t = 1, 2, \dots, \tag{4}$$

with $L = B^k$, $\Phi^{(k)}(L) = \mathbf{I} - \Phi_1^{(k)} L - \dots - \Phi_p^{(k)} L^p$, $\Theta^{(k)}(L) = \mathbf{I} - \Theta_1^{(k)} L - \dots - \Theta_q^{(k)} L^q$, $\mathbf{a}_t^{(k)} \sim WN(0, \Sigma_{(k)})$, $s = p$ and $r \leq [((k-1)p + q)/k]$, where $[a]$ is the integer part of a.

The matrices $\Phi_j^{(k)}$, $j = 1, \dots, p$, of the autoregressive polynomial, are the columns of

$$\Phi_{-k}^{v} \left(\Phi_{-k}^{m}\right)^{-1} \Phi^m - \Phi^v, \tag{5}$$

where

$$\underset{(n\times pnk)}{\Phi^{v}} = \left(\underset{(n\times n)}{\Phi_1}, \ldots, \Phi_p, \mathbf{0}, \ldots, \mathbf{0}\right), \quad \Phi^{m} = \begin{bmatrix} -\mathbf{I} & \Phi_1 & \ldots & \Phi_p & \mathbf{0} & \ldots & \mathbf{0} \\ \mathbf{0} & -\mathbf{I} & \ldots & \Phi_{p-1} & \Phi_p & \ldots & \mathbf{0} \\ \ldots & \ldots & \ldots & \ldots & \ldots & \ldots & \ldots \\ \mathbf{0} & \mathbf{0} & \mathbf{0} & \mathbf{0} & \mathbf{0} & \ldots & \Phi_p \end{bmatrix}, \tag{6}$$

with Φ^m a $[np(k-1) \times pnk]$ matrix, Φ^v_{-k} and Φ^m_{-k} the vector and the matrix of matrices obtained from Φ^v and Φ^m respectively after removing the column kj, for $j = 1, \ldots, p$.

The moving average parameters of $\mathbf{X}_t^{(k)}$ and the $\Sigma_{(k)}$ variance-covariance matrix can be finally obtained from (2) after a proper algebra.

The results in Marcellino (1999), now briefly summarized, suggest some remarks:

R1. Φ^m and Φ^m_{-k} are not defined when $k = 1$, even if in this case it is expected that $\mathbf{X}_t^{(1)} = \mathbf{X}_t$;
R2. If $k \geq 2$ and $|\Phi^m_{-k}| = 0$ the parameters of the aggregated process cannot be defined and so the result (5) cannot be longer applied;
R3. When $k \geq 2$ and $|\Phi^m_{-k}| \neq 0$ the AR parameters obtained from (2) have opposite sign with respect to those given in (5), as discussed with more details in Example 1 and in Sect. 4;
R4. The AR order s of $\mathbf{X}_t^{(k)}$ should be $s \leq p$.

In order to better illustrate these points consider the example below.

Example 1. Given the n variate process $\mathbf{X}_t \sim VARMA(2,0)$, with $n \geq 2$ and aggregation frequency $k = 2$. From (5) the AR parameters of $X_t^{(2)}$ are

$$\Phi_1^{(2)} = -\left(\Phi_1^2 + \Phi_1\Phi_2\Phi_1^{-1} + \Phi_2\right) \quad \text{and} \quad \Phi_2^{(2)} = \left(\Phi_1\Phi_2\Phi_1^{-1}\Phi_2\right).$$

Following the four points before:

E1. To define the parameters of the aggregated process we have fixed $k = 2$;
E2. When $|\Phi_1| = 0$ then the parameters of $\mathbf{X}_t^{(2)}$ cannot be defined. It greatly limits the application of (5), for example when the process $\mathbf{X}_t$ has parameter $\Phi_1 = \mathbf{0}$. Examining the case with $\Phi_1 = \begin{bmatrix} \phi_{11} & 0 \\ \phi_{21} & 0 \end{bmatrix}$ and $\Phi_2 = \begin{bmatrix} \varphi_{11} & 0 \\ 0 & \varphi_{22} \end{bmatrix}$, a solution to the problem under discussion can be given from the generalized inverse $\Phi_1^- = \begin{bmatrix} 1/\phi_{11} & 0 \\ 0 & 0 \end{bmatrix}$ that, even if not unique, allows to obtain the matrices of coefficients

$$\Phi_1^{(2)} = \begin{bmatrix} \phi_{11}^2/(2\varphi_{11}) & 0 \\ \phi_{21}\phi_{11} + \phi_{21}\phi_{11}/\varphi_{22} & \varphi_{22} \end{bmatrix} \quad \text{and} \quad \Phi_2^{(2)} = \begin{bmatrix} \varphi_{11} & 0 \\ \phi_{21}\phi_{11}^2/\phi_{11} & 0 \end{bmatrix};$$

E3. If $\Phi_1 \neq 0$ and $\Phi_2 = 0$, from (5) it follows that

$$\Phi_1^{(2)} = -\Phi_1^2. \tag{7}$$

The same result (7) should be reached if we take advantage of alternative representations of the VARMA process (such as the "markovian" discussed in Sect. 4) or if the covariance matrix (2) is used to define the parameters of the aggregated process.

In this latter case, given:

$$\begin{cases} \Gamma(h) = \Phi_1^h \Gamma(0) \\ \Gamma_{(2)}(h) = \left(\Phi_1^{(2)}\right)^h \Gamma_{(2)}(0), \qquad h = 1, 2, \ldots \end{cases} \tag{8}$$

and recalling that $\Gamma_{(2)}(0) = \Gamma(0)$ and $\Gamma_{(2)}(h) = \Gamma(2h)$, we can assign $h = 2$ and $h = 1$ to the first and second equations in (8) respectively, such that they become:

$$\begin{cases} \Gamma(2) = \Phi_1^2 \Gamma(0) \\ \Gamma(2) = \Phi_1^{(2)} \Gamma(0), \end{cases} \tag{9}$$

and so $\Phi_1^{(2)} = \Phi_1^2$. It differs from (7) where an opposite sign is obtained for the parameters of the aggregated process (the same result obtained from (9) are established in Sect. 4);

E4. When $\Phi_1 = 0$ the process $\mathbf{X}_t$ at time $2t$ becomes $\mathbf{X}_{2t} = \Phi_2 \mathbf{X}_{2t-2} + \mathbf{a}_{2t}$ and the aggregated process, with $k = 2$, is

$$\mathbf{X}_t^{(2)} = \Phi_2 \mathbf{X}_{t-1}^{(2)} + \mathbf{a}_t^{(2)}, \qquad \text{with} \quad \mathbf{a}_t^{(2)} = \mathbf{a}_{2t} \sim WN(0, \Sigma), \tag{10}$$

where the AR order $s = 1$ confirms what stated before in [R4.]. □

4 "Markovian" Representation

The results related to the time aggregation can be further appreciated, and in some case even simplified in their presentation, if we consider the markovian[1] specification of the VARMA model.

Given model (3), its markovian representation is

$$\underset{(np\times n)}{\mathbf{X}_t} = \underset{(np\times np)}{\Phi} \mathbf{X}_{t-1} + \underset{(np\times p)}{\mathbf{1}} \mathbf{u}_t, \tag{11}$$

[1] In the following the adjective "markovian" is only related to the form of the model equation (that looks like a Markov process) and not to the stochastic properties of the generating process.

where

$$\mathbf{u}_t = \Theta(B)\mathbf{a}_t, \quad \Phi = \begin{bmatrix} \Phi_1 & \Phi_2 & \dots & \Phi_{p-1} & \Phi_p \\ \mathbf{I} & \mathbf{0} & \dots & \mathbf{0} & \mathbf{0} \\ \vdots & \vdots & \vdots & \vdots & \vdots \\ \mathbf{0} & \mathbf{0} & \dots & \mathbf{I} & \mathbf{0} \end{bmatrix}, \quad \mathbf{1u}_t = \begin{bmatrix} \mathbf{u}_t \\ \mathbf{0} \\ \vdots \\ \mathbf{0} \end{bmatrix},$$

$$\mathbf{X}_t = \begin{bmatrix} \mathbf{X}_t \\ \mathbf{X}_{t-1} \\ \vdots \\ \mathbf{X}_{t-p} \end{bmatrix}.$$

After $k-1$ iterations, model (11) at time kt becomes

$$\mathbf{X}_{kt} = \Phi^k \mathbf{X}_{k(t-1)} + \sum_{j=0}^{k-1} \Phi^j B^j \mathbf{1u}_{kt}, \tag{12}$$

with $\Phi^k = \Phi \times \ldots \times \Phi$ (k times) and given Φ^j_{ij}, the matrix belonging to Φ^j in position (i, j) (for $j = 0, 1, \ldots, k$), the equation i in (12) is

$$\mathbf{X}_{kt-i+1} = \sum_{j=1}^{p} \Phi^k_{ij} \mathbf{X}_{k(t-1)-j+1} + \sum_{j=0}^{k-1} \Phi^j_{ij} B^j \mathbf{u}_{kt} \qquad i = 1, 2, \ldots, p. \tag{13}$$

For $i = 1$ the point-in-time sampling process with aggregation frequency k is

$$\mathbf{X}_{kt} = \sum_{j=1}^{p} \Phi^k_{1j} \mathbf{X}_{k(t-1)-j+1} + \sum_{j=0}^{k-1} \Phi^j_{1j} B^j \mathbf{u}_{kt}, \tag{14}$$

based on the $p-1$ constraints (13), obtained for $i = 2, 3, \ldots, p$ respectively.

From (12) the generating mechanism (intended as lag lengths) of the aggregated process is $\mathbf{X}_{kt} \sim VARMA(s, r)$ with $s \leq kp$ and $r \leq (k-1)p + q$. According to the notation of the original process $\mathbf{X}_t \sim$VARMA(p, q), it follows that $s \leq p$ and $r \leq [((k-1)p + q)/k]$.

In the next example we show how to use (13) to obtain $\mathbf{X}_{kt}$ from $\mathbf{X}_t \sim VARMA$ (p, q) and the conditions under which this process can be intended closed with respect to the point-in-time temporal aggregation.

Example 2. Given $\mathbf{X}_t \sim VARMA(2, 1)$:

$$\mathbf{X}_t = \Phi_1 \mathbf{X}_{t-1} + \Phi_2 \mathbf{X}_{t-2} + \mathbf{a}_t - \Theta_1 \mathbf{a}_{t-1}, \tag{15}$$

with $\mathbf{a}_t \sim WN(0, \Sigma_a)$ and frequency of aggregation $k = 2$. If the process $\mathbf{X}_{2t}$ exists, from Sect. 3 it should be a VARMA(2,1)

$$\mathbf{X}_t^{(2)} = \Phi_1^{(2)}\mathbf{X}_{t-1}^{(2)} + \Phi_2^{(2)}\mathbf{X}_{t-2}^{(2)} + \mathbf{a}_t^{(2)} - \Theta_1^{(2)}\mathbf{a}_{t-1}^{(2)}, \quad \text{with} \quad \mathbf{a}_t^{(2)} \sim WN(0, \Sigma_a^{(2)}). \tag{16}$$

In order to evaluate the parameters of the aggregated process consider the markovian form of model (15):

$$\begin{pmatrix} \mathbf{X}_t \\ \mathbf{X}_{t-1} \end{pmatrix} = \begin{pmatrix} \Phi_1 & \Phi_2 \\ \mathbf{I} & \mathbf{0} \end{pmatrix} \begin{pmatrix} \mathbf{X}_{t-1} \\ \mathbf{X}_{t-2} \end{pmatrix} + \begin{pmatrix} \mathbf{u}_t \\ \mathbf{0} \end{pmatrix}, \qquad \text{where} \quad \mathbf{u}_t = \mathbf{a}_t - \Theta_1 \mathbf{a}_{t-1}, \tag{17}$$

that after the first iteration becomes

$$\begin{pmatrix} \mathbf{X}_t \\ \mathbf{X}_{t-1} \end{pmatrix} = \begin{pmatrix} \Phi_1^2 + \Phi_2 & \Phi_1\Phi_2 \\ \Phi_1 & \Phi_2 \end{pmatrix} \begin{pmatrix} \mathbf{X}_{t-2} \\ \mathbf{X}_{t-3} \end{pmatrix} + \begin{pmatrix} \mathbf{u}_t + \Phi_1 \mathbf{u}_{t-1} \\ \mathbf{u}_{t-1} \end{pmatrix}. \tag{18}$$

From the second equation (18) it follows that

$$\mathbf{X}_{t-3} = \Phi_1^{-}\mathbf{X}_{t-2} - \Phi_1^{-}\Phi_2\mathbf{X}_{t-4} - \Phi_1^{-}\mathbf{u}_{t-2},$$

with Φ_1^{-} the generalized inverse of Φ_1, whereas the first equation (18) at time $t = 2t$ is

$$\begin{aligned} \mathbf{X}_t^{(2)} &= \left(\Phi_1^2 + \Phi_1\Phi_2\Phi_1^{-1} + \Phi_2\right)\mathbf{X}_{t-1}^{(2)} - \left(\Phi_1\Phi_2\Phi_1^{-1}\Phi_2\right)\mathbf{X}_{t-2}^{(2)} \\ &\quad + \mathbf{u}_{2t} + \Phi_1\mathbf{u}_{2t-1} - \Phi_1\Phi_2\Phi_1^{-}\mathbf{u}_{2t-2} \end{aligned}$$

The AR coefficients of the aggregated process so become

$$\Phi_1^{(2)} = \Phi_1^2 + \Phi_1\Phi_2\Phi_1^{-} + \Phi_2, \qquad \Phi_2^{(2)} = -(\Phi_1\Phi_2\Phi_1^{-}\Phi_2), \tag{19}$$

that: when Φ_1 is not singular $\Phi_1^{-} = \Phi_1^{-1}$ and the result (19) agree with (5), except for the sign; when Φ_1 is singular the solution (19) involves a generalized inverse that, as well known, is not unique for the parameters of the aggregated process. The parameters of the aggregated MA component, $\mathbf{u}_{2t} + \Phi_1\mathbf{u}_{2t-1} - \Phi_1\Phi_2\Phi_1^{-1}\mathbf{u}_{2(t-1)}$, are obtained from:

$$\begin{aligned} \mathbf{a}_t^{(2)} - \Theta_1^{(2)}\mathbf{a}_{t-1}^{(2)} &= \mathbf{u}_{2t} + \Phi_1\mathbf{u}_{2t-1} - \Phi_1\Phi_2\Phi_1^{-1}\mathbf{u}_{2(t-1)} \\ &= \mathbf{a}_{2t} + (\Phi_1 - \Theta_1)\mathbf{a}_{2t-1} - \Phi_1(\Theta_1 + \Phi_2\Phi_1^{-1})\mathbf{a}_{2t-2} \\ &\quad + \Phi_1\Phi_2\Phi_1^{-1}\Theta\mathbf{a}_{2t-3}, \end{aligned} \tag{20}$$

and using the shorter notation $\mathbf{A}_t^{(2)}$ and $\mathbf{A}_{2t}$ to refer the first and the second term in (20) respectively, the MA parameters of the aggregated process can be obtained evaluating the equation based on the variance-covariance matrices $\text{cov}(\mathbf{A}_t^{(2)}, \mathbf{A}_{t-h}^{(2)}) = \text{cov}(\mathbf{A}_{2t}, \mathbf{A}_{2(t-h)})$, for $h = 0, 1$. After some algebra, it can be shown that:

$$\begin{aligned} var(\mathbf{A}_t^{(2)}) = var(\mathbf{A}_{2t}) &= \Sigma_a + (\Phi_1 - \Theta_1)\Sigma_a(\Phi_1 - \Theta_1)' + \Phi_1(\Theta_1 + \Phi_2\Phi_1^-)\Sigma_a \\ &\quad (\Theta_1 + \Phi_2\Phi_1^-)'\Phi_1' + \Phi_1\Phi_2\Phi_1^-\Theta_1\Sigma_a\Theta_1'(\Phi_1\Phi_2\Phi_1^-)' \\ \text{cov}(\mathbf{A}_t^{(2)}, \mathbf{A}_{t-1}^{(2)}) &= \text{cov}(\mathbf{A}_{2t}, \mathbf{A}_{2(t-1)}) \\ &= -\Phi_1(\Theta_1 + \Phi_2\Phi_1^-)\Sigma_a + \Phi_1\Phi_2\Phi_1^-\Sigma_a(\Phi_1 - \Theta_1)' \end{aligned}$$

that only in simple cases can be algebraically solved whereas, in the remaining cases, numerical algorithms are needed.

For example if the process under analysis has $\Phi_1 = 0$, the aggregated process becomes

$$\mathbf{X}_t^{(2)} = \Phi_2\mathbf{X}_{t-1}^{(2)} + \mathbf{a}_t^{(2)},$$

where $\mathbf{a}_t^{(2)} = \mathbf{a}_{2t} - \Theta_1\mathbf{a}_{2t-1}$, with $\mathbf{a}^{(2)} \sim WN(0, \Sigma + \Theta_1\Sigma\Theta_1')$, and finally the aggregated process $\mathbf{X}_t^{(2)} \sim VARMA(1,0)$. □

All results presented in the previous pages can be summarized in the following proposition:

Proposition 1. *Given the n-variate stationary process* $\mathbf{X}_t \sim VARMA(p,q)$:

1. *The point-in-time aggregated process* $\mathbf{X}_t^{(k)}$, *obtained from* $\mathbf{X}_t$, *is a* $VARMA(s,r)$ *with* $s \leq p$ *and* $r \leq [((k-1)p + q)/k]$, $k \in \mathcal{N}$, *(with [a] the integer part of a);*
2. *When* $k > 1$ *if the AR parameters* Φ_j *are different from the null matrix at least for one* $j = 1, \ldots, p-1$, *then the AR parameters of the aggregated process are obtained from:*

$$\Phi^v - \Phi^v_{-k}\left(\Phi^m_{-k}\right)^- \Phi^m, \qquad \text{with} \quad k > 1. \qquad \square \tag{21}$$

The two points enumerated in Proposition 1 integrate the results given in literature on temporal aggregation in presence of stationary VARMA processes: the first gives new results on the order of the AR component of the aggregated process whereas the second point gives new issues on the sign of its parameters that can be appraciated comparing (21) and (5). Further (21) makes use of the generalized inverse to face the remark [R2.] discussed in Sect. 3.

References

Breitung, J., & Swanson, N. (2002). Temporal aggregation and spurious instantaneous causality in multiple time series models. *Journal of Time Series Analysis*, *23*, 651–665

Brewer, K. (1973). Some consequences of temporal aggregation and systematic sampling for ARMA and ARMAX models. *Journal of Econometrics*, *1*, 133–154

Granger, C., & Siklos, P. (1995). Systematic sampling temporal aggregation, seasonal adjustment and cointegration theory and evidence. *Journal of Econometrics*, *66*, 357–369

Jordà, O., & Marcellino, M. (2004). Time-scale transformations of discrete time processes. *Journal of Time Series Analysis*, *25*, 873–894

Lütkephol, H. (1987). *Forecasting aggregated vector ARMA processes*. Berlin: Springer

Marcellino, M. (1999). Some consequences of temporal aggregation in empirical analysis. *Journal of Business and Economic Statistics*, *17*, 129–136

McCrorie, J., & Chambers, M. (2006). Granger causality and sampling of economic processes. *Journal of Econometrics*, *132*, 311–336

Reinsel, G. C. (1993). *Elements of multivariate time series analysis*. New York: Springer

Stram, M., & Wei, W. (1986). Temporal aggregation on ARIMA models. *Journal of Time Series Analysis*, *7*, 279–292

Tiao, G. C. (1972). Asymptotic behaviour of temporal aggregates of time series. *Biometrika*, *59*, 525–531

Wei, W. (1981). Effects of systematic sampling on ARIMA models. *Communications in Statistics: Theory and Methods*, **10**, 2389–2398

Weiss, A. (1984). Systematic sampling and temporal aggregation in time series models. *Journal of Econometrics*, *26*, 271–281

An Index for Ranking Financial Portfolios According to Internal Turnover

Laura Attardi and Domenico Vistocco

Abstract Style analysis models are widely used in common financial practice to estimate the composition of a financial portfolio. The models exploit past returns of the financial portfolio and a set of market indexes, the so-called constituents, that reflect the portfolio investment strategy. The classical model is based on a constrained least squares regression model Sharpe (J Portfol Manage, 1992; Investment management review, 2(6), 59–69, Berlin, Physica, 1998) in which the portfolio returns are regressed on the constituent returns. The quantile regression model, originally proposed in Basset and Chen (Portfolio style: Return-based attribution using quantile regression. In Economic applications of quantile regression (Studies in empirical economics), 293–305, 2001) and revisited in Attardi and Vistocco (Statistica Applicata, 18(2), 2006; On estimating portfolio conditional returns distribution through style analysis models. In Quantitative methods for finance and insurance. Berlin, Springer, 2007), provides a useful complement to the standard model, as it allows the discrimination of portfolios that would be otherwise judged equivalent. Indeed different patterns of weights could correspond to the same conditional expectation, whereas the use of regression models estimating different conditional quantile functions should allow this kind of effect to be discriminated. The aim of this paper is to propose an index based on quantile regression estimates for ranking portfolios according to the level of constituent turnover.

1 Introduction

Portfolio turnover is usually defined as a function of active trading decisions. Several indexes have been proposed with the intent to describe portfolio turnover: they are essentially based on the availability of information on purchases and sales the portfolio manager has carried out during a given period. However portfolio past returns are often the only available information for the final investor. They are thus

D. Vistocco (✉)
Dip.to di Scienze Economiche, Università di Cassino, Italy
e-mail: vistocco@unicas.it

F. Palumbo et al. (eds.), *Data Analysis and Classification*,
Studies in Classification, Data Analysis, and Knowledge Organization,
DOI 10.1007/978-3-642-03739-9_50,

typically used to compare portfolios according to risk/return measures. Style analysis models exploit past returns to describe the investment style with respect to a set of investment classes. The classical style analysis model estimates the effect of portfolio style on the conditional mean return. Using a double constrained model, the estimated coefficients can be interpreted as compositional data, each coefficient standing for the quota of the corresponding constituent in the portfolio.

The aim of this paper is to exploit the information provided by the use of quantile regression for style analysis models (for different values of conditional quantiles) in order to assess the effect of portfolio exposure on the full conditional returns distribution. With respect to the classical model, quantile regression offers a set of estimated coefficients corresponding to the different conditional quantiles of the portfolio returns distribution. These coefficients are useful for comparing different portfolios sharing the same investment classes. In particular we use quantile regression estimates to draw conclusions on portfolio turnover, meant in terms of variation in component weights. The estimates are combined to compute an index able to rank portfolios according to their internal turnover. Although different from typical turnover measures, the index can be used whereas no information on trading decisions are available. The method is tested on a set of illustrative portfolios. The portfolios are artificial with respect to composition but real with respect to returns, as they are composed using the Morgan Stanley equities indexes. The strategy used to obtain portfolio weights is described. The index is computed for the artificial portfolios and the results are interpreted according to the real composition of the portfolios.

The paper is organized as follows. In Sect. 2 the style analysis models are briefly introduced, focusing on the different interpretation of the least squares (LS) model and of the quantile regression (QR) models. An index based on QR estimates is proposed in order to rank portfolios according to their internal level of turnover in Sect. 3: the index is then computed on a set of artificial portfolios. Finally some concluding remarks and further developments are provided in Sect. 4.

2 Style Analysis Models

Style analysis models regress portfolio returns on the returns of a variety of investment class returns. The method thus identifies the portfolio style in the time series of its returns and of constituent returns Horst et al. (2004). The use of past returns is a Hobson's choice as typically there is no other information available to external investors.

Let us denote by $\mathbf{r}^{port}$ the vector of portfolio returns along time and by $\mathbf{R}^{const}$ the matrix containing the returns along time of the ith portfolio constituent on the ith column ($i = 1, \ldots, n$). Data are observed over T subsequent time periods. The style analysis model regresses portfolio returns on the returns of the n constituents:

$$\mathbf{r}^{port} = \mathbf{R}^{const}\mathbf{w}^{const} + \mathbf{e} \qquad \text{s.t.: } \mathbf{w}^{const} \geq 0, \mathbf{1}^T\mathbf{w}^{const} = 1.$$

The vector **e** denotes the tracking error of the portfolio. The two constraints allow the coefficients to be exhaustive and non-negative, thereby facilitating interpretation in terms of compositional data: the estimated coefficients stand for the constituent quotas in the portfolio. The $\mathbf{R}^{const}\mathbf{w}^{const}$ term of the equation can be interpreted as the return of a weighted portfolio: the portfolio with optimized weights is thus a portfolio with the same style as the observed portfolio. It differs from the former as estimates about the internal composition are available Conversano and Vistocco (2004, 2009).

Style analysis models can vary with respect to the choice of style indexes as well as the specific location of the response-conditional distribution they are estimating. The classical style analysis model is based on an LS constrained regression model Sharpe (1992, 1998). The use of an LS model focuses on the conditional expectation of portfolio returns distribution: estimated compositions are interpretable in terms of sensitivity of portfolio expected returns to constituent returns. The LS model, indeed, can be formulated as follows:

$$E(\mathbf{r}^{port} \mid \mathbf{R}^{const}) = \mathbf{R}^{const}\mathbf{w}^{const} \qquad \text{s.t.: } \mathbf{w}^{const} \geq 0, \mathbf{1}^T\mathbf{w}^{const} = 1.$$

Using the LS model, portfolio style is determined by estimating the influence of style exposure on expected returns. Extracting information at other points other than the expected value should provide useful insights as the style exposure could affect returns in different ways at different locations of the portfolio returns distribution. Quantile regression as introduced in Koenker and Basset (1978), Koenker (2005) may be viewed as an extension of classical least squares estimation of conditional mean models to the estimation of a set of conditional quantile functions: exploiting QR a more detailed comparison of financial portfolios can be achieved as QR coefficients are interpretable in terms of sensitivity of portfolio conditional quantile returns to constituent returns Basset and Chen (2001). The QR model for a given conditional quantile θ follows:

$$Q_\theta(\mathbf{r}^{port} \mid \mathbf{R}^{const}) = \mathbf{R}^{const}\mathbf{w}^{const}(\theta) \text{ s.t.: } \mathbf{w}^{const}(\theta) \geq 0, \mathbf{1}^T\mathbf{w}^{const}(\theta) = 1, \forall\theta,$$

where θ $(0 < \theta < 1)$ denotes the particular quantile of interest.

In a similar way as for the LS model, the $w^{const_i}(\theta)$ coefficient of the QR model can be interpreted as the rate of change of the θth conditional quantile of the portfolio returns distribution for one unit change in the ith constituent returns holding the values of $R^{const}_{\cdot j, j \neq i}$ constant.

Therefore QR can be used as a complement to standard analysis, allowing discrimination among portfolios that would be otherwise judged equivalent using only conditional expectation Attardi and Vistocco (2006, 2007). The use of QR thus offers a more complete view of relationships among portfolio returns and constituent returns.

3 Ranking Portfolios According to Internal Turnover

The style analysis model was proposed in order to measure the effectiveness of the investor's overall asset allocation. According to Sharpe (1992), the method can be used to determine "how effectively individual fund managers have performed their functions and the extent (if any) to which value has been added through active management". Indeed, the use of the asset class factor model allows us to obtain information on the internal allocation of the portfolio and compare portfolios with similar investment strategies. Essentially, as described in detail by Horst et al. (2004), style analysis is used: *(a)* to estimate the main factor exposure of a financial portfolio, *(b)* in performance measurement as the style portfolio can be used as a benchmark in evaluating portfolio performance, *(c)* in order to predict future portfolio returns, as, from the empirical results in Horst et al. (2004), factor exposures seem to be more relevant than actual portfolio holdings.

The paper sets out to propose a different use of style analysis models to obtain a ranking of financial portfolios according to their internal turnover. In particular, we exploit quantile regression estimates and we summarize them in an index that can be computed both with respect to the different factors and with respect to the whole portfolio. The index aims to catch information on portfolio activeness. It is worth recalling that two different meanings for an active portfolio are allowed: *(a)* if "activeness" is measured according to a benchmark, a completely passive portfolio is managed by trying to perfectly replicate benchmark returns while an active portfolio shares with the benchmark only the investment market without the need to replicate the same investment results; *(b)* measuring "activeness" according to the internal level of turnover, on the other hand, requires focusing on the variability of portfolio weights. A passive portfolio, in this meaning, is a portfolio whose manager has initially set the constituent quotas and he/she makes no change during the investment period. An active portfolio is, instead, characterised by a high level of assets turnover.

QR estimates can be useful to rank portfolios according to this second definition of activeness, as QR coefficients are related to the conditional distribution of portfolio returns: comparing QR estimates provides information on the different level of portfolio turnover.

In order to illustrate this use of QR coefficients, an application on six equity portfolios follows. The portfolios were obtained as a combination of Morgan Stanley (MSCI) indexes: they consist of ten sector equity indexes tracking the performance of the sectors in question, namely: energy (ENR), materials (MAT), industrials (IND), consumer discretionary (CDIS), consumer staples (CSTA), health care (HC), financial (FNCL), information technology (IT), telecommunication services (TEL) and utilities (UTI). The MSCI website (www.mscibarra.com) provides information about the returns of the global index and its constituents. Daily data ($T = 950$) were used from 7 April 2001 to 23 February 2005 to estimate the whole quantile process.

The six portfolios (from P1 to P6) were formed by using different internal compositions but they share the same mean compositions for each of the ten sectors: the average weights for each sector are identical for the six portfolios, while there are

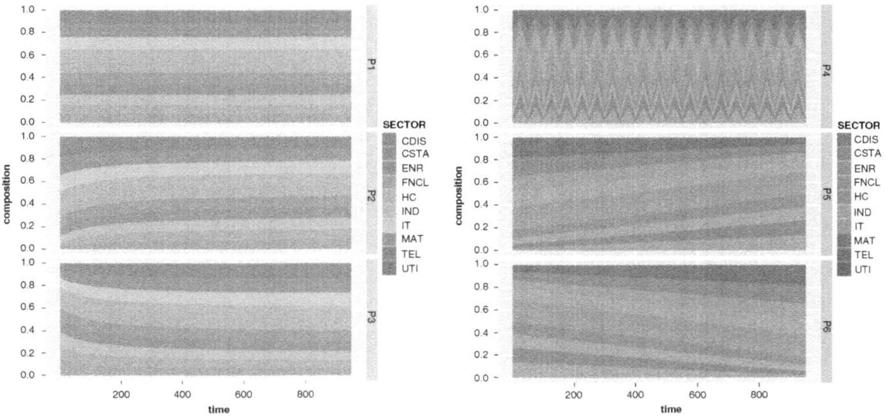

Fig. 1 Real composition of the six portfolios (from P1 to P6)

Table 1 Mean composition and standard deviations of weights for the six portfolios

	ENR	MAT	IND	CDIS	CSTA	HC	FNCL	IT	TEL	UTI
μ	0.079	0.083	0.088	0.093	0.098	0.102	0.107	0.112	0.117	0.121
σ_{P1}	0.000	0.000	0.000	0.000	0.000	0.000	0.000	0.000	0.000	0.000
σ_{P2}	0.015	0.012	0.008	0.005	0.002	0.002	0.005	0.008	0.012	0.015
σ_{P3}	0.015	0.012	0.008	0.005	0.002	0.002	0.005	0.008	0.012	0.015
σ_{P4}	0.036	0.028	0.020	0.012	0.004	0.004	0.012	0.020	0.028	0.036
σ_{P5}	0.034	0.027	0.019	0.011	0.004	0.004	0.011	0.019	0.027	0.034
σ_{P6}	0.034	0.027	0.019	0.011	0.004	0.004	0.011	0.019	0.027	0.034

differences in the interval levels of turnover. Figure 1 illustrates the internal composition of the six portfolios, while means and standard deviations of the weights of the ten sectors are reported in Table 1. Figure 2(a) compares on the same plot the real weights of the six portfolios for the ENR sector (the plot having the same patterns for the other sectors).

It is worth noting that portfolio P1 is completely passive: the manager has set the initial levels of the weights and held them constant over the whole period. Sector turnover is present in the other portfolios. Obviously, the different level of turnover causes the different standard deviations of the related weight distribution (see Table 1). Apart from the case of the completely passive P1 portfolio, the same variability is observed for investment strategies 2 and 3 and for strategies 5 and 6, portfolio P4 denoting a slighty greater variability. From the composition plot of Fig. 1 it is evident that there is the opposite trend in portfolios P2 and P3 as well as in portfolios P5 and P6.

The use of the LS model on the six portfolios provides a single estimate for each sector weight (see Table 2).

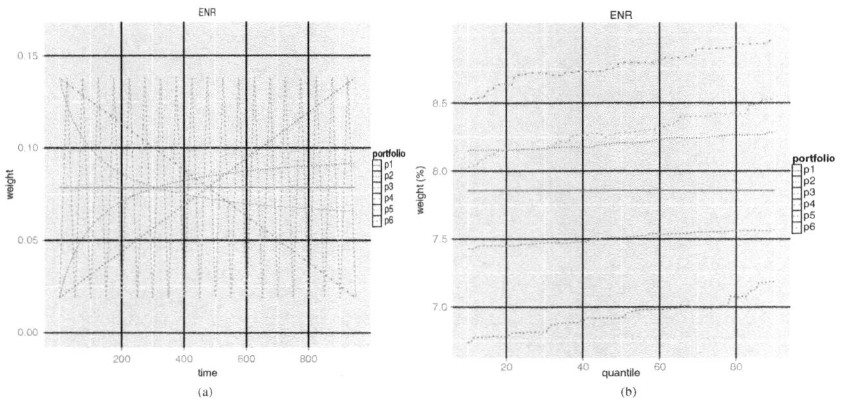

Fig. 2 (**a**) ENR sector weights for the six portfolios (**b**) QR estimates for ENR sector for the six portfolios

Table 2 LS estimates for the six illustrative portfolios

	ENR	MAT	IND	CDIS	CSTA	HC	FNCL	IT	TEL	UTI
p1	0.079	0.083	0.088	0.093	0.098	0.102	0.107	0.112	0.117	0.121
p2	0.077	0.077	0.080	0.092	0.099	0.107	0.112	0.110	0.120	0.125
p3	0.080	0.089	0.096	0.094	0.097	0.098	0.102	0.114	0.113	0.118
p4	0.089	0.076	0.080	0.114	0.092	0.103	0.085	0.111	0.115	0.135
p5	0.075	0.068	0.066	0.090	0.100	0.108	0.105	0.118	0.136	0.132
p6	0.082	0.098	0.110	0.095	0.095	0.097	0.109	0.106	0.097	0.111

Although some slight differences are evident in the LS estimates, it is difficult to draw conclusions on the different variability of the internal composition.

QR estimates allow the six portfolios to be distinguished in terms of the different variability of the composition weights. Figure 2(b) shows the QR estimates of the ENR sector for the six portfolios. Each line in the plot refers to one of the six portfolios. On the x-axis the different quantiles are represented while on the y-axis the corresponding estimates are shown. The QR estimates seem to confirm the ordering of the six portfolios on the basis of internal variability of the ENR weight distributions (compare Table 1). From Fig. 2(b) it is evident that QR estimates detect portfolio P1 as a completely passive portfolio (in the sense of absence of assets turnover) while they couple portfolios P2 with P3 and portfolios P5 with P6. In both cases, the QR models provide mirror estimates with respect to the conditional median. Portfolio P4 shows a different pattern.

In order to summarize the whole QR process, the slope of each line is measured by averaging the absolute values of the differences between each estimate and the previous one (the QR estimates are ordered by increasing value of θ):

$$\bar{\Delta}_{\theta}^{const_i} = \frac{\sum_{\theta} |\Delta(w^{const_i}(\theta))|}{\#(\boldsymbol{\theta}) - 1}$$

Table 3 The $\bar{\Delta}_{\theta}^{const_i}$ index and the corresponding ranking for the ENR sector, portfolio ranking based on the average of the index on all the constituents

	p1	p2	p3	p4	p5	p6
$\bar{\Delta}_{\theta}^{ENR}$	0.000	0.186	0.186	0.663	0.562	0.562
ENR ranking	1.0	2.5	2.5	6.0	4.5	4.5
global ranking	1.0	2.5	2.5	6.0	4.5	4.5

whereas the $\Delta()$ operator is the usual lag operator and $\#()$ denotes the cardinality of a vector. The index is computed for the six illustrative portfolios for each of the constituent sectors (see Table 3): focusing on the first row of the table, reporting the index for the ENR sector, the results shown in Fig. 2(b) are confirmed. For each sector the portfolio can be ranked according to the $\bar{\Delta}_{\theta}^{const_i}$ index (see the second row in Table 3) and a global ranking can be obtained by computing the ranking on the average of the $\bar{\Delta}_{\theta}^{const_i}$ index on all the constituents (see Table 3).

Several tests using a set of simulated portfolios with increasing level of assets turnover were also carried out (results not shown for sake of brevity). The test results suggest the need to further investigate the use of QR estimates to obtain further information on the different levels of assets turnover.

4 Concluding Remarks

By using style analysis models, information may be obtained on the impact of exposure choices on portfolio returns. The classical LS model estimates the effect of style exposure on portfolio expected returns. The QR approach allows information to be extracted at other points other than the expected value, thus providing an appraisal of the influence of exposure choices on the entire conditional returns distribution. Therefore the estimated QR coefficients can be used to discriminate portfolios according to their assets turnover.

An index for ranking portfolios according to internal activeness was proposed. The index was computed on a set of six illustrative portfolios. From the results obtained, further investigation of the use of QR estimates would appear promising. The next step should concern the simulation of a larger set of portfolios according to different investment strategies.

Acknowledgements Authors wish to thank anonymous referee for helpful comments and suggestions on a previous draft of the paper: they helped to improve the final version of the work. This work has been supported by "Laboratorio di Calcolo e Analisi Quantitative", Dipartimento di Scienze Economiche, Università di Cassino.

References

Attardi, L., & Vistocco, D. (2006). Comparing financial portfolio style through quantile regression. *Statistica Applicata, 18*(2)

Attardi, L., & Vistocco, D. (2007). On estimating portfolio conditional returns distribution through style analysis models. In C. Perna & M. Sibillo (eds.), *Quantitative methods for finance and insurance*. Berlin: Springer

Basset, G. W., & Chen, H. L. (2001). Portfolio style: Return-based attribution using quantile regression. In B. Fitzenberger, R. Koenker, & J. A. F. Machado (eds.), *Economic applications of quantile regression (Studies in empirical economics)* (pp. 293–305). Berlin: Physica

Conversano, C., & Vistocco, D. (2004). Model based visualization of portfolio style analysis. In J. Antock (eds.), *Proceedings of the International Conference "COMPSTAT 2004")* (pp. 815–822). Berlin: Physica

Conversano, C., & Vistocco, D. (2009). Analysis of mutual fund management styles: A modeling, ranking and visualizing approach. *Journal of Applied Statistics*, in press.

Horst, J. K. Ter, Nijman, T.H, De Roon, F.A. (2004). Evaluating style analysis. *Journal of Empirical Finance, 11*, 29–51

Koenker, R., & Basset, G. W. (1978). Regression quantiles. *Econometrica, 46*, 33–50

Koenker, R. (2005). *Quantile regression*. Econometric Society Monographs. Cambridge: Cambridge University Press

Koenker, R. (2007). Quantreg: Quantile Regression. R package version 4.10, http://www.r-project.org

R Development Core Team. (2007). R: A Language Environment for Statistical Computing. http://www.r-project.org, R Foundation for Statistical Computing, Vienna, Austria

Sharpe, W. (1992). Asset Allocation: Management Styles and Performance Measurement. *The Journal of Portfolio Management*

Sharpe, W. (1998). Determining a Fund's Effective Asset Mix. *Investment management review*, (Vol. 2(6), pp. 59–69), **December**

Wickham, H. (2007). ggplot2: An Implementation of the Grammar of Graphics. R package version 0.5.6, http://had.co.nz/ggplot2

Bayesian Hidden Markov Models for Financial Data

Rosella Castellano and Luisa Scaccia

Abstract Hidden Markov Models, also known as Markov Switching Models, can be considered an extension of mixture models, allowing for dependent observations. The main problem associated with Hidden Markov Models is represented by the choice of the number of regimes, i.e. the number of generating data processes, which differ one from another just for the value of the parameters. Applying a hierarchical Bayesian framework, we show that Reversible Jump Markov Chain Monte Carlo techniques can be used to estimate the parameters of the model, as well as the number of regimes, and to simulate the posterior predictive densities of future observations. Assuming a mixture of normal distributions, all the parameters of the model are estimated using a well known exchange rate data set.

1 Introduction

A Hidden Markov Model (HMM) or Markov Switching Model is a mixture model whose mixing distribution is a finite state Markov Chain. In practice, given a data set indexed by time, the distribution of each observation is assumed to depend on an unobserved variable, hidden "state" or "regime", whose transition is regulated by a Markov Chain. HMMs have been successfully applied to financial time series: very often financial data show nonlinear dynamics which are possibly due to the existence of two or more regimes, differing one from another only for the value of the parameters. For instance, segmented time-trends in the US dollar exchange rates, Engel and Hamilton (1990), stylized facts about daily returns, Rydén (1998), option prices and stochastic volatilities, Rossi and Gallo (2006), temporal behavior of volatility of daily returns on commodities, Haldrup and Nielsen (2006), have also been modeled via HMMs.

The main problem associated with HMMs is to select the number of regimes (i.e. the number of generating data processes). In a classical perspective, this requires

L. Scaccia (✉)
DIEF, Università di Macerata, Via Crescimbeni, 20, 62100 Macerata, Italy
e-mail: scaccia@unimc.it

F. Palumbo et al. (eds.), *Data Analysis and Classification*,
Studies in Classification, Data Analysis, and Knowledge Organization,
DOI 10.1007/978-3-642-03739-9_51,

hypothesis testing with nuisance parameters, identified only under the alternative. Thus, the regularity conditions for the asymptotic theory to hold are not met and the limiting distribution of the likelihood ratio test must be approximated by simulation, an approach demanding enormous computational efforts. Penalized likelihood methods, as the Akaike and Bayesian information criteria, though are less demanding, do not produce a number quantifying the confidence in the results (i.e. p-values).

In a Bayesian context, several approaches to choose the number of regimes can be listed. A Bayesian non-parametric approach, based on a Dirichlet process (DP) with, a priori, infinite number of regimes is described in Otranto and Gallo (2002). Simulations from the posterior distribution of the process are used to estimate the posterior probabilities of the number of regimes. An alternative approach is based on allocation models: a latent variable is explicitly introduced to allocate each observation to a particular regime, Robert et al. (2000). Then, the Reversible Jump (RJ) algorithm, Green (1995), is used to sample from the posterior joint distribution of all the parameters, including the number of regimes.

In this paper, we prefer to deal with the latter approach for several reasons. From a theoretical point of view, the predictive density of a future observation, based on a DP, assigns to this observation a non-null probability of being exactly equal to one of those already observed. Such a behavior is highly unrealistic if data points are assumed to be drawn from a continuous distribution. Moreover, non-parametric approaches are strongly affected by the influence of the prior distribution on the posterior one, so that the likelihood never dominates the prior and the inferential results are particularly sensitive to prior assumptions. Furthermore, in a DP, a single parameter controls the variability and the clustering, making the prior specification difficult. Finally, the DP is well known to favor, a priori, unequal allocations and this phenomenon becomes more dramatic as soon as the number of observations increases. The unbalance in the prior allocation distribution often persists also a posteriori, Green and Richardson (2001). However, the model proposed in Robert et al. (2000) only allows for regimes being different because of their volatilities. We extend this approach to permit the existence of regimes characterized by different means and/or variances.

The paper is organized as follows: the model and prior assumptions are illustrated in Sect. 2; Sect. 3 deals with computational implementation; Sect. 4 discusses Bayesian inference and forecasting; finally, in Sect. 5 an application is considered.

2 The Model

Let $\boldsymbol{y} = (y_t)_{t=1}^T$ be the observed data, indexed by time. In HMMs, the heterogeneity in the data is represented by a mixture structure, that is, a pair (s_t, y_t), with s_t being an unobserved state variable characterizing the regime of the process at any time t and y_t being independent conditional on the s_t's:

$$y_t | s_t \sim f_{s_t}(y_t) \qquad \text{for} \quad t = 1, 2, \ldots, T , \tag{1}$$

with $f_{s_t}(\cdot)$ being a specified density function. Assuming $\mathcal{S} = \{1, \ldots, k\}$ to be the set of possible regimes, HMMs further postulate that the dynamics of $\boldsymbol{s} = (s_t)_{t=1}^T$ are described by a Markov Chain with transition matrix $\boldsymbol{\Lambda} = (\lambda_{ij})_{i,j=1}^k$. Accordingly, s_t is presumed to depend on the past realizations of $\boldsymbol{y}$ and $\boldsymbol{s}$, only through s_{t-1}:

$$p(s_t = j | s_{t-1} = i) = \lambda_{ij} \ .$$

We study mixtures of normal distributions, so that the model in (1) becomes

$$y_t | \boldsymbol{s}, \boldsymbol{\mu}, \boldsymbol{\sigma} \sim \phi(\cdot; \mu_{s_t}, \sigma_{s_t}^2) \qquad (2)$$

conditional on means $\boldsymbol{\mu} = (\mu_i)_{i=1}^k$ and standard deviations $\boldsymbol{\sigma} = (\sigma_i)_{i=1}^k$, where $\phi(\cdot; \mu_i, \sigma_i^2)$ is the density of the $N(\mu_i, \sigma_i^2)$. Thus, if $s_t = i$, y_t is assumed to be drawn from a $N(\mu_i, \sigma_i^2)$. Notice that, if we let $\boldsymbol{\pi}$ being the stationary vector of the transition matrix, so that $\boldsymbol{\pi}'\boldsymbol{\Lambda} = \boldsymbol{\pi}'$, and we integrate out s_t in (2) using its stationary distribution, the model in (2) can be analogously formalized as

$$y_t | \boldsymbol{\pi}, \boldsymbol{\mu}, \boldsymbol{\sigma} \sim \sum_{i=1}^k \pi_i \phi(\cdot; \mu_i, \sigma_i^2) \qquad \text{for} \quad t = 1, 2, \ldots, T \ .$$

In a classical perspective, the model in (2) can be estimated, conditional on k, by means of EM algorithm, Scott (2002). Then, as already mentioned, the main problem is to choose among different models, characterized by a different number of regimes.

In a Bayesian context, we formalize the uncertainty on the parameters of the model, as well as on the number of regimes, k, using appropriate prior distributions. We choose weakly informative priors, introducing an hyperprior structure, so that $\mu_i | \sigma_i^2 \sim N(\xi, \kappa\sigma_i^2)$ and $\sigma_i^{-2} \sim \text{Ga}(\eta, \zeta)$, independently for each $i = 1, \ldots, k$, with the mean and the variance of the Gamma distribution being η/ζ and η/ζ^2. Then we assume κ to follow an Inverse Gamma distribution with parameters q and r, and ζ to follow a Gamma distribution with parameters f and h. Finally, the rows of the transition matrix have a Dirichlet distribution, so that $\lambda_{ij} \sim D(\boldsymbol{\delta}_j)$, for $i = 1, \ldots, k$ where $\boldsymbol{\delta}_j = (\delta_{ij})_{i=1}^k$, while the number of regimes k is a priori uniform on the values $\{1, 2, \ldots, K\}$, with K being a pre-specified integer corresponding to the maximum hypothesized number of regimes. These settings lead up to the hierarchical model in Fig. 1. The choice of the hyperparameters will be briefly discussed in Sect. 5.

3 Computational Implementation

In order to approximate the posterior joint distribution of all the parameters of the above mixture model, Markov Chain Monte Carlo (MCMC) methods are applied (details can be found in Tierney (1994)). To generate realizations from the posterior

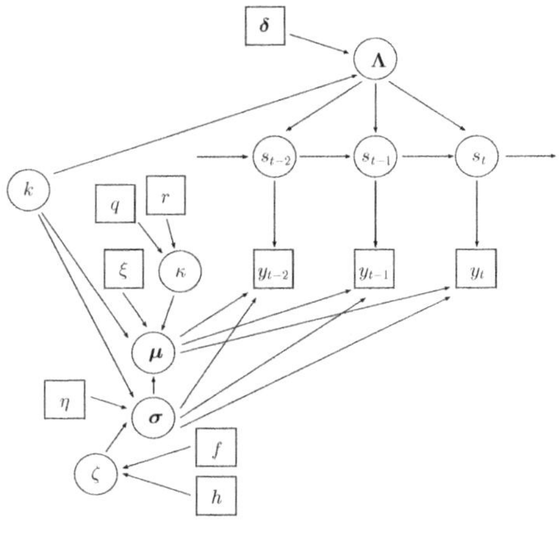

Fig. 1 Directed acyclic graph for the complete hierarchical model

joint distribution, at each sweep of the MCMC algorithm, we update in turn: **(a)** the transition matrix $\mathbf{\Lambda}$, **(b)** the state variable $\boldsymbol{s}$, **(c)** the means $\boldsymbol{\mu}$, **(d)** the standard deviations $\boldsymbol{\sigma}$, **(e)** the hyperparameter κ, **(f)** the hyperparameter ζ, **(g)** the number of regimes k. The first six moves are fairly standard and all performed through Gibbs sampling.

In particular, in **(a)**, the i-th row of $\mathbf{\Lambda}$ is sampled from $D(\delta_{i1}+n_{i1},\ldots,\delta_{ik}+n_{ik})$, where $n_{ij} = \sum_{t=1}^{T-1} I\{s_t = i, s_{t+1} = j\}$ is the number of transitions from regime i to regime j and $I\{\cdot\}$ denotes the indicator function, Robert et al. (1993).

In **(b)**, the standard solution for updating $\boldsymbol{s}$ would be to sample $s_1,\ldots,s_T$ one at a time from $t = 1$ to $t = T$, drawing values from their full conditional distribution $p(s_t = i|\cdots) \propto \lambda_{s_{t-1}i}\phi(y_t;\mu_i,\sigma_i^2)\lambda_{is_{t+1}}$ where "$\cdots$" denotes "all other variables". For a faster mixing algorithm, as in Scott (2002), and Castellano and Scaccia (2007), we instead sample $\boldsymbol{s}$ from $p(\boldsymbol{s}|\boldsymbol{y},\mathbf{\Lambda})$ through a stochastic version of the forward–backward recursion. The forward recursion produces matrices $\boldsymbol{P}_2,\ldots,\boldsymbol{P}_T$, where $\boldsymbol{P}_t = (p_{tij})$ and $p_{tij} = p(s_{t-1} = i, s_t = j|y_1,\ldots,y_t,\mathbf{\Lambda})$. In words, $\boldsymbol{P}_t$ is the joint distribution of $(s_{t-1} = i, s_t = j)$ given parameters and observed data up to time t. $\boldsymbol{P}_t$ is computed from $\boldsymbol{P}_{t-1}$ as $p_{tij} \propto p(s_{t-1} = i, s_t = j, y_t|y_1,\ldots,y_{t-1},\mathbf{\Lambda}) = p(s_{t-1} = i|y_1,\ldots,y_{t-1},\mathbf{\Lambda})\lambda_{ij}\phi(y_t;\mu_j,\sigma_j^2)$ with proportionality reconciled by $\sum_i\sum_j p_{tij} = 1$, where $p(s_{t-1} = i|y_1,\ldots,y_{t-1},\mathbf{\Lambda}) = \sum_j p_{t-1,i,j}$ can be computed once $\boldsymbol{P}_{t-1}$ is known. The recursion starts computing $p(s_1 = i|y_1,\mathbf{\Lambda}) \propto \phi(y_1;\mu_i,\sigma_i^2)\pi_i$ and thus $\boldsymbol{P}_2$. The stochastic backward recursion begins by drawing s_T from $p(s_T|\boldsymbol{y},\mathbf{\Lambda})$, then recursively drawing s_t from the distribution proportional to column s_{t+1} of $\boldsymbol{P}_{t+1}$. In this way, the stochastic backward recursion allows to sample from $p(\boldsymbol{s}|\boldsymbol{y},\mathbf{\Lambda})$, factorizing this distribution as $p(\boldsymbol{s}|\boldsymbol{y},\mathbf{\Lambda}) = p(s_T|\boldsymbol{y},\mathbf{\Lambda})\prod_{t=1}^{T-1} p(s_{T-t}|s_T,\ldots,s_{T-t+1},\boldsymbol{y},\mathbf{\Lambda})$ where $p(s_{T-t} = i|s_T,\ldots,s_{T-t+1},\boldsymbol{y},\mathbf{\Lambda}) = p(s_{T-t} = i|s_{T-t+1},y_1,\ldots,y_{T-t+1},\mathbf{\Lambda}) \propto p_{T-t+1,i,s_{T-t+1}}$.

In **(c)**, for identifiability purpose, we adopt a unique labeling in which the μ_i's are in increasing numerical order, Richardson and Green (1997). Hence, their joint prior distribution is $k!$ times the product of the individual normal densities, restricted to the set $\mu_1 < \mu_2 < \ldots < \mu_k$. The μ_i can be drawn independently from $\mu_i | \cdots \sim N\left(\frac{\kappa \sum_{t:s_t=i} y_t + \xi}{1+\kappa n_i}, \frac{\sigma_i^2 \kappa}{1+\kappa n_i}\right)$, n_i being the number of observations currently allocated in regime i. The move is accepted, provided the invariance of the order.

In **(d)** we update each component of the vector $\boldsymbol{\sigma}^2$ independently, drawing σ_i^{-2} from $\sigma_i^{-2} | \cdots \sim \text{Ga}\left(\eta + \frac{1}{2}(n_i + 1), \zeta + \frac{1}{2} \sum_{t:s_t=i} (y_t - \mu_i)^2 + \frac{1}{2\kappa}(\mu_i - \xi)^2\right)$.

In **(e)** we sample κ^{-1} from $\kappa^{-1} | \cdots \sim \text{Ga}\left(q + \frac{k}{2}, r + \frac{1}{2} \sum_{i=1}^{k} \frac{(\mu_i - \xi)^2}{\sigma_i^2}\right)$ and, finally, in **(f)** we sample ζ from $\zeta | \cdots \sim \text{Ga}\left(f + k\eta, h + \sum_{i=1}^{k} \sigma_i^{-2}\right)$.

Updating k implies a change of dimensionality for $\boldsymbol{\mu}$, $\boldsymbol{\sigma}$ and $\boldsymbol{\Lambda}$. We follow the approach used in Richardson and Green (1997) which consists in a random choice between splitting an existing regime into two, and merging two existing regimes into one. For the *combine proposal* we randomly choose a pair of regimes (i_1, i_2) that are adjacent in terms of the current value of their means. These two regimes are merged into a new one, $i^\star$, reducing k by 1. We then reallocate all the y_t, with $s_t = i_1$ or $s_t = i_2$, to the new regime $i^\star$ and create values for $\mu_{i^\star}$, $\sigma_{i^\star}^2$, $\pi_{i^\star}$ and for the transition probabilities from and to the regimes involved in the move. This is performed in such a way to guarantee that the new HMM and the old one both have the same first and second moments. The *split proposal* starts with the random choice of regime $i^\star$ which is splitted into two new ones, i_1 and i_2, augmenting k by 1. Accordingly, we reallocate the y_t with $s_t = i^\star$ between i_1 and i_2, and create values for $\pi_{i_1}, \pi_{i_2}, \mu_{i_1}, \mu_{i_2}, \sigma_{i_1}, \sigma_{i_2}$ and the transition probabilities for the regimes involved. The aim is to split $i^\star$ in such a way that the dynamics of the Hidden Markov Chain are essentially preserved, Robert et al. (2000). The move is accepted with a probability computed to preserve the *reversibility* between the states of the MCMC algorithm. More details on computational issues can be found in Castellano and Scaccia (2007).

4 Bayesian Inference and Forecasting

After a burn-in period, to guarantee the convergence of the chain to its stationary distribution, the RJ algorithm produces at each sweep n, for $n = 1, \ldots, N$, a draw $(k^{(n)}, \boldsymbol{\Lambda}^{(n)}, \boldsymbol{s}^{(n)}, \boldsymbol{\mu}^{(n)}, \boldsymbol{\sigma}^{(n)}, \kappa^{(n)}, \zeta^{(n)})$ from the joint posterior distribution of all the parameters, including k. The sample obtained after N sweeps can be used to estimate all the quantities of interest. For instance, we can easily estimate the posterior distribution of the number of regimes as the proportion of times each model is visited by the algorithm, i.e. $\hat{p}(k = \ell | \boldsymbol{y}) = \sum_{n=1}^{N} I\{k^{(n)} = \ell\}/N = N_\ell/N$, where N_ℓ is the number of times the model with ℓ regimes is visited.

Conditioning on a particular model, say M_ℓ, the one with ℓ regimes, any other parameter of that model can be estimated, Richardson and Green (1997). Estimating the hidden states $\boldsymbol{s}$ often represents a key question in applied problems. Inference on $\boldsymbol{s}$ derives from its posterior $p(\boldsymbol{s}\,|\,\boldsymbol{y})$, a high-dimensional distribution that must be summarized to be understood. In general, it is sufficient to summarize it through its marginal distributions $p(s_t = i\,|\,\boldsymbol{y})$, whose obvious estimates are $\hat{p}(s_t = i\,|\,\boldsymbol{y}) = \sum_{n:k^{(n)}=\ell} I\{s_t^{(n)} = i\}/N_\ell$. More efficient estimates demand small additional computational effort, as shown in Castellano and Scaccia (2007).

Finally, when historical series are analysed, the main goal is, generally, to forecast future values of the observed variable, on the basis of the information available up to time T. In a Bayesian context, inferences on future observations, i.e. $\boldsymbol{Y} = (y_{T+1}, \ldots, y_{T+G})$, are based on their posterior predictive density, which can be defined into two different ways, depending on what we consider as "information available at time T". If we believe that data are generated by a specific model, say M_ℓ, the information available up to time T will encompass the generating model and the observed data up to time T. Then, the posterior predictive density for $\boldsymbol{Y}$ will be:

$$p(\boldsymbol{Y}\,|\,\boldsymbol{y}, M_\ell) = \int_{\Theta_{M_\ell}} p(\boldsymbol{Y}\,|\,\boldsymbol{y}, \boldsymbol{\theta}_{M_\ell}, M_\ell)\, p(\boldsymbol{\theta}_{M_\ell}\,|\,\boldsymbol{y}, M_\ell)\, d\boldsymbol{\theta}_{M_\ell}, \tag{3}$$

where $\boldsymbol{\theta}_{M_\ell}$ is the vector of all parameters (except k), including the state variable, under the model with ℓ regimes, and Θ_{M_ℓ} is the relative parameter space.

Otherwise, the uncertainty about which one is the true generating model, within a set of possible ones, can be expressed through model averaging, defining the posterior predictive density of $\boldsymbol{Y}$ as:

$$p(\boldsymbol{Y}\,|\,\boldsymbol{y}) = \sum_{k=1}^{K} p(\boldsymbol{Y}\,|\,\boldsymbol{y}, M_k)\, p(M_k\,|\,\boldsymbol{y}) \;, \tag{4}$$

where $p(\boldsymbol{Y}\,|\,\boldsymbol{y}, M_k)$ is defined in (3). Notice that in (4), we consider as available information at time T only the observed data. In both cases, the posterior predictive density can be stimulated as by-product of the MCMC algorithm, Scott (2002).

5 An Application to Financial Data

The proposed model is applied to the exchange rate quarterly returns of the U.S. dollar relative to the French franc, over the period 1973-III to 1988-I, already analysed in Engel and Hamilton (1990) and Otranto and Gallo (2002) through a likelihood ratio and a non-parametric Bayesian approach, respectively. While the approach in Engel and Hamilton (1990) failed to test the model with two regimes against the one with only one regime (more details are reported in Otranto and Gallo (2002)), in Otranto and Gallo (2002) the posterior probability for $k = 2$ was found to be slightly higher than that for $k = 1$. The choice of the data set is motivated by the existence of a benchmark for comparing the results.

For previously unspecified hyperparameters, we let: $\delta_{ij} = 1$, $\forall i, j$; $\xi = 0$; $\eta = 2$; $q = 2$; $r = 2$; $f = 0.95$; $h = 4/R^2$, where R is the data range. All the hyperparameters were chosen in a way that the priors on the parameters would assign large probabilities to a large range of values. Some experimentations highlight that the results are quite robust with respect to reasonable perturbations on the choice of the hyperparameters. Models with a number of regimes up to $K = 5$ were considered. Larger values for K would be unreasonable, given the short time series at hand. Trace plots of the sample parameter values versus iteration, as well as of the sample posterior probability of the number of regimes Fig. 2(b)), were used to control for the stabilization of the simulation. A burn-in of 100,000 sweeps seems sufficient for the convergence to occur. We performed 1,000,000 sweeps of the MCMC algorithm. We observed a quite high acceptance rate ($\approx$17%) for the RJ move, due to the fact that the data set is small and, thus, the posterior distribution of the parameters is not particularly picked, making the algorithm move easily over different models (Fig. 2(a)). After the convergence, the algorithm provides with stable estimates of the posterior model probabilities, given in Fig. 2(b). As in Otranto and Gallo (2002), we get evidence in favor of the two regimes model, the mode of the posterior probability being $k = 2$.

Conditionally on $k = 2$, Fig. 3(b) shows the posterior probability of the U.S. dollar being in a regime of appreciation, with vertical lines representing the switches from one regime to the other. The posterior probability, as well as the switching points, resemble closely the results obtained in Engel and Hamilton (1990). Furthermore, the process seems to stay in the same regime for a while, as confirmed by the estimated transition matrix

$$\hat{\Lambda} = \begin{pmatrix} 0.676 & 0.324 \\ 0.350 & 0.650 \end{pmatrix}$$

showing higher probabilities to persist in the same regime compared to those of switching to the other one (i.e.: the so called *long swings* of the U.S. dollar).

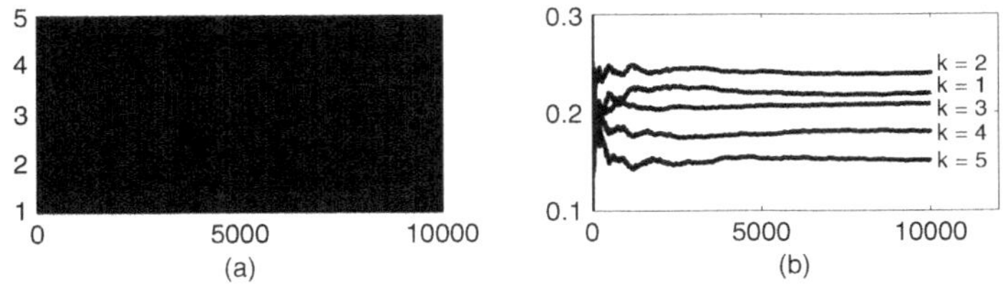

Fig. 2 **(a)** Last 10,000 values of k. **(b)** Estimated posterior distribution of k as a function of number of sweeps, plotted every 100*th* sweep

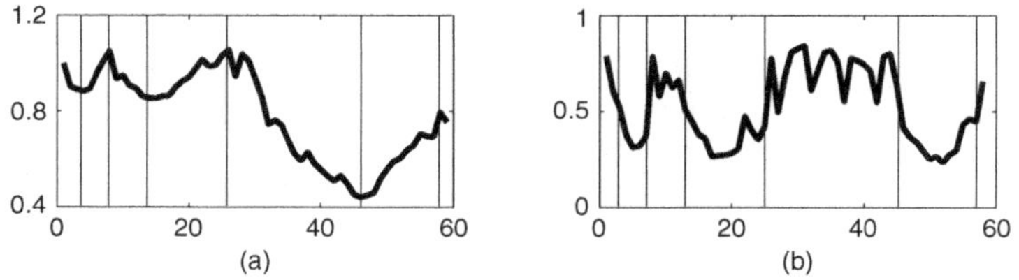

Fig. 3 **(a)** U.S. dollar/French franc exchange rate. **(b)** Estimated posterior probability of the U.S. dollar being in a appreciation regime as a function of t

6 Conclusions

In this paper Bayesian inference for HMMs with an unknown number of regimes and its application to financial time series is illustrated. We considered a hierarchical model which allows to make vague *a priori* assumptions on the parameters. The analytically untractable joint posterior distribution of all the parameters and the unknown number of regimes was simulated through MCMC methods and RJ algorithm.

Future developments could encompass the design of RJ moves visiting a larger set of models, in which some regimes may have equal variances but different means or equal means but different variances. The approach can also be adapted to any extension of HMMs, such as time-varying transition probabilities, Markov switching heteroskedasticity, multiple regime Smooth Transition or Threshold AR models.

References

Castellano, R., & Scaccia, L. (2007). Bayesian hidden Markov Models with an unknown number of regimes. Quaderni del Dipartimento di Istituzioni Economiche e Finanziarie, 43. Università di Macerata

Engel, C., & Hamilton, J. D. (1990). Long swings in the dollar: Are they in the data and do markets know it? *The American Economic Review*, *80*, 689–713

Green, P. J. (1995). Reversible jump Markov chain Monte Carlo computation and Bayesian model determination. *Biometrika*, *82*, 711–732

Green, P. J., & Richardson, S. (2001). Modelling heterogeneity with and without the Dirichlet process. *Scandinavian Journal of Statistics*, *28*, 355–375

Haldrup, N., & Nielsen, M. O. (2006). A regime switching long memory model for electricity prices. *Journal of Econometrics*, *135*, 349–376

Otranto, E., & Gallo, G. (2002). A nonparametric Bayesian approach to detect the number of regimes in Markov switching models. *Econometric Reviews*, *21*, 477–496

Richardson, S., & Green, P. J. (1997). On Bayesian analysis of mixtures with an unknown number of components (with discussion). *Journal of the Royal Statistical Society – Series B: Statistical Methodology*, *59*, 731–792

Robert, C. P., Celeux, G., & Diebolt, J. (1993). Bayesian estimation of hidden Markov chains: a stochastic implementation. *Statistics & Probability Letters*, *16*, 77–83

Robert, C. P., Rydén, T., & Titterington, D. M. (2000). Bayesian inference in hidden Markov models through the reversible jump Markov chain Monte Carlo method. *Journal of the Royal Statistical Society - Series B: Statistical Methodology*, *62*, 57–75

Rossi, A., & Gallo, G. (2006). Volatility estimation via hidden Markov models. *Journal of Empirical Finance*, *13*, 203–230

Rydén, T., Teräsvirta, T., & Åsbrink, S. (1998). Stylized facts of daily return series and the hidden Markov model. *Journal of Applied Economics*, *13*, 217–244

Scott, S. L. (2002). Bayesian methods for hidden Markov models: Recursive computing in the 21st century. *Journal of the American Statistical Association*, *97*, 337–351

Tierney, L. (1994). Markov chains for exploring posterior distributions. *Annals of Statistics*, *22*, 1701–1764

Part XI
Missing Values

Regression Imputation for Space-Time Datasets with Missing Values

Antonella Plaia and Anna Lisa Bondì

Abstract Data consisting in repeated observations on a series of fixed units are very common in different context like biological, environmental and social sciences, and different terminology is often used to indicate this kind of data: panel data, longitudinal data, time series-cross section data (TSCS), spatio-temporal data. Missing information are inevitable in longitudinal studies, and can produce biased estimates and loss of powers. The aim of this paper is to propose a new regression (single) imputation method that, considering the particular structure and characteristics of the data set, creates a "complete" data set that can be analyzed by any researcher on different occasions and using different techniques. Simulated incomplete data from a PM_{10} dataset recorded in Palermo in 2003 have been generated, in order to evaluate the performance of the imputation method by using suitable performance indicators.

1 Introduction

Data consisting in repeated observations on a series of fixed units are very common in different context like biological, environmental and social sciences, and different terminology is often used to indicate this kind of data: panel data, longitudinal data, time series-cross section data (TSCS, Beck 2001), spatio-temporal data (when there is a spatial relation among units). Even if we can always refer to the generic element of the dataset with x_{it}, $i = 1, \ldots, N$, $t = 1, \ldots, T$, usually panels have large number of cross-sections (big N) with each unit observed only a few times (small T), while TSCS or longitudinal data have reasonable sized T and not very large N. Missing information are inevitable in longitudinal studies, and can produce biased estimates and loss of powers. Statistical methods are available that take the missing data into account at the time of analysis. These methods include

A. Plaia (✉)
Department of Statistical and Mathematical Sciences "S. Vianelli", University of Palermo, viale delle Scienze - ed. 13, 90128 Palermo, Italy
e-mail: plaia@unipa.it

F. Palumbo et al. (eds.), *Data Analysis and Classification*,
Studies in Classification, Data Analysis, and Knowledge Organization,
DOI 10.1007/978-3-642-03739-9_52,

likelihood-based approaches, such as generalized linear models and the expectation-maximization (E-M) algorithm when data are "missing at random" (the probability that a value is missing can depend on observed values but does not depend on the missing value itself). A different approach consists in imputing the missing values so that the resulting data set is "complete". This possibility is to be chosen if the data set will be used for many different types of analysis by a number of researchers.

Methods available for creating complete data matrices can be divided into two main categories: single imputation (SI) and multiple imputation (MI) methods. Single imputation methods fill in one value for each missing one; they have many appealing features, because standard complete-data methods can be applied directly and because imputation need to be carried out only once. Multiple imputation methods generate multiple simulated values for each missing value, in order to reflect the uncertainty attached to missing data.

Among the single imputation methods for longitudinal data, we can distinguish methods based on the information on the same subject (e.g. *last observation carried forward*, *next observation carried backward*, *last & next* (Engels and Diehr 2003)), methods that borrow information from other subjects (*row mean/median*) and methods that use both pieces of information (e.g. *conditional mean imputation*, *hot-deck imputation*).

In the present paper we will propose a new single imputation method of the last type, and we will show how it can be transformed into a multiple imputation one. In order to evaluate its performance we will refer to a dataset consisting in PM_{10} concentrations measured every 2 h by eight monitoring stations distributed over the urban area of Palermo, Sicily, during 2003 (4380 observations (T) in eight (N) sites, Table 1).

In this typical space-time dataset, it is particularly important to consider both the row and column information (in Table 1) to impute a missing value (that is to consider both the spatial (row) and temporal (column) information brought by the data). Actually, differently from what can happen for a climatological data set (that can shows the same structure as our data set), here it results to be a main point to consider site specific effects. For example, denoting the generic element of the data set (Table 1) as x_{swdh}, where s refers to the monitoring site ($s = 1, 2, \ldots S$), w to

Table 1 Data structure

W[a]	W-D	H	St1	St2	St3	St4	St5	St6	St7	St8
1	3	2	23.32	12.47	108.02	30.46	NA	4.29	45.05	50.91
1	3	4	34.68	27.09	28.17	9.34	20.10	26.66	NA	26.36
1	3	6	26.85	11.10	34.70	24.17	29.76	21.94	NA	44.46
1	3	8	19.80	24.62	31.51	27.23	34.96	26.20	NA	45.89
1	3	10	24.38	14.84	23.91	18.50	15.21	20.10	20.94	NA
1	3	12	22.47	16.46	30.72	38.54	35.87	16.76	15.20	NA
...	...	...	...	...	...	...	...	...	...	...
53	3	24	29,95	26,39	58,56	26,55	23,73	26,427	26,37	22,31

[a] W refers to the Week $(1, 2, \ldots, 53)$, W-D to the Week-Day $(1, 2, \ldots 7)$ and H to the Hour $(2, 4, \ldots 24)$

the Week ($w = 1, 2, \ldots, 53$), d to the Week-Day ($d = 1, 2, \ldots 7$) and h to the Hour ($h = 2, 4, \ldots 24$), we can consider an hour site mean matrix, which is a 12×8 matrix whose generic element $\bar{x}_{s \cdot\cdot h}$ is the mean of the values observed on day hour h ($h = 2, 4, \ldots, 24$) in site s. As a consequence, it is possible to compute a specific day-hour effect of site s, defined as the difference between $\bar{x}_{s \cdot\cdot h}$ and its marginal row mean $\sum_{s=1}^{S} \frac{\bar{x}_{s \cdot\cdot h}}{S}$. This site-specific effect is usually different from site to site, due to particular sources of pollution located next to the monitoring site.

In (Plaia and Bondì 2006) the authors proposed a single imputation method, SDEM, that considers explicitly a week effect, a day effect and an hour effect (all site-dependent), assuming their additivity. According to SDEM, a missing value will be estimated by:

$$
\begin{aligned}
\hat{x}_{swdh} &= \bar{x}_{\cdot wdh} + \frac{1}{2}\left(\bar{x}_{sw\cdot\cdot} - \sum_{s=1}^{S} \frac{\bar{x}_{sw\cdot\cdot}}{S}\right) \\
&+ \frac{1}{2}\left(\bar{x}_{s \cdot d \cdot} - \sum_{s=1}^{S} \frac{\bar{x}_{s \cdot d \cdot}}{S}\right) + \frac{1}{2}\left(\bar{x}_{s \cdot\cdot h} - \sum_{s=1}^{S} \frac{\bar{x}_{s \cdot\cdot h}}{S}\right). \qquad (1)
\end{aligned}
$$

Here the coefficients "1/2" represent a good compromise, widely tested by the authors, between overestimation and underestimation in the imputed data. The aim of this paper is to transform SDEM into a *regression* (single) imputation method that, considering the particular structure and characteristics of the data set, creates a "complete" data set that can be analyzed by any researcher on different occasions and using different techniques. Moreover we will show how the new method can be converted into a multiple imputation one.

2 A New Regression Single Imputation Method

According to (1), SDEM imputes a missing value through a linear combination of the row mean in Table 1 and the three effects (week, day and hour), with fixed coefficient vector $\beta = [1; 0.5; 0.5; 0.5]$. These values come from a previous analysis, as a compromise between the *Row-Mean* method (Plaia and Bondì 2006) and a previous version of SDEM that, without prior information, considered a fixed coefficient vector $\beta = [1; 1; 1; 1]$. A straightforward transformation of SDEM is towards a real regression imputation method:

$$
\begin{aligned}
\hat{x}_{swdh} &= \beta_0 + \beta_1 \bar{x}_{\cdot wdh} + \beta_2 \left(\bar{x}_{sw\cdot\cdot} - \sum_{s=1}^{S} \frac{\bar{x}_{sw\cdot\cdot}}{S}\right) + \beta_3 \left(\bar{x}_{s \cdot d \cdot} - \sum_{s=1}^{S} \frac{\bar{x}_{s \cdot d \cdot}}{S}\right) \\
&+ \beta_4 \left(\bar{x}_{s \cdot\cdot h} - \sum_{s=1}^{S} \frac{\bar{x}_{s \cdot\cdot h}}{S}\right). \qquad (2)
\end{aligned}
$$

where $\beta = [\beta_0; \beta_1; \beta_2; \beta_3; \beta_4]$ need to be estimated.

Assuming a normal distributed stochastic component (the opportunity of a more appropriate density function for the error is being studied), here we compare three different approaches to estimate β:

- a single linear regression (OLS estimation): therefore a single vector of parameters;
- S site-dependent linear regressions (OLS estimations): therefore a different vector β_s for each monitoring site;
- a multilevel linear model, with monitoring sites as second level units and single measures as first level units, where, as usual in multilevel linear models (Snijders and Bosker 1999), $\beta = \gamma + \mathbf{U}$ where $\mathbf{U}$ reflects the unexplained variation between sites (ML estimation).

Two performance indicators have been considered to assess the goodness of imputation, and to compare it with SDEM' and Row-mean' ones. Both of them are based on the differences between observed and imputed values, as these are the most reliable type of methods to test an imputation technique. Denoting with O_i the $i-th$ observed data point, with $\overline{O}$ the average of observed data, with P_i the $i-th$ imputed data point, with $\overline{P}$ the average of imputed data, with σ_O the standard deviation of the observed data and σ_P the standard deviation of the imputed data, and finally with n the number of imputations (that is the number of missing data), we use (Junninen et al. 2004):

1. *the coefficient of correlation (ρ)* between observed and imputed:

$$\rho = \left[\frac{1}{n} \frac{\sum_{i=1}^{n}[(P_i - \overline{P})(O_i - \overline{O})]}{\sigma_P \sigma_O} \right] \quad \rho \in [-1; 1]; \tag{3}$$

2. *an index of agreement (d)*

$$d = 1 - \left[\frac{\sum_{i=1}^{n}(P_i - O_i)^2}{\sum_{i=1}^{n}(|P_i - \overline{O}| + |O_i - \overline{O}|)^2} \right] \quad d \in [0; 1]. \tag{4}$$

The index of agreement, with respect to the coefficient of correlation, is related to the sizes of the discrepancies between predicted and observed values.

3 Missing Data Simulation

In order to evaluate the performance of the imputation methods, simulated incomplete data have been generated from Palermo dataset, after which the methods have been applied to the data and the performance indicators computed. Actually, the imputation performance does depend both on the amount of missing data and on the missing data pattern: therefore the real missing data pattern has been analysed in order to reproduce it as better as possible.

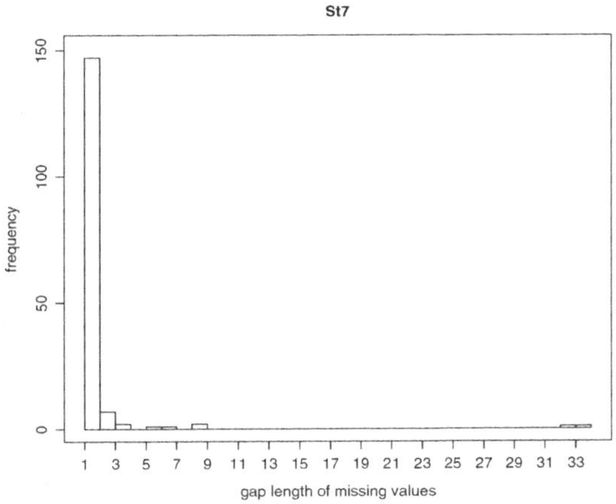

Fig. 1 Distribution of the gap length in a monitoring site

Figure 1 shows a typical distribution of the gap length in a monitoring site (all the monitoring sites show a similar behaviour): as it is possible to notice, the distribution is caracterized by many short gaps (most of them are length-one gaps), but also a few long gaps, of about thirty consecutive missing values (almost 3 days). While simulating incomplete data, we tried to reproduce this pattern as better as possible.

Referring to Table 1, we will consider four different missing data patterns that differ for the total percentage of missing data in the table, for the distribution of the gap length, and for the maximum number of missing values per row. Two different total amounts of missing data have been considered: about 5% and about 15%. Two maximum numbers of missing values per row (in Table 1) have been considered, four and eight, this means that we have supposed to have up to four or up to eight monitoring sites out of order at the same time. For each of the four missing data patterns, 100 missing data indicator matrices M have been generated, drawing the *gap length* from a mixture of two distributions, an Exponential of parameter $\lambda = 0.5$ and a Uniform with parameters (20, 70) and (40, 120) for the 5 and 15% missing data patterns respectively, the *starting point of each gap* from a Uniform (1, 4380) and the number of missing values per row (in Table 1) from a Uniform (0,4) or (0,8).

Matrices M applied to the observed data set create "artificially" missing data (actually real values are known) and this allows to compute the value of the performance indicators to assess the goodness of the imputation methods.

All the analysis, together with the generation of the 400 missing data indicator matrices M, have been carried out using the free software R (R Development Core Team 2007).

Table 2 Performance indices: comparison of methods versus missing data patterns (mean over the 100 matrices M and relative standard deviation)

	Single-regression		S-regressions		Multilevel		SDEM		Row-Mean	
	r	d	r	d	r	d	r	d	r	d
5% 0–4	0.75	0.85	0.76	0.86	0.75	0.85	0.75	0.85	0.65	0.79
	(0.03)	(0.02)	(0.03)	(0.02)	(0.03)	(0.02)	(0.02)	(0.02)	(0.03)	(0.02)
5% 0–8	0.73	0.83	0.74	0.84	0.74	0.83	0.73	0.83	0.64	0.77
	(0.11)	(0.12)	(0.11)	(0.12)	(0.11)	(0.12)	(0.11)	(0.12)	(0.10)	(0.11)
15% 0–4	0.70	0.82	0.71	0.83	0.70	0.82	0.72	0.84	0.65	0.79
	(0.03)	(0.02)	(0.03)	(0.02)	(0.04)	(0.02)	(0.02)	(0.01)	(0.02)	(0.01)
15% 0–8	0.71	0.83	0.72	0.83	0.71	0.83	0.73	0.84	0.65	0.79
	(0.03)	(0.02)	(0.03)	(0.02)	(0.03)	(0.02)	(0.02)	(0.01)	(0.02)	(0.01)

4 Results

Table 2 shows the values of both the correlation coefficient and the index of agreement gained by the five compared methods, according to the four missing data patterns (averaged over the 100 simulated incomplete data sets), together with the corresponding standard deviations (in brackets). As it is possible to notice, all the three approaches to estimate the regression parameters β produce good and comparable results: the single-regression approach, being less computer intensive, results to be preferable. But we can also notice that SDEM' performance is always comparable with regression approaches' one, and indeed outperforms regression as soon as the percentage of missing data increases. As it was foreseeable, regression methods result to be more sensitive to the increases in the percentage of missing, due to the necessity of estimating the regression parameters, on the contrary, with SDEM we don't estimate parameters (as they are assumed to be known), causing the method to be more robust.

The index of agreement (which is considered more informative than the correlation coefficient) plot (Fig. 2) shows the superiority of SDEM as soon as the percentage of missing data increases, while Row-Mean imputation appears not to be competitive at all.

5 Toward a Multiple Imputation Method

Both SDEM and the (single) regression imputation method can be transformed into a multiple imputation method (Schenker and Taylor 1996). Drawing the parameters from their posterior distribution under the Jeffreys prior (Box and Tiao 1973), which is a diffuse prior, the multiple imputation acts according to the following steps:

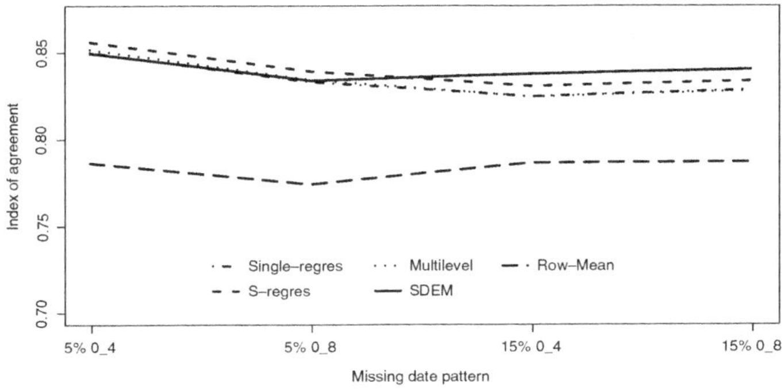

Fig. 2 Index of agreement versus the four missing data patterns

MI-SDEM

1. SDEM is applied to the observed data;
2. a value σ^* is drawn with $\sigma^{*2} \sim SSE/\chi^2_{n_{obs}}$ where SSE is computed from

$$r_{swdh} = x_{swdh} - \bar{x}_{\cdot wdh} - \frac{1}{2}\left(\bar{x}_{sw\cdot\cdot} - \sum_{s=1}^{S}\frac{\bar{x}_{sw\cdot\cdot}}{S}\right) - \frac{1}{2}\left(\bar{x}_{s\cdot d\cdot} - \sum_{s=1}^{S}\frac{\bar{x}_{s\cdot d\cdot}}{S}\right) - \frac{1}{2}\left(\bar{x}_{s\cdot\cdot h} - \sum_{s=1}^{S}\frac{\bar{x}_{s\cdot\cdot h}}{S}\right)$$

   (the sum is extended to the imputed values only);
3. a missing value x_{swdh} is imputed drawing $\hat{x}_{swdh}$ from a $N(Z^T\beta, \sigma^{*2})$

MI-reg (Starting from a linear regression model)

1. A linear model is fitted to the observed data;
2. a value σ^* is drawn with $\sigma^{*2} \sim SSE/\chi^2_{n_{obs}-p-1}$ where SSE is the residual sum of squares from the least-squares fit (χ^2 degrees of freedom are decreased, with respect to MI-SDEM, by the number of parameters in the linear model);
3. a value β^* is drawn with $\beta^* \sim N(\hat{\beta}, \sigma^{*2}(Z^TZ)^{-1})$ where Z is the design matrix (site-dependent week effect, day effect and hour effect);
4. a missing value x_{swdh} is imputed drawing $\hat{x}_{swdh}$ from a $N(Z^T\beta^*, \sigma^{*2})$.

In order to evaluate the procedures, it has been replicated independently $m = 5$ times for both MI-reg and MI-SDEM.

Figure 3 shows the PM_{10} daily means computed both from the observed bi-hour values and from the imputed ones. The data refer to Station 1 taken as example, and a randomly selected matrix M with a pattern of 5% missing and up to eight monitoring sites out of order at the same time. Daily means has been considered here as all the standards for PM_{10}, in directives or guidelines, refer to daily mean (or annual means). As it is possible to notice, both SDEM and the Single regression methods perfectly reproduce the observed data. The peaks are perfectly estimated, and this is a desirable characteristic of an imputation method for pollution data, as European and Italian directives (as well as USA ones) specify the annual number of exceedences permitted.

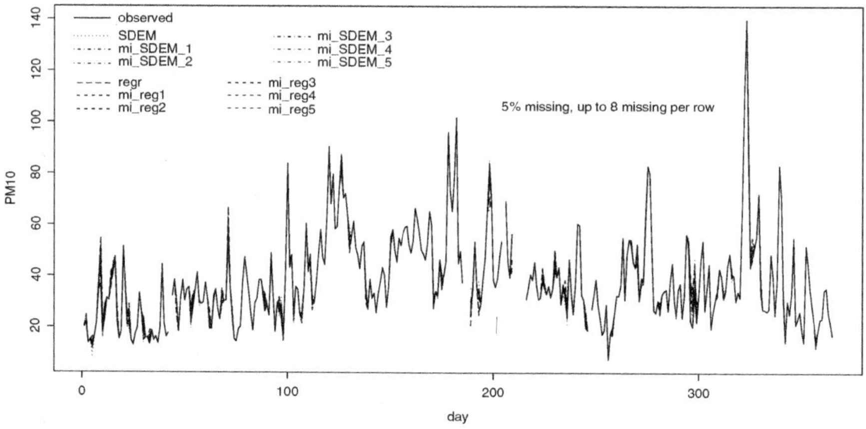

Fig. 3 Daily means: observed and imputed values (Station 1)

Acknowledgements This work is partially funded by the 2006 MIUR grant n.2006131039_003 "Analisi e valutazione di rischi ambientali mediante modelli temporali, spaziali e spazio-temporali".

References

Beck, N. (2001). Time-series-cross-section data. *Statistica Neerlandica, 55*, 111–133

Box, G. E. P., & Tiao, G. C. (1973). *Bayesian inference in statistical analysis*. Reading: Addison-Wesley

Engels, J. M., & Diehr, P. (2003). Imputation of missing longitudinal data: a comparison of methods. *Journal of Clinical Epidemiology, 56*, 968–976

Junninen, H., Niska, H., Tuppurainen, K., Ruuskanen, J., & Kolehmainen, M. (2004). Methods for imputation of missing values in air quality data sets. *Atmospheric Environment, 38*, 2895–2907

Little, R. J. A., & Rubin, D. B. (2002). *Statistical analysis with missing data* (2nd ed.). New York: Wiley

Plaia, A., & Bondì, A. L. (2006). Single imputation method of missing values in environmental pollution data sets. *Atmospheric Environment, 40*, 7316–7330

Snijders, T. A. B., & Bosker, R. J. (1999). Multilevel analysis: An introduction to basic and advanced multilevel modeling. London: Sage

R Development Core Team. (2007). R: A language and environment for statistical computing. R Foundation for Statistical Computing, Vienna, Austria. ISBN 3-900051-07-0 http://www.R-project.org.

Schenker, N., & Taylor, J. M. G. (1996). Partially parametric techniques for multiple imputation. *Computational Statistics & Data Analysis, 22*, 425–446

A Multiple Imputation Approach in a Survey on University Teaching Evaluation

Isabella Sulis and Mariano Porcu

Abstract Missing data is a problem frequently met in many surveys on the evaluation of university teaching. The method proposed in this work uses *multiple imputation by stochastic regression* (MISR) in order to recover partially observed units in surveys where multi-item Likert-type scale are used to measure a latent attribute, namely the *quality of university teaching*. The accuracy of the method has been tested simulating missing values in a benchmark data set *completely at random* (MCR) and *at random* (MAR). A simulation analysis has been carried out in order to assess the accuracy of the imputation procedure according to two standard criteria: accuracy in "distribution" and in "estimation". The procedure has been compared with others widely applied missing data handling methods: *multiple imputation by chained equation* (MICE) and *complete cases analysis* (CCA).

1 Introduction

In the survey on the evaluation of university teaching quality the solution of limiting the analysis to units not affected by missingness could produce a selection bias keeping in the analysis only those students who provide more attention in answering questionnaires. This work proposes a multiple imputation analysis (MIA) (Little and Rubin- 2002) based on stochastic regression (MISR) to cope with missing values in surveys where variables measured on Likert-type scale (with the same number of response categories) define the same latent attribute. The MISR approach replaces missing values with random draws from a distribution whose parameters have been estimated by parametric regressions. The method uses both the information provided by the observed values of the variables affected by missingness and the multivariate structure of the data. The application tackles with data sets affected by different rates of missingness and the mechanisms which generate missing values are of two different types: *Missing Completely at Random* (MCAR) and *Missing at Random*

I. Sulis (✉)
Dip. Ric. Economiche e Sociali - Univ. di Cagliari, Viale S. Ignazio 78, Italy
e-mail: isulis@unica.it

F. Palumbo et al. (eds.), *Data Analysis and Classification*,
Studies in Classification, Data Analysis, and Knowledge Organization,
DOI 10.1007/978-3-642-03739-9_53,

(MAR) (Rubin 1976). The procedure has been validated according to the criteria of *Accuracy in Distribution* (AD) and *Accuracy in Estimation* (AE) (Chambers 2001; Berrington and Borgonini 2004). Data sets with values simulated missing have been imputed using *multiple imputation by stochastic regression* (MISR) and *multiple imputation by chained equations* (MICE). Results arisen from both missing data multiple imputation methods have been compared with the *complete cases analysis* (CCA).

2 An Imputation Procedure to Recover for Missingness

This MISR procedure works in two steps. Let us define a data matrix with n units and p items. For the sake of simplicity, the method is described by supposing that items are questions on teaching attributes and responses are recorded on a 4 (K) category Likert-type scale: 1 = *Definitely No - DN*, 2 = *More No than Yes – MN*, 3 = *More Yes than No – MY*, 4 = *Definitely Yes – DY*. Table 1 (for $i = 3$ obs. and $p = 7$ items) shows the first three units.

Step 1

The procedure starts by building up for each unit i the distribution of the relative frequencies of ratings in each of the K response categories. From Table 1 arises that the rate of response for unit ♯1 is: $DN = \frac{3}{5} = 0.60$; $MN = \frac{2}{5} = 0.40$; $MY = DY = \frac{0}{5} = 0.00$. Unobserved items for each unit i are replaced by drawing values from a *Multinomial* distribution with parameters set equal to the relative frequencies of ratings in each category. For each unobserved record M values are drawn from a *Multinomial* $(\pi_1, \pi_2, \pi_3, \pi_4)$ distribution. The first step uses just the information on subject's pattern of responses in order to generate values. As a result of the multiple imputation procedure M data sets are generated; in each data set simulated values fill in the unobserved records.

Step 2

Let us consider one of the M imputed data set obtained as described in Step 1. In this second step a stochastic regression approach is used: p regression equations are specified. Each of the p items is considered once as a *response variable* (Y) whose values depends upon the set of $p-1$ remaining predictors $\boldsymbol{x}$ and $p-1$ times as predictor. For instance responses to I_0 are assumed to depend upon responses

Table 1 Example of a data matrix affected by missingness

unit	items						
♯	I_0	I_1	I_2	I_3	I_4	I_5	I_6
1	•	1	2	2	1	1	•
2	4	4	4	•	3	4	4
3	•	3	•	4	2	•	3

provided to $I_1 - I_6$, whereas responses to I_1 are assumed to depend upon responses provided to $I_0, I_2 - I_6$ and so forth.

By adopting a *proportional odds* logistic regression model (Agresti 2002) for ordered categorical variables the *logit* of the probability that subject i answers a category lower rather than greater than k for item I_j [$\boldsymbol{y}_i = (y_{i1}, \dots, y_{iK})$] is specified as a function of the pattern of responses provided in the remaining $p-1$ items [$\boldsymbol{x}_i = (x_{i1_1}, \dots, x_{ip_K})$]

$$\text{logit}[P(Y \leq k|\boldsymbol{x})] = \alpha_k + \boldsymbol{\beta}'\boldsymbol{x} \quad k = 1, \dots, K-1. \tag{1}$$

Parameters $\hat{\alpha}_k$s and $\hat{\boldsymbol{\beta}}$s are estimated for any item by applying model 1 to each of the M imputed data set generated in Step 1. Each unobserved unit imputed in step 1 is now replaced by a random draw from a *Multinomial*$(\hat{\pi}_1(\boldsymbol{x}), \dots, \hat{\pi}_K(\boldsymbol{x}))$. The probability to score each of the K categories of an item I_j is expressed as function of responses provided to the remaining $p-1$ items

$$\pi_k = \left[\frac{\exp(\alpha_k + \boldsymbol{\beta}'\boldsymbol{x})}{1 + \exp(\alpha_k + \boldsymbol{\beta}'\boldsymbol{x})} - \frac{\exp(\alpha_{k-1} + \boldsymbol{\beta}'\boldsymbol{x})}{1 + \exp(\alpha_{k-1} + \boldsymbol{\beta}'\boldsymbol{x})}\right]. \tag{2}$$

with vector of parameters $\boldsymbol{\beta}$ estimated using equation (1).

In this step plausible values for missing observations are generated by considering both the information provided by the observed values of the variables affected by missingness and the multivariate structure of the data.

3 An Application to the Data on the Evaluation of University Teaching

MISR has been tested on data provided by a survey on university course quality carried out in an Italian university. Specifically the data set concerns questionnaires gathered at the first level degree scheme at the Faculty of Engineering of Cagliari in 2004–2005 academic year. The analysis has been applied to eight items measured on a four-category Likert scale: *Definitely No*, *More No than Yes*, *More Yes than No*, *Definitely Yes*. The study evaluates the accuracy of the MISR procedure by assessing the extent to which the method fulfills the two standard criteria of AD and AE (Chambers 2001). In order to validate the accuracy of the method the procedure has been applied on a data set where observations have been simulated missing according to two missing data generating processes. The data set contains 1725 records on 24 courses and eight items: seven items concern student's evaluation of lecturer ($L_1 - L_7$) and one students' global satisfaction (S) (Table 2). In the following sections we will refer to it as *complete dataset* (CD). Each of the 24 courses have been evaluated by at least fifty students.

Table 2 Item considered for the application

Item	Contents
L_1	*Lecturer ability on motivating students*
L_2	*Lecturer highlights topics*
L_3	*Lecturer answers questions during the class*
L_4	*Lecturer clarifies goals of the course*
L_5	*Lecturer clearly explains topics*
L_6	*Lecturer suggests how to study*
L_7	*Lecturer gives classes on schedule*
S	*Global satisfaction*
A	*Student's attendance at classes*
I	*Student's interest toward the topics*

Moving from the CD, five data sets with an increasing rate (5, 10, 15, 20, 25%) of missing units have been simulated by setting observations in the CD data set missing according to two different missing data generating processes: MCAR and MAR. The former sets an observation missing independently from any students' characteristic. The observation in the CD is set missing if the result of a random draw from a *Bernoulli* (π) distribution with parameter π is equal to 1.

The MAR fixes the probability to set an observation missing on the bases of the values observed for students' covariates. Two covariates are supposed to influence the probability to observe a missing observation: *Students' attendance at classes* (1 = *Always*; 4 = *Very rarely*) and *Students' interest toward the topic* (1 = *Definitely No*; 4 = *Definitely Yes*).

The cross-classification of units according to the two covariates provides 16 clusters of students; each of them is characterized by a different probability (π) to skip an item. The probability that student i skips a question

$$\pi(\boldsymbol{x}_i) = \exp(\boldsymbol{\beta}'\boldsymbol{x}_i)/(1 + \exp(\boldsymbol{\beta}'\boldsymbol{x}_i))$$

is function of student' covariates $\boldsymbol{x}_i$ (six binary indicators). Students who share the same pattern of covariates belong to the same cluster and are characterized by an equal probability to generate non-responses. Coefficient parameters $\boldsymbol{\beta}$ are set in order to simulate an hypothetical situation in which the lowest probability $\hat{\pi}$ to skip a response is predicted for those students who say to be *definitely yes* interested on the topic and who *always* attend classes, instead the highest $\hat{\pi}$ is predicted for those students who are *not at all* interested and who have *rarely* attended classes. The MAR mechanism set a unit in the matrix missing if the result of a random draw from a *Bernoulli* $(\hat{\pi}(\boldsymbol{x}_i))$ is 1. Ten data sets have been simulated affected by a different rate of missingness: five data sets with a rate of missingness respectively equals to 5, 10, 15, 20 % and missing data generating process MCAR and five with the same rate of missingness but with missing data generating process MAR. The ten data sets have been imputed using MISR and MICE procedures. The MICE procedure has been implemented by Van Buuren and Oudshoorn in the mice package for the R environment (Van Buuren and Oudshoorn 2000). MICE generates multiple imputations for incomplete multivariate data by Gibbs Sampling (Van Buuren and Oudshoorn 2004; Schafen 1997). The algorithm imputes an incomplete variable by generating appropriate imputation values given other variables in the data matrix. In this application the predictors are the set of the remaining columns in the data. The imputation function specified is polyreg, which is the default method for polytomous variables.

3.1 AD

The AD has been assessed by comparing the agreement between the marginal distribution of each item in the CD data set with the marginal distributions in each of the $M = 100$ randomly imputed data sets (Sulis 2007). The *Dissimilarity index* z' (Leti 1983) for ordinal variables has been used to measure the discrepancy. By indicating with F_A and F_B the cumulated probability distributions of two categorical ordered variables 'A' and 'B' with K categories

$$z' = \frac{1}{K-1} \sum_{k=1}^{K-1} |F_{A_k} - F_{B_k}| \tag{3}$$

z' assumes values between [0,1]; 0 when the distributions are similar, 1 in case of maximum dissimilarity. For each item, 100 comparisons have been made. Table 3 shows the average values of the index taken over the 100 data sets. For each of the five rates of missingness, the *dissimilarity index* exhibits better performances when the MCAR assumption holds. Under both MCAR and MAR none of the average indexes calculated on the the data sets imputed using MICE and MISR signal a sensible departure form the benchmark distribution. This highlights a good performance of both the imputation procedures in terms of AD. The overall degree of agrement is very high also when the rate of missingess in the data matrix is equal to 25% (the highest value assumed by the index is 0.02). However, even though for any rate of missingness MICE seems to perform slightly better than MISR, differences in absolute terms may be considered no relevant.

3.2 AE

The AE has been assessed by comparing the parameters of a *random intercept* logit model estimated on the CD with the one obtained as a synthesis (Rubin 1987) of the corresponding estimates observed in the 100 randomly imputed data sets. *Multiple results* have been summarized in a single inferential statement using formula provided by Rubin (1987). By defining $\hat{\theta}_m$ the estimate for θ in data set m, the final estimate for θ is the mean of $\hat{\theta}_m$ taken over M data sets: $\bar{\theta} = M^{-1} \sum_{m=1}^{M} \hat{\theta}_m$.

Table 3 Accuracy in distribution: average values taken over M data sets

	Dissimilarity Index							
% miss	L_1	L_2	L_3	L_4	L_5	L_6	L_7	S
				MCAR				
				MISR				
5%	0.0015	0.0019	0.0014	0.0027	0.0016	0.0025	0.0023	0.0013
10%	0.0038	0.0035	0.0036	0.0024	0.0029	0.0041	0.0055	0.0025
15%	0.0081	0.0057	0.0054	0.0026	0.0032	0.0042	0.0078	0.0026
20%	0.0111	0.0046	0.0075	0.0041	0.0030	0.0080	0.0115	0.0038
25%	0.0157	0.0077	0.0091	0.0067	0.0038	0.0152	0.0199	0.0057
				MICE				
5%	0.0018	0.0016	0.0013	0.0023	0.0016	0.0031	0.0016	0.0015
10%	0.0028	0.0025	0.0022	0.0021	0.0032	0.0049	0.0019	0.0029
15%	0.0048	0.0039	0.0024	0.0028	0.0041	0.0057	0.0022	0.0029
20%	0.0053	0.0035	0.0024	0.0031	0.0031	0.0044	0.0033	0.0039
25%	0.0067	0.0053	0.0028	0.0037	0.0028	0.0049	0.0038	0.0071
				MAR				
				MISR				
5%	0.0018	0.0019	0.0016	0.0023	0.0013	0.0029	0.0020	0.0016
10%	0.0049	0.0031	0.0027	0.0024	0.0021	0.0038	0.0039	0.0041
15%	0.0120	0.0065	0.0052	0.0042	0.0037	0.0049	0.0072	0.0062
20%	0.0119	0.0044	0.0064	0.0056	0.0038	0.0102	0.0110	0.0068
25%	0.0128	0.0108	0.0106	0.0079	0.0039	0.0143	0.0217	0.0067
				MICE				
5%	0.0015	0.0022	0.0012	0.0019	0.0014	0.0034	0.0017	0.0015
10%	0.0033	0.0027	0.0019	0.0023	0.0024	0.0031	0.0019	0.0044
15%	0.0084	0.0040	0.0025	0.0037	0.0042	0.0040	0.0034	0.0056
20%	0.0081	0.0030	0.0025	0.0031	0.0044	0.0033	0.0044	0.0077
25%	0.0075	0.0042	0.0040	0.0029	0.0033	0.0038	0.0045	0.0057

The overall variance of θ is a combination of the *Within* and the *Between* imputation variance: $O = Within + (1 + M^{-1})Between$. Both response and predictor variables have been previously dichotomized. The model specifies the probability to be globally satisfied (item S) as a function of items $L_1 - L_7$

$$\text{logit}[Pr(Y_{ig} = 1|u_g)] = \alpha + \sum_{j=1}^{p} \beta_j x_{ij} + u_g \tag{4}$$

where $i = 1, \ldots, n_g$ are students' evaluations for the g_{th} course and u_g is the random intercept ($u_g \sim N(0, \sigma^2)$) which takes into account that level-1 units (students) are nested in level two-units (courses). Results depicted in Table 4 show that MISR method produces satisfactory estimates of coefficients regression parameters under both MCAR and MAR assumptions for rates of missing records in the matrix not over 10%.

Table 4 Coefficient parameters for the random intercept *logit* model (SE in brackets)

% miss	$\hat{\alpha}$	$\hat{\beta}_{L_1}$	$\hat{\beta}_{L_2}$	$\hat{\beta}_{L_3}$	$\hat{\beta}_{L_4}$	$\hat{\beta}_{L_5}$	$\hat{\beta}_{L_6}$	$\hat{\beta}_{L_7}$	$\hat{\sigma}$
MCAR									
MISR									
5%	−4.048 (.415)	1.484 (.184)	0.599 (.203)	.770 (.235)	.507 (.202)	1.702 (.182)	.734 (.189)	1.713 (.345)	0.559 (.133)
10%	−3.942 (.423)	1.399 (.192)	0.630 (.210)	.839 (.243)	.592 (.209)	1.683 (.191)	.707 (.196)	1.503 (.346)	0.571 (.138)
15%	−3.469 (.398)	1.369 (.193)	0.632 (.225)	.727 (.252)	.465 (.221)	1.649 (.193)	.667 (.204)	1.219 (.340)	0.553 (.137)
20%	−3.155 (.391)	1.274 (.198)	0.590 (.240)	.712 (.270)	.555 (.220)	1.492 (.195)	.742 (.205)	1.002 (.330)	0.552 (.140)
25%	−2.961 (.394)	1.156 (.207)	0.679 (.214)	.599 (.261)	.453 (.228)	1.404 (.202)	.899 (.201)	0.994 (.336)	0.532 (.140)
MICE									
5%	−4.131 (.418)	1.490 (.183)	.587 (.205)	.745 (.239)	.527 (.203)	1.739 (.184)	.709 (.189)	1.806 (.344)	.561 (.134)
10%	−4.294 (.447)	1.409 (.196)	.622 (.218)	.857 (.248)	.605 (.217)	1.785 (.196)	.673 (.203)	1.788 (.375)	.566 (.141)
15%	−3.918 (.444)	1.434 (.204)	.579 (.240)	.750 (.257)	.482 (.229)	1.781 (.205)	.611 (.211)	1.613 (.373)	.554 (.142)
20%	−3.633 (.449)	1.404 (.218)	.488 (.251)	.841 (.288)	.489 (.242)	1.694 (.211)	.644 (.214)	1.382 (.390)	.565 (.139)
25%	−3.627 (.463)	1.282 (.221)	.637 (.243)	.765 (.286)	.265 (.259)	1.684 (.214)	.852 (.239)	1.555 (.397)	.524 (.138)
CCA									
5%	−4.009 (.438)	1.606 (.212)	.660 (.235)	.510 (.270)	639 (.230)	1.672 (.210)	.610 (.215)	1.710 (.367)	.428 (.142)
10%	−4.370 (.571)	1.630 (.274)	.710 (.317)	.516 (.357)	671 (.301)	1.905 (.261)	.612 (.278)	1.775 (.484)	.359 (.206)
15%	−3.294 (.616)	1.756 (.334)	.733 (.383)	.141 (.442)	360 (.377)	1.946 (.324)	.636 (.341)	1.140 (.580)	.441 (.263)
20%	−3.071 (.728)	1.254 (.429)	.678 (.521)	.085 (.554)	549 (.474)	2.008 (.426)	.977 (.445)	.887 (.693)	.000 (.344)
25%	−3.492 (1.003)	1.078 (.542)	.954 (.626)	−.009 (.681)	333 (.621)	1.609 (.565)	1.017 (.523)	1.492 (.978)	.000 (.358)

(*continued*)

Table 4 continued

% miss	$\hat{\alpha}$	$\hat{\beta}_{L_1}$	$\hat{\beta}_{L_2}$	$\hat{\beta}_{L_3}$	$\hat{\beta}_{L_4}$	$\hat{\beta}_{L_5}$	$\hat{\beta}_{L_6}$	$\hat{\beta}_{L_7}$	$\hat{\sigma}$
					MAR				
					MISR				
5%	−4.109 (.415)	1.481 (.183)	.572 (.206)	.774 (.235)	.566 (.200)	1.752 (.183)	.715 (.187)	1.733 (.347)	.558 (.137)
10%	−3.726 (.421)	1.436 (.189)	.733 (.212)	.725 (.249)	.489 (.217)	1.683 (.197)	.759 (.196)	1.341 (.364)	.521 (.136)
15%	−3.421 (.415)	1.342 (.192)	.727 (.225)	.697 (.253)	.433 (.218)	1.537 (.207)	.689 (.207)	1.264 (.356)	.539 (.144)
20%	−3.256 (.405)	1.310 (.201)	.724 (.230)	.679 (.257)	.462 (.232)	1.482 (.200)	.778 (.206)	1.111 (.353)	.536 (.144)
25%	−3.086 (.392)	1.189 (.207)	.792 (.227)	.726 (.258)	.408 (.231)	1.479 (.199)	.769 (.212)	0.967 (.343)	.533 (.146)
					MICE				
5%	−4.175 (.419)	1.502 (.182)	0.535 (.205)	0.755 (.239)	0.584 (.202)	1.783 (.180)	0.705 (.189)	1.809 (.349)	0.554 (.135)
10%	−3.853 (.429)	1.486 (.191)	0.660 (.215)	0.681 (.249)	0.535 (.224)	1.785 (.191)	0.687 (.198)	1.486 (.373)	0.538 (.136)
15%	−4.067 (.467)	1.399 (.200)	0.610 (.243)	0.762 (.273)	0.399 (.239)	1.735 (.203)	0.549 (.209)	1.897 (.412)	0.519 (.139)
20%	−3.894 (.465)	1.408 (.209)	0.546 (.251)	0.721 (.289)	0.416 (.256)	1.743 (.211)	0.719 (.227)	1.761 (.400)	0.546 (.143)
25%	−3.597 (.470)	1.312 (.205)	0.838 (.281)	0.701 (.274)	0.290 (.262)	1.712 (.223)	0.749 (.236)	1.373 (.423)	0.544 (.148)
					CCA				
5%	−3.977 (.438)	1.578 (.203)	.570 (.231)	.518 (.268)	.496 (.226)	1.731 (.204)	.703 (.212)	1.817 (.375)	.438 (.144)
10%	−3.678 (.495)	1.632 (.241)	.813 (.276)	.422 (.320)	.561 (.269)	1.668 (.240)	.556 (.248)	1.382 (.436)	.290 (.200)
15%	−3.673 (.612)	1.894 (.306)	1.080 (.381)	.118 (.415)	.393 (.362)	2.013 (.302)	.415 (.321)	1.182 (.523)	.001 (.170)
20%	−4.187 (.731)	1.696 (.351)	1.248 (.467)	−.059 (.477)	.723 (.421)	2.409 (.347)	.635 (.359)	1.093 (.586)	.000 (.181)
25%	−4.803 (.925)	1.759 (.435)	1.533 (.585)	−.015 (.583)	.355 (.529)	2.490 (.438)	.745 (.443)	1.690 (.721)	.001 (.303)
					CD				
–	−4.254 (.417)	1.494 (.177)	0.666 (.197)	0.658 (.230)	0.575 (.195)	1.742 (.176)	0.767 (.182)	1.871 (.347)	0.599 (.138)

A comparison with the results obtained imputing with chained equations highlights that the advantage of adopting MICE in respect of MISR seem to be higher under the MCAR than MAR. Moreover, the convenience increases as the rate of missingess in the data set becomes severe. From the simulation study arises that both multiple imputation methods do not perform well in estimating the intercept parameter ($\hat{\alpha}$) and the parameter ($\hat{\beta}_{L_7}$). Nevertheless, the latter is better estimated by MICE. The estimates of $\hat{\alpha}$ become strongly unreliable when the rate of missing units increases. Both procedures show a better performance in respect of the CCA results; the latter leads to strongly bias and inefficient estimates of many parameters in data sets heavily affected by missingness (Table 4). In the MAR data sets, the two multiple imputation procedures provide results quite similar for rates of missingness under 15% . Table 4 shows that, as could be expected, MICE and MISR both produce accurate estimates of standard errors. In the *Mixed effect model* framework the greatest advantage of the MICE and MISR approaches is the accuracy in the estimation of the variance of random term (see last column in Table 4).

4 Some Final Remarks

In this article a multiple imputation approach based on stochastic regression models has been described, implemented and evaluated in respect of the widely validated MICE approach. The proposed MISR procedure is an *ad hoc* method to recover for missingness in data where items measured on Likert-type scale define the same latent trait. The examination of subject pattern of responses provides information on the way students score categories and help us to learn if subject tends to use high, low or middle categories. This motivates the first step of the procedure. MISR seems to produce unbiased and efficient estimates of many coefficient parameters in *Mixed effect models* framework. Estimates provided by MISR under MAR assumption do not seems to show a remarkable departure from the one obtained using MICE library, at least when the rate of missingness in the data matrix does not become severe.

References

Agresti, A. (2002). *Categorical data analysis*. Hoboken NJ: Wiley

Berrington, A., & Borgoni, R. (2004). A tree based procedure for multivariate imputation. In *Atti XLII Convegno della Società Italiana di Statistica*. Cleup, Padova

Chambers, R. (2001). Evaluation criteria for statistical editing and imputation. *National Statistics Methodological Series, 28*, 1–41

Leti, G. (1983). *Statistica descrittiva*. Bologna: il Mulino

Little, R. J. A., & Rubin, D. B. (2002). *Statistical analysis with missing data* (2nd ed.), New York: Wiley

Rubin, D. B. (1976). Inference and missing data. *Biometrika, 63*, 581–592

Rubin, D. B. (1987). *Multiple imputation for nonresponses in surveys*. New York: Wiley

Schafer, J. (1997). *Analysis of incomplete multivariate data*. London: Chapman & Hall
Sulis, I. (2007). *Measuring students' assessments of 'university course quality' using mixed-effect models*. PhD Thesis, Università degli Studi di Palermo
Van Buuren, S., & Oudshoorn, C. G. M. (2000). Multivariate imputation by Chained equations: Mice v1.0 users manual. *Technical Report PG/VGZ/00.038*, TNO Prevention and Health, Leiden
Van Buuren, S., & Oudshoorn, C. G. M. (2004). Mice: Multivariate imputation by Chained equations. *R-package* http://cran.r-project.org/web/packages/mice/mice.pdf

GPSR Compliance

The European Union's (EU) General Product Safety Regulation (GPSR) is a set of rules that requires consumer products to be safe and our obligations to ensure this.

If you have any concerns about our products, you can contact us on ProductSafety@springernature.com

In case Publisher is established outside the EU, the EU authorized representative is:

Springer Nature Customer Service Center GmbH
Europaplatz 3
69115 Heidelberg, Germany

Batch number: 10761572

Printed by Printforce, the Netherlands